D1675554

Naminosuke Kubota

Propellants and Explosives

1807–2007 Knowledge for Generations

Each generation has its unique needs and aspirations. When Charles Wiley first opened his small printing shop in lower Manhattan in 1807, it was a generation of boundless potential searching for an identity. And we were there, helping to define a new American literary tradition. Over half a century later, in the midst of the Second Industrial Revolution, it was a generation focused on building the future. Once again, we were there, supplying the critical scientific, technical, and engineering knowledge that helped frame the world. Throughout the 20th Century, and into the new millennium, nations began to reach out beyond their own borders and a new international community was born. Wiley was there, expanding its operations around the world to enable a global exchange of ideas, opinions, and know-how.

For 200 years, Wiley has been an integral part of each generation's journey, enabling the flow of information and understanding necessary to meet their needs and fulfill their aspirations. Today, bold new technologies are changing the way we live and learn. Wiley will be there, providing you the must-have knowledge you need to imagine new worlds, new possibilities, and new opportunities.

Generations come and go, but you can always count on Wiley to provide you the knowledge you need, when and where you need it!

William J. Pesce
President and Chief Executive Officer

Peter Booth Wiley
Chairman of the Board

Naminosuke Kubota

Propellants and Explosives

Thermochemical Aspects of Combustion

Second, Completely Revised and Extended Edition

WILEY-VCH Verlag GmbH & Co. KGaA

The Author

Prof. Dr. Naminosuke Kubota
Asahi Kasei Chemicals
Propellant Combustion Laboratory
Arca East, Kinshi 3-2-1, Sumidaku
Tokyo 130-6591, Japan

First Edition 2001

Library of Congress Card No.:
applied for

British Library Cataloguing-in-Publication Data
A catalogue record for this book is available from the British Library.

Bibliographic information published by the Deutsche Nationalbibliothek
The Deutsche Nationalbibliothek lists this publication in the Deutsche Nationalbibliografie; detailed bibliographic data are available in the Internet at http://dnb.d-nb.de.

Typesetting primustype Robert Hurler GmbH
Printing betz-Druck GmbH, Darmstadt
Binding Litges & Dopf Buchbinderei GmbH, Heppenheim
Cover Design Grafik-Design Schulz, Fußgönheim

Printed in the Federal Republic of Germany
Printed on acid-free paper

ISBN: 978-3-527-31424-9

Table of Contents

Propellants and Explosives. Naminosuke Kubota

ISBN: 978-3-527-31424-9

Preface to the First Edition

Propellants and explosives are composed of energetic materials that produce high temperature and pressure through combustion phenomena. The combustion phenomena include complex physicochemical changes from solid to liquid and to gas, which accompany the rapid, exothermic reactions. A number of books related to combustion have been published, such as an excellent theoretical book, Combustion Theory, 2nd Edition, by F. A. Williams, Benjamin/Cummings, New York (1985), and an instructive book for the graduate student, Combustion, by I. Glassman, Academic Press, New York (1977). However, no instructive books related to the combustion of solid energetic materials have been published. Therefore, this book is intended as an introductory text on the combustion of energetic materials for the reader engaged in rocketry or in explosives technology.

This book is divided into four parts. The first part (Chapters 1–3) provides brief reviews of the fundamental aspects relevant to the conversion from chemical energy to aerothermal energy. References listed in each chapter should prove useful to the reader for better understanding of the physical bases of the energy conversion process; energy formation, supersonic flow, shock wave, detonation, and defl agration. The second part (Chapter 4) deals with the energetics of chemical compounds used as propellants and explosives, such as heat of formation, heat of explosion, adiabatic flame temperature, and specific impulse.

The third part (Chapters 5–8) deals with the results of measurements on the burning rate behavior of various types of chemical compounds, propellants, and explosives. The combustion wave structures and the heat feedback processes from the gas phase to the condensed phase are also discussed to aid in the understanding of the relevant combustion mechanisms. The experimental and analytical data described in these chapters are mostly derived from results previously presented by the author. Descriptions of the detailed thermal decomposition mechanisms from solid phase to liquid phase or to gasphase are not included in this book. The fourth part (Chapter 9) describes the combustion phenomena encountered during rocket motor operation, covering such to pics as the stability criterion of the rocket motor, temperature sensitivity, ignition transients, erosive burning, and combustion oscillations. The fundamental principle of variable-flow ducted rockets is also presented. The combustion characteristics and energetics of the gas-generating propellants used in ducted rockets are discussed.

Since numerous kinds of energetic materials are used as propellants and explosives, it is not possible to present an entire overview of the combustion processes of these materials. In this book, the combustion processes of typical energetic crystalline and polymeri c materials and of varioustypes of propellants are presented so as to provide an informative, generalized approach to understanding their combustion mechanisms.

Kamakura, Japan
March 2001

Naminosuke Kubota

Preface to the Second Edition

The combustion phenomena of propellants and explosives are described on the basis of pyrodynamics, which concerns thermochemical changes generating heat and reaction products. The high-temperature combustion products generated by propellants and explosives are converted into propulsive forces, destructive forces, and various types of mechanical forces. Similar to propellants and explosives, pyrolants are also energetic materials composed of oxidizer and fuel components. Pyrolants react to generate high-temperature condensed and/or gaseous products when they burn. Propellants are used for rockets and guns to generate propulsive forces through deflagration phenomena and explosives are used for warheads, bombs, and mines to generate destructive forces through detonation phenomena. On the other hand, pyrolants are used for pyrotechnic systems such as ducted rockets, gas-hybrid rockets, and igniters and flares. This Second Edition includes the thermochemical processes of pyrolants in order to extend their application potential to propellants and explosives.

The burning characteristics of propellants, explosives, and pyrolants are largely dependent on various physicochemical parameters, such as the energetics, the mixture ratio of fuel and oxidizer components, the particle size of crystalline oxidizers, and the decomposition process of fuel components. Though metal particles are high-energy fuel components and important ingredients of pyrolants, their oxidation and combustion processes with oxidizers are complex and difficult to understand.

Similar to the First Edition, the first half of the Second Edition is an introductory text on pyrodynamics describing fundamental aspects of the combustion of energetic materials. The second half highlights applications of energetic materials as propellants, explosives, and pyrolants. In particular, transient combustion, oscillatory burning, ignition transients, and erosive burning phenomena occurring in rocket motors are presented and discussed. Ducted rockets represent a new propulsion system in which combustion performance is significantly increased by the use of pyrolants.

Heat and mass transfer through the boundary layer flow over the burning surface of propellants dominates the burning process for effective rocket motor operation. Shock wave formation at the inlet flow of ducted rockets is an important process for achieving high propulsion performance. Thus, a brief overview of the fundamentals of aerodynamics and heat transfer is provided in Appendices B–D as a prerequisite for the study of pyrodynamics.

Tokyo, Japan
September 2006

Naminosuke Kubota

1
Foundations of Pyrodynamics

Pyrodynamics describes the process of energy conversion from chemical energy to mechanical energy through combustion phenomena, including thermodynamic and fluid dynamic changes. Propellants and explosives are energetic condensed materials composed of oxidizer-fuel components that produce high-temperature molecules. Propellants are used to generate high-temperature and low-molecular combustion products that are converted into propulsive forces. Explosives are used to generate high-pressure combustion products accompanied by a shock wave that yield destructive forces. This chapter presents the fundamentals of thermodynamics and fluid dynamics needed to understand the pyrodynamics of propellants and explosives.

1.1
Heat and Pressure

1.1.1
First Law of Thermodynamics

The first law of thermodynamics relates the energy conversion produced by chemical reaction of an energetic material to the work acting on a propulsive or explosive system. The heat produced by chemical reaction (q) is converted into the internal energy of the reaction product (e) and the work done to the system (w) according to

$$dq = de + dw \tag{1.1}$$

The work is done by the expansion of the reaction product, as given by

$$dw = pdv \quad \text{or} \quad dw = pd\,(1/\rho) \tag{1.2}$$

where p is the pressure, v is the specific volume (volume per unit mass) of the reaction product, and ρ is the density defined in $v = 1/\rho$. Enthalpy h is defined by

$$dh = de + d\,(pv) \tag{1.3}$$

Propellants and Explosives. Naminosuke Kubota

ISBN: 978-3-527-31424-9

Substituting Eqs. (1.1) and (1.2) into Eq. (1.3), one gets

$$dh = dq + vdp \tag{1.4}$$

The equation of state for one mole of a perfect gas is represented by

$$pv = R_g T \quad \text{or} \quad p = \rho R_g T \tag{1.5}$$

where T is the absolute temperature and R_g is the gas constant. The gas constant is given by

$$R_g = \mathrm{R}/M_g \tag{1.6}$$

where M_g is the molecular mass, and R is the universal gas constant, R = 8.314472 J mol^{-1} K^{-1}. In the case of n moles of a perfect gas, the equation of state is represented by

$$pv = nR_g T \quad \text{or} \quad p = n\rho R_g T \tag{1.5 a}$$

1.1.2
Specific Heat

Specific heat is defined according to

$$c_v = (de/dT)_v \quad c_p = (dh/dT)_p \tag{1.7}$$

where c_v is the specific heat at constant volume and c_p is the specific heat at constant pressure. Both specific heats represent conversion parameters between energy and temperature. Using Eqs. (1.3) and (1.5), one obtains the relationship

$$c_p - c_v = R_g \tag{1.8}$$

The specific heat ratio γ is defined by

$$\gamma = c_p/c_v \tag{1.9}$$

Using Eq. (1.9), one obtains the relationships

$$c_v = R_g/(\gamma - 1) \quad c_p = \gamma R_g/(\gamma - 1) \tag{1.10}$$

Specific heat is an important parameter for energy conversion from heat energy to mechanical energy through temperature, as defined in Eqs. (1.7) and (1.4). Hence, the specific heat of gases is discussed to understand the fundamental physics of the energy of molecules based on kinetic theory.[1,2] The energy of a single molecule, ε_m, is given by the sum of the internal energies, which comprise translational energy,

ε_t, rotational energy, ε_r, vibrational energy, ε_v, electronic energy, ε_e, and their interaction energy, ε_i:

$$\varepsilon_m = \varepsilon_t + \varepsilon_r + \varepsilon_v + \varepsilon_e + \varepsilon_i$$

A molecule containing n atoms has $3n$ degrees of freedom of motion in space:

molecular structure	degrees of freedom		translational		rotational		vibrational
monatomic	3	=	3				
diatomic	6	=	3	+	2	+	1
polyatomic linear	$3n$	=	3	+	2	+	$(3n–5)$
polyatomic nonlinear	$3n$	=	3	+	3	+	$(3n–6)$

A statistical theorem on the equipartition of energy shows that an energy amounting to $kT/2$ is given to each degree of freedom of translational and rotational modes, and that an energy of kT is given to each degree of freedom of vibrational modes. The Boltzmann constant k is 1.38065×10^{-23} J K^{-1}. The universal gas constant R defined in Eq. (1.6) is given by $R = k\zeta$, where ζ is Avogadro's number, $\zeta = 6.02214 \times 10^{23}$ mol^{-1}.

When the temperature of a molecule is increased, rotational and vibrational modes are excited and the internal energy is increased. The excitation of each degree of freedom as a function of temperature can be calculated by way of statistical mechanics. Though the translational and rotational modes of a molecule are fully excited at low temperatures, the vibrational modes only become excited above room temperature. The excitation of electrons and interaction modes usually only occurs at well above combustion temperatures. Nevertheless, dissociation and ionization of molecules can occur when the combustion temperature is very high.

When the translational, rotational, and vibrational modes of monatomic, diatomic, and polyatomic molecules are fully excited, the energies of the molecules are given by

$$\varepsilon_m = \varepsilon_t + \varepsilon_r + \varepsilon_v$$

$$\varepsilon_m = 3 \times kT/2 = 3\ kT/2 \text{ for monatomic molecules}$$

$$\varepsilon_m = 3 \times kT/2 + 2 \times kT/2 + 1 \times kT = 7\ kT/2 \text{ for diatomic molecules}$$

$$\varepsilon_m = 3 \times kT/2 + 2 \times kT/2 + (3n - 5) \times kT = (6n - 5)\ kT/2 \text{ for linear molecules}$$

$$\varepsilon_m = 3 \times kT/2 + 3 \times kT/2 + (3n - 6) \times kT = 3(n - 1)\ kT \text{ for nonlinear molecules}$$

Since the specific heat at constant volume is given by the temperature derivative of the internal energy as defined in Eq. (1.7), the specific heat of a molecule, $c_{v,m}$, is represented by

$$c_{v,m} = d_{\varepsilon m}/dT = d\varepsilon_t/dT + d\varepsilon_r/dT + d\varepsilon_v/dT + d\varepsilon_e/dT + d\varepsilon_i/dT \qquad \text{J molecule}^{-1}\ \text{K}^{-1}$$

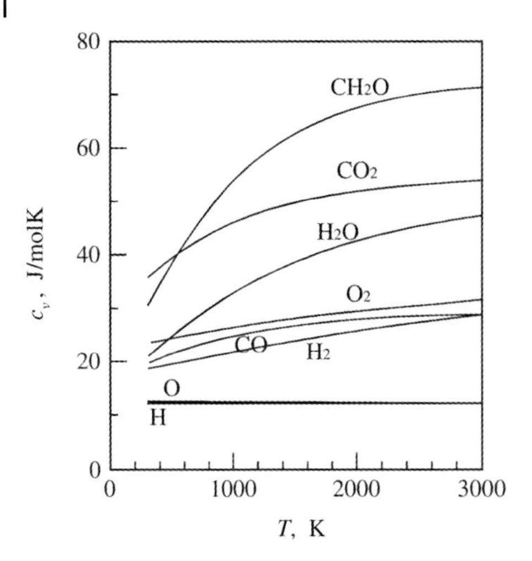

Fig. 1.1 Specific heats of gases at constant volume as a function of temperature.

Thus, one obtains the specific heats of gases composed of monatomic, diatomic, and polyatomic molecules as follows:

$c_v = 3R/2 = 12.47$ J mol^{-1} K^{-1} for monatomic molecules

$c_v = 7R/2 = 29.10$ J mol^{-1} K^{-1} for diatomic molecules

$c_v = (6n - 5)R/2$ J mol^{-1} K^{-1} for linear molecules

$c_v = 3(n - 1)R$ J mol^{-1} K^{-1} for nonlinear molecules

The specific heat ratio defined by Eq. (1.9) is 5/3 for monatomic molecules; 9/7 for diatomic molecules. Since the excitations of rotational and vibrational modes only occur at certain temperatures, the specific heats determined by kinetic theory are different from those determined experimentally. Nevertheless, the theoretical results are valuable for understanding the behavior of molecules and the process of energy conversion in the thermochemistry of combustion. Fig. 1.1 shows the specific heats of real gases encountered in combustion as a function of temperature.[3] The specific heats of monatomic gases remain constant with increasing temperature, as determined by kinetic theory. However, the specific heats of diatomic and polyatomic gases are increased with increasing temperature as the rotational and vibrational modes are excited.

1.1.3 Entropy Change

Entropy s is defined according to

$$ds \equiv dq/T \tag{1.11}$$

Substituting Eqs. (1.4), (1.5), and (1.7) into Eq. (1.11), one gets

$$ds = c_p\, dT/T - R_g dp/p \qquad (1.12)$$

In the case of isentropic change, $ds = 0$, Eq. (1.12) is integrated as

$$p/p_1 = (T/T_1)^{cp/Rg} \qquad (1.13)$$

where the subscript 1 indicates the initial state 1. Using Eqs. (1.10), (1.5), and (1.13), one gets

$$p/p_1 = (T/T_1)^{\gamma/(\gamma-1)} \quad \text{and} \quad p\,(1/\rho)^{\gamma} = p_1\,(1/\rho_1)^{\gamma} \qquad (1.14)$$

When a system involves dissipative effects such as friction caused by molecular collisions or turbulence caused by a non-uniform molecular distribution, even under adiabatic conditions, ds becomes a positive value, and then Eqs. (1.13) and (1.14) are no longer valid. However, when these physical effects are very small and heat loss from the system or heat gain by the system are also small, the system is considered to undergo an isentropic change.

1.2 Thermodynamics in a Flow Field

1.2.1 One-Dimensional Steady-State Flow

1.2.1.1 Sonic Velocity and Mach Number

The sonic velocity propagating in a perfect gas, a, is given by

$$a = (\partial p/\partial \rho)_s^{1/2} \qquad (1.15)$$

Using the equation of state, Eq. (1.8), and the expression for adiabatic change, Eq. (1.14), one gets

$$a = (\gamma R_g T)^{1/2} \qquad (1.16)$$

Mach number M is defined according to

$$M = u/a \qquad (1.17)$$

where u is the local flow velocity in a flow field. Mach number is an important parameter in characterizing a flow field.

1.2.1.2 Conservation Equations in a Flow Field

Let us consider a simplified flow, that is, a one-dimensional steady-state flow-without viscous stress or a gravitational force. The conservation equations of continuity, momentum, and energy are represented by:
rate of mass in – rate of mass out = 0

$$d(\rho u) = 0 \tag{1.18}$$

rate of momentum gain by convection + pressure difference acting on flow = 0

$$\rho u du + dp = 0 \tag{1.19}$$

rate of energy input by conduction + rate of energy input by convection = 0

$$d\,(h + u^2/2) = 0 \tag{1.20}$$

Combining Eqs. (1.20) and Eq. (1.4), one obtains the relationship for the enthalpy change due to a change of flow velocity as

$$dh = dq - u\,du \tag{1.21}$$

1.2.1.3 Stagnation Point

If one can assume that the process in the flow field is adiabatic and that dissipative effects are negligibly small, the flow in the system is isentropic ($ds = 0$), and then Eq. (1.21) becomes

$$dh = -u\,du \tag{1.22}$$

Integration of Eq. (1.22) gives

$$h_0 = h + u^2/2 \tag{1.23}$$

where h_0 is the stagnation enthalpy at $u = 0$ of a stagnation flow point. Substituting Eq. (1.7) into Eq. (1.23), one gets

$$c_p T_0 = c_p T + u^2/2 \tag{1.24}$$

where T_0 is the stagnation temperature at $u = 0$.

The changes in temperature, pressure, and density in a flow field are expressed as a function of Mach number as follows:

$$\frac{T_0}{T} = 1 + \frac{\gamma - 1}{2} M^2 \tag{1.25}$$

$$\frac{p_0}{p} = \left(1 + \frac{\gamma - 1}{2} M^2\right)^{\frac{\gamma}{\gamma - 1}} \tag{1.26}$$

$$\frac{\rho_0}{\rho} = \left(1 + \frac{\gamma - 1}{2} M^2\right)^{\frac{1}{\gamma - 1}} \tag{1.27}$$

1.2.2
Formation of Shock Waves

One assumes that a discontinuous flow occurs between regions 1 and 2, as shown in Fig. 1.2. The flow is also assumed to be one-dimensional and in a steady state, and not subject to a viscous force, an external force, or a chemical reaction.
The mass continuity equation is given by

$$\rho_1 u_1 = \rho_2 u_2 = m \tag{1.28}$$

The momentum equation is represented by

$$p_1 + m u_1{}^2 = p_2 + m u_2{}^2 \tag{1.29}$$

The energy equation is represented by the use of Eq. (1.20) as

$$c_p T_1 + u_1{}^2/2 = c_p T_2 + u_2{}^2/2 \tag{1.30}$$

where m is the mass flux in a duct of constant area, and the subscripts 1 and 2 indicate the upstream and the downstream of the discontinuity, respectively. Substituting Eq. (1.29) into Eq. (1.30), one gets

$$p_1 + \rho_1 u_1{}^2 = p_2 + \rho_2 u_2{}^2 \tag{1.31}$$

Using Eq. (1.25), the temperature ratio in regions 2 and 1 is represented by the Mach number in 2 and 1 according to

$$\frac{T_2}{T_1} = \frac{1 + \frac{\gamma - 1}{2} M_1^2}{1 + \frac{\gamma - 1}{2} M_2^2} \tag{1.32}$$

Using Eqs. (1.5), (1.17), and (1.28), one gets

$$\frac{T_2}{T_1} = \left(\frac{M_2}{M_1}\right)^2 \left(\frac{p_2}{p_1}\right)^2 \tag{1.33}$$

Combining Eqs. (1.31) and (1.32), the pressure ratio is obtained as a function of M_1 and M_2:

$$\frac{p_2}{p_1} = \frac{M_1}{M_2} \frac{\sqrt{1 + \frac{\gamma - 1}{2} M_1^2}}{\sqrt{1 + \frac{\gamma - 1}{2} M_2^2}} \tag{1.34}$$

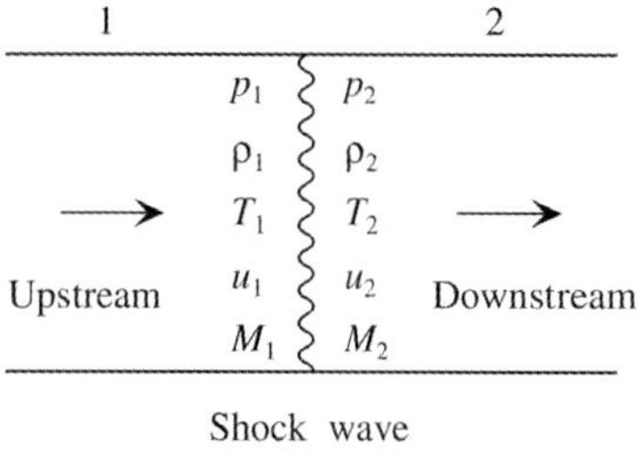

Fig. 1.2 Shock wave propagation.

Combining Eqs. (1.33) and (1.34), the Mach number relationship in the upstream 1 and downstream 2 is obtained as

$$\frac{M_1\sqrt{1+\frac{\gamma-1}{2}M_1^2}}{1+\gamma M_1^2}=\frac{M_2\sqrt{1+\frac{\gamma-1}{2}M_2^2}}{1+\gamma M_2^2} \tag{1.35}$$

One obtains two solutions from Eq. (1.35):

$$M_2 = M_1 \tag{1.36}$$

$$M_2=\left[\frac{\frac{2}{\gamma-1}+M_1^2}{\frac{2\gamma}{\gamma-1}M_1^2-1}\right]^{\frac{1}{2}} \tag{1.37}$$

The solution expressed by Eq. (1.36) indicates that there is no discontinuous flow between the upstream 1 and the downstream 2. However, the solution given by Eq. (1.37) indicates the existence of a discontinuity of pressure, density, and temperature between 1 and 2. This discontinuity is called a "normal shock wave", which is set-up in a flow field perpendicular to the flow direction. Discussions on the structures of normal shock waves and supersonic flow fields can be found in the relevant monographs.[4,5]

Substituting Eq. (1.37) into Eq. (1.34), one obtains the pressure ratio as

$$\frac{p_2}{p_1}=\frac{2\gamma}{\gamma+1}M_1^2-\frac{\gamma-1}{\gamma+1} \tag{1.38}$$

Substituting Eq. (1.37) into Eq. (1.33), one also obtains the temperature ratio as

$$\frac{T_2}{T_1}=\frac{1}{M_1^2}\frac{2(\gamma-1)}{(\gamma+1)^2}\left(1+\frac{\gamma-1}{2}M_1^2\right)\left(\frac{2\gamma}{\gamma-1}M_1^2-1\right) \tag{1.39}$$

The density ratio is obtained by the use of Eqs. (1.38), (1.39), and (1.8) as

$$\frac{\rho_2}{\rho_1}=\frac{p_2}{p_1}\frac{T_2}{T_1} \tag{1.40}$$

Using Eq. (1.24) for the upstream and the downstream and Eq. (1.38), one obtains the ratio of stagnation pressure as

$$\frac{p_{02}}{p_{01}}=\left(\frac{\gamma+1}{2}M_1^2\right)^{\frac{\gamma}{\gamma-1}}\left(1+\frac{\gamma-1}{2}M_1^2\right)^{\frac{\gamma}{1-\gamma}}\left(\frac{2\gamma}{\gamma+1}M_1^2-\frac{\gamma-1}{\gamma+1}\right)^{\frac{1}{1-\gamma}} \tag{1.41}$$

The ratios of temperature, pressure, and density in the downstream and upstream are expressed by the following relationships:

$$\frac{T_2}{T_1}=\frac{p_2}{p_1}\left(1+\frac{1}{\xi}\frac{p_2}{p_1}\right)\Big/\left(\frac{1}{\xi}+\frac{p_2}{p_1}\right) \tag{1.42}$$

$$\frac{p_2}{p_1}=\left(\xi\frac{\rho_2}{\rho_1}-1\right)\Big/\left(\xi-\frac{\rho_2}{\rho_1}\right) \tag{1.43}$$

$$\frac{\rho_2}{\rho_1} = \left(\xi \frac{p_2}{p_1} + 1 \right) \bigg/ \left(\xi + \frac{p_2}{p_1} \right) \tag{1.44}$$

where $\zeta = (\gamma + 1)/(\gamma - 1)$. The set of Eqs. (1.42), (1.43), and (1.44) is known as the Rankine–Hugoniot equation for a shock wave without any chemical reactions. The relationship of p_2/p_1 and ρ_2/ρ_1 at $\gamma = 1.4$ (for example, in the case of air) shows that the pressure of the downstream increases infinitely when the density of the downstream is increased approximately six times. This is evident from Eq. (1.43), as when $\rho_2/\rho_1 \to \zeta$, then $p_2/p_1 \to \infty$.

Though the form of the Rankine–Hugoniot equation, Eqs. (1.42)–(1.44), is obtained when a stationary shock wave is created in a moving coordinate system, the same relationship is obtained for a moving shock wave in a stationary coordinate system. In a stationary coordinate system, the velocity of the moving shock wave is u_1 and the particle velocity u_p is given by $u_p = u_1 - u_2$. The ratios of temperature, pressure, and density are the same for both moving and stationary coordinates.

A shock wave is characterized by the entropy change across it. Using the equation of state for a perfect gas shown in Eq. (1.5), the entropy change is represented by

$$s_2 - s_1 = c_p \ln(T_2/T_1) - R_g \ln(p_2/p_1) \tag{1.45}$$

Substituting Eqs. (1.38) and (1.39) into Eq. (1.45), one gets

$$s_2 - s_1 = c_p \ln\left[\frac{2}{(\gamma+1)M_1^2} + \frac{1}{\xi} \right] + \frac{c_p}{\gamma} \ln\left[\frac{2\gamma}{\gamma+1} M_1^2 - \frac{1}{\xi} \right] \tag{1.46}$$

It is obvious that the entropy change will be positive in the region $M_1 > 1$ and negative in the region $M_1 < 1$ for gases with $1 < \gamma < 1.67$. Thus, Eq. (1.46) is valid only when M_1 is greater than unity. In other words, a discontinuous flow is formed only when $M_1 > 1$. This discontinuous surface perpendicular to the flow direction is the normal shock wave. The downstream Mach number, M_1, is always < 1, i. e. subsonic flow, and the stagnation pressure ratio is obtained as a function of M_1 by Eqs. (1.37) and (1.41). The ratios of temperature, pressure, and density across the shock wave are obtained as a function of M_1 by the use of Eqs. (1.38)–(1.40) and Eqs. (1.25)–(1.27). The characteristics of a normal shock wave are summarized as follows:

	Front	← Shock wave ←	Behind
Velocity	u_1	$>$	u_2
Pressure	p_1	$<$	p_2
Density	ρ_1	$<$	ρ_2
Temperature	T_1	$<$	T_2
Mach number	M_1	$>$	M_2
Stagnation pressure	p_{01}	$>$	p_{02}
Stagnation density	ρ_{01}	$>$	ρ_{02}
Stagnation temperature	T_{01}	$=$	T_{02}
Entropy	s_1	$<$	s_2

1.2.3 Supersonic Nozzle Flow

When gas flows from stagnation conditions through a nozzle, thereby undergoing an isoentropic change, the enthalpy change is represented by Eq. (1.23). The flow velocity is obtained by substitution of Eq. (1.14) into Eq. (1.24) as

$$u^2 = 2c_p T_0\{1 - (p/p_0)^{Rg/cp}\} \tag{1.47}$$

Substitution of Eqs. (1.10) and (1.47) gives the following relationship:

$$u = \left[\frac{2\gamma}{\gamma-1} R_g T_0 \left\{1 - \left(\frac{p}{p_0}\right)^{\frac{\gamma-1}{\gamma}}\right\}\right]^{\frac{1}{2}} \tag{1.48 a}$$

The flow velocity at the nozzle exit is represented by

$$u_e = \left[\frac{2\gamma}{\gamma-1} R_g T_0 \left\{1 - \left(\frac{p_e}{p_0}\right)^{\frac{\gamma-1}{\gamma}}\right\}\right]^{\frac{1}{2}} \tag{1.48 b}$$

where the subscript e denotes the exit of the nozzle. The mass flow rate is given by the law of mass conservation for a steady-state one-dimensional flow as

$$\dot{m} = \rho u A \tag{1.49}$$

where $\dot{m}$ is the mass flow rate in the nozzle, ρ is the gas density, and A is the cross-sectional area of the nozzle. Substituting Eqs. (1.48), (1.5), and (1.14) into Eq. (1.49), one obtains

$$\dot{m} = p_0 A \left[\frac{2\gamma}{\gamma-1} \frac{1}{R_g T_0} \left(\frac{p}{p_0}\right)^{\frac{2}{\gamma}} \left\{1 - \left(\frac{p}{p_0}\right)^{\frac{\gamma-1}{\gamma}}\right\}\right]^{\frac{1}{2}} \tag{1.50}$$

Thus, the mass flux defined in $\dot{m}/A$ is given by

$$\frac{\dot{m}}{A} = p_0 \left[\frac{2\gamma}{\gamma-1} \frac{1}{R_g T_0} \left(\frac{p}{p_0}\right)^{\frac{2}{\gamma}} \left\{1 - \left(\frac{p}{p_0}\right)^{\frac{\gamma-1}{\gamma}}\right\}\right]^{\frac{1}{2}} \tag{1.51 a}$$

The mass flux can also be expressed as a function of Mach number using Eqs. (1.25) and (1.26) as follows:

$$\begin{aligned} \frac{\dot{m}}{A} &= \rho u = \frac{pu}{R_g T} \\ &= \sqrt{\frac{\gamma}{R_g T_0}}\, p M \left(1 + \frac{\gamma-1}{2} M^2\right)^{\frac{1}{2}} \\ &= \sqrt{\frac{\gamma}{R_g T_0}}\, p_0 M \left(1 + \frac{\gamma-1}{2} M^2\right)^{\frac{\zeta}{2}} \end{aligned} \tag{1.51 b}$$

Differentiation of Eq. (1.50) yields

$$\frac{d}{dM}\left(\frac{\dot{m}}{A}\right)=\sqrt{\frac{\gamma}{R_g T_0}}\,p_0(1-M^2)\left(1+\frac{\gamma-1}{2}M^2\right)^{\frac{1-3\gamma}{2(\gamma-1)}} \tag{1.51c}$$

It is evident that $\dot{m}$ is maximal at $M = 1$. The maximum mass flux, $(\dot{m}/A)_{max}$, is obtained when the cross-sectional area is A^* as

$$\left(\frac{\dot{m}}{A^*}\right)_{max}=\sqrt{\frac{\gamma}{R_g T_0}}\,p_0\left(\frac{2}{\gamma+1}\right)^{\frac{\zeta}{2}} \tag{1.52}$$

Thus, the area ratio, A/A^*, is obtained as

$$\frac{A}{A^*}=\frac{1}{M}\left\{\frac{2}{\gamma+1}\left(1+\frac{\gamma-1}{2}M^2\right)\right\}^{\frac{\zeta}{2}} \tag{1.53}$$

The flow Mach number at A is obtained by the use of Eq. (1.53) when $\dot{m}$, T_0, p_0, R_g, and γ are given. In addition, T, p, and ρ are obtained by the use of Eqs. (1.25), (1.26), and (1.27). Differentiation of Eq. (1.53) with respect to Mach number yields Eq. (1.54):

$$\frac{d}{dM}\left(\frac{A}{A^*}\right)=\frac{M^2-1}{M^2}\frac{2}{\gamma+1}\left\{\frac{2}{\gamma+1}\left(1+\frac{\gamma-1}{2}M^2\right)\right\}^{\frac{2}{\gamma-1}-\frac{\zeta}{2}} \tag{1.54}$$

Equation (1.54) indicates that A/A^* becomes minimal at $M = 1$. The flow Mach number increases as A/A^* decreases when $M < 1$, and also increases as A/A^* increases when $M > 1$. When $M = 1$, the relationship $A = A^*$ is obtained and is independent of γ. It is evident that A^* is the minimum cross-sectional area of the nozzle flow, the so-called "nozzle throat", in which the flow velocity becomes the sonic velocity. Furthermore, it is evident that the velocity increases in the subsonic flow of a convergent part and also increases in the supersonic flow of a divergent part.

The velocity u^*, temperature T^*, pressure p^*, and density ρ^* in the nozzle throat are obtained by the use of Eqs. (1.16), (1.18), (1.19), and (1.20), respectively:

$$u^*=\sqrt{\gamma R T^*} \tag{1.55}$$

$$\frac{T^*}{T_0}=\frac{2}{\gamma+1} \tag{1.56}$$

$$\frac{p^*}{p_0}=\left(\frac{2}{\gamma+1}\right)^{\frac{\gamma}{\gamma-1}} \tag{1.57}$$

$$\frac{\rho^*}{\rho_0}=\left(\frac{2}{\gamma+1}\right)^{\frac{1}{\gamma-1}} \tag{1.58}$$

For example, $T^*/T_0 = 0.833$, $p^*/p_0 = 0.528$, and $\rho^*/\rho_0 = 0.664$ are obtained when $\gamma = 1.4$. The temperature T_0 at the stagnation condition decreases by 17 % and the pressure p_0 decreases by 50 % in the nozzle throat. The pressure decrease is more rapid than the temperature decrease when the flow expands through a convergent nozzle. The maximum flow velocity is obtained at the exit of the divergent part of the nozzle. When the pressure at the nozzle exit is a vacuum, the maximum velocity is obtained by the use of Eqs. (1.48) and (1.6) as

$$u_{e,max} = \sqrt{\frac{2\gamma}{\gamma-1}\frac{R}{M_g}T_0} \tag{1.59}$$

This maximum velocity depends on the molecular mass M_g, the specific heat γ, and the stagnation temperature T_0. The velocity increases as γ and M_g decrease, and as T_0 increases. Based on Eq. (1.52), a simplified expression for mass flow rate in terms of the nozzle throat area A_t (= A^*) and the chamber pressure p_c (= p_0) is given by

$$\dot{m} = c_D A_t p_c \tag{1.60}$$

where c_D is the nozzle discharge coefficient given by

$$c_D = \sqrt{\frac{M_g}{T_0}}\sqrt{\frac{\gamma}{R}\left(\frac{2}{\gamma+1}\right)^{\zeta}} \tag{1.61}$$

1.3
Formation of Propulsive Forces

1.3.1
Momentum Change and Thrust

One assumes a propulsion engine operated in the atmosphere, as shown in Fig. 1.3. Air enters in the front end i, passes through the combustion chamber c, and is expelled from the exit e. The heat generated by the combustion of an energetic material is transferred to the combustion chamber. The momentum balance to generate thrust F is represented by the terms:

$$F + p_a\,(A_e - A_i) = (\dot{m}_e\,u_e - p_e\,A_e) - (\dot{m}_i\,u_i + p_i\,A_i) \tag{1.62}$$

$\dot{m}_i\,u_i$ = incoming momentum at i

$\dot{m}_e\,u_e$ = outgoing momentum at e

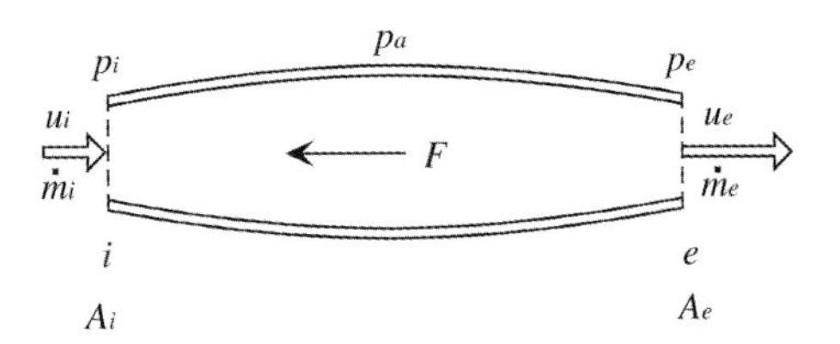

Fig. 1.3 Momentum change for propulsion.

$p_i\ A_i$ = pressure force acting at i

$p_e\ A_e$ = pressure force acting at e

$F + p_a\ (A_e - A_i)$ = force acting on the outer surface of engine

where u is the flow velocity, $\dot{m}$ is the mass flow rate, A is the area, and the subscripts i, e, and a denote inlet, exit, and ambient atmosphere, respectively. The mass flow rate of the energetic material supplied to the combustion chamber $\dot{m}_p$ is given by the difference in the mass flow rates at the exit and the inlet, $\dot{m}_e - \dot{m}_i$. In the case of rocket propulsion, the front end is closed ($A_i = 0$) and there is no influx of mass to the combustion chamber ($\dot{m}_i = 0$). Thus, the thrust for rocket propulsion is represented by

$$F = \dot{m}_e\ u_e + A_e\ (p_e - p_a) \tag{1.63}$$

where $\dot{m}_p = \dot{m}_g$. Thus, the thrust is determined by the flow velocity and pressure at the exit when $\dot{m}_e$, A_e, and p_a are given.

Differentiation of Eq. (1.63) with respect to A_e gives

$$dF/dA_e = u_e\ d\dot{m}_g/dA_e + \dot{m}_g\ du_e/dA_e + A_e\ dp_e/dA_e + p_e - p_a \tag{1.64}$$

The momentum equation at the nozzle exit is represented by $\dot{m}_g\ du_e = -A_e\ dp_e$, and $d\dot{m}_g = 0$ for a steady-state flow at the nozzle. Thus, from Eq. (1.64), one obtains the relationship

$$dF/dA_e = p_e - p_a \tag{1.65}$$

The maximum thrust is obtained at $p_e = p_a$, i. e., when the pressure at the nozzle exit is equal to the ambient pressure.

However, it must be noted that Eq. (1.62) is applicable for ramjet propulsion, as in ducted rockets and solid-fuel ramjets, because in these cases air enters through the inlet and a pressure difference between the inlet and the exit is set up. The mass flow rate from the inlet $\dot{m}_i$ plays a significant role in the generation of thrust in the case of ramjet propulsion.

1.3.2 Rocket Propulsion

Fig. 1.4 shows a schematic drawing of a rocket motor composed of propellant, combustion chamber, and nozzle. The nozzle is a convergent–divergent nozzle designed to accelerate the combustion gas from subsonic to supersonic flow through the nozzle throat. The thermodynamic process in a rocket motor is shown in Fig. 1.4 by a pressure–volume diagram and an enthalpy–entropy diagram.[6] The propellant contained in the chamber burns and generates combustion products, and this increases the temperature from T_i to T_c at a constant pressure, p_c. The com-

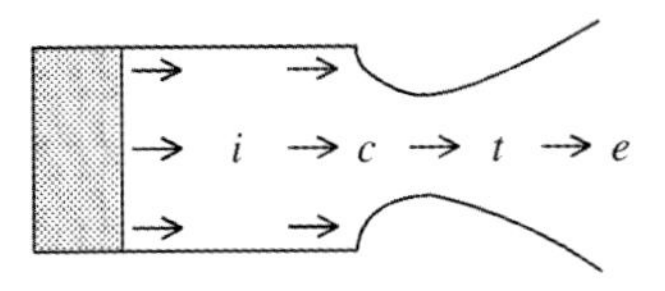

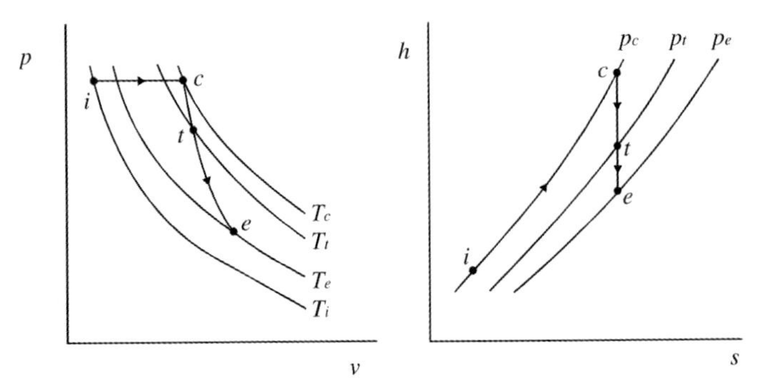

Fig. 1.4 Pressure–volume and enthalpy–entropy diagrams for rocket propulsion.

bustion products expand through the convergent nozzle to give pressure p_t and temperature T_t at the nozzle throat. The combustion products continue to expand through a divergent nozzle to give pressure p_e and temperature T_e at the nozzle exit.

If one can assume that (1) the flow is one-dimensional and in a steady-state, (2) the flow is an isentropic process, and (3) the combustion gas is an ideal gas and the specific heat ratio is constant, the plots of p vs. v and of h vs. s are uniquely determined.[6–9] The enthalpy change due to the combustion of the propellant is given by

$$\Delta h = c_p\,(T_c - T_i) \tag{1.66}$$

where Δh is the heat of reaction of propellant per unit mass. The expansion process $c \rightarrow t \rightarrow e$ shown in Fig. 1.4 follows the thermodynamic process described in Section 1.2.3.

1.3.2.1 **Thrust Coefficient**

The thrust generated by a rocket motor is represented by Eq. (1.63). Substituting Eqs. (1.48 b) and (1.52) into Eq. (1.63), one gets

$$F = A_t p_c \left[\frac{2\gamma^2}{\gamma-1}\left(\frac{2}{\gamma+1}\right)^{\frac{\gamma+1}{\gamma-1}}\left\{1-\left(\frac{p_e}{p_c}\right)^{\frac{\gamma-1}{\gamma}}\right\}\right]^{\frac{1}{2}} + (p_e - p_a)\,A_e \tag{1.67}$$

As shown by Eq. (1.65), the maximum thrust F_{max} is obtained when $p_e = p_a$ at a given specific heat ratio of the combustion gas:

$$F_{max} = A_t p_c \left[\frac{2\gamma^2}{\gamma-1} \left(\frac{2}{\gamma+1} \right)^{\frac{\gamma+1}{\gamma-1}} \left\{ 1 - \left(\frac{p_e}{p_c} \right)^{\frac{\gamma-1}{\gamma}} \right\} \right]^{\frac{1}{2}} \tag{1.68}$$

Equation (1.68) can be represented by a simplified expression for thrust in terms of nozzle throat area and chamber pressure:

$$F = c_F A_t p_c \tag{1.69}$$

where c_F is the thrust coefficient and is given by

$$c_F = \left[\frac{2\gamma^2}{\gamma-1} \left(\frac{2}{\gamma+1} \right)^{\frac{\gamma+1}{\gamma-1}} \left\{ 1 - \left(\frac{p_e}{p_c} \right)^{\frac{\gamma-1}{\gamma}} \right\} \right]^{\frac{1}{2}} + \frac{p_e - p_a}{p_c} \frac{A_e}{A_t} \tag{1.70}$$

The maximum thrust coefficient $c_{F,max}$ is then given by

$$c_{F,max} = \left[\frac{2\gamma^2}{\gamma-1} \left(\frac{2}{\gamma+1} \right)^{\frac{\gamma+1}{\gamma-1}} \left\{ 1 - \left(\frac{p_e}{p_c} \right)^{\frac{\gamma-1}{\gamma}} \right\} \right]^{\frac{1}{2}} \tag{1.71}$$

When the nozzle expansion ratio becomes infinity, the pressure ratio p_c/p_a also becomes infinity. The maximum thrust coefficient $c_{F,max}$ then becomes

$$c_{F,max} = \left[\frac{2\gamma^2}{\gamma-1} \left(\frac{2}{\gamma+1} \right)^{\frac{\gamma+1}{\gamma-1}} \right]^{\frac{1}{2}} \tag{1.72}$$

For example, $c_{F,max}$ is 2.246 for $\gamma = 1.20$ and 1.812 for $\gamma = 1.40$.

1.3.2.2 **Characteristic Velocity**

The characteristic velocity c^* is defined according to

$$c^* = \frac{A_t p_c}{\dot{m}_g} \tag{1.73}$$

Substituting Eq. (1.52) into Eq. (1.73), one gets

$$c^* = \sqrt{\frac{RT_c}{\gamma M_g} \left(\frac{2}{\gamma+1} \right)^{-\frac{\zeta}{2}}} \tag{1.74}$$

It can be shown that c^* is dependent only on T_g, M_g, and γ and that it is independent of the pressure and the physical dimensions of the combustion chamber and exhaust nozzle; c^*, as defined in Eq. (1.74), is a parameter used to describe the energetics of combustion.

1.3.2.3 Specific Impulse

Specific impulse I_{sp} is a parameter used to describe the energy efficiency of propellant combustion, which is represented by

$$I_{sp} = F/\dot{m}_g\, g \tag{1.75}$$

where g is the gravitational acceleration, 9.80665 m s^{-2}, and hence specific impulse is expressed in terms of seconds. Thermodynamically, specific impulse is the effective time required to generate thrust that can sustain the propellant mass against the gravitational force through energy conversion. Since the mass flow rate $\dot{m}_g$ is given by Eq. (1.50) and F is given by Eq. (1.67), I_{sp} is represented by

$$I_{sp} = \frac{1}{g}\left[\frac{2\gamma}{\gamma-1}\frac{R}{M_g}T_c\left\{1-\left(\frac{p_e}{p_c}\right)^{\frac{\gamma-1}{\gamma}}\right\}\right]^{\frac{1}{2}} + \frac{1}{g}\left(\frac{\gamma+1}{2}\right)^{\frac{\zeta}{2}}\sqrt{\frac{RT_c}{\gamma M_g}}\left(\frac{p_e-p_a}{p_c}\right)\frac{A_e}{A_t} \tag{1.76}$$

$$\sim (T_g/M_g)^{1/2} \tag{1.77}$$

where T_g is the combustion temperature and M_g is the molecular mass of the combustion products. Though $I_{sp,max}$ is also a function of the specific heat ratio γ of the combustion products, γ varies relatively little among propellants. It is evident from Eq. (1.77) that an energetic material that produces high-T_g and high-M_g combustion products is not always a useful propellant. A propellant that generates low-T_g can also be useful if M_g is sufficiently low. Similar to F_{max} and $c_{F,max}$, the maximum specific impulse $I_{sp,max}$ is obtained when $p_e = p_a$:

$$I_{sp,max} = \frac{1}{g}\left[\frac{2\gamma}{\gamma-1}\frac{R}{M_g}T_g\left\{1-\left(\frac{p_a}{p_c}\right)^{\frac{\gamma-1}{\gamma}}\right\}\right]^{\frac{1}{2}} \tag{1.78}$$

In addition, the specific impulse is given by the thrust coefficient and the characteristic velocity according to

$$I_{sp} = c_F\, c^*/g \tag{1.79}$$

Since c_F indicates the efficiency of the expansion process in the nozzle flow and c^* indicates the efficiency of the combustion process in the chamber, I_{sp} gives an indication of the overall efficiency of a rocket motor.

1.3.3 Gun Propulsion

1.3.3.1 Thermochemical Process of Gun Propulsion

Gun propellants burn under conditions of non-constant volume and non-constant pressure. The rate of gas generation changes rapidly with time and the temperature changes simultaneously because of the displacement of the projectile in the combustion chamber of the gun barrel.[10–12] Though the pressure change is rapid, the linear burning rate is assumed to be expressed by a pressure exponent law, the so-called Vieille's law, i. e.:

$$r = ap^n \tag{1.80}$$

where r is the burning rate (mm s^{-1}), p is the pressure (MPa), n is a constant dependent on the composition of the propellant, and a is a constant dependent on the initial chemical composition and temperature of the propellant.

The fundamental difference between gun propellants and rocket propellants lies in the magnitude of the burning pressure. Since the burning pressure in guns is extremely high, more than 100 MPa, the parameters of the above equation are empirically determined. Though rocket propellant burns at below 20 MPa, in general, the burning rate expression of gun propellants appears to be similar to that of rocket propellants. The mass burning rate of the propellant is also dependent on the burning surface area of the propellant, which increases or decreases as the burning proceeds. The change in the burning surface area is determined by the shape and dimensions of the propellant grains used.

The effective work done by a gun propellant is the pressure force that acts on the base of the projectile. Thus, the work done by propellant combustion is expressed in terms of the thermodynamic energy, f, which is represented by

$$f = pv = RT_g/M_g = p_0 v_0 T_g/T_0 \tag{1.81}$$

where p_0, v_0, and T_0 are the pressure, volume, and temperature, respectively, generated by the combustion of unit mass of propellant in the standard state. The thermodynamic energy f is expressed in units of MJ kg^{-1}. It is evident that a higher f value is favorable for a gun propellant, similar to I_{sp} used to evaluate the thermodynamic energy of rocket propellants.

The thermal energy generated by propellant combustion is distributed to various non-effective energies.[10] The energy losses of a caliber gun are approximately as follows:

Sensible heat of combustion gas	42 %
Kinetic energy of combustion gas	3 %
Heat loss to gun barrel and projectile	20 %
Mechanical losses	3 %

The remaining part of the energy, 32 %, is used to accelerate the projectile. It is obvious that the major energy loss is the heat released from the gun barrel. This is an unavoidable heat loss based on the laws of thermodynamics: the pressure in the gun barrel can only be expended by the cooling of the combustion gas to the atmospheric temperature.

1.3.3.2 **Internal Ballistics**

The one-dimensional momentum equation for the internal ballistics of a gun is represented by:[10–12]

$$M_w\, du/dt = M_w\, u\, du/dx = pA_{bi} \tag{1.82}$$

where M_w is the mass of the projectile, u is its velocity, x is the distance travelled, t is time, p is pressure, and A_{bi} is the cross-sectional area of the gun barrel. Integration of Eq. (1.82) from 0 to L_b gives

$$u_{be} = \sqrt{\frac{2[p]L_b A_{bi}}{M_w}} \tag{1.83}$$

where u_{be} is the velocity at the barrel exit and L_b is the effective length of the barrel used to accelerate the projectile. If one assumes an averaged pressure in the barrel, $[p]$, given by

$$f = pv = \mathrm{R}\frac{T_g}{M_g} \tag{1.84}$$

the velocity of the projectile is given by

$$u_{be} = (2[p]\ A_{bi}\ L_b/M_w)^{1/2} \tag{1.85}$$

With fixed physical dimensions of a gun barrel, the thermodynamic efficiency of a gun propellant is expressed by its ability to produce as high a pressure in the barrel as possible from a given propellant mass within a limited time.

In general, the internal pressure in a gun barrel exceeds 200 MPa, and the pressure exponent, n, of the propellant burning rate given by Eq. (1.80) is 1. When $n = 1$, the burning rate of a gun propellant is represented by

$$r = ap \tag{1.86}$$

where r is the burning rate, p is the pressure, and a is constant dependent on the chemical ingredients and the initial temperature of the propellant grain. The volumetric burning rate of a propellant grain is represented by $S(t)r$, where $S(t)$ is the surface area of the propellant grain at time t. The volumetric burning change of the propellant grain is defined by

$$dz/dt = V(t)/V_0$$

$$\frac{S_0}{V_0}\frac{S(t)}{S_0}r(t) \tag{1.83}$$

where V_0 is the initial volume of the propellant grain, $V(t)$ is the volume of the propellant grain at time t, and z is a geometric function of the grain. The surface area ratio change, termed the "form function", φ, is defined according to

$$\varphi(z) = S(t)/S_0 \tag{1.88}$$

Table 1.1 Form functions for various types of propellant grain.

Grain Shape	φ(z)
spherical, cubic grain	$(1 - z)^{2/3}$
disk, square, strand	$(1 - z)^{1/2}$
short column	$(1 - z)^{3/5}$
short tubular	$(1 - 0.57z)^{1/2}$
center perforated disk	$(1 - 0.33z)^{1/2}$
long tubular	1
seven-holes short tubular	$(1 + z)^{1/2}$

Table 1.1 shows form functions for several types of propellant grains. Substituting Eqs. (1.80), (1.86), and (1.88) into Eq. (1.87), one obtains a simplified expression for the volumetric burning rate change:

$$dz/dt = a\sigma\varphi(z)p \tag{1.89}$$

Substituting Eq. (1.89) into Eq. (1.82), the velocity change of the projectile is determined by

$$du = (A_{bi}/M_w)(1/a\sigma)dz/\varphi(z) \tag{1.90}$$

The velocity of the projectile is obtained by integration of Eq. (1.90) from the initial stage to the stage z_1:

$$u = (A_{bi}/M_w)(1/a\sigma)\int_0^{z1} dz/\phi(z) \tag{1.91}$$

where u is the velocity when $z = z_1$. In general, the projectile starts to move when the pressure in the barrel reaches a certain initial pressure, p_c, due to the action of the shot resistance between the projectile and the barrel. The velocity of the projectile is then represented by

$$u = (A_{bi}/M_w)(1/a\sigma)\left\{\int_0^{z1} dz/\phi(z) - \int_0^{z0} dz/\phi(z)\right\} \tag{1.92}$$

where z_0 is the volumetric burning change at p_c. After the propellant grain is completely consumed, the pressure in the barrel changes isentropically according to

$$p = \rho_g p^*/\rho_g^* \tag{1.93}$$

where p^* and ρ_g^* denote the pressure and density, respectively, when burning is complete.

1.4
Formation of Destructive Forces

1.4.1
Pressure and Shock Wave

When a propellant grain burns in a closed chamber, a large number of gaseous molecules are produced. The pressure generated by these molecules acts on the inner surface the chamber. The pressure increases slowly due to the continuous burning of the propellant. When the pressure exceeds the mechanical strength of the chamber wall, mechanical breakage occurs at the weakest portion of the chamber wall. The force acting on the chamber wall is caused by the static pressure of the combustion gas.

When an explosive detonates in a closed container, a shock wave is formed. The shock wave travels toward the inner surface of the chamber and acts on the chamber wall. The pressure wave is caused by this shock wave, rather than the pressure created by the detonated burned gases. The shock wave travels first through the air in the chamber and the burned gas follows somewhat later. When the shock wave reaches the inner surface of the chamber wall, the chamber will be damaged if the mechanical strength of the chamber wall is lower than the mechanical force created by the shock wave. Though the time for which the shock wave acts on the wall is very short, in contrast to the static pressure built up by combustion gases, the impulsive force caused at the wall leads to destructive damage. Though no pressure is formed when a propellant grain burns outside of the chamber, a shock wave is still formed when an explosive detonates externally. When this shock wave reaches the outer surface of the chamber, the chamber wall may well be damaged.

1.4.2
Shock Wave Propagation and Reflection in Solid Materials

When a shock wave travels in a solid wall from one end to the other, a compressive force is created at the front end of the shock wave. When the shock wave reaches the other end, a reflection wave is formed, which travels back in the reverse direction. This reflection wave forms an expansion force that acts on the wall.

There are two general modes for the destruction of solid materials, namely ductile fracture and brittle fracture. These modes are dependent on the type of material and on the type of forces acting on the material. The mechanical force created by a shock wave is similar to the force created by an impact stress. The breakage mechanism of materials is dependent on the action of the mechanical force. When a shock wave travels in a concrete wall from one end to the other, it generates a compressive stress, and no damage is observed. However, when the shock wave is reflected at the other end of the wall, a reflection wave is formed, accompanied by an expansion stress. Since the compression strength of the concrete is sufficient to endure the compressive stress created by the shock wave, no mechanical damage results from the shock wave itself. However, when the concrete wall is subjected to the tensile stress created by the expansion wave, the expansion force exceeds the tensile strength of the wall, thus leading to its breakage.

References

1 Jeans, J., Introduction to the Kinetic Theory of Gases, The University Press, Cambridge (1959).
2 Dickerson, R. E., Molecular Thermodynamics, W. A. Benjamin, New York (1969), Chapter 5.
3. JANAF Thermochemical Tables, Dow Chemical Co., Midland, Michigan (1960–1970).
4 Liepmann, H. W., and Roshko, A., Elements of Gas Dynamics, John Wiley & Sons, New York (1957), Chapter 2.
5 Shapiro, A. H., The Dynamics and Thermodynamics of Compressible Fluid Flow, The Ronald Press Company, New York (1953), Chapter 5.
6 Summerfield, M., The Liquid Propellant Rocket Engine, Jet Propulsion Engines, Princeton University Press, New Jersey (1959), pp. 439–520.
7 Glassman, I., and Sawyer, F., The Performance of Chemical Propellants, Circa Publications, New York (1970), Chapter 2.
8 Sutton, G. P., Rocket Propulsion Elements, 6th edition, John Wiley & Sons, Inc., New York (1992), Chapter 3.
9 Kubota, N., Rocket Combustion, Nikkan Kogyo Press, Tokyo (1995), Chapter 2.
10 Weapons Systems Fundamentals – Analysis of Weapons, U.S. Navy Weapons Systems, NAVWEPS Operating Report 3000, Vol. 2 , 1963.
11 Krier, H., and Adams, M. J., An Introduction to Gun Interior Ballistics and a Simplified Ballistic Code, Interior Ballistics of Guns (Eds.: Krier, H., and Summerfield, M.), Progress in Astronautics and Aeronautics, Vol. 66, AIAA, New York (1979).
12 Gun Propulsion Technology (Ed.: Stiefel, L.), Progress in Astronautics and Aeronautics, Vol. 109, AIAA, Washington DC (1988).

2
Thermochemistry of Combustion

2.1
Generation of Heat Energy

2.1.1
Chemical Bond Energy

All materials are composed of atoms, which are tightly bonded to each other to form molecules of the materials. When chemical reaction occurs, the materials change to other materials through bond breakage and new bond formation. The chemical bond energy is dependent on the type of atoms and their energetic state pbased on the electrons of each atom. Though the exact energetic state of materials should strictly be evaluated by a quantum mechanical treatment, an overall energetic state is evaluated from the experimentally determined chemical bond energies.

Typical chemical bonds, as found in energetic materials and their combustion products, can be classified as covalent or ionic. Gaseous reactants such as hydrogen, oxygen, nitrogen, and hydrocarbons contain covalent bonds, which may be single, double, and/or triple bonds. Many energetic crystalline materials are formed by ionic bonds. The status of the bonds determines the energetics and physicochemical properties of the materials.

The total chemical bond energy between each constituent atom of a molecule is the energy of the molecule. Bond breakage in a gas molecule can be induced by collision between molecules or by externally provided photochemical energy. When a molecule dissociates into atoms, energy is needed to break the chemical bonds and to separate the atoms. The chemical bond energy of a molecule A–B is equal to the dissociation energy to form atoms A and B. For example, an H_2O molecule dissociates to form two H atoms and one O atom according to $H_2O \rightarrow 2H + O$ (-912 kJ mol^{-1}). If the two O–H bonds are broken according to $H_2O \rightarrow H + OH$ (-494 kJ mol^{-1}) followed by $OH \rightarrow O + H$ (-419 kJ mol^{-1}), the first O–H bond breakage would require 494 kJ mol^{-1} and the second 419 kJ mol^{-1}, giving an apparent difference in O–H bond energies of 75 kJ mol^{-1}. However, the two dissociation processes occur simultaneously for H_2O molecules and the averaged O–H bond energy appears to be 457 kJ mol^{-1}.

Propellants and Explosives. Naminosuke Kubota

ISBN: 978-3-527-31424-9

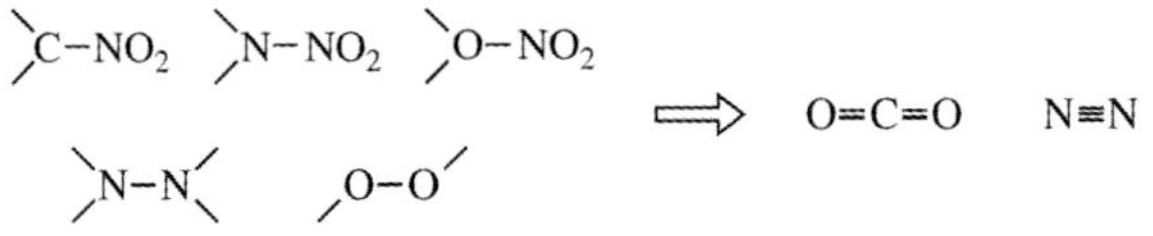

$>C-NO_2$ $>N-NO_2$ $>O-NO_2$ ⟹ $O=C=O$ $N \equiv N$

$>N-N<$ $-O-O-$

Fig. 2.1 Generation of heat by bond breakage in energetic materials and formation of CO_2 and N_2.

Typical chemical bonds used to formulate energetic materials are $C-NO_2$, $N-NO_2$, $O-NO_2$, N–N, and O–O bonds. When such bonds are broken in molecules due to thermal decomposition or reactions with other molecules, molecules of the gases CO_2 and N_2 are formed, as shown in Fig. 2.1. The difference between the bond energy of the energetic material and that of the gas molecules is released as heat energy. The chemical bond energies of typical bonds found in molecules related to combustion are shown in Table 2.1.[1–3]

Table 2.1 Chemical bond energies of the constituent bonds of energetic materials.

Chemical Bond	Bond Energy, kJ mol^{-1}
C – H	411
C – C	358
C = C	599
C ≡ C	812
C – O	350
C = O (ketone)	766
C = O (formaldehyde)	699
C = O (acetaldehyde)	720
C ≡ N	883
H – H	435
O – H	465
O – O	138
N – H	368
N – O	255
N = O	601
N = N	443
N ≡ N	947

2.1.2 Heat of Formation and Heat of Explosion

When reactant R of an energetic material reacts to generate product P, heat is released (or absorbed). Since the chemical bond energy of R is different from that of P, the energy difference between R and P appears as heat. The rearrangement of the molecular structure of R changes the chemical potential. The heat of reaction at constant pressure, represented by Q_p, is equal to the enthalpy change of the chemical reaction:

$$\Delta H = Q_p \tag{2.1}$$

where H is the enthalpy, ΔH is the enthalpy change of the reaction, and the subscript p indicates the condition of constant pressure.

The heat produced by a chemical reaction is expressed by the "heat of explosion", H_{exp}. H_{exp} is determined by the difference between the heat of formation of the reactants, $\Delta H_{f,\mathrm{R}}$, and the heat of formation of the products, $\Delta H_{f,\mathrm{p}}$, as represented by

$$H_{exp} = \Delta H_{f,\mathrm{R}} - \Delta H_{f,\mathrm{p}} \tag{2.2}$$

The heats of formation, ΔH_f, are dependent on the chemical structures and chemical bond energies of the constituent molecules of the reactants and products. Equation (2.2) indicates that the higher the value of $\Delta H_{f,\mathrm{R}}$ for the reactants and the lower the value of $\Delta H_{f,\mathrm{p}}$ for the products, the higher the H_{exp} that will be obtained.

2.1.3 Thermal Equilibrium

Though combustion is a very fast exothermic chemical reaction compared with other chemical reactions, the reaction time is finite and the combustion products are formed after a large number of molecular collisions, which also produce a large number of intermediate molecules. When the time-averaged numbers of molecules reach a constant level and the temperature becomes constant, the reaction system is said to be in a state of thermal equilibrium.[1,2,4]

The Gibbs free energy F for one mole of an ideal gas is defined according to

$$F = h - Ts = e + pv - Ts \tag{2.3}$$

Substituting Eq. (1.3) into Eq. (2.3) and then differentiating, one gets

$$dF = v\,dp - s\,dT \tag{2.4}$$

Substituting the equation of state for one mole of an ideal gas, Eq. (1.5), into Eq. (2.4), one gets

$$dF = (\mathrm{R}T/p)dp - s\,dT \tag{2.5}$$

When the temperature of the gas remains unchanged, Eq. (2.5) is represented by

$$dF = \mathrm{R}T\,d\ln p \tag{2.6}$$

Integrating Eq. (2.6), one gets

$$F - F_0 = \mathrm{R}T\ln p \tag{2.7}$$

where F_0 is the standard free energy at temperature T and a pressure of 1 MPa.

Considering a reversible reaction, one assumes that the chemical species A, B, C, and D are in a thermal equilibrium state according to:

$$\mathrm{aA + bB \rightleftharpoons cC + dD} \tag{2.8}$$

where a, b, c, and d are the stoichiometric coefficients of the reversible reaction in the thermal equilibrium state. Based on Eq. (2.7), the change in free energy dF associated with the reaction is represented by

$$\Delta F - \Delta F_0 = \mathrm{R}T\,[(\mathrm{c}\ln p_\mathrm{c} + \mathrm{d}\ln p_\mathrm{d}) - (\mathrm{a}\ln p_\mathrm{a} + \mathrm{b}\ln p_\mathrm{b})]$$

$$= \mathrm{R}T \ln (p_\mathrm{c}^{\mathrm{c}}\, p_\mathrm{d}^{\mathrm{d}} / p_\mathrm{a}^{\mathrm{a}}\, p_\mathrm{b}^{\mathrm{b}}) \tag{2.9}$$

where p_a, p_b, p_c, and p_d are the partial pressures of the chemical species A, B, C, and D, respectively, and ΔF_0, is the standard free energy change of the reaction represented by Eq. (2.8). When the reaction is in thermodynamic equilibrium, the free energy change becomes zero, $\Delta F = 0$, and Eq. (2.9) becomes

$$\Delta F_0 = -\mathrm{R}T \ln (p_\mathrm{c}^{\mathrm{c}}\, p_\mathrm{d}^{\mathrm{d}} / p_\mathrm{a}^{\mathrm{a}}\, p_\mathrm{b}^{\mathrm{b}}) \tag{2.10}$$

$$= -\mathrm{R}T \ln K_p \tag{2.11}$$

where K_p is the equilibrium constant defined by

$$K_p = p_\mathrm{c}^{\mathrm{c}}\, p_\mathrm{d}^{\mathrm{d}} / p_\mathrm{a}^{\mathrm{a}}\, p_\mathrm{b}^{\mathrm{b}} \tag{2.12}$$

Substituting Eq. (2.11) into Eq. (2.3), one gets

$$dh_0 - Tds_0 = -\mathrm{R}T \ln K_p \tag{2.13}$$

where dh_0 and ds_0 are the changes in enthalpy and entropy between standard pressure (1 MPa) and a pressure p at temperature T. Thus, the left-hand side of Eq. (2.13) is the free energy change between standard pressure (1 MPa) and pressure p at temperature T. A K_p value can be obtained for any reaction of the type shown in Eq. (2.8). Since the free energy change is dependent on temperature, and not on pressure, K_p is also dependent only on temperature. The equilibrium constants of combustion products are tabulated in the JANAF Thermochemical Tables.[6] If the reaction is of the type a + b = c + d, the number of molecules involved is not changed by the reaction and the position of the equilibrium appears to be pressure insensitive.

2.2 Adiabatic Flame Temperature

The heat of reaction of a process is given by the difference in the heats of formation of the reactants and products. On going from a reactant temperature of T_0 to a product temperature of T_1, the energy change due to a chemical reaction is represented by the energy conservation law according to[4,5,7]

$$\sum_i n_i[(H_{T1} - H_0^0) - (H_{T0} - H_0^0) + (\Delta H_f)_{T0}]_i$$
$$= \sum_j n_j[(H_{T2} - H_0^0) - (H_{T0} - H_0^0) + (\Delta H_f)_{T0}]_j + \sum_j n_j Q_j \tag{2.14}$$

where $H^0{}_0$ is the standard enthalpy at 0 K, n is the stoichiometry of the chemical species involved in the reaction, the subscripts T_1 and T_2 indicate the temperatures of the reactant and product, respectively, i denotes the reactant, and j denotes the product. This expression indicates that the energy change is not dependent on the reaction pathway from the reactant to the product.

When the temperature of the reactants is T_1 and the temperature of the product is T_2, Eq. (2.14) is represented by

$$\sum_i n_i[(H_{T1} - H_{T0}) + (\Delta H_f)_{T0}]_i$$
$$= \sum_j n_j[(H_{T2} - H_{T0}) + (\Delta H_f)_{T0}]_j + \sum_j n_j Q_j \tag{2.15}$$

where T_0 is the temperature at which the heat of formation is defined, and H is the enthalpy given by

$$H_T - H_{T0} = \int_{T0}^{T} c_p dT \tag{2.16}$$

When T_0 = 298.15 K is used as the standard temperature, Eq. (2.16) becomes

$$\sum_i n_i[(H_{T1} - H^0) + \Delta H_f^0]_i$$
$$= \sum_j n_j[(H_{T2} - H^0) + \Delta H_f^0]_j + \sum_j n_j Q_j \tag{2.17}$$

where H^0 is the standard enthalpy at 298.15 K. The JANAF Thermochemical Tablesshow $H_T - H_0$ values as a function of T for various materials.

When the reaction is conducted under adiabatic conditions, all of the heat generated, n_jQ_j, is converted to the enthalpy of the products; T_2 then becomes T_f, which is defined as the adiabatic flame temperature. Equation (2.17) becomes

$$\sum_i n_i[(H_{T1} - H^0) + (\Delta H_f)_{T0}]_i = \sum_j n_j[(H_{Tf} - H^0) + (\Delta H_f)_{T0}]_j \tag{2.18}$$

Equation (2.18) can be split into two separate equations as follows:

$$Q_1 = \sum_i n_i (\Delta H_f)_{T0,i} - \sum_j n_j (\Delta H_f)_{T0,j} \tag{2.19}$$

$$Q_2 = \sum_j n_j (H_{Tf} - H_{T0})_j - \sum_i n_i (H_{T1} - H_{T0})_i \tag{2.20}$$

T_f and n_j are determined at $Q_1 = Q_2$. In the case of $T_0 = 298.15$ K, Eqs. (2.19) and (2.20) become

$$Q_1^0 = \sum_i n_i (\Delta H_f^0)_i - \sum_j n_j (\Delta H_f^0)_j \tag{2.21}$$

$$Q_2^0 = \sum_j n_j (H_{Tf} - H^0)_j - \sum_i n_i (H_{T1} - H^0)_i \tag{2.22}$$

T_f and n_j are also determined at $Q_1^0 = Q_2^0$. When the temperature of the reactants is 298.15 K, $H_{T1} = H^0$, and then

$$Q_2^0 = \sum_j n_j \int_{298}^{Tf} c_{p,j}\, dT \tag{2.23}$$

is obtained by the use of Eqs. (2.22) and (2.16).

Though the above equations are nonlinear and complex, T_f and n_j may be computed for any combustion reaction for which thermochemical data are available. In the following, the reaction $3H_2 + O_2$ at 2 MPa is used to demonstrate a representative computation, illustrating the procedure for the determination of T_f and n_j and reiterating the principles of thermochemical equilibrium and adiabatic flame temperature. First, the following reaction scheme and products are assumed:

$$3H_2 + O_2 \rightarrow n_{H2O}H_2O + n_{H2}H_2 + n_H H + n_O O + n_{O2}O_2 + n_{OH}OH \tag{2.24}$$

The mass of the each chemical species is conserved before and after reaction. For the number of hydrogen atoms:

$$3H_2 = 2n_{H2O} + 2n_{H2} + n_H + n_{OH} = 6 \tag{2.25}$$

For the number of oxygen atoms:

$$O_2 = n_{H2O} + n_O + 2n_{O2} + n_{OH} = 2 \tag{2.26}$$

The equation of state for each species is given as

$$p_j v = n_j RT \quad \text{and} \quad pv = nRT \tag{2.27}$$

Thus, one gets

$$n_j = p_j v/RT = p_j n/p \tag{2.28}$$

where n is the total number of moles of products. Substituting Eq. (2.28) into Eq. (2.25), one gets

$$2p_{H2O} + 2p_{H2} + p_H + p_{OH} = 6(RT/v) \tag{2.29}$$

$$p_{H2O} + p_O + 2p_{O2} + p_{OH} = 2(RT/v) \tag{2.30}$$

The chemical species of the products are all in chemical equilibrium according to:

$$1/2O_2 \rightleftharpoons O \tag{2.31}$$

$$1/2H_2 \rightleftharpoons H \tag{2.32}$$

$$1/2O_2 + 1/2H_2 \rightleftharpoons OH \tag{2.33}$$

$$H_2 + 1/2O_2 \rightleftharpoons H_2O \tag{2.34}$$

As defined in Eq. (2.12), the equilibrium constants for each chemical equilibrium state are given as:

$$K_{p,1} = p/p_{O2}{}^{1/2} \tag{2.35}$$

$$K_{p,2} = p_H/p_{H2}{}^{1/2} \tag{2.36}$$

$$K_{p,3} = p_{OH}/(p_{O2}{}^{1/2}\, p_{H2}{}^{1/2}) \tag{2.37}$$

$$K_{p,4} = p_{H2O}/(p_{H2}\, p_{O2}{}^{1/2}) \tag{2.38}$$

These equilibrium constants are determined by means of Eq. (2.13) as a function of temperature (see the JANAF Thermochemical Tables[6]). There are seven unknown parameters (T_f, p_{H2O}, p_{H2}, p_O, p_H, p_{O2}, and p_{OH}), and seven equations (Eqs. (2.18), (2.29), (2.30), and (2.35)–(2.38)). The procedure for the computations is as follows:

1) Determine $H_{T1} - H_{T0}$ at T_0, T_1, and p from the Thermochemical Tables.
2) Determine $H_{f,T0}$ from the Thermochemical Tables.
3) Determine $H_{Tf,1} - H_{T0}$ and $K_{p,j}$ at $T_{f,1}$ (assumed) from the Thermochemical Tables.
4) Determine p_j (or n_j) at $T_{f,1}$ and p.
5) Determine Q_1 and Q_2.
6) If $Q_1 < Q_2$ or $Q_1 > Q_2$, assume $T_{f,2}$. Repeat steps 1–6 iteratively until $Q_1 = Q_2$.
7) If $Q_1 = Q_2$ at $T_{f,m}$, then $T_{f,m}$ is the adiabatic flame temperature, T_f.

The computational process may indicate, for example, that Q_1 decreases monotonically with T and that Q_2 increases monotonically with T. The adiabatic flame temperature is determined by interpolation of the computed results. Practical computations are carried out with the aid of computer programs.[8] The results of this example are as follows:

$$T_f = 3360 \text{ K}$$

$$n_{H2O} = 1.837,\ n_{H2} = 1.013,\ n_O = 0.013,\ n_H = 0.168,\ n_{O2} = 0.009, \text{ and } n_{OH} = 0.131$$

Adiabatic flame temperatures and product mole fractions obtained from the reaction of hydrogen and oxygen in different mixture ratios are shown in Table 2.2. The maximum adiabatic flame temperature is obtained from the mixture of $2H_2 + O_2$, which is the stoichiometric ratio for the reaction.

Table 2.2 Adiabatic flame temperatures and mole fractions of the combustion products of H_2 and O_2 mixtures at 2 MPa.

Reactants	T_f (K)	Mole fractions					
		H_2O	H_2	O	H	O_2	OH
$H_2 + O_2$	3214	0.555	0.021	0.027	0.009	0.290	0.098
$2H_2 + O_2$	3498	0.645	0.136	0.022	0.048	0.042	0.107
$3H_2 + O_2$	3358	0.579	0.319	0.004	0.053	0.003	0.042
$4H_2 + O_2$	3064	0.479	0.481	0.000	0.030	0.000	0.010

If a condensed material is formed as a combustion product, no equilibrium constant as defined by Eq. (2.12) is obtained. For example, the reaction of solid carbon and oxygen produces carbon dioxide according to

$$C_{(s)} + O_{2(g)} \rightarrow CO_{2(g)} \tag{2.39}$$

Since solid carbon vaporizes as a result of its vapor pressure, $p_{vp,C}$, the chemical equilibrium constant $K_{p,6}$ is defined as

$$K_{p,6} = p_{CO2}/(p_{vp,C}\, p_{O2}) \tag{2.40}$$

However, since the vapor pressure is independent of the reaction system, it can be regarded as a characteristic value of the thermodynamics. The equilibrium constant $K_{p,7}$ is defined as

$$K_{p,7} = K_{p,6} \times p_{vp,C}$$

$$= p_{CO2}/p_{O2} \tag{2.41}$$

Thus, $K_{p,7}$ is determined by the vapor pressure and the standard free energy. In the first step of the computations, it is assumed that no solid species are formed. If the partial pressure is higher than the vapor pressure, the next step of the computations should include liquid or solid materials. The computation is then continued until the conditions are matched.

2.3 Chemical Reaction

2.3.1 Thermal Dissociation

When high-temperature products are in an equilibrium state, many of the constituent molecules dissociate thermally. For example, the rotational and vibrational modes of carbon dioxide are excited and their motions become very intense. As the temperature is increased, the chemical bonds between the carbon and oxygen atoms are broken. This kind of bond breakage is called thermal dissociation. The dissociation of H_2O becomes evident at about 2000 K and produces H_2, OH, O_2, H, and O at 0.1 MPa. About 50% of H_2O is dissociated at 3200 K, rising to 90% at 3700 K. The products H_2, O_2, and OH dissociate to H and O as the temperature is increased further. The fraction of thermally dissociated molecules is suppressed as the pressure is increased at constant temperature.

2.3.2 Reaction Rate

Since chemical reactions occur through molecular collisions, the reaction of a chemical species M_j can be represented by

$$\sum_{j=1}^{n} \nu_j' \, M_j \rightarrow \sum_{j=1}^{n} \nu_j'' \, M_j \tag{2.42}$$

where ν' and ν'' are the stoichiometric coefficients of the reactant and product, respectively, and n is the number of chemical species involved in the reaction. The reaction rate Ω is expressed by the law of mass action[2,4] as

$$\Omega = k_f \prod_{j=1}^{n} [M_j]^{\nu_j'} \tag{2.43}$$

where $[M_j]$ is the concentration of species M_j and k_f is the reaction rate constant for the reaction shown in Eq. (2.42). The reaction rate constant k_f is determined by statistical theory and is expressed by

$$k_f = Z_c \exp(-E/RT) \tag{2.44}$$

where E is the activation energy and Z_c is a pre-exponential factor determined by collision theory. The exponential term in Eq. (2.44) is the Boltzmann factor, which indicates the fraction of M_j having energy higher than the activation energy E.

The rate of production of the species M_i is expressed by

$$d[M_i]/dt = (\nu_i'' - \nu_i')\,\Omega = (\nu_i'' - \nu_i')\,k_f \prod_{j=1}^{n} [M_j]^{\nu_j'} \tag{2.45}$$

where $\nu_i'' - \nu_i'$ is the stoichiometric factor and t is time. The overall order of a reaction, m, is defined according to:

$$m = \sum_{j=1}^{n} \nu_j' \tag{2.46}$$

When $m = 1$, the reaction is called a first-order reaction, when $m = 2$ it is a second-order reaction, and when $m = 3$ it is a third-order reaction, etc. When an elementary chemical reaction expressed as

$$A + 2B \rightarrow 3P$$

produces 3 moles of P from 1 mole of A and 2 moles of B, the reaction rate in moles $m^{-3}\,s^{-1}$ is expressed by

$$d[P]/dt = -3d[A]/dt = -3/2\ d[B]/dt$$

2.4 Evaluation of Chemical Energy

Since the energy density of materials is determined by their intermolecular structure, the energy density of polymeric materials is limited due to the relatively long chemical bonds that separate the atoms. Furthermore, the density of polymeric materials is also limited by the molecular structures involved. On the other hand, the density of crystalline materials is high due to the three-dimensionally arranged atoms in the molecular structures. Since the distances between the atoms in a crystalline structure are relatively short, the bond energy between the atoms is high. Combination of the different favorable properties of polymeric materials and crystalline materials in mixtures has led to the development of high energy density materials used as propellants or explosives. Thus, the selection of polymeric materials and crystalline materials is central to the formulation of high-energy propellants or explosives.

When metal particles are oxidized, a large amount of heat is generated. However, the oxidized products are agglomerated condensed-phase particles and very little gaseous product is formed. Therefore, the potential use of metal particles as fuel components of propellants and explosives is limited. When organic materials composed of C, H, N, and O atoms are oxidized, CO_2, H_2O, N_2, and other hydrocarbon gases are produced and a large amount of heat is generated.

Typical crystalline materials used as oxidizers are perchlorates, nitrates, nitro compounds, nitramines, and metal azides. The polymeric materials used as fuel components are divided into nitrate esters, inert polymers, and azide polymers. Optimized combinations of these oxidizer and fuel components yield the desired ballistic characteristics of propellants or explosives.

2.4.1 Heats of Formation of Reactants and Products

Propellants, explosives, and pyrolants are all energetic materials designed to have maximum energy density within a limited domain of mechanical sensitivities, manufacturing requirements, physical properties, and combustion characteristics. As shown by Eq. (2.2), $\Delta H_{f,R}$ needs to be as high as possible and $\Delta H_{f,P}$ as low as possible in order to gain high H_{exp}. Tables 2.3 and 2.4 show $\Delta H_{f,R}$ values of typical ingredients used to formulate energetic materials and $\Delta H_{f,P}$ values of typical combustion products of energetic materials at 298 K, respectively.[1–9]

Table 2.3 Heats of formation of energetic materials.

Reactant	$\Delta H_{f,R}$ (MJ kg^{-1})
NG	−1.70
NC	−2.60
DEGDN	−2.21
TEGDN	−2.53
TMETN	−1.61
DBP	−3.03
DEP	−7.37
TA	−5.61
2NDPA	−0.01
HMX	+0.25
RDX	+0.27
AP	−2.52
NP	+0.23
KP	−3.12
AN	−4.56
TAGN	−0.281
ADN	−1.22
HNF	−0.39
HNIW	+0.96
KN	−4.87
Ammonium picrate	−1.50
Diazodinitrophenol	−1.46
Diethyleneglycol	−2.10
Nitroglycol	−1.51
NQ	−0.77

Table 2.3 Continued.

Reactant	$\Delta H_{f,R}$ (MJ kg^{-1})
NIBGTN	–0.80
NM	–1.73
Hydrazine nitrate	–2.45
PETN	–1.59
TNT	–0.185
Trinitroanisole	–0.548
TNB	–0.097
TNChloroB	+0.169
Methyl nitrate	–1.91
Tetryl	+0.196
Picric acid	–0.874
Lead azide	+1.66
CTPB	–0.89
HTPB	–0.31
GAP	+0.96
BAMO	+2.46
Cubane	+5.47
B	0
C	0
Al	0
Mg	0
Ti	0
Zr	0

Table 2.4 Heats of formation of combustion products.

Product	$\Delta H_{f,P}$ (MJ kg^{-1})
CO	–3.94
CO_2	–8.94
H_2	0
$H_2O_{(g)}$	–13.4
N_2	0
Al_2O_3	–16.4
B_2O_3	–18.3
MgO	–14.9

Typical materials containing oxygen and nitrogen atoms are nitrate esters, such as nitrocellulose (NC) and nitroglycerin (NG). Nitrate esters contain -O–NO_2 chemical bonds in their structures. The oxidizer component is represented by the oxygen atoms and the fuel components are the carbon and hydrogen atoms. The oxidized combustion products are CO_2 and H_2O, the $\Delta H_{f,P}$ values of which are –8.94 MJ

kg^{-1} and −13.42 MJ kg^{-1}, respectively, as shown in Table 2.4.[9–17] The nitrogen atoms in the reactants produce nitrogen gas, for which $\Delta H_{f,P}$ is zero (Table 2.4).

Table 2.5 shows H_{exp} values and nitrogen concentrations, N (%), of typical energetic materials used as major components of propellants and explosives.[9–17] In order to obtain higher H_{exp} values of propellants and explosives, various types of chemicals are admixed, such as plasticizers, stabilizers, and reaction rate modifiers. The major chemicals are fuels and oxidizers. The fuels react with the oxidizers to produce heat and gaseous products. Even when the $\Delta H_{f,R}$ values of both the fuel

Table 2.5 Heats of explosion and nitrogen concentrations of energetic materials.

Material	Chemical formula	H_{exp} (MJ kg^{-1})	N (%)
NG	$(ONO_2)_3(CH_2)_2CH$	6.32	18.50
NC	$C_{12}H_{14}N_6O_{22}$	4.13	14.14
DEGDN	$(CH_2)_4O(ONO_2)_2$	4.85	14.29
TEGDN	$(CH_2)_6O_2(ONO_2)_2$	3.14	11.67
TMETN	$CH_3C(CH_2)_3(ONO_2)_3$	5.53	16.46
AP	NH_4ClO_4	1.11	11.04
AN	NH_4NO_3	1.60	35.00
ADN	$NH_4N(NO_2)_2$		45.16
NQ	$CH_4N_4O_2$	2.88	53.83
TAGN	$CH_9N_7O_3$	3.67	58.68
HMX	$(NNO_2)_4(CH_2)_4$	5.36	37.83
RDX	$(NNO_2)_3(CH_2)_3$	5.40	37.84
HNIW	$(NNO_2)_6(CH)_6$	6.80	38.45
Ammonium picrate	$NH_4OC_6H_2(NO_2)_3$	4.28	22.77
Diazodinitrophenol	$C_6H_2N_2O(NO_2)_2$		26.67
Diethyleneglycol dinitrate	$(ONO_2)_2(CH_2)_4O$	4.85	14.29
Nitroglycol	$(ONO_2)_2(CH_2)_2$	6.83	18.42
Nitroisobutylglycerol trinitrate	$(ONO_2)_3NO_2C(CH_2)_3$	7.15	19.58
NM	CH_3NO_2	4.54	22.96
Hydrazine nitrate	$(NH_2)_2HNO_3$	3.87	44.20
HNF	$N_2H_5C(NO_2)_3$		38.25
PETN	$(ONO_2)_4(CH_2)_4C$	5.90	17.72
TNT	$(NO_2)_3C_7H_5$	5.07	18.50
Trinitroanisole	$(NO_2)_3C_7H_5O$	4.62	17.29
TNB	$(NO_2)_3C_6H_3$	5.34	19.72
TNChloroB	$(NO_2)_3C_6H_2Cl$		16.98
Methyl nitrate	CH_3ONO_2	6.12	18.19
Tetryl	$(NO_2)_4C_7H_5N$	5.53	24.39
Picric acid	$(NO_2)_3C_6H_2OH$	5.03	18.37
Lead azide	$Pb(N_3)_2$		28.85

and oxidizer are low, a higher H_{exp} is obtained if the oxidizer has the potential to oxidize the fuel effectively. The oxidation reaction, i. e. combustion, then produces combustion products with lower $\Delta H_{f,P}$. It is also apparent that H_{exp} is high for materials with high N (%).

2.4.2 Oxygen Balance

The concentration of oxygen atoms within an oxidizer represented by its "oxygen balance", denoted "[OB]", is an important parameter indicating the potential of oxidizers. Oxygen balance expresses the number of oxygen molecules remaining after oxidation of H, C, Mg, Al, etc., to produce H_2O, CO_2, MgO_2, Al_2O_3, etc. If excess oxygen molecules are remaining after the oxidation reaction, the oxidizer is said to have a "positive" oxygen balance. If the oxygen molecules are completely consumed and excess fuel molecules remain, the oxidizer is said to have a "negative" oxygen balance.

Let us consider the reaction of an oxidizer with the composition $C_aH_bN_cO_dCl_eS_f$, as represented by

$$C_aH_bN_cO_dCl_eS_f \rightarrow$$

$$aCO_2 + 1/2(b - e)H_2O + c/2N_2 + eHCl + fSO_2 - \{(a + f) + 1/4(b - e) - d/2\}O_2$$

The oxygen balance expressed in mass percent is given by

$$[OB] = -\{(a + f) + 1/4(b - e) - d/2\} \times 32/(\text{molecular mass of material}) \times 100\ (\%)$$

For example, NG produces excess oxygen molecules among its combustion products according to

$$C_3H_5N_3O_9 \rightarrow 3\ CO_2 + 5/2\ H_2O + 3/2\ N_2 + 1/4\ O_2$$

The oxygen balance of NG is given by

$$[OB]_{NG} = +1/4 \times 32/227 \times 100 = +3.52\%$$

The oxygen balance of any specified mixture of oxidizers can be obtained by assuming the nature of the oxidized products. Table 2.6 shows the oxygen balances [OB] and densities ρ of some energetic materials.

2.4.3 Thermodynamic Energy

The chemical energy generated by the combustion of energetic materials is converted to thermodynamic energy used for propulsion and explosion. As described in the preceding sections of this chapter, the amount of stored chemical energy is

Table 2.6 Oxygen balances and densities of energetic materials.

Material	[OB] (%)	ϱ (kg m^{-3})
NG	+3.5	1590
NC	−28.7	1670
DEGDN	−40.8	1380
TEGDN	−66.7	1340
TMETN	−34.5	1470
DBP	−224.2	1050
TA	−139.0	1150
AP	+34.0	1950
NP	+71.5	2220
AN	+20.0	1720
ADN	+25.8	1720
HNF	+25.0	1860
NQ	−30.7	1710
TAGN	−33.5	1500
HMX	−21.6	1900
RDX	−21.6	1820
HNIW	−10.9	2040
Ammonium picrate	−52.0	1720
Diazodinitrophenol	−60.9	1630
Diethyleneglycol	−40.8	1380
Nitroglycol	0	1480
NIBGTN	0	1680
NM	−39.3	1140
Hydrazine nitrate	+8.6	1640
PETN	−10.1	1760
TNT	−73.9	1650
Trinanisole	−62.5	1610
TNB	−56.3	1760
TNChloroB	−45.3	1800
Methyl nitrate	−10.4	1220
Tetryl	−47.4	1730
Picric acid	−45.4	1770
Lead azide	−5.5	4600

determined by the chemical structure of the molecules of the energetic materials. However, the available thermodynamic energy is determined by the conversion of the heat and combustion products into pressure as described in Chapter 1. The characteristic velocity c^* defined in Eq. (1.74), the specific impulse I_{sp} defined in Eq. (1.76), and the heat of explosion H_{exp} defined in Eq. (2.2) are used to evaluate the potential thermodynamic energy of materials. Of these, c^* is used to evaluate the energetics in rocket motors, I_{sp} is used to evaluate overall energetics, including those pertaining to nozzle expansion processes, and H_{exp} is used to evaluate enthalpic potential. In addition, the ratio of the combustion temperature, T_g, to the molecular mass of the combustion products, M_g, defined in $\Theta = T_g/M_g$, is also used to evaluate the energetics of materials.

Though the thermodynamic energies of propellants and explosives are not determined by the thermodynamic energies of their individual components, it is important to recognize the thermochemical properties through the thermodynamic energy of each component. Table 2.7 shows T_g, M_g, Θ, I_{sp}, and the combustion products of the major components used in propellants and explosives, as obtained by computations with a NASA program.[8]

Table 2.7 Thermochemical properties of energetic chemicals (10 MPa).

	T_g (K)	M_g (kg kmol^{-1})	Θ (kmol K kg^{-1})	I_{sp} (s)
NC (12.6%N)	2600	24.7	105	233
NG	3300	28.9	114	247
TMETN	2910	23.1	126	256
TEGDN	1390	19.0	73	186
DEGDN	2520	21.8	116	244
AP	1420	27.9	51	160
AN	1260	22.9	55	164
ADN	2060	24.8	83	206
HNF	3120	26.4	118	265
HNIW	3640	27.5	132	281
NP	610	36.4	17	88
RDX	3300	24.3	136	269
HMX	3290	24.3	135	269
TAGN	2310	18.6	124	251

Major combustion products (mol/mol)	O_2	H_2O	CO	CO_2	H_2	N_2	OH	HCl	Cl_2
NC (12.6%N)		0.225	0.147	0.128	0.116	0.111			
NG	0.069	0.280	0.107	0.275	0.014	0.181	0.041		
TMETN		0.263	0.357	0.096	0.140	0.136			
TEGDN		0.110	0.397	0.063	0.335	0.079			
DEGDN		0.253	0.365	0.079	0.190	0.111			
AP	0.287	0.377				0.119		0.197	0.020
NP	0.750					0.125			0.125
AN	0.143	0.571				0.286			
RDX		0.226	0.246	0.082	0.089	0.326			
HMX		0.227	0.246	0.082	0.089	0.326			
TAGN		0.209	0.098	0.013	0.290	0.389			
HNF	0.098	0.337	0.002	0.125		0.348	0.003		
ADN	0.196	0.339				0.397			
HNIW	0.018	0.137	0.235	0.142	0.028	0.367	0.033		

References

1 Dickerson, R. E., Molecular Thermodynamics, W. A. Benjamin, New York (1969), Chapter 5.
2 Laidler, K. J., Chemical Kinetics, Second Edition, McGraw-Hill, New York (1969), Chapter 4.
3 Sarner, S. F., Propellant Chemistry, Reinold Publishing Corporation, New York (1966), Chapter 4.
4 Penner, S. S., Chemistry Problems in Jet Propulsion, Pergamon Press, New York (1957), Chapters 12 and 13.
5 Wilkins, R. L., Theoretical Evaluation of Chemical Propellants, Prentice-Hall, Englewood Cliffs (1963), Chapters 3–5.
6 JANAF Thermochemical Tables, The Clearing House for Federal Scientific and Technical Information, U.S. Department of Commerce, Springfield, Virginia.
7 Glassman, I., Combustion, Academic Press, New York (1977), Chapter 1.
8 Gordon, S., and McBridge, B. J., Computer Program for Calculation of Complex Chemical Equilibrium Compositions, Rocket Performance, Incident and Reflected Shocks, and Chapman–Jouguet Detonations, NASA SP-273, 1971.
9 Meyer, R., Explosives, Verlag Chemie, Weinheim, 1977.
10 Sarner, S. F., Propellant Chemistry, Reinhold Publishing Corporation, New York, 1966.
11 Japan Explosives Society, Energetic Materials Handbook, Kyoritsu Shuppan, 1999.
12 Chan, M. L., Reed, Jr., R., and Ciaramitaro, D. A., Advances in Solid Propellant Formulations, Solid Propellant Chemistry, Combustion, and Motor Interior Ballistics (Eds.: Yang, V., Brill, T. B., and Ren, W.-Z.), *Progress in Astronautics and Aeronautics*, AIAA, Vol. 185, 2000, Chapter 1.7.
13 Doriath, G., Available Propellants, Solid Rocket Technical Committee Lecture Series, AIAA Aerospace Sciences Meeting, Reno, Nevada, 1994.
14 Miller, R. R., and Guimont, J. M., Ammonium Dinitramide Based Propellants, Solid Rocket Technical Committee Lecture Series, AIAA Aerospace Sciences Meeting, Reno, Nevada, 1994.
15 Kubota, N., Propellant Chemistry, Journal of Pyrotechnics, Inc., 2000, pp. 25–45.
16 Sanderson, A., New Ingredients and Propellants, Solid Rocket Technical Committee Lecture Series, AIAA Aerospace Sciences Meeting, Reno, Nevada, 1994.
17 Miller, R., Advancing Technologies: Oxidizers, Polymers, and Processing, Solid Rocket Technical Committee Lecture Series, AIAA Aerospace Sciences Meeting, Nevada, 1994.

3
Combustion Wave Propagation

3.1
Combustion Reactions

3.1.1
Ignition and Combustion

Combustion phenomena have been studied extensively and a number of instructive books have been published, covering both experimental[1–4] and theoretical aspects.[5] In several respects, the definition of combustion is not clear. The combustion of gaseous materials produces heat accompanied by emission from luminous reaction products. However, an energetic polymer burns very fast by itself and produces heat without luminous emission. When a reactive gas is heated by an external energy source, chemical reactions occur between the constituent molecules of the gas. This initiation process is an exothermic reaction and it forms high-temperature products. Such a process is described as ignition, a key part of the combustion phenomenon. When the heat produced by this exothermic reaction-serves to heat up the unreacted portion of the reactive gas, a successive ignition process is established without external heating. This process is said to be self-sustaining combustion, that is, what we commonly think of as combustion. The ignited region between the unburned and burned regions is called a combustion wave, which propagates toward the unburned region.

When heat is supplied to the surface of an energetic solid material, both the surface temperature and the subsurface temperature are increased. When the surface temperature reaches the decomposition or gasification temperature, endothermicand/or exothermic reactions occur on and above the surface. The gases given off react to form decomposition products accompanied by a high heat release, and the temperature in the gas phase increases. This process is the ignition of the energetic solid material. If this reaction process is maintained after removal of the heat supply to the surface, then combustion has been established. On the other hand, if the exothermic and gasification reactions are terminated after removal of the heat supply to the surface, then ignition has failed and combustion has not been established.

External heating is needed for ignition, and thereafter successive heating from the high-temperature burned portion to the low-temperature unburned portion is

Propellants and Explosives. Naminosuke Kubota

ISBN: 978-3-527-31424-9

needed for combustion. The ignition and combustion of reactive gases and energetic solid materials are fundamentally the same. However, additional physicochemical processes, such as a phase transition from solid to liquid and/or gas, are needed for energetic solid materials. In the combustion wave, melting, decomposition, sublimation, and/or gasification processes are involved.

3.1.2 Premixed and Diffusion Flames

Any reactants capable of forming combustion products are composed of a mixture of oxidizer and fuel components, and the flame is produced by the reaction of the mixture. Two types of flames can be formed when the mixture burns: a premixed flame or a diffusion flame.[1,2] A premixed flame is formed by the combustion of the two components when they are premixed prior to burning in the combustion zone. The intermingled oxidizer and fuel component molecules in the premixed reactants then react homogeneously. The temperature and the concentration of the products increase uniformly in the combustion zone.

When the oxidizer and fuel components are physically separated and allowed to diffuse into each other in the combustion zone, a diffusion flame is formed. Since the molecular distributions of the oxidizer and fuel components are not uniform, the temperature and combustion products are also not uniformly distributed in the combustion zone. Thus, the rate of the reaction generating the combustion products is low when compared to that in a premixed flame because an additional diffusional process is needed to form the diffusion flame.

3.1.3 Laminar and Turbulent Flames

The characteristics of a reactive gas (a premixed gas) are dependent not only on the type of reactants, pressure, and temperature, but also on the flow conditions. When the flame front of a combustion wave is flat and one-dimensional in shape, the flame is said to be a laminar flame. When the flame front is composed of a large number of eddies, which are three-dimensional in shape, the flame is said to be a turbulent flame. In contrast to a laminar flame, the combustion wave of a turbulent flame is no longer one-dimensional and the reaction surface of the combustion wave is significantly increased by the eddies induced by the dynamics of the fluid flow.

When the same chemical compositions of the reactants are used to generate both types of flame, the chemical reaction rate is considered to be the same in both cases. However, the reaction surface area of the turbulent flame is increased due to the nature of eddies and the overall reaction rate at the combustion wave appears to be much higher than that in the case of the laminar flame. Furthermore, the heat transfer process from the burned gas to the unburned gas in the combustion wave is different because of the thermophysical properties; specifically, the thermal diffusivity is higher for the turbulent flame than for the laminar flame. Thus, the flame speed of a turbulent flame appears to be much higher than that of a laminar flame.

The creation of eddies in a combustion zone is dependent on the nature of the flow of the unburned gas, i. e., the Reynolds number. If the upstream flow is turbulent, the combustion zone tends to be turbulent. However, since the transport properties, such as viscosity, density, and heat conductivity, are changed by the increased temperature and the force acting on the combustion zone, a laminar upstream flow tends to generate eddies in the combustion zone and here again the flame becomes a turbulent one. Furthermore, in some cases, a turbulent flame accompanied by large-scale eddies that exceed the thickness of the combustion wave is formed. Though the local combustion zone seems to be laminar and one-dimensional in nature, the overall characteristics of the flame are not those of a laminar flame.

3.2 Combustion Wave of a Premixed Gas

3.2.1 Governing Equations for the Combustion Wave

The combustion wave of a premixed gas propagates with a certain velocity into the unburned region (with flow speed = 0). The velocity is sustained by virtue of thermodynamic and thermochemical characteristics of the premixed gas. Figure 3.1 illustrates a combustion wave that propagates into the unburned gas at velocity u_1, one-dimensionally under steady-state conditions. If one assumes that the observer of the combustion wave is moving at the same speed, u_1, then the combustion wave appears to be stationary and the unburned gas flows into the combustion wave at the velocity $-u_1$. The burned gas is expelled downstream at a velocity of $-u_2$ with respect to the combustion wave. The thermodynamic characteristics of the combustion wave are described by the velocity (u), pressure (p), density (ρ), and temperature (T) of the unburned gas (denoted by the subscript 1) and of the burned gas (denoted by the subscript 2), as illustrated in Fig. 3.1.

The governing equations of the combustion wave are the set of equations for the conservation of mass, momentum, and energy:

$$\rho_1 u_1 = \rho_2 u_2 = m \tag{3.1}$$

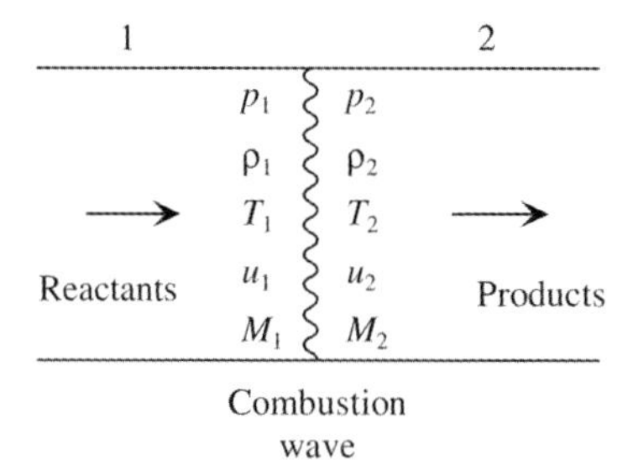

Fig. 3.1 Combustion wave propagation.

$$p_1 + \rho_1 u_1^2 = p_2 + \rho_2 u_2^2 \tag{3.2}$$

$$c_p T_1 + u_1^2/2 + q = c_p T_2 + u_2^2/2 \tag{3.3}$$

where m is the mass flux and q is the heat of reaction per unit mass. These equations are equivalent to Eqs. (1.28), (1.29), and (1.30) for a shock wave, except for the additional q term in Eq. (3.3). The heat of reaction produced in the combustion wave is given by:

$$q = h^0{}_1 - h^0{}_2 \tag{3.4}$$

The enthalpy is given by the sum of the sensible enthalpy and chemical enthalpy:

$$h = c_p T + h^0 \tag{3.5}$$

$$c_p T = e + p/\rho \tag{3.6}$$

where h and h^0 are the heats of formation per unit mass at temperature T and in the standard state, respectively.

Equations (3.1) and (3.2) give the velocities u_1 and u_2 as:

$$u_1^2 = \frac{1}{\rho_1^2}\left(\frac{p_2 - p_1}{1/\rho_1 - 1/\rho_2}\right) \tag{3.7}$$

$$u_2^2 = \frac{1}{\rho_2^2}\left(\frac{p_2 - p_1}{1/\rho_1 - 1/\rho_2}\right) \tag{3.8}$$

3.2.2
Rankine–Hugoniot Relationships

Combining Eqs. (3.1)–(3.4), one obtains:

$$h_2 - h_1 = \frac{1}{2}(p_2 - p_1)(1/\rho_1 + 1/\rho_2) \tag{3.9}$$

which is termed the Rankine–Hugoniot equation. Since the internal energy e is represented by Eq. (3.6), the Rankine–Hugoniot equation can also be expressed as:

$$e_2 - e_1 = \frac{1}{2}(p_1 + p_2)(1/\rho_1 - 1/\rho_2) + q \tag{3.10}$$

If the reactant and product are assumed to be in thermodynamic equilibrium, e_1 and e_2 can be expressed by known functions of pressure and density:

$$e_1 = e(p_1, \rho_1) \quad e_2 = e(p_2, \rho_2) \tag{3.11}$$

The Rankine–Hugoniot relationship expressed by Eq. (3.10) or Eq. (3.9) is shown in Fig. 3.2 as a function of $1/\rho$ and p, and such a plot is called the Hugoniot curve. The Hugoniot curve for $q = 0$, i. e., no chemical reaction, passes through the initial point

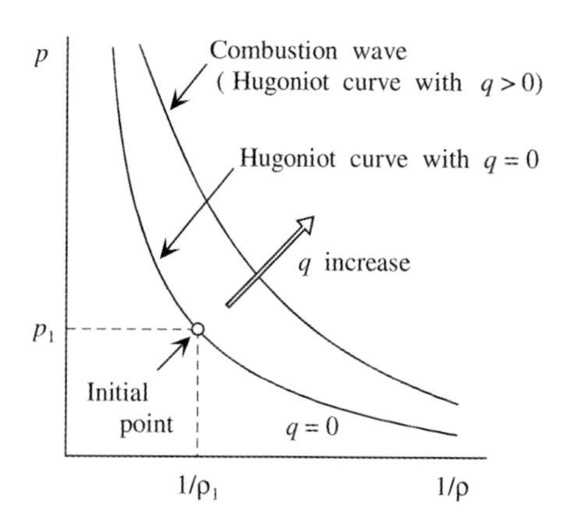

Fig. 3.2 Hugoniot curve for a combustion wave.

$(1/\rho_1, p_1)$ and is exactly equivalent to the shock wave described in Chapter 1. When heat q is produced in the combustion wave, the Hugoniot curve shifts in the direction indicated by the arrow in Fig. 3.2. It is evident that two different types of combustion are possible on the Hugoniot curve: (1) a detonation, in which pressure and density increase, and (2) a deflagration, in which pressure and density decrease.

Equations (3.1) and (3.2) yield the following relationship, the so-called Rayleigh equation:

$$(p_2 - p_1)/(1/\rho_2 - 1/\rho_1) = -m^2 \tag{3.12}$$

As shown in Fig. 3.3, tangents from the initial point 1 $(1/\rho_1, p_1)$ to the points J and K $(1/\rho_2, p_2)$ on the Hugoniot curve represent the Rayleigh lines, which are expressed by the following equations:[6]

$$[(p_2 - p_1)/(1/\rho_2 - 1/\rho_1)]_J = [\partial p/\partial(1/\rho)]_J = \tan\theta_J \tag{3.13}$$

$$[(p_2 - p_1)/(1/\rho_2 - 1/\rho_1)]_K = [\partial p/\partial(1/\rho)]_K = \tan\theta_K \tag{3.14}$$

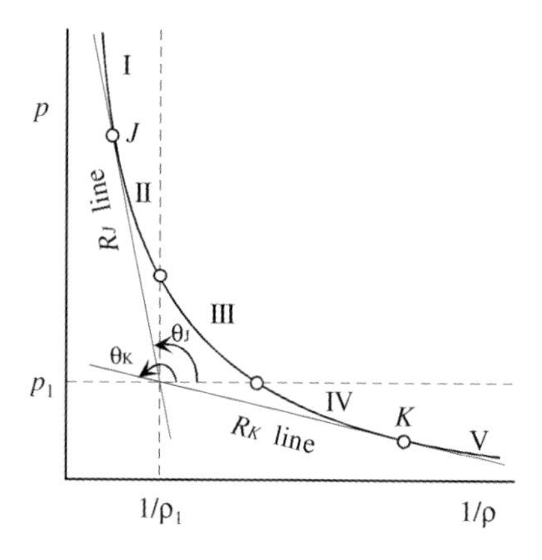

Fig. 3.3 Definition of the Hugoniot curve for a combustion wave.

The mass flow rate in Eq. (3.1) and the velocity u_2 at point J or K are given by:

$$m = \sqrt{-\tan\theta_J} = \sqrt{-\tan\theta_K} \tag{3.15}$$

$$u_2 = \frac{1}{\rho_2}\sqrt{-\tan\theta_J} \tag{3.16}$$

$$u_2 = \frac{1}{\rho_2}\sqrt{-\tan\theta_K} \tag{3.17}$$

3.2.3
Chapman–Jouguet Points

As illustrated in Fig. 3.3, the Hugoniot curve is divided into five regions.[1–4] Since the velocity u_2 expressed by Eq. (3.16) becomes an imaginary value in region III because $\tan\theta_J$ has a positive value, a combustion wave is not physically possible in this region. Points J and K are called the Chapman–Jouguet points. Chapman–Jouguet theory states that the velocities at J and K on the Hugoniot curve represent the minimum and maximum values, respectively, of the propagation velocity u_D, relative to the unburned gas. The entropy of the reaction products varies along the Hugoniot curve. Instructive descriptions of the Chapman–Jouguet relationship are available in the cited monographs.[3–5]

Since the Rayleigh line through point J or K is a tangent to the Hugoniot curve, it is also a tangent to the line of constant entropy through J or K. Thus, the slope of the line of constant entropy is exactly the slope of the Hugoniot curve at J or K.[6] Differentiation of Eq. (3.10) gives:

$$de = -\frac{1}{2}(p_1 + p_2)d(1/\rho_2) + \frac{1}{2}(1/\rho_1 - 1/\rho_2)d(p_2) \tag{3.18}$$

The entropy curve is expressed by

$$Tds = de + p_1 d(1/\rho) \tag{3.19}$$

Combining Eqs. (3.15), (3.18), and (3.19), one obtains:

$$T\left(\frac{\partial s}{\partial p}\right)_{\mathrm{H}} = \left(\frac{1}{\rho_1} - \frac{1}{\rho_2}\right)\left[1 + \frac{m^2}{\partial p/\partial(1/\rho)}\right]_{\mathrm{H}} \tag{3.20}$$

along the Hugoniot curve (H). Since the Rayleigh line (R) is tangential to the Hugoniot curve at points J and K, $\{\partial p/\partial(1/\rho)\}_{\mathrm{H}} = \{\partial p/\partial(1/\rho)\}_{\mathrm{R}}$, the relationship

$$\left[\frac{\partial p}{\partial(1/\rho)}\right]_{\mathrm{H}} = -m^2 \tag{3.21}$$

is obtained based on Eqs. (3.13) and (3.15). Substituting Eq. (3.21) into Eq. (3.20), the relationship $(\partial s/\partial p)_{\mathrm{H}} = 0$ on the Hugoniot curve is obtained, and so $ds = 0$ at points J and K.

The velocity of sound in the burned gas is expressed by:

$$a^2 = \left(\frac{\partial p}{\partial \rho}\right)_S = -\frac{1}{\rho^2}\left[\frac{\partial p}{\partial(1/\rho)}\right]_S \tag{3.22}$$

Using Eq. (3.13), one gets

$$(a_2^2)_J = \left[\frac{1}{\rho_2^2}\frac{p_2 - p_1}{1/\rho_1 - 1/\rho_2}\right]_J \tag{3.23}$$

Thus, the following relationship is obtained:

$$(u_2)_J = (c_2)_J \quad \text{or } M_2 = 1 \text{ at point } J$$

The velocity of the burned gas (u_2) at point J is equal to the speed of sound in the burned gas (a_2). A similar result is obtained at point K, as

$$M_2 = 1 \text{ at point } K$$

The velocity of the burned gas relative to a stationary observer, *up*, defined as "particle velocity", is given by

$$u_1 = u_D = u_2 + u_p \tag{3.24}$$

Using Eq. (3.1), one gets

$$u_p = u_1(1 - \rho_1/\rho_2) \tag{3.25}$$

Since $\rho_1 < \rho_2$ for a detonation and $\rho_1 > \rho_2$ for a deflagration, the flow field becomes $0 < u_p < u_1$ for detonation, and $u_p < 0$ for deflagration. In the case of detonation, the velocity of the combustion products is less than the detonation wave velocity. In the case of deflagration, the combustion products are expelled in the opposite direction to the deflagration wave.

The line of constant entropy represented by Eq. (1.14), $p\,(1/\rho)^\gamma$ = constant, rises more steeply than the Rayleigh line and less steeply than the Hugoniot curve with increasing pressure in regions I and V, and these are called the strong detonation branch and the strong deflagration branch, respectively. On the other hand, the line of constant entropy rises less steeply than the Rayleigh line and more steeply than the Hugoniot curve with increasing pressure in regions II and IV, and these are called the weak detonation branch and the weak deflagration branch, respectively. The velocity of the reaction products relative to the reaction front is subsonic behind the combustion wave of regions I and IV. The velocity is sonic behind a Chapman–Jouguet detonation at J or deflagration at K, and is supersonic behind the combustion wave of regions II and V. These characteristics are shown in Fig. 3.4 and are summarized as follows:[3–6]

Region I	$p_2 > p_J$	supersonic flow to subsonic flow, strong detonation
Region II	$p_2 < p_J$	supersonic flow to supersonic flow, weak detonation
Region III		physically invalidated flow
Region IV	$p_2 > p_K$	subsonic flow to subsonic flow, weak deflagration
Region V	$p_2 < p_K$	supersonic flow to supersonic flow, strong deflagration

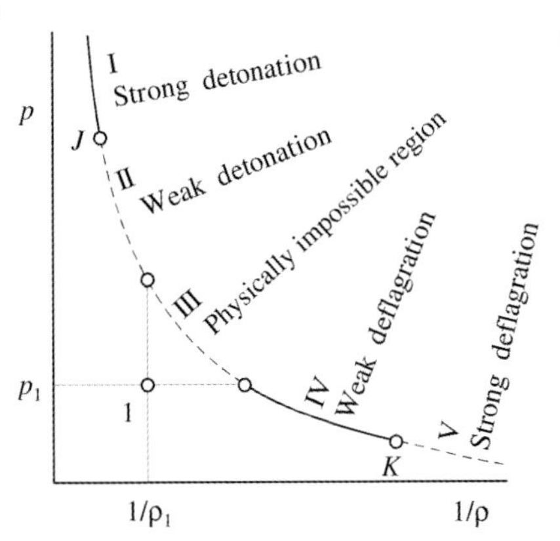

Fig. 3.4 Regions of detonation and deflagration on the Hugoniot curve.

However, region II and region V are regions wherein no physical processes are attainable, as described in the standard monographs.[1–4] Experimentally, most deflagration waves are observed in region IV, i. e. the weak deflagration branch, and most detonation waves are observed in region I, i. e. the strong detonation branch. It should be noted that the relationships of the Hugoniot curve are obtained by considering idealized conditions: the combustion wave is assumed to be formed under one-dimensional and steady-state flow, and the reactant and product gases are also assumed to be perfect gases. In reality, the phenomenon of combustion wave propagation is more complex and is accompanied by thermophysical and thermochemical effects. Nevertheless, the general characteristics of detonation and deflagration waves can be summarized as shown in Table 3.1.

Table 3.1 Thermophysical characteristics of deflagration and detonation waves.

	Deflagration	**Detonation**
p_2/p_1	< 1	> 1
ϱ_2/ϱ_1	< 1	> 1
T_2/T_1	> 1	> 1
u_2/u_1	> 1	< 1
M_1	< 1	> 1
M_2	< 1	< 1

In general, the wave propagation velocity in the deflagration branch is termed the flame speed, while that in the detonation branch is termed the detonation velocity.

3.3 Structures of Combustion Waves

3.3.1 Detonation Wave

A detonation wave formed by a reactive gas under one-dimensional steady-state flow conditions is shown in Fig. 3.5. The Hugoniot curve indicates that the pressure, density, and temperature increase rapidly at the front of the detonation wave because of the passage of the shock wave. Non-equilibrium molecular collisions lead to the conversion of translational energy to rotational energy and to vibrational energy in the shock wave. The increased temperature initiates an exothermic chemical reaction of the reactive gas behind the shock wave, which further increases the temperature. The pressure behind the shock wave decreases over a relaxation time and reaches a steady-state condition corresponding to the Chapman–Jouguet point J, the so-called CJ point, as shown in Fig. 3.4. In general, the increases in pressure and temperature resulting from a detonation wave are of the order of $p_2/p_1 = 15$–50 and $T_2/T_1 = 10$–20. However, the increase in density is only of the order of $\rho_2/\rho_1 = 1.5$–2.5, as shown in Fig. 3.5. A detonation temperature is approximately 400–800 K higher than a deflagration temperature because kinetic energy is converted into pressure and then into heat in the case of a detonation wave.

The structural model of a detonation wave proposed by Zeldovich, von Neumann, and Döring (ZND model) involves the pressure at the shock front increasing along the Hugoniot curve without chemical reaction until it attains the value at the point of intersection of the Rayleigh line and the Hugoniot curve,

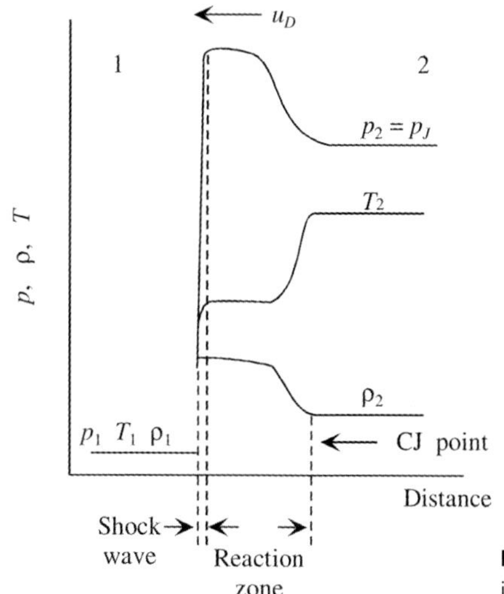

Fig. 3.5 Structure of a detonation wave showing the Chapman–Jouguet (CJ) point.

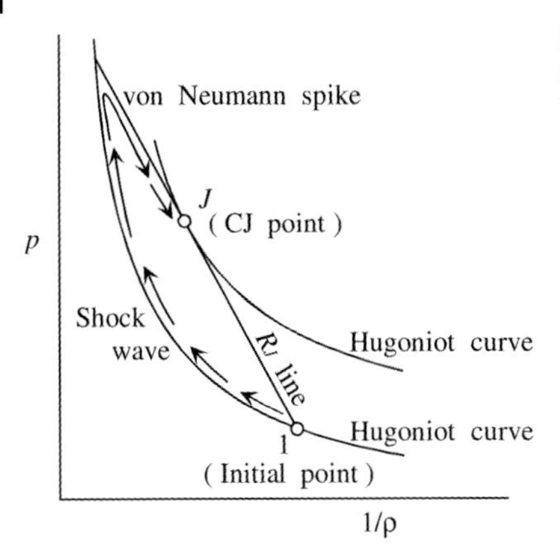

Fig. 3.6 Detonation wave formation from shock wave to von Neumann spike, and then to the Chapman–Jouguet point.

thereby defining the so-called von Neumann spike[3–6] as shown in Fig. 3.6. The pressure then decreases along the Rayleigh line to the Chapman–Jouguet point J, at which the detonation velocity reaches the speed of sound M_2. Experimental observations indicate that the pressure of the detonation front is higher than the pressure at J, but lower than the pressure at the point of intersection of the Rayleigh line and the Hugoniot curve, and finally reaches the pressure at J shown in Fig. 3.6.

The procedure for the computation of detonation speed is as follows:

1) Assume a reaction scheme (reactants to products).
2) Assume T_2.
3) Assume $1/\rho_2$ and determine p_2 by use of the equation of state.
4) Determine p_j (or n_j) at T_2 and p_2 by the same procedure as used for the determination of T_f described in Section 2.2 of Chapter 2.
5) Repeat steps (3) and (4) until Eq. (3.9) is satisfied.
6) Determine u_1 by means of Eq. (3.7)
7) Assume a different T_2
8) Repeat the procedure from (3) to (7) until the minimum value of u_1 is determined.
9) The minimum value of u_1 is the detonation velocity u_D at point J (Chapman–Jouguet detonation velocity).

Computations of the thermochemical values of various combinations of oxidizers and fuels can be found in the JANAF tables.[7] Practical computations are carried out by the use of computer programs such as the cited NASA program.[8] Table 3.2 shows an example of a computation comparing the detonation and deflagration characteristics of the gaseous mixture $2H_2 + O_2$.

Since the detonation velocity is equal to the speed of sound at the CJ point, u_D is determined by means of Eqs. (3.24) and (3.25). The temperature of detonation at the CJ point is higher than the temperature of deflagration because of the shock wave compression on the detonation wave.

Table 3.2 Detonation and deflagration characteristics of $2H_2 + O_2$.

Initial conditions	Detonation	Deflagration
$p_1 = 0.1$ MPa	$p_J = 1.88$ MPa	$p_2 = 0.1$ MPa
$T_1 = 298$ K	$T_2 = 3680$ K	$T_2 = 3500$ K
$a_1 = 538$ m s^{-1}	$a_2 = 1550$ m s^{-1}	$a_2 = 1380$ m s^{-1}
	$M_1 = 5.28$	
	$u_D = 2840$ m s^{-1}	

3.3.2 Deflagration Wave

A deflagration wave formed by a reactive gas under one-dimensional steady-state flow conditions is illustrated in Fig. 3.7. In the combustion wave, the temperature increases from the initial temperature of the unburned gas to the ignition temperature and then reaches the flame temperature. The heat generated in the reaction zone is transferred back to the unburned gas zone.

The thermal balance of the heat flux transferred back from the reaction zone to the unburned gas zone and the heat supplied to the unburned gas to increase its temperature from the initial temperature T_0 to the ignition temperature T_b is represented by

$$\lambda(dT/dx)_b = c_p\rho u\ (T_b - T_0) \tag{3.26}$$

where T is temperature, x is distance, u is the flow velocity, which is equal to the laminar flame speed, ρ is density, c_p is specific heat, and λ is thermal conductivity. The subscripts 0 and b indicate the initial condition and the onset location of the

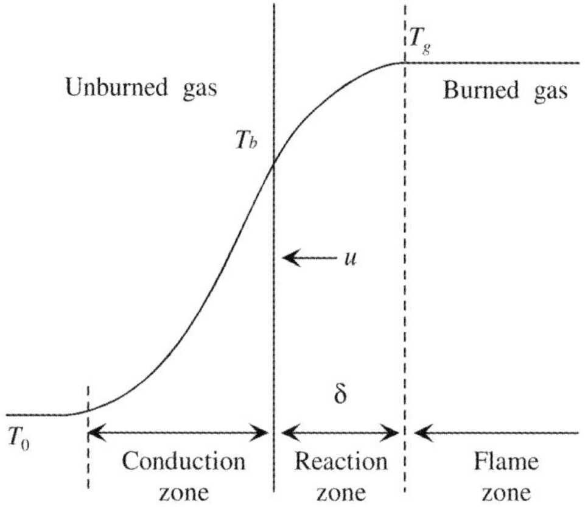

Fig. 3.7 Thermal structure of a deflagration wave.

chemical reaction, respectively. If the temperature in the reaction zone increases linearly, the temperature gradient in this zone is given by

$$(dT/dx)_b = (T_g - T_b)/\delta \tag{3.27}$$

where δ is the thickness of the reaction zone, and T_g is the flame temperature. Combining Eqs. (3.26) and (3.27), one gets

$$\lambda(T_g - T_b)/\delta = c_p \rho u(T_b - T_0) \tag{3.28}$$

and then

$$u = \frac{\lambda}{c_p \rho \delta} \frac{T_g - T_b}{T_b - T_0} \tag{3.29}$$

The reaction rate in the combustion wave is given by

$$[\omega]\delta = \int_0^\delta \omega dx \tag{3.30}$$

and then one obtains the relationship $\rho u = [\omega]\delta$, where ω is the reaction rate and $[\omega]$ is an averaged value of the reaction rate in the combustion wave. Thus, one gets

$$u = \frac{1}{\rho} \sqrt{\frac{\lambda[\omega]}{c_p} \frac{T_g - T_b}{T_b - T_0}} \tag{3.31}$$

The thickness of the combustion wave is determined according to

$$\delta = \lambda/(c_p \rho u) \tag{3.32}$$

The reaction rate in the combustion wave is expressed by

$$\omega = \rho^m [\varepsilon]^m Z \exp(-E/RT) \tag{3.33}$$

where ε is the mole fraction of the reactant, $[\varepsilon]$ is an averaged value of the mole fraction in the combustion wave, m is the order of the chemical reaction, E is the activation energy, R is the universal gas constant, and Z is a constant. Substituting Eqs. (3.33) and (1.5) into Eq. (3.31), the laminar flame speed is represented by

$$u \sim p^{m/2 - 1} \exp(-E/2RT) \tag{3.34}$$

where p is pressure, and T is assumed to be an averaged value of T_b and T_g. Though the analysis is a simplified one, the characteristics of a laminar flame speed are given. In general, gas-phase reactions are bimolecular and so the order of the chemical reaction is approximately 2. The laminar flame speed appears to be independent of pressure. When the mixture of fuel and oxidizer components of a premixed gas is in the stoichiometric ratio, the flame temperature T_g is maximized, and then the laminar flame speed is also maximized. The laminar flame speed is also increased if the initial temperature T_0 of a premixed gas is increased. The laminar flame speed given by Eq. (3.31) is confirmed by experimental results. For example, with $u = 350$ mm s^{-1}, the thickness of the reaction zone of a propane/air mixture is obtained from Eq. (3.32) as $\delta = 1.6$ mm.

3.4 Ignition Reactions

3.4.1 The Ignition Process

When heat is supplied to a gaseous mixture of oxidizer and fuel components, i. e., a premixed gas, an exothermic reaction occurs and the temperature increases. The reaction may continue and proceed into the unreacted portion of the mixture even after the source of the heat is removed. The amount of heat that has to be supplied to the mixture to achieve this is defined as the ignition energy. If, however, the reaction terminates after removal of the heat source, ignition of the mixture has failed. This is because the heat generated in the combustion zone is not sufficient to heat the unreacted portion of the mixture from the initial temperature to the ignition temperature.

Ignition is dependent on various physicochemical parameters, such as the type of reactants, reaction rate, pressure, the heat transfer process from the external heat source to the reactants, and the size or mass of the reactants. The rate of heat production is dependent on the heats of formation of the reactants and products, the temperature, and the activation energy. As the process of ignition includes an external heating and an exothermic reaction of the reactants, there is a non-steady heat balance during these phases.

3.4.2 Thermal Theory of Ignition

One assumes that a reactive gas is kept in a container and that an exothermic reaction is initiated. The rate of heat generation by the reactive material, q_R (self-heating by the exothermic reaction), is represented, according to Eq. (3.33), by

$$
\begin{aligned}
q_R &= QV\omega = \rho^m [\varepsilon]^m Z \exp(-E/RT) \\
&= QVA\, p^m \exp(-E/RT) \qquad (3.35)
\end{aligned}
$$

where V is the volume of the container, Q is the heat of reaction, and A is a constant. The system will be subject to a heat loss, q_L, proportional to the temperature difference between the container surface and the surroundings, as given by

$$q_L = h_g\, S\, (T - T_0) \qquad (3.36)$$

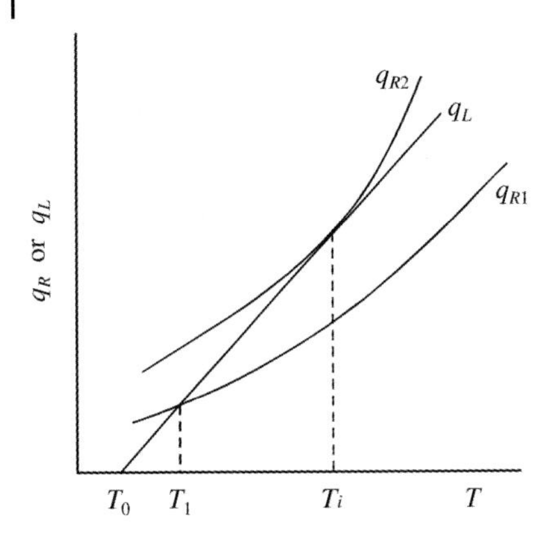

Fig. 3.8 Ignition criterion in terms of heat loss and heat gain.

where S is the surface area of the container, h_g is the heat transfer coefficient from the container surface to the surroundings, and T_0 is the temperature of the surroundings and also the initial temperature of the reactive gas.

Equations (3.35) and (3.36) are illustrated in Fig. 3.8 as a function of temperature. The temperature T_i indicates the ignition temperature of the reactive gas. If the rate of heat generation by the reactant is given by the curve q_{R1}, self-heating occurs at T_0 without heat loss and temperature increases toward T_1. However, when the temperature reaches T_1, the rate of heat loss q_L becomes equal to q_{R1}, and so the temperature increase by self-heating is stopped at T_1. Thus, the temperature never reaches the ignition temperature T_i. On the other hand, if the rate of heat generation by the reactant is given by the curve q_{R2}, the self-heating increases the temperature up to T_i because q_{R2} is larger than q_L in the temperature range between T_0 and T_i. Since q_{R2} is also larger than q_L in the temperature range above T_i, the self-heating continues, and then ignition occurs.

The ignition criteria of the thermal theory of ignition are then represented by

$$q_L = q_{R2} \quad \text{and} \quad d_{qL}/dT = dq_{R2}/dT \quad \text{at } T = T_i \tag{3.37}$$

Substituting Eqs. (3.35) and (3.36) into Eq. (3.37), one obtains the simplified relationship

$$T - T_0 = RT_0^2/E \tag{3.38}$$

It is evident from Eqs. (3.35) and (3.38) that the ignition temperature decreases as the density of the reactants increases, i. e., as pressure increases.

3.4.3
Flammability Limit

The rate of heat generation by a mixture depends on the mixture ratio of oxidizer and fuel components. As the mixture ratio becomes fuel-rich or oxidizer-rich, the rate of heat generation decreases, as does the reaction rate. At a certain mixture ratio, no combustion occurs even when excess ignition energy is supplied to the mixture. This is a combustion limit, the so-called flammability limit. Thus, there exist two flammability limits, a lower limit and an upper limit, for a given reactant consisting of different mixture ratios of the same oxidizer and fuel components. For example, mixtures of hydrogen and air burn in mixture ratios of 0.04 to 0.74 by hydrogen volume. The lower limit is 0.04 and the upper limit is 0.74 by volume. The maximum reaction rate and the highest temperature are obtained at the mixture ratio of 0.292, that is, the stoichiometric ratio of hydrogen by volume. In the case of hydrogen and oxygen mixtures, the lower limit is 0.04 and the upper limit is 0.94 by volume. The stoichiometric ratio for this mixture is 0.667.

3.5 Combustion Waves of Energetic Materials

3.5.1 Thermal Theory of Burning Rate

3.5.1.1 Thermal Model of Combustion Wave Structure

A schematic representation of the combustion wave structure of a typical energetic material is shown in Fig. 3.9 and the heat transfer process as a function of the burning distance and temperature is shown in Fig. 3.10. In zone I (solid-phase zone or condensed-phase zone), no chemical reactions occur and the temperature increases from the initial temperature (T_0) to the decomposition temperature (T_u). In zone II (condensed-phase reaction zone), in which there is a phase change from solid to liquid and/or to gas and reactive gaseous species are formed in endothermic or exothermic reactions, the temperature increases from T_u to the burning surface temperature (T_s). In zone III (gas-phase reaction zone), in which exothermic gas-phase reactions occur, the temperature increases rapidly from T_s to the flame temperature (T_g).

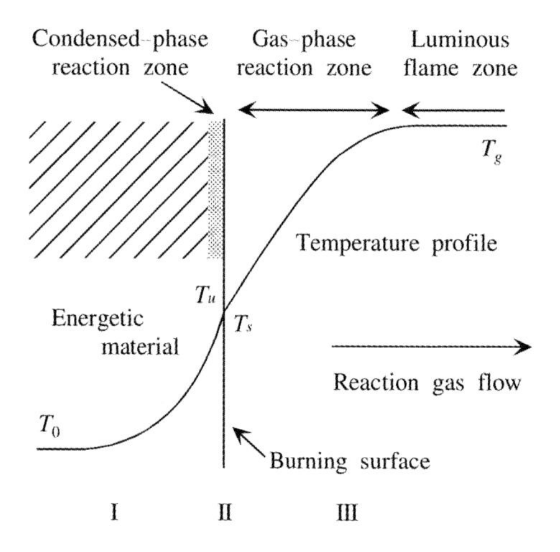

Fig. 3.9 Combustion wave structure of an energetic material.

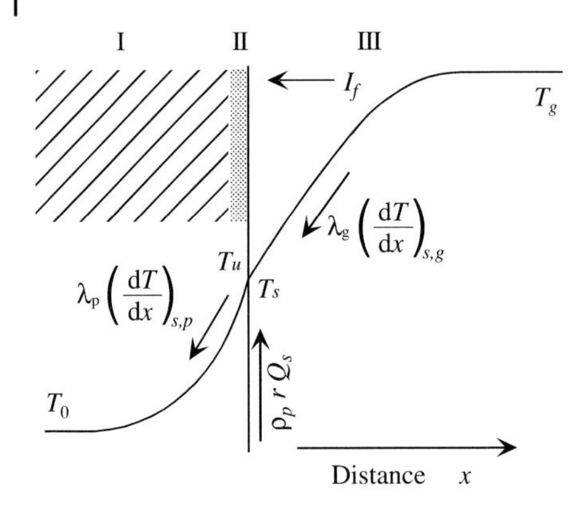

Fig. 3.10 Thermal structure of a combustion wave and heat feedback processes therein.

The basic assumptions in the following description of the burning rate modelare:

- One-dimensional burning
- Steady-state burning at a fixed pressure
- Radiative energy from the gas phase is absorbed at the burning surface

With reference to Figs. 3.9 and 3.10, the energy and the species equations are:
condensed-phase energy equation:

$$\frac{d}{dx}\left(\lambda_p \frac{dT}{dx}\right) - \rho_p r c_p \frac{dT}{dx} + \omega_p Q_p = 0 \tag{3.39}$$

condensed-phase species equation for species j:

$$\frac{d}{dx}\left(\rho_p D_{p,j} \frac{d\varepsilon_j}{dx}\right) - \rho_p r \frac{d\varepsilon_j}{dx} - \omega_{p,j} = 0 \tag{3.40}$$

gas-phase energy equation:

$$\frac{d}{dx}\left(\lambda_g \frac{dT}{dx}\right) - \rho_g u_g c_g \frac{dT}{dx} + \omega_g Q_g = 0 \tag{3.41}$$

gas-phase species equation for species i:

$$\frac{d}{dx}\left(\rho_g D_{g,i} \frac{d\varepsilon_i}{dx}\right) - \rho_g u_g \frac{d\varepsilon_i}{dx} - \omega_{g,i} = 0 \tag{3.42}$$

where r is the burning rate, Q is the heat of reaction, D is the diffusion coefficient, the subscripts p and g denote the condensed phase and the gas phase, respectively, and j and i refer to the species in the condensed phase and the gas phase, respectively.

The heat transfer in the gas phase and in the condensed phase can be viewed schematically as shown in Fig. 3.11. The heat flux transferred from the high-temperature zone, i. e., the flame zone, to the condensed phase through the burning surface is determined by the sum of the heat produced by the conductive heat $d/dx(\lambda dT/dx)$, by the convective heat $-\rho rc\, dT/dx$, and by the chemical reaction ωQ.

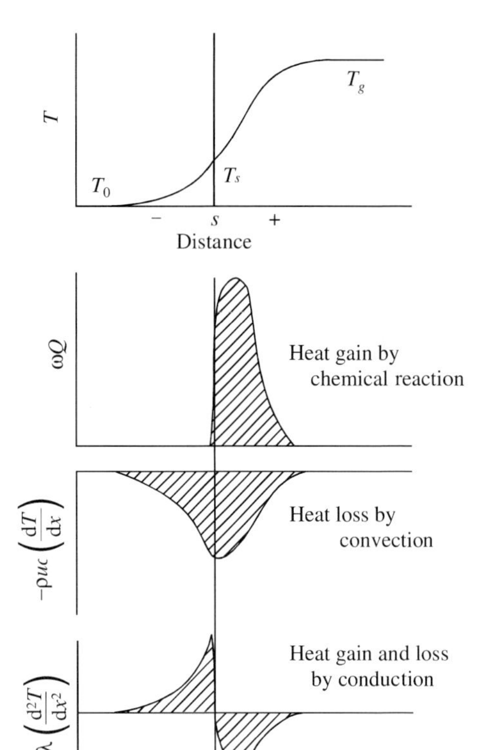

Fig. 3.11 Schematic representation of heat feedback processes in a combustion wave.

The temperature in the combustion wave increases from the initial temperature T_0 to the burning surface temperature T_s and then to the flame temperature T_g. The conductive heat $d/dx\ (\lambda dT/dx)$ in the gas phase decreases with increasing distance from the surface and reaches a minimum at some distance from the burning surface, before increasing once more at some point further downstream. The conductive heat in the condensed phase also decreases with increasing distance from the burning surface. Both the convective heat $-\rho rc\ dT/dx$ in the gas phase and the convective heat in the condensed phase increase with increasing distance from the burning surface. On the other hand, the heat release rate by the chemical reaction ωQ first increases downstream of the burning surface, reaches a maximum value at a certain distance, and then tails off to zero at a distant point. The heat transfer in the gas phase terminates at a distance far from the burning surface, where the temperature reaches a maximum and the final combustion products are formed.

3.5.1.2 Thermal Structure in the Condensed Phase

In order to describe the energy transfer process in the condensed phase, several additional assumptions are applied to the above equations:[9,10] (1) no endothermic or exothermic reaction is involved within the condensed phase (below the burning surface), (2) the luminous flame zone does not contribute to the conductive heat feedback from the gas phase to the burning surface, and (3) there is no diffusion of any species in the condensed phase or in the gas phase. Equations (3.39) and (3.40) are then simplified as follows:

$$\frac{d}{dx}\left(\lambda_p \frac{dT}{dx}\right) - \rho_p r c_p \frac{dT}{dx} = 0 \tag{3.43}$$

$$-\rho_p r \frac{d\varepsilon_j}{dx} - \omega_{p,i} = 0 \tag{3.44}$$

Integration of Eq. (3.43) under the boundary conditions of

$T = T_0$ at $x = -\infty$

$T = T_s$ at $x = 0$

leads to

$$T(x) - T_0 = (T_s - T_0) \exp(\mathrm{r}x/\alpha_\mathrm{p}) \tag{3.45}$$

where α_p is the thermal diffusivity of the condensed phase, defined according to $\alpha_p = \lambda_p/\rho_p c_p$, which is assumed to be independent of temperature. Figure 3.12 shows the temperature profiles in the condensed phase at r = 2 mm s^{-1}, 10 mm s^{-1}, and 50 mm s^{-1} under the assumptions of T_0 = 325 K, T_s = 600 K, λ_p = 2.10 × 10^{-4} kJ s^{-1} m^{-1} K^{-1}, ρ_p = 1600 kg m^{-3}, c_p = 1.47 kJ kg^{-1} K^{-1}, and $\alpha_p = \lambda_p/\rho_p c_p$ = 8.93 × 10^{-8} m^2 s^{-1}. The thermal wave thickness in the condensed phase, defined according to $\delta_p = \alpha_p/r$, is 45 μm, 9 μm, and 1.8 μm at r = 2 mm s^{-1}, 10 mm s^{-1}, and 50 mm s^{-1}, respectively. The temperature gradient in the condensed phase increases as the burning rate increases, i. e., δ_p decreases as the burning rate increases.

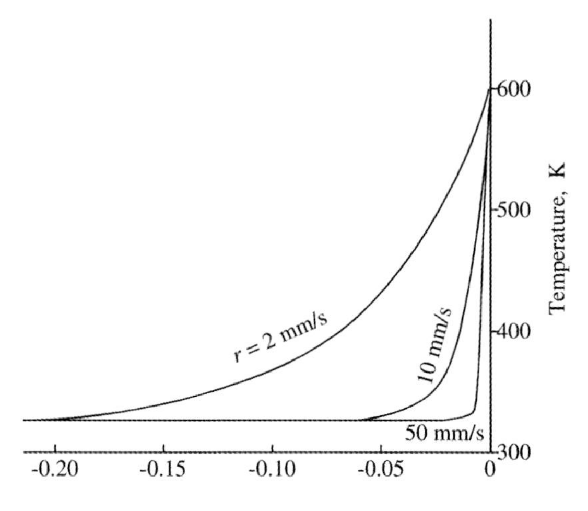

Fig. 3.12 Temperature profile in the condensed phase of an energetic material.

3.5.1.3 **Thermal Structure in the Gas Phase**

Since energetic materials are composed of several chemical ingredients and each molecular structure is complex, numerous gaseous species are produced at the burning surface and their reaction schemes are also complex. Thus, the determination of the gas-phase reaction rate for each species *j* during burning is highly difficult. A treatment of the gas-phase reaction that involves determining the temperature gradient just above the burning surface, $\phi = (dT/dx)_{s,g}$, and the heat feedback from the gas phase to the condensed phase, $\Lambda_g = \lambda_g \phi$, provides the basic foundation that underpins the burning rate equation. In order to gain a fundamental understanding of the heat feedback process in the gas phase, a heat release in the gas phase is assumed to have a positive constant value, $Q_g \omega_g$. The model represents the heat flux feedback from the gas phase to the condensed phase by integration of Eq. (3.41) with the boundary condition that the heat flux at infinity must be zero; one then obtains:

$$\Lambda_g = Q_g \int_0^{\infty} \exp(-\rho_g c_g u_g x/\lambda_g) \omega_g dx \tag{3.46 a}$$

Similar to the expression for the condensed phase, the thermal diffusivity in the gas phase is given by $\alpha_g = \lambda_g/\rho_g c_g$ and is assumed to be independent of temperature. The thermal wave thickness in the gas phase δ_g is defined according to $\delta_g = \alpha_g/u_g$. Then, Eq. (3.46 a) can be written as:

$$\Lambda_g = Q_g \int_0^{\infty} \exp(-u_g x/\alpha_g) \omega_g dx \tag{3.46 b}$$

$$= Q_g \int_0^{\infty} \exp(-x/\delta_g) \omega_g dx \tag{3.46 c}$$

In general, reaction rates are strongly dependent on the temperature when the activation energy is high. Thus, it is assumed that the two major effects of the reaction rate, temperature and concentration of the reactants, tend to cancel each other out; as the reaction proceeds the temperature increases but the concentration of the reactants decreases. Furthermore, it is assumed that the resulting constant rate of reaction occurs only in a limited region, that is, between $x = x_i$ and $x = x_g$ in Fig. 3.13, and that the reaction rate in the gas phase can be expressed by a step function. Thus, integrating Eq. (3.46 c), one gets

$$\Lambda_g = (\delta_g[\omega_g]Q_g)\{\exp(-x_i/\delta_g) - \exp(-x_g/\delta_g)\} \tag{3.46 d}$$

where $[\omega_g]$ is a positive constant for $x_i < x \le x_g$ and is zero elsewhere; $[\omega_g]$ should be considered an average value for the real reaction rate occurring in the gas phase.

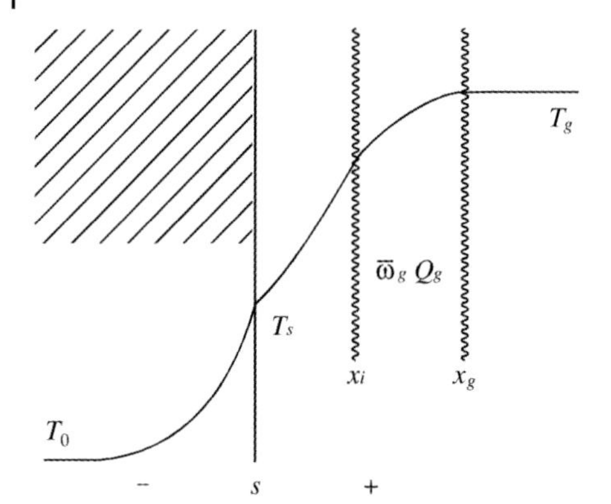

Fig. 3.13 Step function model for the gas-phase reaction that starts at $x = x_i$ and ends at $x = x_g$.

If one can assume that the chemical reaction represented by a step function starts at $x_i = 0$, i. e. the burning surface, and ends at $x = x_g$ in the gas phase, as shown in Fig. 3.14, Eq. (3.46 d) can be rewritten as

$$\Lambda_g = (\delta_g[\omega_g]Q_g)\{1 - \exp(-x_g/\delta_g)\} \tag{3.47}$$

When the term in the exponent $x_g/\delta_g >> 1$, a simplified expression for the heat flux feedback from the gas phase to the solid phase is obtained:[8]

$$\Lambda_g = \delta_g[\omega_g]Q_g \tag{3.48}$$

This asymptote occurs whenever the heat transferred back to the burning surface is small compared to the heat released in the gas phase, which implies that x_g/δ_g is sufficiently large. Though Eq. (3.48) is not precise, it indicates the burning rate behavior without introducing mathematical complexities and reaction parameters.

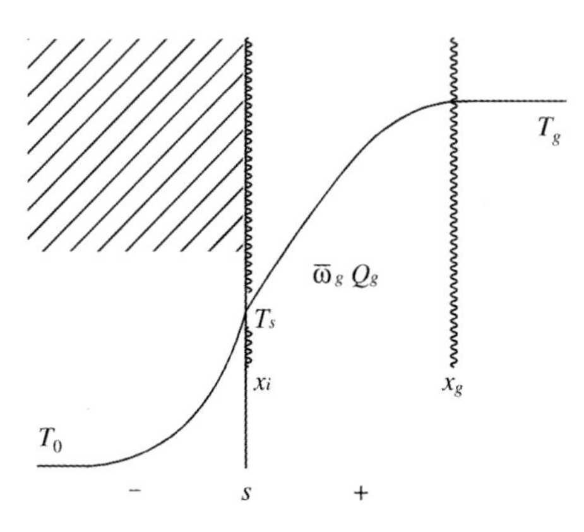

Fig. 3.14 Step function model for the gas-phase reaction that starts at $x_i = 0$ and ends at $x = x_g$.

3.5.1.4 **Burning Rate Model**

The heat flux feedback from zone II to zone I by conduction heat transfer, Λ_p, is given by

$$\Lambda_p = \lambda_p\,(dT/dx)_{s,p} = \rho_p\, c_p\, r(T_s - T_0) \tag{3.49}$$

and the heat flux generated in the condensed-phase zone at the burning surface, Γ_s, is given by

$$\Gamma_s = \rho_p\, r\, Q_s \tag{3.50}$$

where the subscript s denotes the burning surface and the subscript p denotes the condensed-phase reaction zone at the burning surface. The respective boundary conditions for the energy equation in the condensed phase and at the burning surface are given by

$$T = T_0 \text{ at } x = -\infty$$

$$\Lambda_p = \Lambda_g + \Gamma_s \quad \text{at } x = 0 \tag{3.51}$$

Generally, the reaction rate ω_j of a one-step reaction

$$\sum_{i=1}^{N} \nu_i'\, \mathrm{M}_i \xrightarrow{k_g} \sum_{i=1}^{N} \nu_i''\, \mathrm{M}_i$$

is represented by

$$\omega_j = \rho_g \frac{d\varepsilon_j}{dt} = \rho_g\, u_g \frac{d\varepsilon_j}{dx} = (\nu_i'' - \nu_i')\, k_g \prod_{h=1}^{N} (\rho_g\, \varepsilon_h)^{\nu_h'} \tag{3.52}$$

where M denotes an arbitrary chemical species, N is the number of parallel reaction paths in the gas phase, k_g is the reaction rate constant, and ν_i' and ν_i'' are the stoichiometric coefficients for species i as a reactant and species j as a product, respectively. Therefore, the heat feedback from the gas phase to the burning surface is obtained from Eqs. (3.42), (3.48), and (3.52):

$$\Lambda_g = \delta_g \sum_{i=1}^{N} Q_{g,i}(\nu_i'' - \nu_i')\, k_g \prod_{h=1}^{N} (\rho_g \varepsilon_h)^{\nu_h'} \tag{3.53}$$

Let us assume that the thermal diffusivity at the interface between the condensed phase and the gas phase, i. e., at the burning surface, is given by $\alpha_0 = \lambda_g/\rho_p c_g$. Using the mass continuity equation between the condensed phase and the gas phase, $\rho_p r = \rho_g u_g$, represented by Eq. (3.1), the thermal wave thickness, δ_g, is given by $\alpha_0/r = \delta_g$. Then, combining Eq. (3.51) with Eq. (3.53) and solving for r, one gets:

$$r = \left[\frac{\alpha_0}{\rho_p c_p (T_s - T_0 - Q_s/c_p)} \sum_{i=1}^{N} Q_{g,i}(\nu_i'' - \nu_i')\, k_g \prod_{h=1}^{N} (\rho_g \varepsilon_h)^{\nu_h'} \right]^{\frac{1}{2}} \tag{3.54}$$

Equation (3.54) is the simplified burning rate equation. If the reaction rates in the gas phase are known, the burning rate is given in terms of gas density (pressure), burning surface temperature, initial propellant temperature, and physical properties of the energetic material.

The burning surface temperature is related to the burning rate by an Arrhenius equation, which assumes a first-order decomposition reaction for each reaction species at the burning surface.

$$r = \sum_{j=1}^{K} \varepsilon_j Z_{s,j} \exp(-E_{s,j}/RT_s) \tag{3.55}$$

where K is the number of independent gasification reaction paths in the condensed phase, which are assumed to be parallel. Combination of Eqs. (3.54) and (3.55) allows the burning rate and burning surface temperature to be obtained for any given set of conditions.

Generally, the gas-phase reactions in both flame models for premixed gases and the burning of energetic materials are assumed to be bimolecular and hence of second order. Eq. (3.54) can then be expressed as

$$r = [\{\alpha_0 Q_g(\varepsilon_g \rho_g)^2 k_g\}/\{\rho_p c_p\ (T_s - T_0 - Q_s/c_p)\}]^{1/2} \tag{3.56}$$

The reaction rate constant, k_g, is a function of temperature and is expressed as

$$k_g = Z_g \exp\ (-E_g/RT_g) \tag{3.57}$$

The perfect gas law is also used to relate the assumed spatially constant density to p and Tg:

$$\rho_g = p/R_g T_g \tag{3.58}$$

where R_g is the gas constant in the reaction zone. Substituting Eqs. (3.57) and (3.58) into Eq. (3.56), one obtains the burning rate equation for energetic materials under the aforementioned assumptions:

$$r = p\left[\frac{\alpha_0}{\rho_p c_p}\frac{1}{(R_g T_g)^2}\frac{Q_g \varepsilon_g^2 Z_g \exp(-E_g/RT_g)}{(T_s - T_0 - Q_s/c_p)}\right]^{\frac{1}{2}} \tag{3.59}$$

where T_g is given as

$$T_g = T_0 + Q_s/c_p + Q_g/c_g \tag{3.60}$$

The burning surface decomposition rate, i. e., burning rate, is given from Eq. (3.55) as

$$r = Z_s \exp\ (-E_s/RT_s) \tag{3.61}$$

The nonlinear character of the algebraic equations, Eqs. (3.59) and (3.61), implies the need for an iteration solution.

3.5.2 Flame Stand-Off Distance

If the reactive gas produced at the burning surface of an energetic material reacts slowly in the gas phase and generates a luminous flame, the distance L_g between the burning surface and the luminous flame front is termed the flame stand-off distance. In the gas phase shown in Fig. 3.9, the temperature gradient appears to be small and the temperature increases relatively slowly. In this case, heat flux by conduction, the first term in Eq. (3.41), is neglected. Similarly, the rate of mass diffusion, the first term in Eq. (3.42), is assumed to be small compared with the rate of mass convection, the second term in Eq. (3.42). Thus, one gets

$$-\rho_g\, u_g\, c_g\, dT/dx + \omega_g Q_g = 0 \qquad (3.62)$$

$$-\rho_g\, u_g\, d\varepsilon_i/dx - \omega_i = 0 \qquad (3.63)$$

The reaction rate for an m th order reaction (ignoring the temperature dependence of ρ_g) is given by

$$\omega_g = \varepsilon_g{}^m \rho_g{}^m\, Z_g \exp(-E_g/RT_g) \qquad (3.64)$$

Combining Eqs. (3.62) and (3.64), one obtains

$$dT/dx = (1/c_g\, u_g) Q_g \varepsilon_g{}^m \rho_g{}^{m-1}\, Z_g \exp\,(-E_g/RT_g) \qquad (3.65)$$

The mass flow continuity relationship between the gas phase and solid is

$$u_g = r\rho_p/\rho_g \qquad (3.66)$$

Combining Eqs. (3.65) and (3.66) and the perfect gas law gives

$$dT/dx = (1/c_g\rho_p r) Q_g \varepsilon^m\, (R_g T_g)^{-m}\, p^m\, Z_g \exp\,(-E_g/RT_g) \qquad (3.67)$$

The burning rate of an energetic material is represented by Vieille's law (Saint Robert's law) according to

$$r = ap^n \qquad (3.68)$$

where n is the pressure exponent of the burning rate and a is a constant dependent on the chemical composition and the initial propellant temperature. Substituting Eq. (3.68) into Eq. (3.67), one gets

$$dT/dx = (1/c_g\rho_p a) Q_g \varepsilon^m\, (R_g T_g)^{-m}\, p^{m-n}\, Z_g \exp\,(-E_g/RT_g) \qquad (3.69)$$

The temperature gradient, dT/dx, in the gas phase is approximately equal to $\Delta T_g/L_g$, where ΔT_g is the temperature change across the gas-phase zone. Thus, the flame stand-off distance L_g is represented by

$$L_g = p^{n-m} \frac{\Delta T_g c_g \rho_p a (R_g T_g)^m}{Q_g \varepsilon_g^m Z_g \exp(-E_g / R T_g)} \tag{3.70}$$

$$\sim p^{n-m} = p^d \tag{3.70 a}$$

3.5.3 Burning Rate Characteristics of Energetic Materials

3.5.3.1 Pressure Exponent of Burning Rate

In general, the burning rate of an energetic material is seen to increase linearly with increasing pressure in an ln p versus ln r plot represented by Eq. (3.68) at constant initial temperature T_0. Thus, the pressure sensitivity of burning rate at a constant initial temperature, n, is defined by

$$n = \left(\frac{\partial \ln r}{\partial \ln p} \right)_{T_0} \tag{3.71}$$

3.5.3.2 Temperature Sensitivity of Burning Rate

The burning rate of an energetic material also depends on its initial temperature, T_0, even when the burning pressure is kept constant. The temperature sensitivity of burning rate, σ_p, is defined by the change in burning rate when T_0 is changed according to

$$\sigma_p = \frac{1}{r} \frac{r_1 - r_0}{T_1 - T_0} \tag{3.72a}$$

where r_0 and r_1 are the burning rates at T_0 and T_1, respectively, and r is the averaged burning rate between T_0 and T_1. Thus, the unit of σ_p appears to be K^{-1}. The differential form of Eq. (3.72 a) is expressed by

$$\sigma_p = \frac{1}{r} \left(\frac{\partial r}{\partial T_0} \right)_p = \left(\frac{\partial \ln r}{\partial T_0} \right)_p \tag{3.72b}$$

Using Eq. (3.68), one gets

$$\sigma_p = \left[\frac{\partial \ln (a p^n)}{\partial T_0} \right]_p = \frac{1}{a} \left(\frac{\partial a}{\partial T_0} \right)_p \tag{3.72c}$$

The temperature sensitivity of the burning rate defined in Eq. (3.72) is a parameter of considerable relevance in energetic materials.

3.5.4
Analysis of Temperature Sensitivity of Burning Rate

In order to understand the fundamental concept of the cause of temperature sensitivity, in the analysis described in this section it is assumed that the combustion wave is homogeneous and that it consists of steady-state, one-dimensionally successive reaction zones. The gas-phase reaction occurs with a one-step temperature rise from the burning surface temperature to the maximum flame temperature. The heat transfer in the combustion wave structure of an energetic material is illustrated in Fig. 3.10. The heat flux feedback from zone III to zone II by conductive heat transfer, $\Lambda_g = \lambda_g\ (dT/dx)_{s,g}$, is given by Eq. (3.46), and the heat flux feedback from zone II to zone I by conduction heat transfer, $\Lambda_p = \lambda_p\ (dT/dx)_{s,p}$, is given by Eq. (3.49). Using the integrated energy equation, Eq. (3.51), at the burning surface, the burning rate is represented by

$$r = \alpha_s \phi/\varphi \tag{3.73}$$

$$\phi = (dT/dx)_{s,g} \tag{3.74}$$

$$\varphi = T_s - T_0 - Q_s/c_p \tag{3.75}$$

$$\alpha_s = \lambda_g/\rho_p c_p \tag{3.76}$$

where α_s is the thermal diffusivity at the burning surface. Equation (3.73) indicates that the burning rate of an energetic material is determined by two parameters: the gas-phase parameter ϕ, which is determined by the physical and chemical properties of the gas phase, and the condensed-phase parameter φ, which is determined by the physical and chemical properties of the condensed phase. When the initial temperature is increased from T_0 to $T_0 + \Delta T_0$, the temperature profile is as shown by the dashed line in Fig. 3.15. The burning surface temperature T_s is also increased to $T_s + \Delta T_s$, and the final combustion temperature is increased from T_g to $T_g + \Delta T_g$.

When the logarithmic form of the burning rate equation given by Eq. (3.73) is differentiated with respect to the initial temperature of the energetic material at a constant pressure, the following is derived:

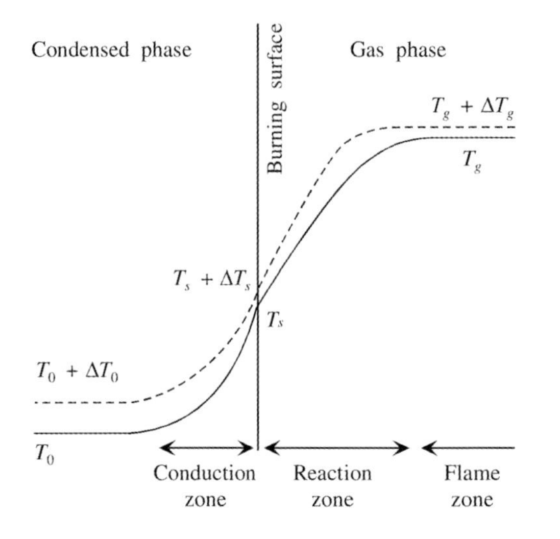

Fig. 3.15 Temperature profiles in combustion waves at different initial temperatures of an energetic material.

$$(\partial \ln r/\partial T_0)_p = \{\partial \ln \lambda_g (dT/dx)_{s,g}/\partial T_0\}_p - \{\partial \ln \rho_p c_p (T_s - T_0 - Q_s/c_p)/\partial T_0\}_p \quad (3.77)$$

Hence, the physical property α_s is assumed to be independent of T_0, and Eq. (3.77) is written as

$$\sigma_p = (\partial \ln \phi/\partial T_0)_p - (\partial \ln \varphi/\partial T_0)_p$$

$$= \Phi + \Psi \quad (3.78)$$

$$\text{where } \Phi = (\partial \ln \phi/\partial T_0)_p \quad (3.79)$$

$$\text{and } \Psi = -(\partial \ln \varphi/\partial T_0)_p \quad (3.80)$$

From Eq. (3.78), it can be seen that the temperature sensitivity depends on two parameters,[11] Φ and Ψ: Φ is the so-called "temperature sensitivity of the gas phase", which is determined by the parameters of the gas phase, and Ψ is the so-called "temperature sensitivity of the condensed phase", which is determined by the parameters of the condensed phase.

If one assumes that Eq. (3.48) is applicable to the reaction in the gas phase, the heat flux feedback from the gas phase to the burning surface is given by

$$\lambda_g \phi = \alpha_0 [\omega_g] Q_g / r \quad (3.81)$$

Differentiating the logarithmic form of Eq. (3.81) with respect to T_0 at a constant pressure, one gets

$$\Phi = (\partial \ln [\omega_g] Q_g/\partial T_0)_p - (\partial \ln r/\partial T_0)_p$$

$$= (\partial \ln [\omega_g]/\partial T_0)_p + (\partial \ln Q_g/\partial T_0)_p - \sigma_p$$

$$= \Omega + \Theta - \sigma_p \quad (3.82)$$

where

$$\Omega = (\partial \ln [\omega_g]/\partial T_0)_p \quad (3.83)$$

$$\Theta = (\partial \ln Q_g/\partial T_0)_p \quad (3.84)$$

Thus, one obtains the temperature sensitivity expression:

$$\sigma_p = \Omega/2 + \Theta/2 + \Psi/2 \quad (3.85)$$

If one assumes that the reaction rate in the gas phase is given by a one-step, kth-order Arrhenius-type equation and substitutes this in Eq. (3.83), one gets

$$\Omega = (E_g/RT_g^2)(\partial T_g/\partial T_0)_p \quad (3.86)$$

The heat generated in the gas phase is then given by

$$Q_g = c_g(T_g - T_s) \tag{3.87}$$

Substituting Eq. (3.87) into Eq. (3.84), one gets

$$\Theta = (\partial T_g/\partial T_0 - \partial T_s/\partial T_0)_p/(T_g - T_s) \tag{3.88}$$

Substituting Eqs. (3.86), (3.88), and (3.80) into Eq. (3.85), one gets

$$\sigma_p = (E_g/2RT_g^2)(\partial T_g/\partial T_0)_p + (\partial T_g/\partial T_0 - \partial T_s/\partial T_0)_p/2(T_g - T_s)$$
$$- \{\partial T_s/\partial T_0 - 1 - (\partial Q_s/\partial T_0)/c_p\}_p/2(T_s - T_0 - Q_s/c_p) \tag{3.89}$$

Equation (3.89) is the expression for the temperature sensitivity of an energetic material based on the analysis of a one-dimensional, one-step reaction in the combustion wave.

References

1 Lewis, B., and von Elbe, G., Combustion, Flames and Explosions of Gases, Academic Press, New York (1951).
2 Gaydon, A. G., and Wolfhard, H. G., Flames: Their Structure, Radiation and Temperature, Chapman and Hall, London (1960).
3 Strehlow, R. A., Fundamentals of Combustion, International Textbook Company, Scranton, Pennsylvania (1968), Chapter 5.
4 Glassman, I., Combustion, Academic Press, New York (1977), Chapter 5.
5 Williams, F. A., Combustion Theory, 2nd edition, Benjamin/Cummings, New York (1985), Chapters 6 and 7, pp. 182–246.
6 Zucrow, M. J., and Hoffman, J. D., Gas Dynamics, John Wiley & Sons, New York (1976), Chapter 9.
7 JANAF Thermochemical Tables, Dow Chemical Co., Midland, Michigan (1960–1970).
8 Gordon, S., and McBridge, B. J., Computer Program for Calculation of Complex Chemical Equilibrium Compositions, Rocket Performance, Incident and Reflected Shocks, and Chapman–Jouguet Detonations, NASA SP-273, 1971.
9 Kubota, N., Ohlemiller T. J., Caveny, L. H., and Summerfield, M., The Mechanism of Super-Rate Burning of Catalyzed Double-Base Propellants, AMS Report No. 1087, Aerospace and Mechanical Sciences, Princeton University, Princeton, NJ (1973).
10 Kubota, N., "Survey of Rocket Propellants and Their Combustion Characteristics", Fundamentals of Solid-Propellant Combustion (Eds.: Kuo, K. K., Summerfield, M.), Progress in Astronautics and Aeronautics, Vol. 90, Chapter 1, AIAA, Washington DC (1984).
11 Kubota, N., "Temperature Sensitivity of Solid Propellants and Affecting Factors: Experimental Results", Nonsteady Burning and Combustion Stability of Solid Propellants (Eds.: De Luca, L., Price, E. W., and Summerfield, M.), Progress in Astronautics and Aeronautics, Vol. 143, Chapter 4, AIAA, Washington DC (1990).

4
Energetics of Propellants and Explosives

4.1
Crystalline Materials

4.1.1
Physicochemical Properties of Crystalline Materials

Energetic materials are composed of fuel and oxidizer components, which are incorporated into their chemical structures. Fuel components are mostly hydrocarbon structures made up of hydrogen and carbon atoms. As shown in Fig. 2.1, NO_2 is a typical oxidizer component, which is attached to carbon, nitrogen, or oxygen atoms in hydrocarbon structures to form C–NO_2, N–NO_2, or O–NO_2 bonds. Breakage of an N–N or O–O bond generates heat with the formation of CO_2 or N_2 as a reaction product. Crystalline materials used as oxidizer components of propellants and explosives are decomposed thermally to produce gaseous oxidizer fragments. On the other hand, hydrocarbon polymers used as fuel components such as polyurethane and polybutadiene are decomposed endothermically to produce hydrogen, solid carbon, and other hydrocarbon fragments. Mixtures of these hydrocarbon polymers with the crystalline materials constitute energetic materials that are gasified upon heating to generate both fuel and oxidizer fragments simultaneously. These fragments react exothermically and produce high-temperature combustion products. The combustion process of the energetic materials is dependent on various physical and chemical characteristics, such as the individual properties of both the fuel and oxidizer components, the ratio in which they are mixed, the particle size of crystalline oxidizers, the presence of additives such as catalysts and modifiers, as well as the combustion pressure and the initial temperature.

Extensive studies on the formation of energetic materials have been carried out to formulate high-energy propellants and explosives.[1–15] High-density energetic materials are made by the formation of three-dimensional structures with nitrogen-nitrogen bonds. Nitrogen-nitrogen single bonds produce significant heat when they are cleaved and dinitrogen molecules containing nitrogen-nitrogen triple bonds are formed. Computational molecular design predicts the possibility of the formation of energetic nitrogen crystals composed of -N–N- single bonds such as N_4, N_6, N_7, N_{20}, and N_{60} molecular structures as shown in Fig. 4.1. These molecules are akin to carbon molecular structures such as diamond, carbon nanotubes, and

Propellants and Explosives. Naminosuke Kubota

ISBN: 978-3-527-31424-9

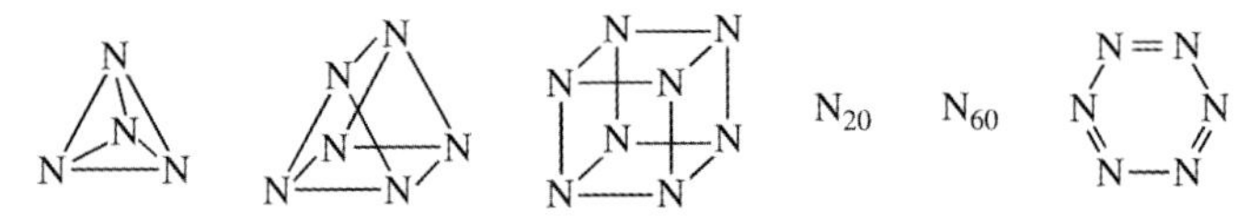

Fig. 4.1 Molecular structures of energetic nitrogen crystals.

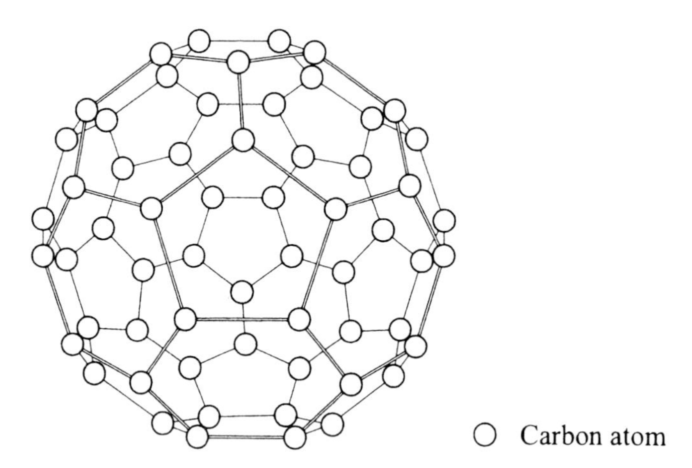

Fig. 4.2 Molecular structure of fullerene, C_{60}.

fullerene (C_{60}).[16] The size of C_{60} is about 4×10^{-10} m and the molecular structure is shown in Fig. 4.2. These are used as high-density fuel components.

Hexanitrohexaazaisowurtzitane (HNIW) is a typical high energy density material that incorporates six -N–NO_2 bonds and twelve -C–N- bonds in one molecular structure. Since the number of oxygen atoms in the molecular structure of HNIW is not sufficient to fulfil a role as an oxidizer, the heat released on its decomposition is extremely high due to the process of bond breakage and the formation of dinitrogen molecules. This chemical process is equivalent to the decomposition and heat release processes of HMX and RDX. Similar chemical bond formations yielding more complex high energy density materials can be envisaged by means of computational molecular design. Though most of them are thermally unstable and sensitive to mechanical shock, the high energy density materials shown in Fig. 4.3 are considered to be useful for propellants and explosives.[17] Based on the fundamental understanding of energetic materials,[18–20] many other polymeric and crystalline energetic materials have been synthesized and processed.[21–27]

4.1.2
Perchlorates

Perchlorates are characterized by a ClO_4 fragment/anion in their molecular structures and are crystalline materials used in propellants and explosives.[18–21] The oxy-

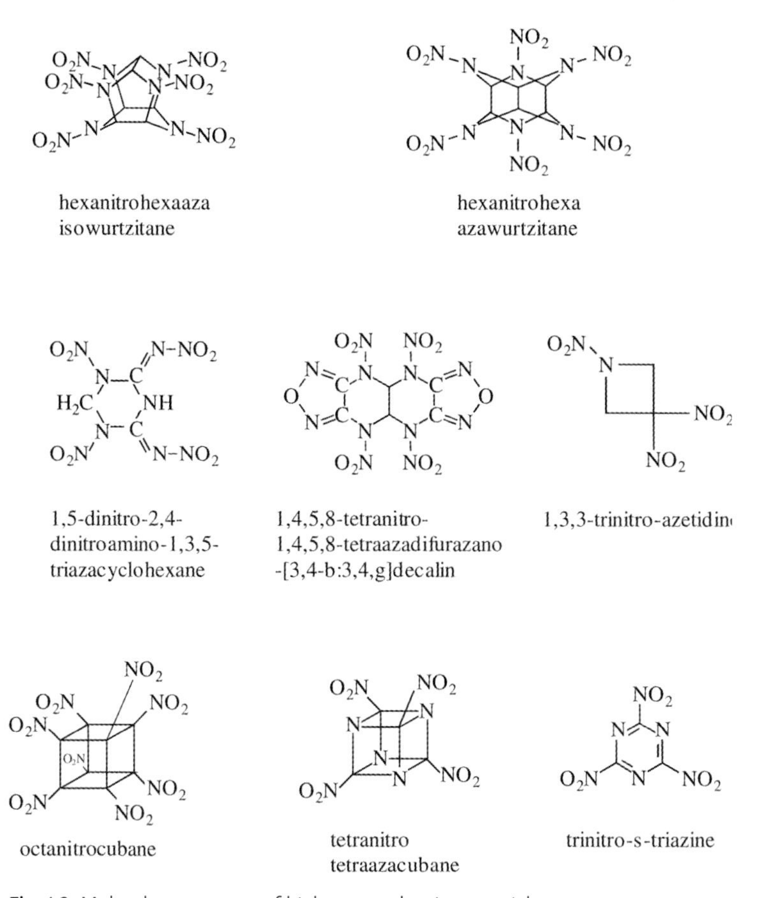

Fig. 4.3 Molecular structures of high energy density materials.

gen atoms in the ClO_4 fragment act as an oxidizer. Typical perchlorates are ammonium perchlorate (AP: NH_4ClO_4), nitronium perchlorate (NP: NO_2ClO_4), and potassium perchlorate (KP: $KClO_4$). Since AP contains no metal atoms and the molecular mass of the combustion products is low, it is the major crystalline oxidizer used for composite rocket propellants.

4.1.2.1 **Ammonium Perchlorate**

Ammonium perchlorate (AP: NH_4ClO_4) is a white, crystalline material, the crystal structure of which changes from orthorhombic to cubic at 513 K, which affects the decomposition process. AP is not hygroscopic in the atmosphere and the mass fraction of oxygen is 0.545. It is used as an oxidizer in various types of propellants and pyrolants. A rapid sublimation occurs between 670 K and 710 K at low pressures. Under slow heating, AP starts to decompose at about 470 K according to:

$$4NH_4ClO_4 \rightarrow 2Cl_2 + 3O_2 + 8H_2O + 2N_2O$$

At about 620 K, AP decomposes according to:

$$2NH_4ClO_4 \rightarrow Cl_2 + O_2 + 4H_2O + 2NO$$

When the heating rate is high, the overall reaction process is

$$NH_4ClO_4 \rightarrow NH_3 + HClO_4$$

$$HClO_4 \rightarrow HCl + 2O_2$$

This reaction is exothermic and produces excess oxygen as an oxidizer.

The oxygen molecules produced by the decomposition act as an oxidizer when AP particles are mixed with a polymeric fuel component, for example:

$$NH_4ClO_4 + C_mH_n \text{ (hydrocarbon polymer)} \rightarrow CO_2 + H_2O + N_2 + HCl$$

This reaction produces significant amounts of heat and gaseous molecules, which yield a high I_{sp} as defined in Eq. (1.76).

Though AP is relatively stable to mechanical shocks, an excessive shock will cause it to detonate with a velocity of 3400 m s^{-1}. A mixture of AP and ammonium nitrate (AN) with small amounts of silicon and iron is used as an industrial explosive.

4.1.2.2 **Nitronium Perchlorate**

The oxygen content of NP is higher than that of most other crystalline materials. Thus, its chemical potential as an oxidizer is high when combined with a fuel component. Since its theoretical density is 2220 kg m^{-3} and its heat of formation is a positive value, +33.6 kJ mol^{-1}, NP is an ideal material to serve as an oxidizer. However, NP is extremely hygroscopic and its hydrolysis forms nitric acid and perchloric acid according to:

$$NO_2ClO_4 + H_2O \rightarrow HNO_3 + HClO_4$$

Moreover, NO_2 is formed by the thermal decomposition of NP at about 360 K, a temperature too low to permit its use as a practical oxidizer ingredient.

4.1.2.3 **Potassium Perchlorate**

Potassium perchlorate (KP: $KClO_4$) is a well-known oxidizer, used as an oxidizer component of black powder. Since KP produces potassium oxides and condensed products, the high molecular mass M_g of the combustion products is not favorable for its use as an oxidizer in rocket propellants. A mixture of 75% KP with 25% asphalt pitch was used as a rocket propellant named Galcit, which was the original prototype of a composite propellant in the 1940s. Potassium chlorate ($KClO_3$) is also a crystalline oxidizer, and although it has a lower oxygen content compared with KP, it is more sensitive to mechanical shock and easier to ignite, and is also easier to detonate.

4.1.3
Nitrates

Typical crystalline nitrates used in propellants and explosives are ammonium nitrate (AN: NH_4NO_3), potassium nitrate (KN: KNO_3), sodium nitrate (SN: $NaNO_3$), pentaerythrol tetranitrate (PETN: $C_5H_8(ONO_2)_4$), and triaminoguanidine nitrate (TAGN: $CH_9N_7O_3$).

4.1.3.1 Ammonium Nitrate

Ammonium nitrate (AN: NH_4NO_3) is a white, crystalline material, the crystal structure of which varies with temperature.[18–21] Its melting point is 442 K and its heat of fusion is 71.4 kJ kg^{-1}. Though the mass fraction of oxygen of AN is 0.5996, it is highly hygroscopic and absorbs moisture from the atmosphere to form liquid AN acid. This limits its application in propellants and pyrolants. However, AN is widely used as an oxidizer of explosives such as slurry explosives and ANFO (ammonium nitrate fuel oil) explosives.

Table 4.1 Physicochemical properties of ammonium nitrate, AN.

Phase	Melt	I	II	III	IV	V
Transition temperature, K	442	398	357	305	256	
Density, kg m^{-3}		1550	1600	1570	1700	1750
Heat of crystal transition, kJ kg^{-1}		50.0	22.3	21.0	6.7	
Volume change, $\times 10^{-6}$ m^3 kg^{-1}		+13	−8	+22	−16	

The thermal decomposition of NH_4NO_3 depends on the heating process. At low temperatures, i. e. about 430 K, the decomposition reaction is expressed by

$$NH_4NO_3 \rightarrow NH_3 + HNO_3$$

and is a reversible reaction. The gasification reaction is endothermic by 181 kJ mol^{-1}. At high temperatures, i. e. about 550 K, the decomposition reaction is expressed by

$$NH_4NO_3 \rightarrow N_2O + 2H_2O$$

and is exothermic by 36.6 kJ mol^{-1}. This reaction is followed by the decomposition of N_2O, and so the overall reaction is expressed by

$$NH_4NO_3 \rightarrow N_2 + 2H_2O + 1/2O_2$$

which is exothermic by 119 kJ mol^{-1}.

The disadvantages of using AN in rocket propellants and pyrolants are its hygroscopic nature and its crystal structure transitions. The crystal transformations from phase I to phase V occur with decreasing temperature as follows:

V		IV		III		II		I
	256 K		305 K		357 K		398 K	
tetragonal		orthorhombic		disordered orthorhombic		tetragonal		cubic

Accompanying density and volumetric changes occur as shown in Table 4.1. The volumetric changes associated with these transitions affect internal mechanical stresses of the propellant grains. The transition from orthorhombic to disordered orthorhombic at 305 K is accompanied by a significant density decrease, which can sometimes lead to damage of the propellant and pyrolant grains under temperature cycling conditions. During propellant and pyrolant processing, humidity control is unavoidable due to the hygroscopic nature of AN. In order to suppress the phase transitions of AN crystals and to obtain phase-stabilized ammonium nitrate (PSAN), metal salts are incorporated into the AN lattice. Copper- and nickel salts are typical metal salts that have a beneficial effect on the phase transitions. On the other hand, the hygroscopic nature and crystal phase transitions of AN impose no limitations on its use in dynamites, emulsion explosives or slurry explosives.

4.1.3.2 **Potassium Nitrate and Sodium Nitrate**

Though the oxidation potentials of potassium nitrate (KN: KNO_3) and sodium nitrate (SN: $NaNO_3$) are high, both metal nitrates generate combustion products of high M_g. Thus, the specific impulse becomes low when KN or SN is used in a rocket propellant. KN and SN are used as major ingredients of explosives and in pyrotechnics. KN is a well-known material as a major component of black powder.

4.1.3.3 **Pentaerythrol Tetranitrate**

In contrast to other crystalline nitrates, pentaerythrol tetranitrate (PETN: $C_5H_8(ONO_2)_4$) is a crystalline nitrate ester similar to NG and NC. Though PETN is one of the most powerful energetic materials used in explosives, no excess oxidizer fragments are formed when it decomposes. Thus, PETN is not used as an oxidizer of propellants.

4.1.3.4 Triaminoguanidine Nitrate

Triaminoguanidine nitrate (TAGN: $CH_9N_7O_3$) contains a relatively high mole fraction of hydrogen, and its oxidizer fragment (HNO_3) is attached by an ionic bond in the molecular structure. The molecular mass of the combustion products of TAGN is low due to the high concentration of hydrogen.

4.1.4 Nitro Compounds

Energetic materials containing -C–NO_2 bonds in their molecular structures come under the heading of nitro compounds. Similar to nitrates, when nitro compoundsare decomposed thermally, NO_2 molecules are formed and act as an oxidizer component. The NO_2 molecules react exothermically with remaining hydrocarbon fragments and generate a large number of high-temperature combustion products. Typical nitro compounds used in explosives are dinitrotoluene (DNT: $C_7H_6N_2O_4$), trinitrotoluene (TNT: $C_7H_5N_3O_6$), hexanitrostilbene (HNS: $C_{14}H_6N_6O_{12}$), diaminotrinitrobenzene (DATB: $C_6H_5N_5O_6$), triaminotrinitrobenzene (TATB: $C_6H_6N_6O_6$), diazodinitrophenol (DDNP: $C_6H_2N_4O_5$), hydrazinium nitroformate (HNF: $N_2H_5C(NO_2)_3$), hexanitroazobenzene (HNB: $C_{12}H_4N_8O_{12}$), 2,4,6-trinitrophenylmethylnitramine (tetryl: $C_7H_5N_5O_8$), and 2,4,6-trinitrophenol (picric acid: $C_6H_3N_3O_7$). Computational molecular design predicts several possible high energy density nitro compounds, such as octanitrocubane, tetranitrotetraazacubane, trinitro-*s*-triazine, and hexanitrohexaazawurtzitane. The chemical structures of these energetic materials are shown in Fig. 4.3.

In general, the energy density of nitro compounds is high and their susceptibility to detonation is also high. Thus, nitro compounds are used as major components of explosives but are not used as components of propellants. However, since the energy density of HNF is higher than that of AN or ADN, this salt is used as an oxidizer in rocket propellants. In contrast to AP propellants, HNF propellants generate halogen-free combustion products and are used as eco-friendly propellants. The molecular structure of HNF is $N_2H_5^+ \ ^-C(NO_2)_3$, made by acid-base reaction between nitroform and hydrazine. Like AN and ADN, HNF is hygroscopic, necessitating humidity control during processing.

4.1.5 Nitramines

Nitramines are characterized by -N–NO_2 chemical bonds attached to hydrocarbon structures. N–N bond breakage produces NO_2, which acts as an oxidizer. The remaining hydrocarbon fragments act as fuel components. Typical nitramines are cyclo-1,3,5-trimethylene-2,4,6-trinitramine (RDX: $C_3H_6N_6O_6$), cyclo-1,3,5,7-tetramethylene-2,4,6,8-tetranitramine (HMX: $C_4H_8N_8O_8$), nitroguanidine (NQ:$CH_4N_4O_2$), hexanitrohexaazaisowurtzitane (HNIW: $(NNO_2)_6(CH)_6$), and ammonium dinitramide (ADN: $NH_4N(NO_2)_2$). Other high energy density nitramines, such as hexanitrohexaazawurtzitane (HNHAW) and hexanitrohexaazaadamantane(HNHAA), are predicted by computational molecular design to be potential ingredients of propellants and explosives.

RDX and HMX are also known as hexogen and octogen, respectively. The physicochemical properties of RDX and HMX are shown in Tabs. 2.3, 2.5, and 2.6. Though the densities, heats of formation, and heats of explosion are approximately the same for both RDX and HMX, the melting point temperature of HMX is much higher than that of RDX. RDX was synthesized in an effort to obtain higher energy density than that of nitroglycerin in order to improve explosive power, and takes its name from "**r**esearch and **d**evelopment e**x**plosives". HMX was synthesized to obtain a higher melting point than that of RDX, and takes its name from "**h**igh **m**elting point e**x**plosives".

The decomposition reactions of RDX and HMX are stoichiometrically balanced when it is assumed that CO, rather than CO_2, is formed as a combustion product:

$$\text{RDX:} \quad C_3H_6O_6N_6 \rightarrow 3CO + 3H_2O + 3N_2$$

$$\text{HMX:} \quad C_4H_8O_8N_8 \rightarrow 4CO + 4H_2O + 4N_2$$

Though the adiabatic flame temperatures are 3300 K for RDX and 3290 K for HMX at 10 MPa, no excess oxidizer fragments are produced. Thus, RDX and HMX are not used as oxidizers in propellants.

The overall initial decomposition reaction of HMX is represented by

$$3(CH_2NNO_2)_4 \rightarrow 4NO_2 + 4N_2O + 6N_2 + 12CH_2O$$

NO_2 and N_2O act as oxidizers and CH_2O acts as the fuel component. Since nitrogen dioxide reacts quite rapidly with formaldehyde, the gas-phase reaction

$$7NO_2 + 5CH_2O \rightarrow 7NO + 3CO + 2CO_2 + 5H_2O$$

is probably the dominant reaction immediately after the decomposition reaction. The NO thus produced oxidizes the remaining fuel fragments such as H_2 and CO. However, this oxidation by NO is reported to give the final combustion products only slowly. The dominant gas-phase reaction that dictates the burning rate of HMX is the oxidation by NO_2. A similar combustion process is envisaged for RDX.

Nitroguanidine (NQ) is a nitramine compound containing one $N-NO_2$ group in its molecular structure. In contrast to cyclic nitramines such as HMX and RDX, its density is low and its heat of explosion is also comparatively low. However, the *Mg* of its combustion products is low because of the high mass fraction of hydrogen contained within the molecule. Incorporating NQ particles into a double-base propellant forms a composite propellant termed a triple-base propellant, as used in guns.

Ammonium dinitramide (ADN) is a crystalline oxidizer with the formula $NH_4N(NO_2)_2$, that is, it is composed of ionically bonded ammonium cations, NH_4^+, and dinitramide anions, $^-N(NO_2)_2$. Though ADN is crystalline and has a high oxygen content, similar to AP and KP, it has no halogen or metal atoms within its structure. ADN is used as an oxidizer in smokeless composite propellants, similar to AN and HNF. It melts at about 364 K, accompanied by the latent heat of fusion.

The onset temperature of the exothermic decomposition is about 432 K, and the reaction is complete at about 480 K without a residue. The activation energy for the exothermic decomposition process ranges from 117 kJ mol^{-1} to 151 kJ mol^{-1}.

4.2 Polymeric Materials

4.2.1 Physicochemical Properties of Polymeric Materials

Polymeric materials that act as fuels and oxidizers are composed of nitrogen, oxygen, carbon, and hydrogen atoms. The hydrocarbon structures act as fuel components, and the oxidizer fragments, such as -C–NO_2, -O–NO_2, -O–NO, or -N-NO_2, are attached to the hydrocarbon structures through covalent chemical bonds.

Polymeric materials composed of hydrocarbon structures and $-N{=}N^{-}{=}N^{+}$ bonds are called azide polymers. Azide polymers generate heat when they are thermally decomposed. Cleavage of the azide bonds results in the formation of gaseous dinitrogen, which is accompanied by the release of much heat by virtue of the high chemical bond energy between the two nitrogen atoms of N_2. Thus, azide polymers produce heat without oxidation reactions. As shown in Table 2.5, a higher H_{exp} is obtained from materials having higher nitrogen concentrations.

Polymeric materials are used as binders to hold solid particles together so as to formulate composite explosives or composite propellants. The polymeric materials also constitute part of the fuel ingredients when the crystalline particles are oxidizer-rich. Various types of hydrocarbon polymers are used as polymeric binders.

The viscosity of a polymeric binder needs to be relatively low during the mixing process with crystalline particles in order to obtain a uniformly dispersed structure. The curing time needs to be long enough to allow for homogeneous mixing. In addition, the elasticity after curing should be sufficient to provide adequate mechanical strength and elongation characteristics of the composite explosives and propellants.

Three types of polymeric materials are used: inert polymers, active polymers, and azide polymers. No exothermic heat is produced when inert polymers are decomposed thermally. On the other hand, exothermic reactions occur when active polymers and azide polymers are decomposed. Self-sustaining burning is possible when active polymers and azide polymers are ignited.

4.2.2 Nitrate Esters

Nitrate esters are characterized by -O–NO_2 bonds in their structures. Typical nitrate esters used in propellants and explosives are nitrocellulose (NC), nitroglycerin(NG), triethyleneglycol dinitrate (TEGDN), trimethylolethane trinitrate (TMETN), diethyleneglycol dinitrate (DEGDN), and nitratomethyl methyl oxetane (NIMO). These nitrate esters are all liquid at room temperature, with the exception of NC, and are used as energetic plasticizers to formulate propellants and explosives. The thermal decomposition of nitrate esters involves O–NO_2 bond breakage with the evolution of NO_2 gas.[19,21] The remaining hydrocarbon structures are also decomposed to produce aldehydes and other fuel fragments, which are oxidized by the NO_2. This oxidation reaction is highly exothermic and generates high-temperature combustion products.

NC is an energetic nitropolymer consisting of a hydrocarbon structure with -O–NO_2 bonds as oxidizer fragments. In general, NC is produced from the cellulose, $\{C_6H_7O_2(OH)_3\}_n$, of cotton or wood, which is nitrated using nitric acid (HNO_3) to introduce -O–NO_2 bonds into its structure.

$$\{C_6H_7O_2(OH)_3\}_n + xHNO_3 \rightarrow (C_6H_7O_2)_n(OH)_{3n\text{-}x}(ONO_2)x + xH_2O$$

Through this nitration, OH groups contained within the cellulose are replaced with O–NO_2 groups, and the degree of nitration determines the energy content, i. e., the potential to form high-temperature combustion gases. The maximum degree of nitration of NC is obtained when its nitrogen content becomes 14.14 % by mass. The nitrogen content of the NC used as a conventional ingredient of propellants and explosives ranges from 13.3 % to 11.0 % N. An NC molecule containing 12.6 % N is represented by $C_{2\cdot20}H_{2\cdot77}O_{3\cdot63}N_{0\cdot90}$, for which the heat of formation is ΔH_f = −2.60 MJ kg^{-1}. The heat of formation decreases with decreasing degree of nitration, as shown in Table 4.2.

Table 4.2 Heats of formation of nitrocelluloses with varying N content.

% N	13.3	13.0	12.5	12.0	11.5	11.0
ΔH_f (MJ kg^{-1})	−2.39	−2.48	−2.61	−2.73	−2.85	−3.01

NC decomposes in an autocatalytic reaction with the evolution of NO_2 as a result of breakage of the weakest bond of -O–NO_2. The reaction between 363 K and 448 K is a first-order process and the activation energy is 196 kJ mol^{-1}. The remaining fragments form aldehydes such as HCHO and CH_3CHO. The reaction between NO_2 and these aldehydes produces heat and combustion gases.

NG has a relatively low molecular mass of 227.1 kg $kmol^{-1}$. It is a liquid at room temperature, but solidifies below 286 K.[20,21] Since NG is shock-sensitive and easy to detonate, desensitizers are admixed to render it safe for practical applications. NG is one of the major ingredients used in propellants and explosives. Typical examples are double-base propellants, in which it is mixed with nitrocellulose, and dynamites, in which it is again mixed with nitrocellulose and/or with other crystalline materials. The autocatalytic decomposition of NG, caused by O–NO_2 bond breakage with the evolution of NO_2, occurs at 418 K and has an activation energy of 109 kJ mol^{-1}. Self-ignition occurs after a critical concentration of NO_2 is achieved at 491 K.

Though nitroglycerin (NG) is a liquid nitrate ester rather than a nitropolymer, it becomes a polymeric material when it is mixed with plasticizers. Like NC, NG is composed of a hydrocarbon structure with -O–NO_2 bonds as oxidizer fragments. The thermal decomposition of NG is fundamentally the same as that of NC, producing NO_2 as an oxidizer and aldehydes as fuel components.

4.2.3 Inert Polymers

Various types of inert polymers are used to formulate propellants and explosives.[21] The polymers serve both as fuel components and as binders between the energetic particles, providing the necessary mechanical properties to prevent break-up of the grains or crack formation during ignition and combustion. Since the mass fraction of an inert polymer needed to formulate an ideal propellant or explosive is 0.12 or less, low polymer viscosity is required during mixing with the energetic particles. In general, polymers become soft at high temperatures and become brittle at low temperatures. High mechanical strength at high temperature and high elongation at low temperature are needed for applications in propellants and explosives.

Inert polymers are classified according to their chemical bond structures. Figure 4.4 shows typical basic polymer units and bond structures.[21] These polymers are based on hydrocarbon structures that have relatively low viscosities during

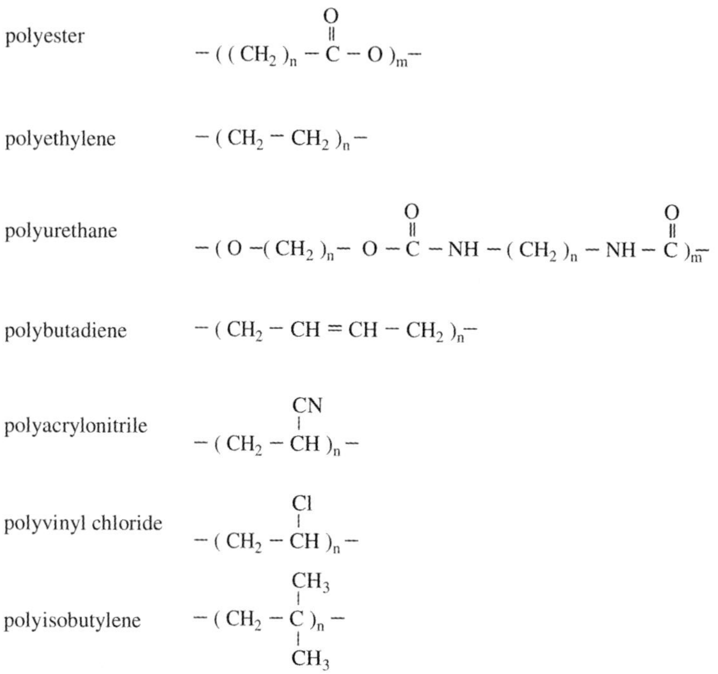

polyester $-((CH_2)_n - \overset{\overset{\displaystyle O}{\|}}{C} - O)_m-$

polyethylene $-(CH_2 - CH_2)_n-$

polyurethane $-(O-(CH_2)_n - O - \overset{\overset{\displaystyle O}{\|}}{C} - NH - (CH_2)_n - NH - \overset{\overset{\displaystyle O}{\|}}{C})_m-$

polybutadiene $-(CH_2 - CH = CH - CH_2)_n-$

polyacrylonitrile $-(CH_2 - \overset{\overset{\displaystyle CN}{|}}{CH})_n-$

polyvinyl chloride $-(CH_2 - \overset{\overset{\displaystyle Cl}{|}}{CH})_n-$

polyisobutylene $-(CH_2 - \underset{\underset{\displaystyle CH_3}{|}}{\overset{\overset{\displaystyle CH_3}{|}}{C}})_n-$

Fig. 4.4 The structures of typical basic polymer units.

the process of mixing with the crystalline oxidizer particles used for propellants and the energetic particles used for explosives. Two types of copolymers are used to formulate modern propellants and explosives: (1) polyurethane copolymer, and (2) polybutadiene copolymer. Polyether and polyester chemical bond structures are present in polyurethane copolymers. Since the oxygen content is relatively high in a polyurethane binder, this class of binder is used to obtain high combustion efficiency with a low concentration of crystalline oxidizer materials. On the other hand, the heat of formation of a polybutadiene copolymer is high and the oxygen content is low when compared with a polyurethane copolymer. This class of binder is used to obtain a high combustion temperature when mixed with crystalline oxidizer particles.

Polybutadiene acrylonitrile (PBAN) is used as the binder of the principal booster propellant deployed in the Space Shuttle. Carboxy-terminated polybutadiene(CTPB) and hydroxy-terminated polybutadiene (HTPB) are widely used in modern composite propellants. CTPB and HTPB form regularly distributed polymer matrices through crosslinking reactions. For example, HTPB polymer, HO–$(CH_2–CH{=}CH–CH_2)_n$–OH, can be cured with isophorone diisocyanate (IPDI) to form a polymeric binder. Using this binder, a high load density of oxidizer particles is obtained. In order to obtain superior mechanical properties of propellant grains, a small amount of a bonding agent is added to adhere each oxidizer particle to the binder.

Hydroxy-terminated polyester (HTPS) is made from diethylene glycol and adipic acid, and hydroxy-terminated polyether (HTPE) is made from propylene glycol. Hydroxy-terminated polyacetylene (HTPA) is synthesized from butynediol and paraformaldehyde and is characterized by acetylenic triple bonds. The terminal OH groups of these polymers are cured with isophorone diisocyanate. Table 4.3 shows the chemical properties of typical polymers and prepolymers used in composite propellants and explosives.[21] All of these polymers are inert, but, with the exception of HTPB, contain relatively high oxygen contents in their molecular structures.

Table 4.3 Chemical properties of polymers and prepolymers used in propellants and explosives.

Polymer	Chemical formula	$\xi(O)$	ΔH_f
HTPS	$C_{4.763}H_{7.505}O_{2.131}N_{0.088}$	34.1	−0.550
HTPE	$C_{5.194}H_{9.840}O_{1.608}N_{0.194}$	25.7	−0.302
HTPA	$C_{4.953}H_{8.184}O_{1.843}N_{0.205}$	29.5	−0.139
HTPB	$C_{7.075}H_{10.65}O_{0.223}N_{0.063}$	3.6	−0.058

$\xi(O)$: Oxygen content, % by mass; ΔH_f: heat of formation (298 K), MJ mol^{-1}

Polyether prepolymer:

$$H{-}(O{-}CH(CH_3){-}CH_2)n{-}O{-}CH_2{-}CH(CH_3){-}O{-}(CH_2{-}CH(CH_3){-}O)n{-}H \quad n = 17$$

Polyester prepolymer:

$$[O{-}\{CH_2CH_2{-}O{-}CH_2CH_2{-}O{-}C({=}O){-}(CH_2)_4{-}C({=}O){-}O\}_{3\text{-}4}{-}CH_2]_3C{-}CH_3$$

Figure 4.5 shows the chemical processes and molecular structures of typical inert binders used in composite propellants and plastic-bonded explosives.[21] Polysulfides are characterized by sulfur atoms in their structures and produce H_2O molecules during the polymerization process. These H_2O molecules should be re-

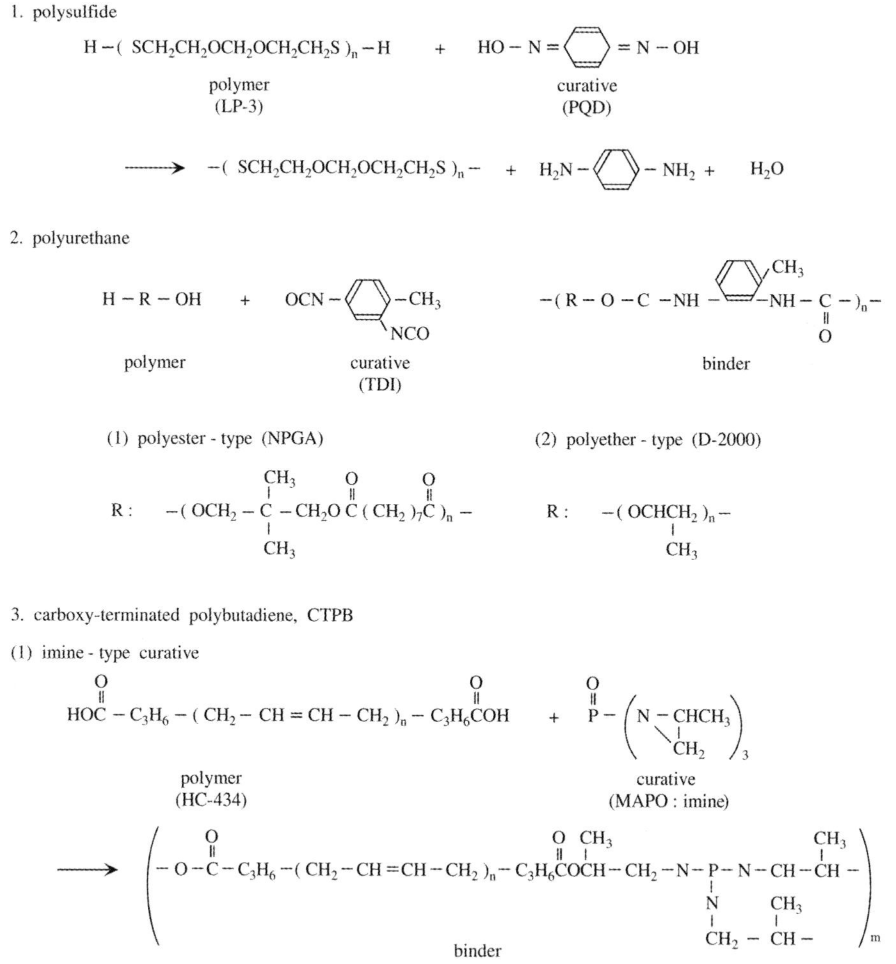

1. polysulfide

$H-(SCH_2CH_2OCH_2OCH_2CH_2S)_n-H$ + HO – N = ⟨ring⟩ = N – OH

polymer (LP-3) — curative (PQD)

⟶ $-(SCH_2CH_2OCH_2OCH_2CH_2S)_n-$ + H_2N–⟨ring⟩–NH_2 + H_2O

2. polyurethane

H – R – OH + OCN–⟨ring⟩–CH_3 (NCO) ⟶ –(R – O – C – NH –⟨ring, CH_3⟩–NH – C(=O) –$)_n$–

polymer — curative (TDI) — binder

(1) polyester - type (NPGA)

R : $-(OCH_2-C(CH_3)_2-CH_2OC(=O)(CH_2)_7C(=O))_n-$

(2) polyether - type (D-2000)

R : $-(OCHCH_2)_n-$ with CH_3 on CH

3. carboxy-terminated polybutadiene, CTPB

(1) imine - type curative

$HOC(=O)-C_3H_6-(CH_2-CH=CH-CH_2)_n-C_3H_6COH(=O)$ + $O{=}P-(N-CHCH_3, CH_2)_3$

polymer (HC-434) — curative (MAPO : imine)

⟶ $\left(-O-C(=O)-C_3H_6-(CH_2-CH=CH-CH_2)_n-C_3H_6C(=O)OCH(CH_3)-CH_2-N-P(N-CH_2-CH(CH_3)-)-N-CH(CH_3)-CH-\right)_m$

binder

Fig. 4.5 Chemical processes and molecular structures of typical binders used in composite propellants.

(2) epoxy - type curative

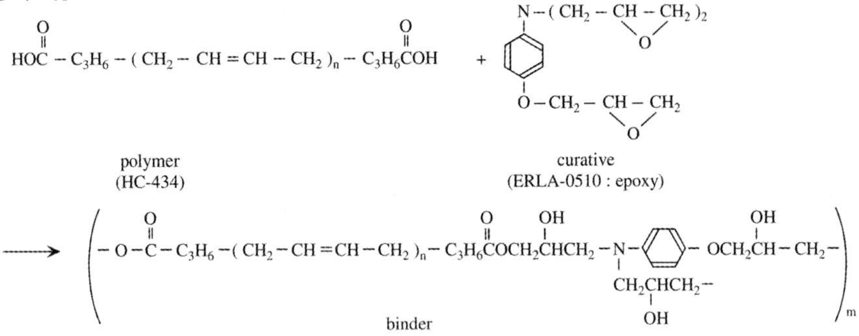

$HOC(=O) - C_3H_6 - (CH_2 - CH = CH - CH_2)_n - C_3H_6COH(=O)$ + $N-(CH_2-CH-CH_2)_2$ (epoxide, O) on benzene ring, $O-CH_2-CH-CH_2$ (epoxide, O)

polymer (HC-434)   curative (ERLA-0510 : epoxy)

$\longrightarrow \left(-O-C(=O)-C_3H_6-(CH_2-CH=CH-CH_2)_n-C_3H_6C(=O)OCH_2CH(OH)CH_2-N(CH_2CH(OH)CH_2-)-C_6H_4-OCH_2CH(OH)-CH_2- \right)_m$

binder

4. hydroxy-terminated polybutadiene, HTPB

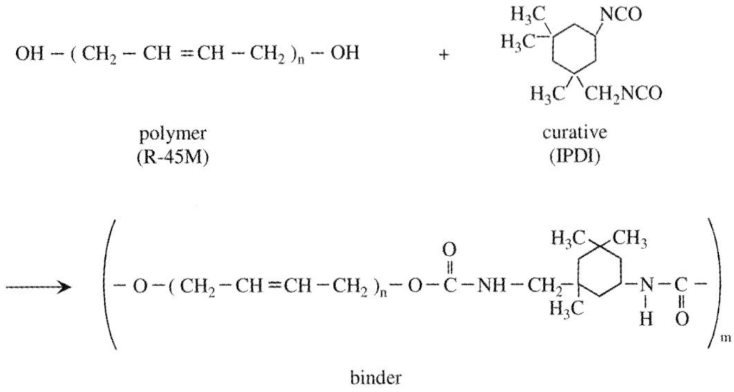

$OH - (CH_2 - CH = CH - CH_2)_n - OH$ + (cyclohexane ring with H_3C, H_3C, NCO, H_3C, CH_2NCO)

polymer (R-45M)   curative (IPDI)

$\longrightarrow \left(-O-(CH_2-CH=CH-CH_2)_n-O-C(=O)-NH-CH_2-[\text{ring}: H_3C, CH_3, H_3C]-N(H)-C(=O)- \right)_m$

binder

Fig. 4.5 Continued.

moved under vacuum in order to avoid bubble formation within the propellants and explosives. Two types of curative process, an imine-type and an epoxy-type, are used for CTPB. Superior mechanical properties at high and low temperatures are obtained when HTPB is used as an inert binder of propellants and explosives.

4.2.4 Azide Polymers

As mentioned in Section 4.2.1, materials containing $-N=N^{-}=N^{+}$ bonds are called azides, and these produce nitrogen gas accompanied by a significant release of heat when they are thermally decomposed. Polymeric materials consisting of hydrocarbon structures with $-N=N^{-}=N^{+}$ bonds are known as azide polymers and are characterized by $-N_3$ units attached to carbon atoms. Glycidyl azide polymer (GAP monomer: $C_3H_5ON_3$), poly(bis-azide methyl oxetane) (BAMO monomer: $C_5H_8ON_6$), and poly(3-azidomethyl-3-methyl oxetane) (AMMO monomer: $C_5H_9ON_3$) are typical energetic azide polymers used as active binders of propellants and explosives.[22,26–33]

The decomposition of $-N_3$ bonds within azide polymers generates a significant amount of heat without oxidation by oxygen atoms. Bond breakage of $-N_3$ is the initial step of the reaction, and is accompanied by melting and gasification processes.

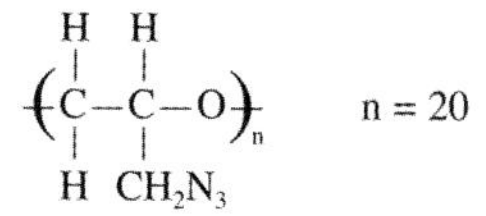

$$-\!\!\left(\begin{matrix} H & H \\ | & | \\ C - & C - O \\ | & | \\ H & CH_2N_3 \end{matrix}\right)_{\!n} \qquad n = 20$$

Fig. 4.6 GAP prepolymer.

Gaseous fragments are eliminated on heating, and numerous chemical species are formed at the reacting surface of the azide polymer. The heat transfer process from the high-temperature zone to the reacting surface determines the burning rate of an azide polymer.

4.2.4.1 GAP

GAP is synthesized by replacing C–Cl bonds of polyepichlorohydrin with C–N_3 bonds.[14] The three nitrogen atoms of the N_3 moiety are attached linearly with ionic and covalent bonds in every GAP monomer unit, as shown in Fig. 4.6. The bond energy of N_3 is reported to be 378 kJ mol^{-1} per azide group. Since GAP is a liquid at room temperature, it is polymerized by allowing the terminal –OH groups to react with hexamethylene diisocyanate (HMDI) so as to formulate GAP copolymer, as shown in Fig. 4.7, and crosslinked with trimethylolpropane (TMP) as shown in Fig. 4.8. The physicochemical properties of GAP prepolymer and GAP copolymer are shown in Table 4.4 and Table 4.5, respectively.[32]

Table 4.4 Chemical properties GAP prepolymer.

Chemical formula	$C_3H_5ON_3$
Molecular mass	1.98 kg mol^{-1}
Heat of formation	0.957 MJ kg^{-1} at 293 K
Adiabatic flame temperature	1470 K at 5 MPa

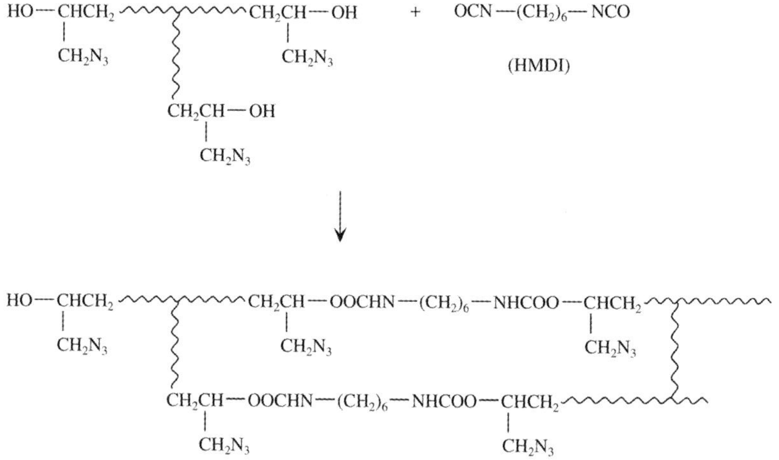

HO—CHCH₂ ~~~~~ CH₂CH—OH (side groups CH₂N₃; branch CH₂CH—OH with CH₂N₃) + OCN—(CH₂)₆—NCO (HMDI)

↓

HO—CHCH₂ ~~~~~ CH₂CH—OOCHN—(CH₂)₆—NHCOO—CHCH₂ ~~~~~ (side groups CH₂N₃); branch CH₂CH—OOCHN—(CH₂)₆—NHCOO—CHCH₂ ~~~~~ (side groups CH₂N₃)

Fig. 4.7 Polymerization of GAP with HMDI.

(GAP) + (TMP) + (HMDI)

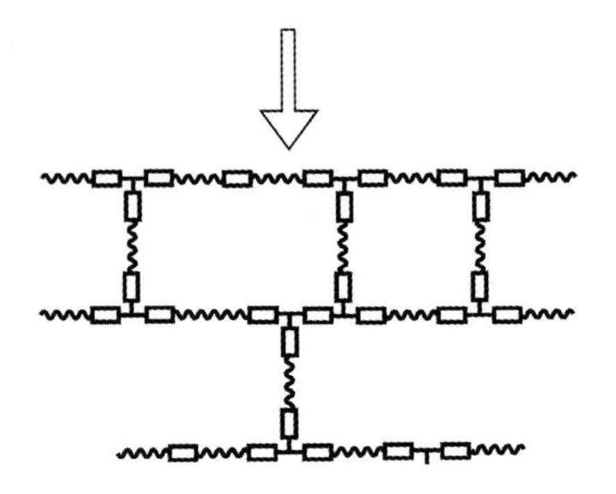

GAP : $HO\text{-}(CH_2\text{—}CH(CH_2N_3)O)_n\text{-}H$

TMP : $CH_3CH_2\text{—}C(CH_2OH)_3$

HMDI : $OCN\text{-}(CH_2CH_2CH_2CH_2CH_2CH_2)\text{-}NCO$

Fig. 4.8 Copolymerization of GAP with HMDI and crosslinking with TMP.

Table 4.5 Physicochemical properties of GAP copolymer.

Chemical formula	$C_{3.3}H_{5.6}O_{1.12}N_{2.63}$
Molecular mass	1.27 kg mol^{-1}
Adiabatic flame temperature	1370 K at 5 MPa

Combustion products (mole fractions) at 5 MPa

N_2	C(s)	CO	CO_2	CH_4	H_2	H_2O
0.190	0.298	0.139	0.004	0.037	0.315	0.016

The adiabatic flame temperature of GAP copolymer is 1370 K at 5 MPa and large amounts of $C_{(s)}$, H_2, and N_2 are formed as initial combustion products. Fuel components such as $C_{(s)}$, CO, and H_2 predominate, with only very small amounts of CO_2 and H_2O being formed.

4.2.4.2 **BAMO**

BAMO monomer is synthesized by replacing the C–Cl bonds of 3,3-bis(chloromethyl) oxetane (BCMO) with C–N_3 bonds.[33] BAMO polymer has two N_3 bonds per BAMO monomer unit. BAMO polymer is obtained by polymerizing the monomer in a similar manner as described for GAP. Table 4.6 shows the physicochemical properties of BAMO polymer. Its heat of formation has a positive value, and its adiabatic flame temperature is higher than that of GAP. The molecular mass distribution of the BAMO polymer spans the range from 300 to 5×10^5 and the main peak is at $M_n = 9600$ and $M_w = 26700$ ($M_w/M_n = 2.8$), where M_n is the number-averaged molecular mass and M_w is the mass-averaged molecular mass.[33]

Table 4.6 Physicochemical properties of BAMO polymer.

Molecular formula	$HO\text{-}(C_5H_8N_6O)_n\text{-}H$
Molecular mass	2.78 kg mol^{-1} ($n = 16.4$)
Density	1300 kg m^{-3}
Melting point	334 K
Glass transition temperature	234 K
Heat of formation	2.46 MJ kg^{-1} at 293 K
Adiabatic flame temperature	2020 K at 10 MPa

Combustion products (mole fraction) at 10 MPa

N_2	$C_{(s)}$	CO	CH_4	H_2	HCN
0.252	0.328	0.084	0.006	0.324	0.005

Since BAMO polymer is a solid at room temperature, BAMO monomer is copolymerized with tetrahydrofuran (THF) in order to formulate a liquid BAMO copolymer that is used as a binder in propellants and explosives, as shown in Fig. 4.9. The terminal OH groups of the BAMO-THF copolymer are cured by reaction with the NCO groups of hexamethylene diisocyanate (HMDI) and then cross-linking is carried out with trimethylolpropane (TMP). The physical properties of such a copolymer with a BAMO/THF composition of 60/40 mol% are shown in Table 4.7.[15]

Table 4.7 Physicochemical properties of a 60/40 mol% BAMO-THF copolymer.

Chemical formula	$HO\text{-}(C_5H_8N_6O)_n\text{-}(C_4H_8O)_m\text{-}H$
Molecular mass	2.24 kg mol^{-1} ($n = 10.4$, $m = 6.9$)
Density	1270 kg m^{-3}
Melting point	249 K
Glass transition temperature	212 K
Heat of formation	1.19 MJ kg^{-1} at 293 K
Adiabatic flame temperature	1520 K at 10 MPa

Combustion products (mole fraction) at 10 MPa

N_2	$C_{(s)}$	CO	CH_4	H_2	HCN
0.186	0.342	0.095	0.037	0.331	0.007

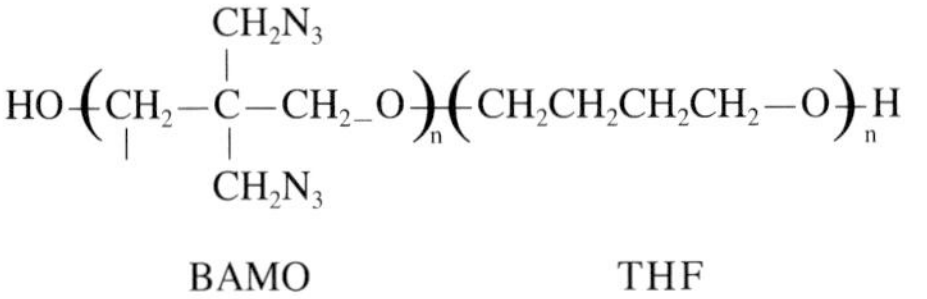

$$HO{-}\!\left(CH_2{-}\overset{\displaystyle CH_2N_3}{\underset{\displaystyle CH_2N_3}{C}}{-}CH_2{-}O\right)_{\!n}\!\left(CH_2CH_2CH_2CH_2{-}O\right)_{\!n}\!H$$

BAMO THF

Fig. 4.9 BAMO/THF copolymer.

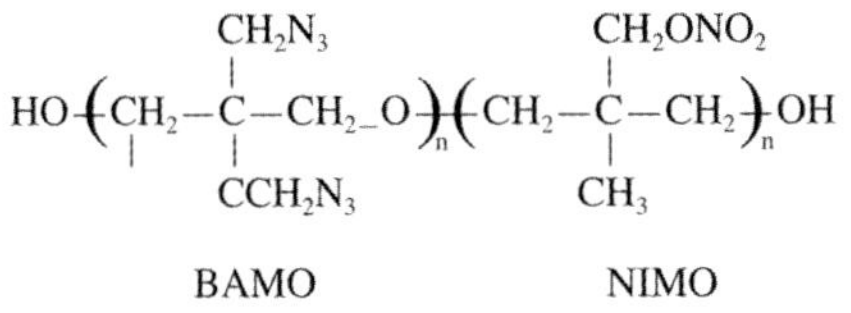

$$HO{-}\!\left(CH_2{-}\overset{\displaystyle CH_2N_3}{\underset{\displaystyle CCH_2N_3}{C}}{-}CH_2{-}O\right)_{\!n}\!\left(CH_2{-}\overset{\displaystyle CH_2ONO_2}{\underset{\displaystyle CH_3}{C}}{-}CH_2\right)_{\!n}\!OH$$

BAMO NIMO

Fig. 4.10 BAMO-NIMO copolymer.

BAMO is also copolymerized with nitratomethyl methyl oxetane (NIMO) to formulate the energetic liquid polymer BAMO-NIMO. Since NIMO is a nitrate ester containing an $-O-NO_2$ bond in its molecular structure, BAMO-NIMO copolymer is more energetic than BAMO-THF copolymer. The chemical structures of BAMO and NIMO are both based on the oxetane structure, and the structure of the BAMO-NIMO copolymer is shown in Fig. 4.10.

The physical properties of the copolymer, such as viscosity, elasticity, and hardness, are tuned by adjusting the molecular mass through appropriate selection of m and n.

4.3 Classification of Propellants and Explosives

Propellants and explosives are both composed of highly energetic materials capable of producing high-temperature gaseous products. Propellants are used to obtain propulsive forces and explosives are used to obtain destructive forces through combustion phenomena. When propellants and explosives are burned in a vessel, the gaseous products in the vessel generate high pressure, which is converted into the propulsive force or destructive force. Though the energetics of both propellants and explosives are fundamentally the same, their combustion phenomena are different due to a difference in the heat release processes. Propellants burn in the deflagration branch of the Hugoniot curve (region IV in Fig. 3.4), whereas explosives burn in the detonation branch (region I in Fig. 3.4).

The energy produced by a unit volume, defined as "energy density", is an important property for both propellants and explosives. For rocket and gun propulsions, the aerodynamic drag during flight in the atmosphere increases as the cross-sectional area of these projectiles increases. It is therefore favorable to reduce this cross-sectional area by reducing the size of the combustion chamber of the projectiles, which, in turn, necessitates increased energy density of the propellant used. In other words, high energy density is required to gain high volumetric propulsive force.

On the other hand, a high pressure build-up in a closed vessel or an open vessel is required for explosives. When the shell wall of a closed vessel is burst by the pressure, a large number of shell wall fragments may be formed if the rate of pressure rise is high enough to lead to non-ductile destruction of the shell wall. This fragment formation process is used in bomb and warhead explosions. When explosives are detonated in a drilled hole in a rock, the high pressure generated in the hole breaks the rock into a large number of small pieces or several blocks. These destructive processes are applied to blasting in mines.

When an energetic material is slowly burned in a closed shell, giving rise to a deflagration wave rather than a detonation wave, the shell does not burst until the pressure reaches the critical limit of the breakage strength of the shell wall. When the shell wall is designed to burst suddenly and to produce fragments, the shock wave generated by the pressure difference between the vessel and the atmosphere propagates into the atmosphere. In this case, the energetic material behaves as an explosive even though it burns in the deflagration branch.

Though high pressure is generated when an energetic material burns and produces high-temperature gas, the high-temperature gas is not the only means of generating high pressure. When the gas consists of molecules of low molecular mass, a high pressure may be generated even when the temperature is low, as described in Chapter 1. The following example highlights another scenario: the reaction between titanium and solid carbon produces solid titanium carbide according to

$$Ti + C \rightarrow TiC + 184 \text{ kJ mol}^{-1}$$

and the combustion temperature reaches about 3460 K without any pressure rise when the reaction is conducted in a closed vessel. This is caused by the reaction occurring without the generation of gaseous products. Energetic materials of this latter type, the so-called "pyrolants", cannot be used as propellants or explosives. However, the heat produced by pyrolants can be used, for example, to initiate ignition of propellants or explosives. Thus, the use of energetic materials falls into two categories: (1) to obtain both high temperature and low molecular mass gaseous products when serving as propellants or explosives, or (2) to obtain high temperature and high molecular mass products when serving as pyrolants.

The materials used as propellants and explosives are classified into two types: (1) energetic materials consisting of chemically bonded oxidizer and fuel components in the same molecule, and (2) energetic composite materials consisting of physically mixed oxidizer and fuel components. One strives for stoichiometrically balanced materials in order to obtain high heat release and low molecular mass combustion products. Nitropolymers consisting of nitro groups and hydrocarbon structures are representative of energetic materials that produce oxidizer and fuel fragments when decomposed. NC is a typical nitropolymer used as a major component of nitropolymer propellants. Though NG is not a nitropolymer, it also consists of nitro groups and a hydrocarbon structure, and it is used in conjunction with NC to formulate nitropolymer propellants, the so-called double-base propellants. Double-base propellants are near-stoichiometrically balanced energetic materials

a)

b)

Fig. 4.11 Rocket flight trajectories assisted by (a) an NC-NG double-base propellant and (b) an aluminized AP composite propellant.

and produce high temperature and low molecular mass combustion products. Figure 4.11 (a) shows a rocket flight assisted by the combustion of a double-base propellant. No smoke signature is seen from the rocket nozzle because the combustion products are mainly CO_2, CO, H_2O, and N_2.

Materials containing either excess oxidizer fragments or excess fuel fragments in their chemical structures are also used to formulate energetic composite materials. Mixing an oxidizer-rich component with a fuel-rich component can form a stoichiometrically balanced material useful as a propellant or explosive. Though ammonium perchlorate (AP) produces oxidizer-rich fragments, its energy density is not high. Polymeric hydrocarbons are also low energy density materials. However, when AP particles and a hydrocarbon polymer are mixed together, a stoichiometrically balanced material named AP composite propellant is formed. In order to increase the specific impulse, aluminum powder is added as a fuel component. Figure 4.11 (b) shows a rocket flight assisted by the combustion of an AP composite propellant. A white smoke signature is seen from the rocket nozzle because the combustion products are mainly HCl, Al_2O_3, CO_2, and H_2O. HCl generates white smoke when combined with H_2O (moisture) in the atmosphere.

The physicochemical properties of explosives are fundamentally equivalent to those of propellants. Explosives are also made of energetic materials such as nitropolymers and composite materials composed of crystalline particles and polymeric materials. TNT, RDX, and HMX are typical energetic crystalline materials used as explosives. Furthermore, when ammonium nitrate (AN) particles are mixed with an oil, an energetic explosive named ANFO (ammonium nitrate fuel oil) is formed. AN with water is also an explosive, named slurry explosive, used in industrial and civil engineering. A difference between the materials used as explosives and propellants is not readily evident. Propellants can be detonated when they are subjected to excess heat energy or mechanical shock. Explosives can be deflagrated steadily without a detonation wave when they are gently heated without mechanical shock.

4.4 Formulation of Propellants

Molecules in which fuel and oxidizer components are chemically bonded within the same structure are suitably predisposed for the formulation of energetic materials. Nitropolymers are composed of O–NO_2 groups and a hydrocarbon structure. The bond breakage of O–NO_2 produces gaseous NO_2, which acts as an oxidizer fragment, and the remaining hydrocarbon structure acts as a fuel fragment. NC is a typical nitropolymer used as a major component of propellants. The propellants composed of NC are termed "nitropolymer propellants".

Crystalline particles that produce gaseous oxidizer fragments are used as oxidizer components and hydrocarbon polymers that produce gaseous fuel fragments are used as fuel components. Mixtures of these crystalline particles and hydrocarbon polymers form energetic materials that are termed "composite propellants". The oxidizer and fuel components produced at the burning surface of each component mix together to form a stoichiometrically balanced reactive gas in the gas phase.

The polymeric hydrocarbon also acts as a binder of the particles, holding them together so as to formulate a propellant grain. Ammonium perchlorate (AP) is a typical crystalline oxidizer and hydroxy-terminated polybutadiene (HTPB) is a typical polymeric fuel. When AP and HTPB are decomposed thermally on the propellant surface, oxidizer and fuel gases are produced, which diffuse into each other and react to produce high-temperature combustion gases.

An energetic material composed of granulated energetic particles is called a "granulated propellant". Granulated propellants are used as gun propellants and in pyrotechnics. For example, a mixture of crystalline potassium nitrate particles, sulfur, and charcoal forms an energetic granulated material known as black powder. Granulated NC particles are used as a single-base propellant for guns. Granulated single-base, double-base, and triple-base propellants are used as gun propellants. Though the linear burning rate (the burning velocity in the direction perpendicular to the propellant burning surface) of granulated propellants is not high, the mass burning rate (mass generation rate from the entire burning surface of the grains) is very high because of the large burning surface area of the grains. The high mass burning rate of these propellants is exploited in gun propulsion to create a short burn time and high pressure in the barrels. The burning times of granulated propellants in gun tubes are of the order of 10 ms to 100 ms. The mass burning rates of these grains are high because the web thickness is very thin compared with that of rocket propellants, and the resulting burning pressure is of the order of 100 MPa to 1000 MPa. Though the physical structures of single-base and double-base granulated propellants are essentially homogeneous, the grains burn independently when they are burned in combustion chambers. Thus, the flame structures appear to be heterogeneous in nature. The shape of each grain is designed to obtain an adequate pressure versus time relationship during burning.

The mechanical properties of propellants are important with regard to the formulation of the desired propellant grains. During pressure build-up processes in a rocket motor, such as the ignition transient or unstable burning in the chamber, or under the very high pressures of more than 1 GPa generated in gun tubes, the grains are subjected to very high mechanical stresses. If the internal grain shape is complicated, increased chamber pressure can lead to cracking of the grains, thereby increasing the burning surface area. The increased burning surface area due to such unexpected cracking increases the chamber pressure, which can cause a catastrophic explosion of the rocket motor or the gun tube. In general, the mechanical properties of propellants are dependent on the ambient temperature. The elongation properties of propellants become poor at low temperatures, below about 200 K. As a result, deep crack formation occurs when a mechanical stress is applied to the grains at low temperatures. On the other hand, mechanical strength becomes poor at high temperatures, approximately above 330 K. This leads to deformation of the grain when it is subjected to an external force such as an acceleration force or gravitational force. Accordingly, the selection of propellant ingredients to formulate the desired propellant grain shape not only requires consideration of the combustion performance, but also the mechanical properties of the formulated propellant grain. The design criteria of gun propellants are different from those of rocket propellants. The size and mass of each grain of a gun propellant are much smaller than those of a rocket propellant. The burning surface area per unit mass of propellant is much larger for a gun propellant. Furthermore, the operational combustion pressure is of the order of 1 GPa for a gun propellant and 1–10 MPa for a rocket propellant.

4.5 Nitropolymer Propellants

4.5.1 Single-Base Propellants

Single-base propellants are made from NC that is gelatinized with the solvents ethanol or diethyl ether. A small amount of diphenylamine, $(C_6H_5)_2NH$, is also added as a chemical stabilizer of NC. In some cases, a small amount of K_2SO_4 or KNO_3 is added as a flame-suppressor. Ethanol (C_2H_5OH) or diethyl ether ($C_2H_5OC_2H_5$) is mixed with the NC in order to make it soft and to obtain an adequate size and shape of the propellant grains. The grain surface is then coated with carbon black to keep it smooth. Table 4.8 shows the chemical composition, flame temperature, and combustion products of a typical single-base propellant.

Table 4.8 Physicochemical properties of a single-base propellant.

Composition (% by mass)				T_f (K)	Combustion products (mole fractions)				
NC	DNT	DBP	DPA		CO_2	CO	H_2O	H_2	N_2
85.0	10.0	5.0	1.0	1590	0.052	0.508	0.130	0.212	0.098

4.5.2 Double-Base Propellants

Double-base propellants are formed from NC that is gelatinized with energetic nitrate esters such as NG, DEGDN, TEGDN, or TMETN. These liquid nitrate esters are used to obtain a rigid gel network of plasticized NC and to form double-base propellants having a homogeneous physical structure. For example, liquid NG is absorbed by solid NC, producing a homogeneous gelatinized material. Though both materials burn by themselves, propellants composed of NC and NG maintain the desired grain shapes used in rockets and guns and generate the desired temperature and combustion products when they burn.

Dibutyl phthalate (DBP), diethyl phthalate (DEP), and triacetin (TA) are the typical plasticizers and stabilizers used for double-base propellants. These chemicals are used to obtain superior characteristics of propellant grain formation, and to improve mechanical properties, shock sensitivities, and chemical stability. Several types of amines are used as anti-ageing agents in double-base propellants. Amines are known to react with gaseous NO_2, which is generated by $O–NO_2$ bond breakage in nitrate esters. This reaction prevents gas formation in the propellant grain, thereby preventing mechanical destruction of the grain. A small amount of 2-nitrodiphenylamine (2NDPA) is commonly added to nitropolymer propellants. The heats of formation of DBP, TA, and 2NDPA are shown in Table 2.3.

4.5.2.1 NC-NG Propellants

These double-base propellants are best known as smokeless propellants used in guns and rockets. Two major ingredients are used to formulate the double-base propellant grains: NC and NG. As covered in Section 4.2.2, NG is a nitrate ester characterized by $–O–NO_2$ groups, and is known as a high explosive. Since NG is a liquid at room temperature, it is absorbed by NC and serves to gelatinize the latter to form

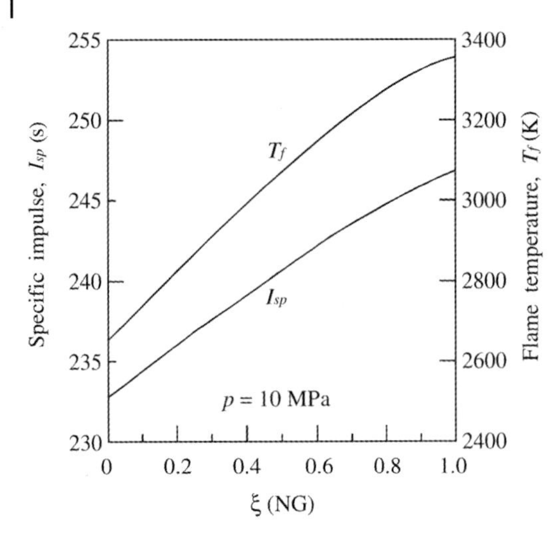

Fig. 4.12 Specific impulse and adiabatic flame temperature of NC-NG double-base propellants as a function of ξ(NG).

double-base propellant grains. Table 4.9 shows the chemical compositions and energetic properties of two typical double-base propellants.

NC and NG have oxygen available in the form of $O{-}NO_2$, which is attached to the organic moiety (e. g., cellulose) from which they are derived. The physicochemical properties of double-base propellants, such as energy density, mechanical strength, and chemical stability, are dependent on the proportions of NC, NG, stabilizers, plasticizers, and catalysts used to formulate them. Though the energy density is increased by increasing the fraction of NG, the mechanical strength and chemical stability are decreased.

Two types of production method are used to formulate double-base propellant grains used in rockets and guns: (1) an extrusion method using an external mechanical press, and (2) a cast method using finely divided NC or NC-NG powders. The extrusion method is used for small-sized grains of less than 0.1 kg, as used in guns and pyrotechnics. The cast method is used for large-sized grains weighing more than 1 kg, as used in booster and sustainer rockets. In Fig. 4.12, specific impulse I_{sp} of double-base propellants composed of NC and NG is shown as a function of the mass fraction of NG, ξ(NG). When ξ(NG) = 1.0, the theoretical maximum I_{sp} of 247 s is obtained on combustion at 10 MPa with an ideal expansion process to 0.1 MPa. However, NG is a liquid at room temperature and is sensitive to mechanical shock. To formulate practical double-base propellants, plasticizers and/or stabilizers always have to be mixed with NG. Since NG is a detonable explosive, ξ(NG) is approximately 0.5 or lower in conventional double-base propellants. Furthermore, various types of chemicals, such as burning rate catalysts, modifiers, and anti-ageing agents, are added to NC-NG mixtures in order to obtain superior mechanical properties at high and low environmental temperatures and to improve burning rate characteristics.

The mechanical properties and shock sensitivities of these double-base propellants are highly dependent on the mixture ratio of NC and NG. Though the specific impulse of the double-base propellant increases with increasing NG concentration, its mechanical strength decreases. If ξ(NG) approaches 0.6, it becomes difficult to maintain grain shape at room temperature. If, on the other hand, ξ(NG) is less than 0.4, the elongation property of the double-base propellant becomes poor. Thus, double-base propellants need to contain desensitizers, stabilizers, and chemicals to improve mechanical properties. In order to obtain superior mechanical properties, plasticizers and stabilizers such as dibutyl phthalate (DBP: $C_{16}H_{22}O_4$), triacetin (TA: $C_9H_{14}O_6$), ethyl centralite (EC: $CO\{N(C_6H_5)(C_2H_5)\}_2$), or diethyl phthalate (DEP: $C_{12}H_{14}O_4$) are added. A typical chemical composition and the thermochemical properties of an NC-NG double-base propellant are shown in Table 4.9.

4.5.2.2 **NC-TMETN Propellants**

Since NG is highly shock-sensitive, other types of nitrate esters can be used to formulate non-NG double-base propellants. DEGDN, TEGDN, and TMETN are typical examples of energetic nitrate esters that can be mixed with NC. These nitrate esters are less energetic than NG, and their sensitivities to friction and mechanical shock are accordingly lower than those of NG. Thus, the mass fraction of desensitizer used in propellant formulation can be lower than when NG is involved. The physicochemical properties of these nitrate esters are shown in Tabs. 2.3 and 2.5–2.7.

TMETN is a liquid at room temperature and the production process of NC-TMETN propellants is the same as that described for NC-NG propellants. The shock sensitivity of TMETN is sufficiently lower than that of NG that no desensitizers are needed for NC-TMETN propellants. Instead of the DEP or TA used as low energy density plasticizers and stabilizers of NC-NG propellants, TMETN is mixed with TEGDN, which is a dinitrate ester and hence a relatively high energy density material. Thus, the overall energy density of double-base propellants composed of NC-TMETN is equivalent to or even higher than that of NC-NG double-base propellants.

The chemical compositions and thermochemical properties of representative NC-NG and NC-TMETN double-base propellants are compared in Table 4.9. Though the NC/NG mass ratio of 0.80 is much smaller than the NC/TMETM mass ratio of 1.38, the combustion performance in terms of T_f and M_g is seen to be similar, and Θ is 109 kmol K kg^{-1} for both propellants. In the case of rocket motor operation, I_{sp} and ρ_p are also approximately equivalent for both propellants.

4.5.2.3 **Nitro-Azide Polymer Propellants**

Double-base propellants containing azide polymers are termed nitro-azide polymer propellants. DEP used as a plasticizer of double-base propellants is replaced with azide polymers in order to increase energy density. The compatibility of GAP prepolymer with NG serves to suitably desensitize the mechanical sensitivity of NG and gives superior mechanical properties in the formulation of rocket propellant grains.

Table 4.9 Chemical compositions and thermochemical properties of NC-NG and NC-TMETN double-base propellants (10 MPa).

% by mass	NC-NG	NC-TMETN
NC	39.6	53.8
NG	49.4	–
TMETN	–	39.1
DEP	10.0	–
REGDN	–	7.0
EC	1.0	0.1
ρ_p (kg m^{-3})	1550	1550
T_f (K)	2690	2570
M_g (kg kmol^{-1})	24.6	23.6
Θ (kmol K kg^{-1})	109	109
I_{sp} (s)	242	240
Combustion products (% by mol)		
CO	39.7	39.8
CO_2	12.4	19.4
H_2	11.5	14.3
H_2O	23.8	23.6
N_2	12.4	11.8
OH	0.1	0.0
H	0.2	0.1

Table 4.10 shows a comparison of the theoretical combustion properties of NC-NG-DEP and NC-NG-GAP propellants at 10 MPa. Though the molecular mass of the combustion products, M_g, remains relatively unchanged by the replacement of DEP with GAP, the adiabatic flame temperature is increased from 2557 K to 2964 K when 12.5 % DEP is replaced with 12.5 % GAP. Thus, the specific impulse I_{sp} is increased from 237 s to 253 s. The density of a propellant, ρ_p, is also an important parameter in evaluating its thermodynamic performance. The density is increased from 1530 kg m^{-3} to 1590 kg m^{-3} by the replacement of DEP with GAP. Since GAP is also compatible with DEP, double-base propellants composed of four major ingredients, NC, NG, DEP, and GAP, are also formulated.

4.5.2.4 **Chemical Materials of Double-Base Propellants**

Though the major components of double-base propellants are NC-NG or NC-TMETN, various additives such as plasticizers, burning rate modifiers, or combustion instability suppressants are needed. Table 4.11 shows the materials used to formulate double-base propellants.

Table 4.10 Chemical compositions and thermochemical properties of NC-NG-DEP and NC-NG-GAP propellants (10 MPa).

% by mass	NC-NG-DEP	NC-NG-GAP
NC	37.5	37.5
NG	50.0	50.0
DEP	12.5	–
GAP	–	12.5
ϱ_p (kg m^{-3})	1530	1590
T_f (K)	2557	2964
M_g (kg kmol^{-1})	24.0	25.0
Θ (kmol K kg^{-1})	107	118
I_{sp} (s)	237	253
Combustion products (% by mol)		
CO	41.5	33.7
CO_2	11.0	13.4
H_2	13.4	9.1
H_2O	22.2	25.9
N_2	11.9	16.9

Table 4.11 Chemical materials used to formulate double-base propellants.

Plasticizer (oxidizer and fuel)	NG, TMETN, TEGDN, DNT
Plasticizer (fuel)	DEP, DBP, TA, PU
Stabilizer	EC, 2NDPA, DPA
Plasticizer (energetic fuel)	GAP, BAMO, AMMO
Binder (fuel and oxidizer)	NC
Burning rate catalyst	PbSa, PbSt, Pb_2EH, CuSa, CuSt, LiF
Burning rate catalyst modifier	C (carbon black, graphite)
Combustion instability suppressant	Al, Zr, ZrC
Opecifier	C (carbon black, graphite)
Flame suppressant	KNO_3, K_2SO_4

4.6 Composite Propellants

Crystalline materials such as KNO_3, NH_4NO_3, and NH_4ClO_4 are used as oxidizers due to the high oxygen contents in their molecular structures. These materials generate high concentrations of gaseous oxidizing fragments when they are thermally decomposed. On the other hand, hydrocarbon-based polymeric materials such as polyurethane and polybutadiene generate high concentrations of gaseous fuel fragments when they are thermally decomposed. The two types of gaseous fragments diffuse into each other and react to generate heat and combustion products. Thus, mixtures of the crystalline materials and the polymeric materials form highly combustible energetic materials of heterogeneous physical structure. The energetic materials thus formed are termed composite propellants or heterogeneous propellants. Accordingly, the ballistic properties of these systems, such as burning rate and pressure sensitivity, are not only dependent on the chemical ingredients of the oxidizer and fuel components, but also on the shape and size of the oxidizer particles.

The mechanical properties of composite propellant grains are dependent on the physical and chemical properties of the polymeric materials and on the presence of additives such as bonding agents, surfactants, crosslinkers, and curing agents. Physical properties such as mechanical strength and elongation are altered by the use of different types of binder, varying the mass fractions, or changing the oxidizer particle size. Low viscosity of the mixed materials is needed to formulate a stoichiometrically balanced propellant. High concentrations of oxidizer particles and/or aluminum powders are needed to obtain high specific impulse. Burning rate catalysts or modifiers are also added to obtain a wide range of burning rates.

4.6.1 AP Composite Propellants

4.6.1.1 AP-HTPB Propellants

As shown in Figs. 4.13 and 4.14, T_f and I_{sp} at 10 MPa of AP composite propellants are highest at $\xi(AP) = 0.90$. The maximum I_{sp} is 259 s and the maximum T_f is 3020 K. In order to further increase I_{sp}, aluminum particles are added as a fuel component. Though the addition of aluminum particles also increases M_g, the increase in T_f is more pronounced, resulting in a net increase in I_{sp}. The effects of the mass fraction of aluminum particles added, $\xi(Al)$, on I_{sp} and T_f are also shown in Figs. 4.13 and 4.14. Though T_f increases as $\xi(Al)$ increases, I_{sp} reaches a maximum at about $\xi(Al) = 0.20$: $I_{sp} = 270$ s and $T_f = 3890$ K.

When aluminized AP composite propellant burns, a high mole fraction of aluminum oxide is produced as a combustion product, which generates visible smoke. If smoke has to be avoided, e. g. for military purposes or a fireworks display, aluminum particles cannot be added as a component of an AP composite propellant. In addition, a large amount of white smoke is produced even when non-aluminized AP composite propellants burn. This is because the combustion product HCl acts as a nucleus for moisture in the atmosphere and relatively large-sized water drops are formed as a fog or mist. This physical process only occurs when the relative humidity in the atmosphere is above about 60%. If, however, the atmospheric temperature is below 260 K, white smoke is again formed because of the condensation of water vapor with HCl produced as combustion products. If the HCl smoke generated by AP combustion cannot be tolerated, the propellant should be replaced with a double-base propellant or the AP particles should be replaced with another

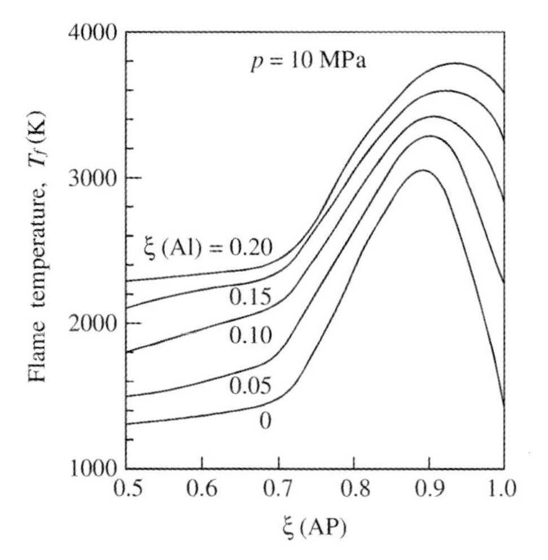

Fig. 4.13 Adiabatic flame temperature **vs.** composition of aluminized AP-HTPB composite propellants.

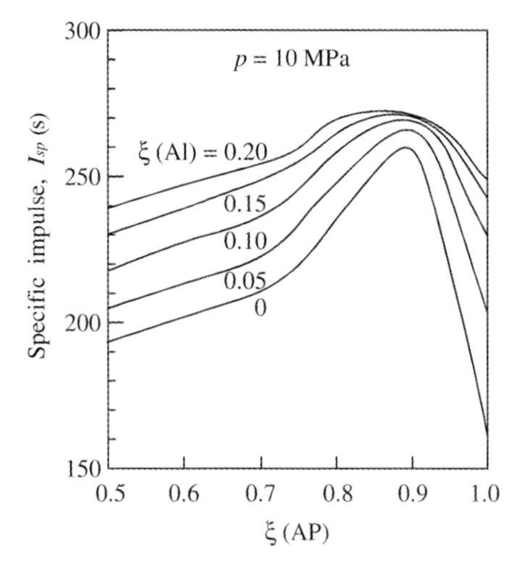

Fig. 4.14 Specific impulse **vs.** composition of aluminized AP-HTPB composite propellants.

non-halogen- or non-metal-containing oxidizer such as AN, RDX, HMX, ADN, or HNF. When aluminum particles are replaced with magnesium particles, the mole fraction of HCl is significantly reduced due to the formation of $MgCl_2$ as a combustion product.

Hydroxy-terminated polybutadiene (HTPB) is considered to be the best binder for obtaining high combustion performance, superior elongation properties at low temperatures, and superior mechanical strength properties at high temperatures. This combination of properties is difficult to achieve in double-base propellants. HTPB is characterized by terminal –OH groups on a butadiene polymer. The other type of butadiene polymer used is carboxy-terminated polybutadiene (CTPB), which is cured with an imine or an epoxy resin. It should be noted that CTPB is somewhat sensitive to humidity, which has an adverse effect on its ageing charac-

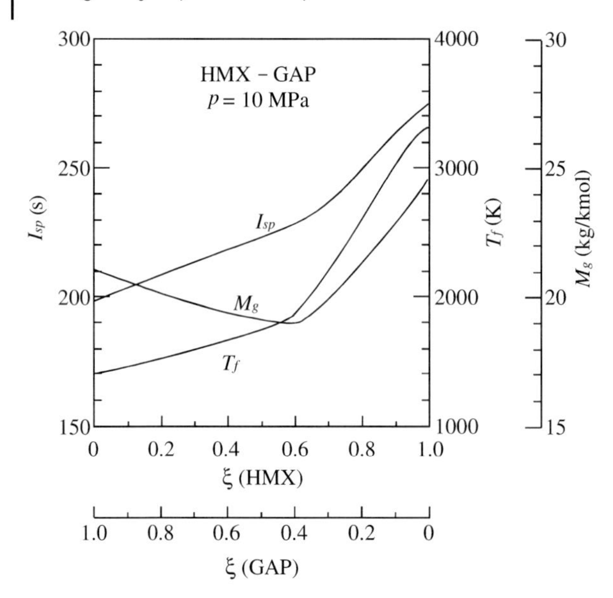

Fig. 4.15 Specific impulse, adiabatic flame temperature, and molecular mass of the combustion products for AP-GAP composite propellants.

teristics. The prepolymer of HTPB is cured and crosslinked with isophorone diisocyanate (IPDI) to formulate HTPB polymer used as a binder of oxidizer particles. The functionality of HTPB prepolymer is also an important chemical parameter during the process of curing and crosslinking with regard to obtaining superior mechanical properties of the HTPB binder. This also affects the ageing characteristics of AP-HTPB propellants.

4.6.1.2 AP-GAP Propellants

Azide polymers such as GAP and BAMO are also used to formulate AP composite propellants in order to give improved specific impulses compared with those of the above-mentioned AP-HTPB propellants. Since azide polymers are energetic materials that burn by themselves, the use of azide polymers as binders of AP particles, with or without aluminum particles, increases the specific impulse compared to those of AP-HTPB propellants. As shown in Fig. 4.15, the maximum I_{sp} of 260 s is obtained at ξ(AP) = 0.80 and is approximately 12 % higher than that of an AP-HTPB propellant because the maximum loading density of AP particles is obtained at about ξ(AP) = 0.86 in the formulation of AP composite propellants. Since the molecular mass of the combustion products, M_g, remains relatively unchanged in the region above ξ(AP) = 0.8, I_{sp} decreases rapidly as ξ(AP) increases.

4.6.1.3 Chemical Materials of AP Composite Propellants

As in the case of double-base propellants, various types of materials, such as plasticizers, burning rate modifiers, and combustion instability suppressants, are added to mixtures of AP and a binder. Table 4.12 shows the materials used to formulate AP composite propellants.

Table 4.12 Chemical materials used to formulate AP composite propellants.

Oxidizer	AP
Binder (fuel)	HTPB, CTPB, PBAN, HTPE, HTPS, HTPA, PU, PS, PVC
Binder (energetic fuel)	GAP, BAMO, AMMO, NIMO
Curing and/or crosslinking agent	IPDI, TDI, PQD, HMDI, MAPO
Bonding agent	MAPO, TEA, MT-4
Plasticizer	DOA, IDP, DOP
Burning rate catalyst	Fe_2O_3, FeO(OH), ferrocene, catocene
Burning rate negative catalyst	LiF, $SrCO_3$
Negative burning rate modifier	OXM
Metal fuel	Al
High-energy additive	RDX, HMX, NQ, HNIW, ADN
Combustion instability suppressant	Al, Zr, ZrC
HCl suppressant	Mg, MgAl, $NaNO_3$

4.6.2
AN Composite Propellants

Ammonium nitrate contains a relatively high concentration of oxidizer fragments, as shown in Table 2.6. In order to maximize I_{sp}, the binder mixed with AN is GAP. The maximum I_{sp} of 238 s and the maximum T_f of 2400 K are obtained at ξ(AN) = 0.85, as shown in Fig. 4.16. However, since AN crystal particles are not wholly compatible with GAP, the practical ξ(AN) is less than 0.8, at which I_{sp} drops to 225 s and T_f drops to 2220 K.

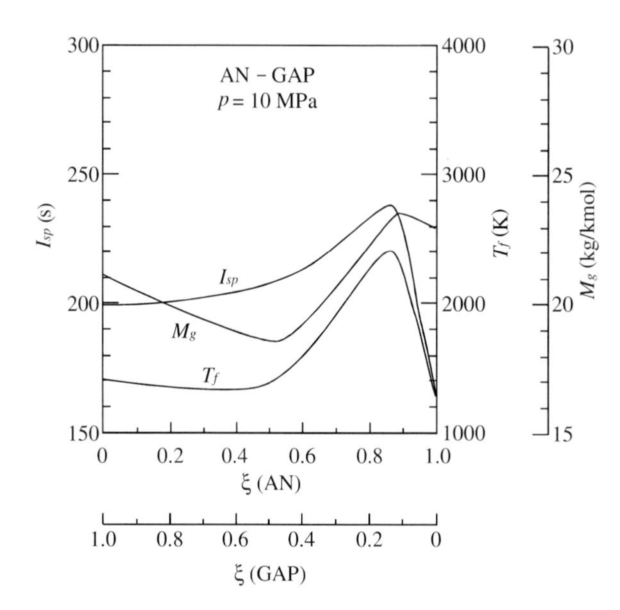

Fig. 4.16 Specific impulse, adiabatic flame temperature, and molecular mass of the combustion products for AN-GAP composite propellants.

4.6.3 Nitramine Composite Propellants

Nitramine composite propellants composed of HMX or RDX particles and polymeric materials offer the advantages of low flame temperature and low molecular mass combustion products, as well as reduced infrared emissions. The reduced infrared emissions result from the elimination of CO_2 and H_2O from the combustion products. To formulate these composite propellants, crystalline nitramine monopropellants such as HMX or RDX are mixed with a polymeric binder. Since both HMX and RDX are stoichiometrically balanced, the polymeric binder acts as a coolant, producing low-temperature, fuel-rich combustion products. This is in contrast to AP composite propellants, in which the binder surrounding the AP particles acts as a fuel to produce high-temperature combustion products.

The polymeric binders used for nitramine composite propellants are similar to those used for AP composite propellants, i. e. HTPB, HTPE, and GAP. The combustion performance and the products of HMX composite propellants are shown in Figs. 4.17 and 4.18, respectively. Here, the binder mixed with the HMX particles is GAP, as in the AN-GAP propellants shown in Fig. 4.16. Though the maximum T_f and I_{sp} are obtained at ξ(HMX) = 1.0, the maximum HMX loading is less than ξ(HMX) = 0.80 for practical HMX-GAP propellants, at which I_{sp} = 250 s and T_f = 2200 K. It is important to note that no H_2O, CO_2, or $C_{(s)}$ are formed as combustion products at ξ(HMX) = 0.60. Though the mole fractions of H_2 and N_2 are relatively high, there is no infrared emission from or absorption by these molecules. The emission from CO is not high compared with that from CO_2, H_2O, or $C_{(s)}$. The use of this class of propellants significantly reduces infrared emission from rocket exhaust gas.

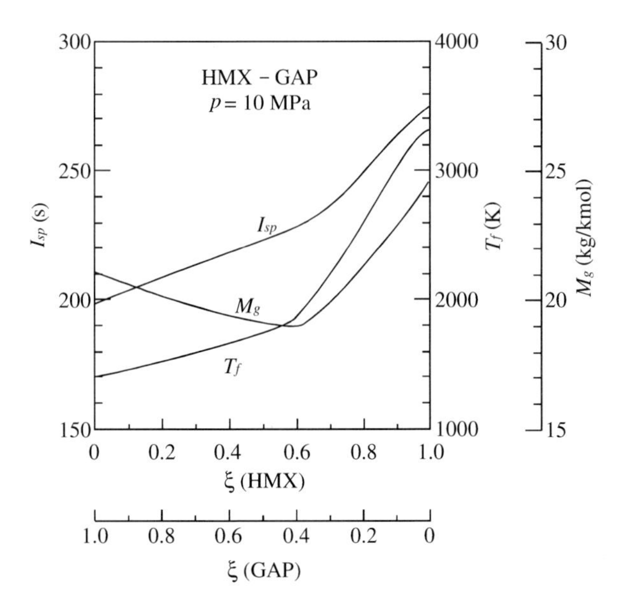

Fig. 4.17 Specific impulse, adiabatic flame temperature, and molecular mass of combustion products for HMX-GAP composite propellants.

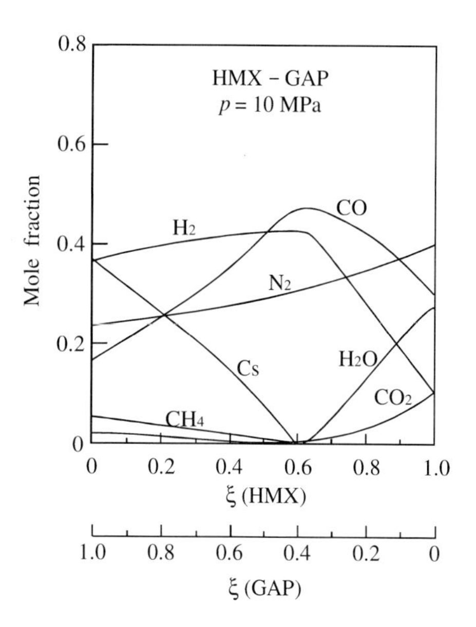

Fig. 4.18 Mole fractions of combustion products from HMX-GAP composite propellants.

Figure 4.19 shows the specific impulses of ADN-GAP and HNIW-GAP propellants as functions of ξ(ADN) and ξ(HNIW). The maximum I_{sp} of 270 s is obtained at ξ(ADN) = 0.87 and the maximum I_{sp} of 280 s is obtained at ξ(HNIW) = 1.00. Since the mass fraction of GAP, ξ(GAP), needed to formulate a practical composite propellant is at least 0.13, I_{sp} = 270 s is possible for ADN-GAP propellant, but only I_{sp} = 260 s is accessible for HNIW-GAP propellant.

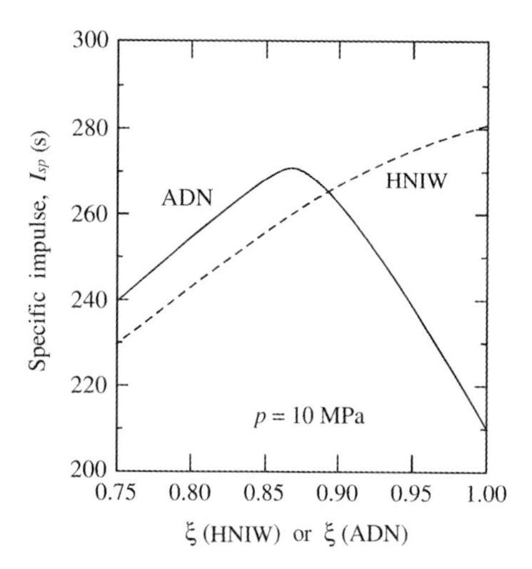

Fig. 4.19 Specific impulses of ADN-GAP and HNIW-GAP composite propellants.

4.6.4
HNF Composite Propellants

Hydrazinium nitroformate (HNF) contains a relatively high concentration of oxidizer fragments, as shown in Table 2.6. When GAP is used as a binder of HNF particles, HNF-GAP composite propellants are made. The maximum I_{sp} of 285 s and the maximum T_f of 3280 K are obtained at ξ(HNF) = 0.90 with an optimum expansion from 10 MPa to 0.1 MPa, as shown in Figs. 4.20 and 4.21, respectively. Since a

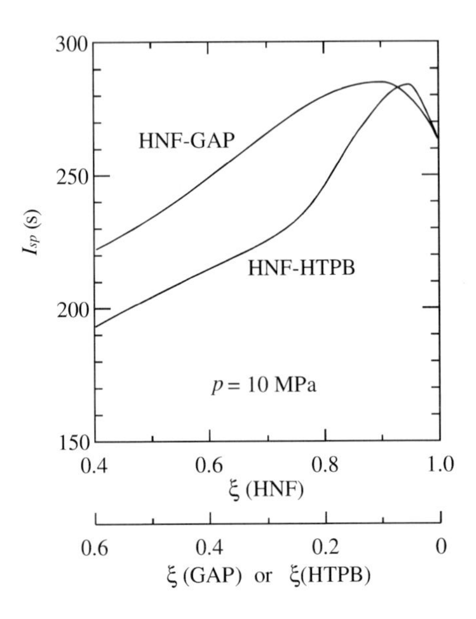

Fig. 4.20 Specific impulses of HNF-GAP and HNF-HTPB composite propellants.

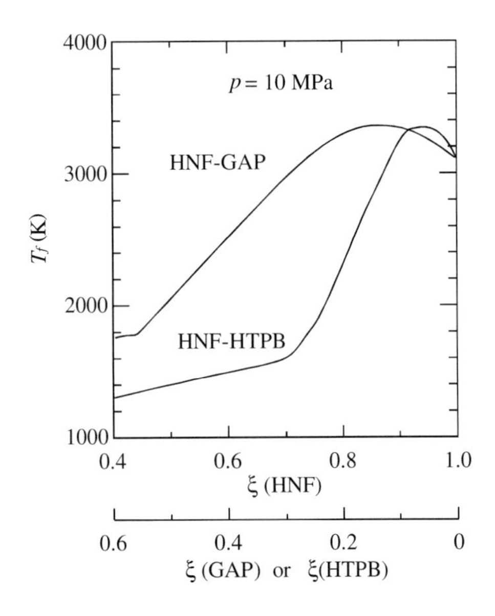

Fig. 4.21 Adiabatic flame temperatures of HNF-GAP and HNF-HTPB composite propellants.

ξ(HNF) of 0.90 is too high to formulate a practical HNF-GAP propellant, a ξ(HNF) of 0.75 is chosen. Though I_{sp} and T_f decrease to 274 s and 3160 K, respectively, the superiority of HNF-GAP propellants is evident when the data are compared with those for HNF-HTPB propellants, as are also shown in Figs. 4.20 and 4.21. However, it is important to note that HNF is highly hygroscopic and that its impact sensitivity and friction sensitivity are high compared to those of other crystalline oxidizers.

4.6.5
TAGN Composite Propellants

TAGN is a unique energetic material that contains a relatively high mole fraction of hydrogen. When TAGN is mixed with a polymeric material, it acts only as a moderate oxidizer and produces fuel-rich combustion products. Figure 4.22 shows I_{sp}, T_f, and M_g of propellants composed of TAGN and GAP, while Fig. 4.23 shows the combustion products as a function of ξ(TAGN) or ξ(GAP). Since T_f is low, I_{sp} is low, and TAGN-GAP propellants are formulated for use as gas-generating propellants in ducted rocket engines or as gun propellants giving reduced gun barrel erosion.

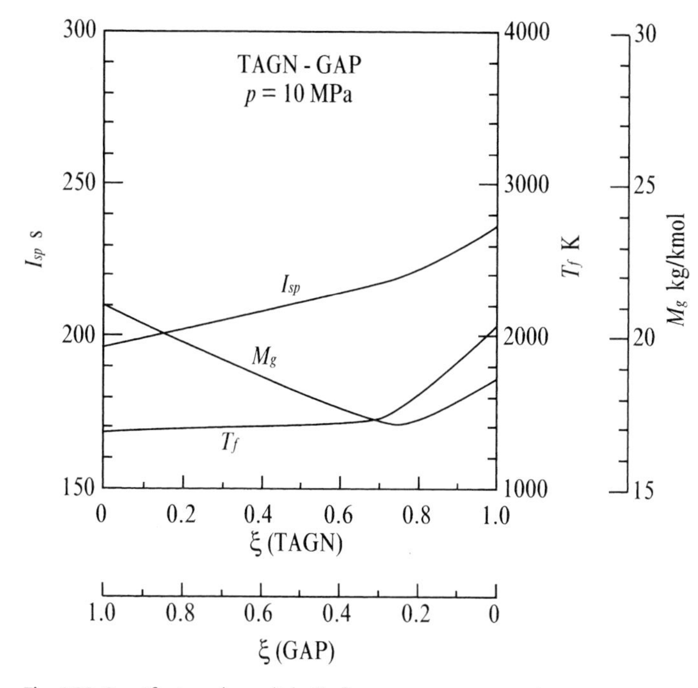

Fig. 4.22 Specific impulse, adiabatic flame temperature, and molecular mass of combustion products for TAGN-GAP composite propellants.

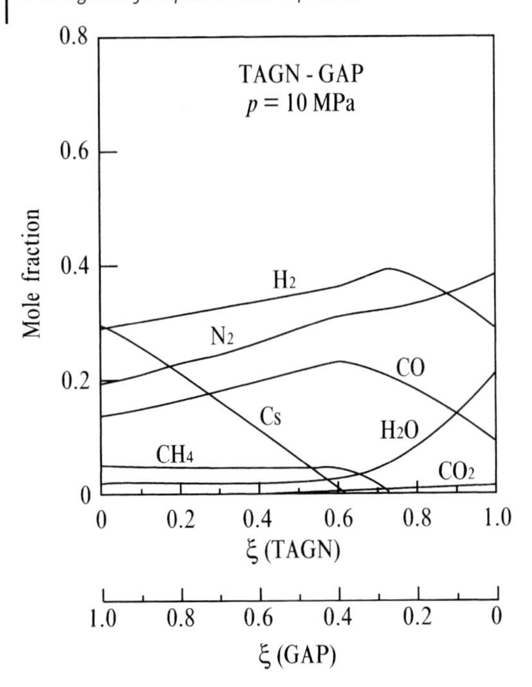

Fig. 4.23 Mole fractions of combustion products of TAGN-GAP composite propellants.

4.7
Composite-Modified Double-Base Propellants

Since the energetics of nitropolymer propellants composed of NC-NG or NC-TMETN are limited due to the limited concentration of oxidizer fragments, some crystalline particles are mixed within these propellants in order to increase the thermodynamic energy or specific impulse. The resulting class of propellants is termed "composite-modified double-base (CMDB) propellants". The physicochemical properties of CMDB propellants are intermediate between those of composite and double-base propellants, and these systems are widely used because of their great potential to produce a high specific impulse and their flexibility of burning rate.

Though the physical structures of CMDB propellants are heterogeneous, similar to those of composite propellants, the base matrix used as a binder burns by itself and the combustion mode of CMDB propellants appears to be different from that of composite propellants and double-base propellants. The burning rate of a CMDB propellant is dependent on the type of crystalline particles incorporated.

4.7.1
AP-CMDB Propellants

When crystalline AP particles are mixed with nitropolymers, ammonium perchlorate composite-modified double-base (AP-CMDB) propellants are formulated. A nitropolymer such as NC-NG or NC-TMETN double-base propellant acts as a

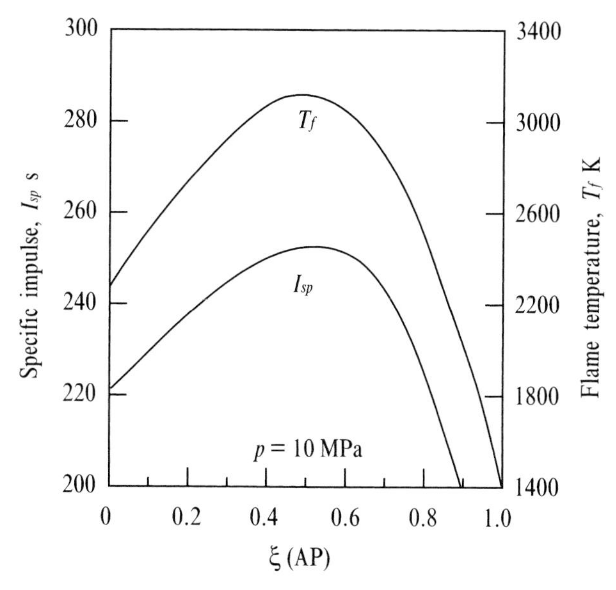

Fig. 4.24 Specific impulse and adiabatic flame temperature of AP-CMDB propellants.

base matrix to bond the AP particles within the propellant. Each AP particle thermally decomposes to produce an oxidizer-rich flame stream, which diffuses into the reactive gases produced by decomposition of the base matrix. The final combustion products are formed somewhat downstream of the gas phase and the combustion temperature is maximized.

Figure 4.24 shows the specific impulse and the flame temperature of AP-CMDB propellants as a function of ξ(AP) at 10 MPa. The base matrix NC-NG propellant has the composition 0.5/0.5. The maximum I_{sp} (253 s) and T_f (3160 K) are obtained at ξ(AP) = 0.5.

4.7.2
Nitramine CMDB Propellants

When nitramine particles such as HMX or RDX particles are mixed with a double-base propellant, nitramine composite-modified double-base propellants are formulated. Since HMX and RDX are stoichiometrically balanced materials, the use of these nitramine particles leads to a somewhat different mode of combustion as compared to AP-CMDB propellants. Since each nitramine particle can burn independently of the base matrix at the burning surface, a monopropellant flamelet is formed in the gas phase from each particle. The monopropellant flamelet diffuses into the reactive gas of the base matrix above the burning surface and a homogeneously mixed gas is formed.

Figure 4.25 shows the adiabatic flame temperatures of mixtures of HMX with an NC-NG double-base propellant as a function of the mass fraction of HMX,

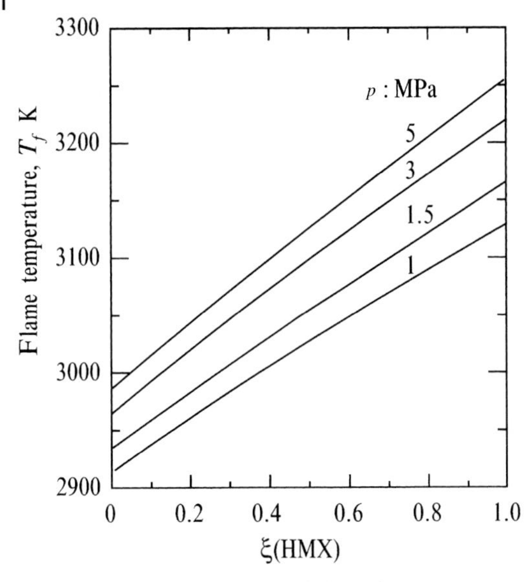

Fig. 4.25 Specific impulse and adiabatic flame temperature of HMX-CMDB propellant.

ξ(HMX). The base matrix of the double-base propellant is composed of NC and NG in the ratio of 0.2/0.8. Since HMX is a stoichiometrically balanced energetic material, it serves as an energetic material rather than as an oxidizer. Accordingly, the flame temperature increases monotonously as ξ(HMX) is increased.

4.7.3
Triple-Base Propellants

Triple-base propellants are made by the addition of crystalline nitroguanidine (NQ) to double-base propellants, similar to the way in which nitramine is added to CMDB propellants as described in the preceding section. Since NQ has a relatively high mole fraction of hydrogen within its molecular structure, the molecular mass of the combustion products becomes low even though the flame temperature is reduced. Table 4.13 shows the chemical composition, adiabatic flame temperature, and thermodynamic energy, f, as defined in Eq. (1.84), of a triple-base propellant at 10 MPa (NC: 12.6% N).

Table 4.13 Chemical composition and properties of a triple-base propellant.

Composition (% by mass)				T_f	f
NC	NG	NQ	EC	K	MJ kg^{-1}
28.0	22.8	47.7	1.5	3050	1.09

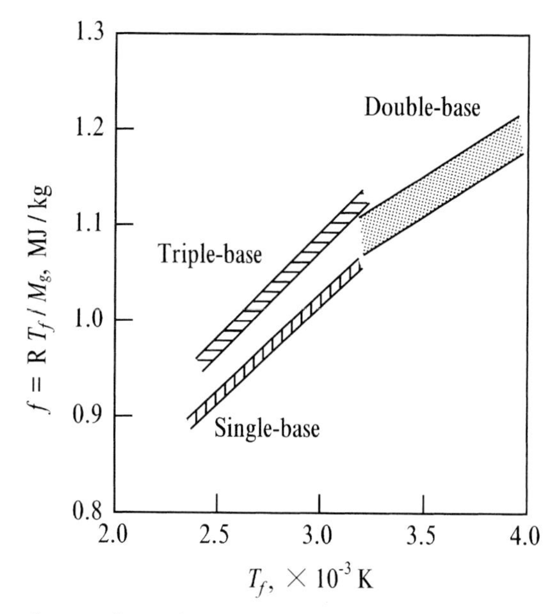

Fig. 4.26 Thermodynamic energies of single-, double-, and triple-base propellants.

Figure 4.26 shows the thermodynamic energy, f, as defined in Eq. (1.84), of single-base, double-base, and triple-base propellants as a function of the combustion temperature, T_g. Though the f values of double-base propellants are high, T_g is also high. In order to suppress gun erosion, T_g needs to be low. Triple-base propellants are thus formulated to reduce erosion while maximizing the f value.

4.8 Black Powder

Granulated propellants consist of a number of powdered energetic materials, and are commonly packed into cartridges. Propellants of this type are used for gun propulsion and for aerial shell ejection in fireworks. Black powder is a typical granulated propellant used for small-scale rockets, guns, and fireworks. Since each particle of black powder is small and has high porosity, the mass burning rate is very high.

Though the theoretical I_{sp} of black powder is less than that of single-base, double-base, and AP-based composite propellants, black powder is useful for short-duration operation of simple propulsive systems. The advantages of the use of black powder are its low cost, its minimal deterioration with age, and simple regulation of the propulsive force by adjustment of the quantities used.

Black powder is a mechanically mixed material consisting of KN powder (60–80 %), charcoal (10–25 %), and sulfur (8–25 %), which is pressed, granulated, and packed into the desired shape for use. When black powder is ignited, the combustion occurs from the entire surface of the grains. Thus, the rate of gas production becomes much higher than in the case of conventional propellants used for rockets. However, this combustion phenomenon is deflagration, not detonation. The burning rate of black powder is not defined like that of rocket propellants because of the distinct nature of the combustion of the granulated material. The overall gas production rate is an important parameter for combustion, which is dependent on each constituent granulated powder and the density of the packed shape. The burning rate of black powder is highly dependent on the particle size of the ingredients. The porosity of the charcoal used is the dominant factor affecting the burning rate when the mixture ratio is fixed. The type of wood used to produce the charcoal is also an important parameter in determining the burning characteristics. Table 4.14 shows the chemical composition and thermochemical parameters of a black powder used for the ejection of the aerial-bursting shells of fireworks.

Table 4.14 Chemical composition and thermochemical properties of black powder.

Ingredient	% by mass
Potassium nitrate	60–80
Charcoal	10–25
Sulfur	8–25
ρ_p (kg m^{-3})	1200–2000
T_f (K)	1400–3200
I_{sp} (s)	60–150

4.9 Formulation of Explosives

Nitroglycerin is used to formulate explosives in combination with nitrocellulose, the same ingredients as used to form double-base propellants. However, the mass fraction of NC is only 0.06–0.08 for explosives, as compared to 0.3–0.6 for double-base propellants. Though both materials have homogeneous physical structures, the energy density of explosives is higher than that of double-base propellants due to the high mass fractions of NG.

Nitroglycol (NGC) has similar physical and chemical properties to those of NG. However, its vapor pressure is too high to permit its use as a major component of propellants and explosives. NGC is used as an additive to reduce the freezing temperature of NG and to formulate explosives. However, the shock-sensitivity of the resulting NG-based explosives is much higher than that of other types of explosives.

The addition of crystalline oxidizers such as ammonium nitrate or potassium nitrate to nitrate esters or energetic polymers leads to the formation of composite explosives of heterogeneous physical structure. The energy density of these composite explosives is higher than that of homogeneous explosives.

When explosives are used in the warheads of gun projectiles and missiles, the acceleration force to which they are subjected is high and aerodynamic heating due to supersonic flight also becomes high. Thus, explosives containing TNT as a major ingredient cannot be adequately protected from the effects of high acceleration and heating. Instead, mixtures of polymeric materials and crystalline energetic particles, such as HMX, PETN, DATB, and HNS, are used as high explosives. The use of polymeric materials improves tensile stress and elongation properties, thereby offering greater protection from the effects of external mechanical shock and heating.

4.9.1 Industrial Explosives

4.9.1.1 ANFO Explosives

A mixture of ammonium nitrate and light oil forms a low-strength explosive, which is used as a blasting compound in mines and in industrial engineering. This class of explosives is named ANFO explosives (ammonium nitrate fuel oil explosives).[3] A typical ANFO explosive is formulated from a mixture of 95 % AN particles and 5 % light oil by mass. Porous AN particles are used to effectively absorb the oil and to make granular ANFO explosives. Since AN is highly hygroscopic, humidity control is needed during the formulation process. The density of ANFO explosives is in the range 8000–9000 kg m^{-3} and the detonation velocity is 2500–3500 m s^{-1}.

4.9.1.2 Slurry Explosives

Slurry explosives consist of saturated aqueous solutions of ammonium nitrate with sensitizing additives.[1,3] Nitrates such as monomethylamine nitrate, ethylene glycol mononitrate, or ethanolamine mononitrate are used as sensitizers. Aluminum powder is also added as an energetic material. Table 4.15 shows a typical chemical composition of a slurry explosive. It is important that so-called micro-bubbles are present within the explosives in order to facilitate the initial detonation and the ensuing detonation wave. These micro-bubbles are made of glass or polymeric materials.

Table 4.15 Chemical composition of a slurry explosive.

Component	wt. %
water	0.10
ammonium nitrate	0.45
potassium nitrate	0.10
monomethylamine nitrate	0.30
aluminum powder	0.02
others	0.03

Being partly aqueous, slurry explosives are not inactivated by water or humidity, and they are essentially insensitive to mechanical shock or heat. The strength of detonation is approximately equal to that of NG-NC-based explosives. Since slurry explosives are composed of a mixture of aqueous AN and an oil, they exist as emulsions and hence are also termed emulsion explosives.

4.9.2 Military Explosives

4.9.2.1 TNT-Based Explosives

Mixtures of TNT, RDX, and/or AN are used as TNT-based explosives. Various additives such as aluminum powder, barium nitrate, and/or some other small amounts of materials are used. Densities are in the range 1450–1810 kg m^{-3}. Aluminum powder is added to obtain bubble energy when used in underwater conditions.

4.9.2.2 Plastic-Bonded Explosives

Since TNT-based explosives have relatively low melting points, their deformation or unexpected ignition can occur when warheads are subjected to a high heat flux as a result of aerodynamic heating in high-supersonic or hypersonic flight. Plastic-bonded explosives (PBX) have been developed based on similar chemical processes to those used to develop composite rocket propellants. Crystalline materials such as RDX and HMX are mixed with liquid copolymers such as polystyrene and polybutadiene prepolymers. The mixtures are cast in the warheads under vacuum conditions in order to remove bubbles, and are then crosslinked and cured to impart rubber-like properties. The mechanical strengths and decomposition temperatures of the resulting explosives are much higher than those of TNT-based "Composition B" (see Section 9.2.3.1).

Various types of crystalline materials and polymers are used to formulate PBX, as summarized in Table 4.16. Though the polymers used are not the same as those used for propellants, the fundamental concepts for the selection of materials for PBX are the same.

Table 4.16 Chemical materials used to formulate PBX.

Energetic materials (oxidizer)	RDX, HMX, TNT, HNIW, AP, AN
Polymeric materials (energetic binder)	GAP-THF, BAMO-AMMO, BAMO-NIMO
Polymeric materials (binder and fuel)	Nylon, Viton, polyester-styrene, HTPB, polyurethane, silicone resin
Plasticizer	fluoronitropolymer, TEGDN
Metallic fuel	Al, Mg, Mg-Na alloy, B, Zr

References

1 Elliot, M. S., Smith, F. J., and Fraser, A. M., Synthetic Procedures Yielding Targeted Nitro and Nitroso Derivatives of the Propellant Stabilisers Diphenylamine, **N**-Methyl-4-nitroaniline, and **N**,**N**′-Diethyl-**N**,**N**′-diphenylurea, **Propellants, Explosives, Pyrotechnics**, Vol. 25, 2000, pp. 31–36.

2 Teipel, U., Heintz, T., and Krause, H. H., Crystallization of Spherical Ammonium Dinitramide (ADN) Particles, **Propellants, Explosives, Pyrotechnics**, Vol. 25, 2000, pp. 81–85.

3 Niehaus, M., Compounding of Glycidyl Azide Polymer with Nitrocellulose and its Influence on the Properties of Propellants, **Propellants, Explosives, Pyrotechnics**, Vol. 25, 2000, pp. 236–240.

4 Beal, R. W., Incarvito, C. D., Rhatigan, B. J., Rheingold, A. L., and Brill, T. B., X-ray Crystal Structures of Five Nitrogen-Bridged Bifurazan Compounds, **Propellants, Explosives, Pyrotechnics**, Vol. 25, 2000, pp. 277–283.

5 Hammerl, A., Klapötke, T. M., Piotrowski, H., Holl, G., and Kaiser, M., Synthesis and Characterization of Hydrazinium Azide Hydrazinate, **Propellants, Explosives, Pyrotechnics**, Vol. 26, 2001, pp. 161–164.

6 Klapötke, T. M., and Ang, H.-G., Estimation of the Crystalline Density of Nitramine (N–NO_2-based) High Energy Density Materials (HEDM), **Propellants, Explosives, Pyrotechnics**, Vol. 26, 2001, pp. 221–224.

7 Simões, P., Pedroso, L., Portugal, A., Carvalheira, P., and Campos, J., New Propellant Component, Part I. Study of 4,6-Dinitroamino-1,3,5-triazine-2(1**H**)-one (DNAM), **Propellants, Explosives, Pyrotechnics**, Vol. 26, 2001, pp. 273–277; Simões, P., Pedroso, L., Portugal, A., Plaksin, I., and Campos, J., New Propellant Component, Part I. Study of a PSAN/DNAM/HTPB Based Formulation, **Propellants, Explosives, Pyrotechnics**, Vol. 26, 2001, pp. 278–283.

8 Eaton, P. E., Zhang, M.-X, Gilardi, R., Gelber, N., Iyer, S., and Surapaneni, R., Octanitrocubane: A New Nitrocarbon, **Propellants, Explosives, Pyrotechnics**, Vol. 27, 2002, pp. 1–6.

9 Bunte, G., Neumann, H., Antes, J., and Krause, H. H., Analysis of ADN, its Precursor and Possible By-Products Using Ion Chromatography, **Propellants, Explosives, Pyrotechnics**, Vol. 27, 2002, pp. 119–124.

10 Teipel, U., and Mikonsaari, I., Size Reduction of Particulate Energetic Material, **Propellants, Explosives, Pyrotechnics**, Vol. 27, 2002, pp. 168–174.

11 Kwok, Q. S. M., Fouchard, R. C., Turcotte, A.-M., Lightfoot, P. D., Bowes, R., and Jones, D. E. G., Characterization of Aluminum Nanopowder Compositions, **Propellants, Explosives, Pyrotechnics**, Vol. 27, 2002, pp. 229–240.

12 Wingborg, N., and Eldsäter, C., 2,2-Dinitro-1,3-bis-nitrooxy-propane (NPN): A New Energetic Plasticizer, **Propellants, Explosives, Pyrotechnics**, Vol. 27, 2002, pp. 314–319.

13 Chen, F.-T., Duo, Y.-Q., Luo, S.-G., Luo, Y.-J., and Tan, H.-M., Novel Segmented Thermoplastic Polyurethane Elastomers Based on Tetrahydrofuran/Ethylene Oxide Copolymers as High Energetic Propellant Binders, **Propellants, Explosives, Pyrotechnics**, Vol. 28, 2003, pp. 7.

14 Spitzer, D., Braun, S., Schäfer, M. R., Ciszek, F., Comparative Crystallization Study of Several Linear Dinitramines in Nitrocellulose-Based Gels, **Propellants, Explosives, Pyrotechnics**, Vol. 28, 2003, pp. 58.

15 Venkatachalam, S., Santhosh, G., and Ninan, K. N., High Energy Oxidisers for Advanced Solid Propellants and Explosives (Eds.: Varma, M., and Chatterjee, A. K.), pp. 87–106, Tata McGraw-Hill Publishing Co., Ltd., India, 2002.

16 Wang, N.-X., Review on the Nitration of [60]Fullerene, **Propellants, Explosives, Pyrotechnics**, Vol. 26, 2001, pp. 109–111.

17 Matsunaga, T., and Fujiwara, S., Material Design of High Energy Density Materials, Explosion, Japan Explosives Society, Vol. 9, No. 2, 1999, pp. 100–110.

18 Meyer, R., Explosives, Verlag Chemie, Weinheim, 1977.

19 Sarner, S. F., Propellant Chemistry, Reinhold Publishing Corporation, New York, 1966.

20 Japan Explosives Society, Energetic Materials Handbook, Kyoritsu Shuppan, 1999.
21 Kubota, N., Propellant Chemistry, Pyrotechnic Chemistry, Journal of Pyrotechnics, Inc., Colorado, 2004, Chapter 12.
22 Chan, M. L., Reed, Jr., R., and Ciaramitaro, D. A., Advances in Solid Propellant Formulations, Solid Propellant Chemistry, Combustion, and Motor Interior Ballistics (Eds.: Yang, V., Brill, T. B., and Ren, W.-Z.), Progress in Astronautics and Aeronautics, Vol. 185, Chapter 1.7, AIAA, Virginia, 2000.
23 Doriath, G., Available Propellants, Solid Rocket Technical Committee Lecture Series, AIAA Aerospace Sciences Meeting, Reno, Nevada, 1994.
24 Miller, R. R., and Guimont, J. M., Ammonium Dinitramide Based Propellants, Solid Rocket Technical Committee Lecture Series, AIAA Aerospace Sciences Meeting, Reno, Nevada, 1994.
25 Kubota, N., Combustion of Energetic Azide Polymer, **J. Propulsion and Power**, Vol. 11, No. 4, 1995, pp. 677–682.
26 Sanderson, A., New Ingredients and Propellants, Solid Rocket Technical Committee Lecture Series, AIAA Aerospace Sciences Meeting, Reno, Nevada, 1994.
27 Miller, R., Advancing Technologies: Oxidizers, Polymers, and Processing, Solid Rocket Technical Committee Lecture Series, AIAA Aerospace Sciences Meeting, Nevada, 1994.
28 Kuwahara, T., Takizuka, M., Onda, T., and Kubota, N., Combustion of GAP Based Energetic Pyrolants, **Propellants, Explosives, Pyrotechnics**, Vol. 25, 2000, pp. 112–116.
29 Beckstead, M. W., Overview of Combustion Mechanisms and Flame Structures for Advanced Solid Propellants, Solid Propellant Chemistry, Combustion, and Motor Interior Ballistics (Eds.: Yang, V., Brill, T. B., and Ren, W.-Z.), Progress in Astronautics and Aeronautics, Vol. 185, Chapter 2.1, AIAA, Virginia, 2000.
30 Bazaki, H., Combustion Mechanism of 3-Azidomethyl-3-methyloxetane (AMMO) Composite Propellants, Solid Propellant Chemistry, Combustion, and Motor Interior Ballistics (Eds.: Yang, V., Brill, T. B., and Ren, W.-Z.), Progress in Astronautics and Aeronautics, Vol. 185, Chapter 2.8, AIAA, 2000.
31 Komai, I., Kobayashi, K., and Kato, K., Burning Rate Characteristics of Glycidyl Azide Polymer (GAP) Fuels and Propellants, Solid Propellant Chemistry, Combustion, and Motor Interior Ballistics (Ed.: Yang, V., Brill, T. B., and Ren, W.-Z.), Progress in Astronautics and Aeronautics, Vol. 185, Chapter 2.9, AIAA, Virginia, 2000.
32 Kubota, N., and Sonobe, T., Combustion Mechanism of Azide Polymer, **Propellants, Explosives, Pyrotechnics**, Vol. 13, 1988, pp. 172–177.
33 Miyazaki, T., and Kubota, N., Energetics of BAMO, **Propellants, Explosives, Pyrotechnics**, Vol. 17, 1992, pp. 5–9.

5
Combustion of Crystalline and Polymeric Materials

5.1
Combustion of Crystalline Materials

5.1.1
Ammonium Perchlorate (AP)

5.1.1.1 Thermal Decomposition

Experimental studies on the thermal decomposition and combustion processes of AP have been carried out and their detailed mechanisms have been reported.[1–11] Fig. 5.1 shows the thermal decomposition of AP as measured by differential thermal analysis (DTA) and thermal gravimetry (TG) at a heating rate of 0.33 K s^{-1}. An endothermic peak is seen at 520 K, corresponding to an orthorhombic to cubic lattice crystal structure phase transition, the heat of reaction for which amounts to

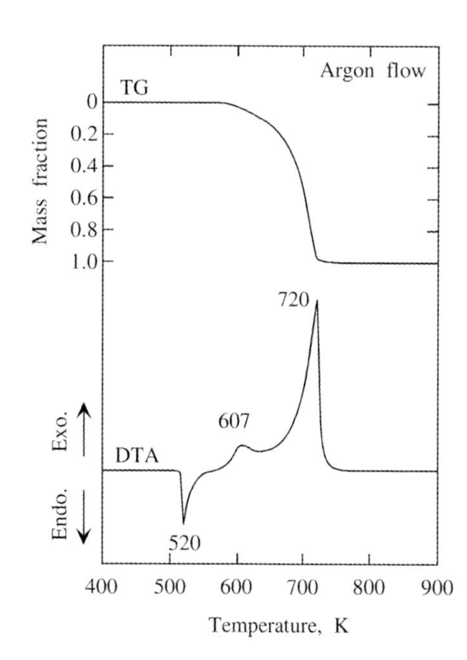

Fig. 5.1 Thermal decomposition process of AP measured by thermal gravimetry (TG) and by differential thermal analysis (DTA).

Propellants and Explosives. Naminosuke Kubota

ISBN: 978-3-527-31424-9

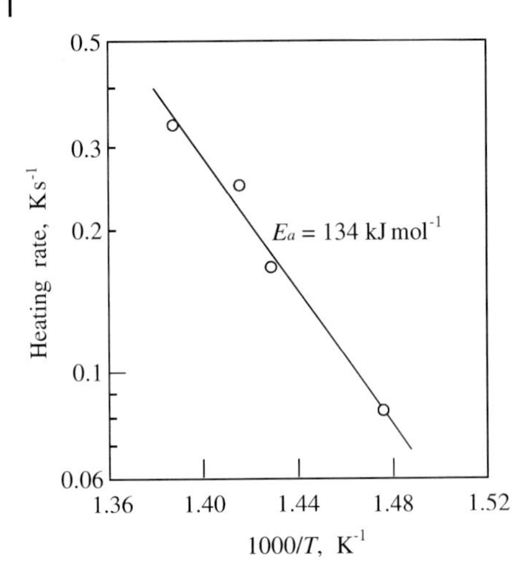

Fig. 5.2 Determination of the activation energy of AP decomposition.

–85 kJ kg^{-1} without mass loss. An exothermic reaction occurs between 607 K and 720 K accompanied by mass loss. This exothermic reaction occurs through the following overall reaction scheme:[1,2]

$$NH_4ClO_4 \rightarrow NH_3 + HClO_4$$
$$HClO_4 \rightarrow HCl + 2O_2$$

This process produces excess oxygen as an oxidizer. The exothermic peak shifts towards higher temperatures as the heating rate is increased. Fig. 5.2 shows that a plot of heating rate versus reciprocal temperature produces a straight line that indicates the activation energy for the exothermic gasification reaction. The activation energy is thereby determined as 134 kJ mol^{-1}. However, the decomposition reaction includes sublimation and melting processes that cannot be identified from the DTA and TG data. Dissociative sublimation occurs when the heating rate is slow compared with the heating rate of normal burning.[5] This sublimation is endothermic by 2.1 MJ kg^{-1} and zeroth order with respect to pressure. Melting of AP occurs in a higher temperature zone, above 725 K, when the heating rate is high.

5.1.1.2 **Burning Rate**

The mode of degradation of AP was first studied by Bircumshaw and Newman.[3] They found that below 570 K only 30% decomposition occurs; the remaining 70% is a porous solid residue chemically identical to the starting AP that does not react further when the pressure and temperature are very low. Above 670 K, no solid residue remains. When the pressure is increased, pure sublimation is suppressed and the decomposition reaction is favored. AP degradation involves dissociative sublimation of a loosely bound $NH_3 \cdot HClO_4$ complex as a first step, liberating gaseous NH_3 and $HClO_4$. The combustion of AP particles is sustained by the heat generated at the burning surface and the transfer back from the gas phase of the heat produced by the reaction between $HClO_4$ and NH_3 molecules. The gas-phase reaction is a second-order process and the adiabatic flame temperature is 1205 K.

The burning rate of a pressed strand of AP as a function of pressure has been dealt with by Arden[1] and by Levy and Friedman.[2] The lower pressure limit of AP burning is about 2.7 MPa and the burning rate increases as the pressure is increased above this lower limit. The thickness of the gas-phase reaction of NH_3/$HClO_4$ is less than 100 µm at 10 MPa and decreases as the pressure is increased, and the reaction time is inversely proportional to the pressure (MPa) represented by $6.5 \times 10^{-7}/p$ seconds.[8]

5.1.1.3 Combustion Wave Structure

When a pressed strand of AP burns, a high-temperature flame is formed in the gas phase due to the exothermic reaction between NH_3 and $HClO_4$. Mitani and Niioka measured the gas-phase structure above the regressing surface of an AP pellet and found a two-stage flame,[12] and an AP deflagration model has been presented by Guirao and Williams.[13] Heat conducted from the flame to the burning surface is used to heat the solid phase from its initial temperature to the surface temperature. At the burning surface, the AP crystals go through an orthorhombic to cubic lattice transition. This transition is endothermic by 80 kJ kg^{-1} and occurs at about 513 K. Tanaka and Beckstead carried out a computational study of the condensed-phase and gas-phase structures of AP by assuming 107 reaction steps and 32 gaseous species.[14] The burning surface temperature and the melt layer thickness were also computed as a function of pressure. The activation energy of the surface reaction was found to be about 63 kJ mol^{-1} and the pressure exponent of burning rate, n, is around 0.77 between 2.7 MPa and 10 MPa.[14]

5.1.2 Ammonium Nitrate (AN)

5.1.2.1 Thermal Decomposition

AN melts at 443 K and begins to gasify above 480 K. The decomposition process of AN is temperature-dependent. At low temperatures, i. e. around 480 K, the gasification process of AN is the endothermic (–178 kJ mol^{-1}) reversible reaction represented by[15]

$$NH_4NO_3 \rightleftharpoons NH_3 + HNO_3$$

The decomposition process shifts to an exothermic (37 kJ mol^{-1}) gasification reaction as the temperature increases:

$$NH_4NO_3 \rightarrow N_2O + 2H_2O$$

and then the overall decomposition reaction of AN appears to be

$$NH_4NO_3 \rightarrow N_2 + 2H_2O + 1/2O_2$$

This reaction is highly exothermic (119 kJ mol^{-1}) and produces oxygen molecules which act as an oxidizer. Though the ignition of AN is difficult due to the initial endothermic reaction, AN becomes highly inflammable in the high pressure region and also becomes detonable when heated beyond 550 K. Furthermore, the inflammable characteristics of AN are dependent on impurities or additives.

5.1.3
HMX

5.1.3.1 Thermal Decomposition

Very detailed literature accounts and discussions of the thermal decomposition of nitramines such as HMX and RDX have been presented by Boggs,[16] and a general picture of the decomposition processes of nitramines can be gleaned from further references.[15–24] When HMX is slowly heated, a single-stage mass-loss process is observed:[17] the mass loss begins at 550 K and a rapid gasification reaction occurs at 553 K. No solid residue remains above 553 K. Two endothermic peaks and one exothermic peak are seen: the first endothermic peak at 463 K corresponds to the crystal transformation from β to δ and the second endothermic peak at 550 K corresponds to the phase change from solid to liquid. The exothermic peak at 553 K is caused by the reaction and the accompanying gas-phase reaction.

A thermally degraded HMX sample, as obtained by decomposition interrupted at the 50% mass-loss condition (the heating is stopped at 552 K and the sample is cooled to room temperature, 293 K) is identified as a recrystallized material.[17] Gasification of the thermally degraded HMX begins at 550 K and rapid decomposition occurs at 553 K, which is equivalent to the thermal decomposition process of non-degraded HMX. However, the endothermic peak observed at 463 K is not seen for the degraded HMX. The results of an infrared (IR) analysis of β-HMX, δ-HMX, and the degraded HMX show that the degraded HMX is equivalent to δ-HMX, which implies that the endothermic process of the phase change from solid to liquid observed at 550 K can be ascribed to δ-HMX.

5.1.3.2 Burning Rate

Since HMX is composed of fine, crystalline particles, it is difficult to measure its linear burning rate. When a large-sized HMX single crystal (approximately 10 mm × 10 mm × 20 mm) is ignited by means of an electrically heated wire attached to the top of the crystal, the crystal immediately breaks into fragments following ignition because of the thermal stress created therein. As a result, no steady burning of the crystal is possible, and no linear burning rate exists for HMX because the heat conduction rate from the ignited surface to the interior of the crystal is faster than the surface regressing rate (burning rate). A breakage of the HMX crystal structure occurs due to the thermal stress caused by the temperature difference in the crystal.

When a pressed pellet made of HMX particles is ignited from the top, it burns steadily without breakage of the pellet. The thermal stress created within the pressed pellet is dissipated at the interfaces of the HMX particles within the pellet. Furthermore, the density of the pressed pellet is at most 95% of the theoretical HMX den-

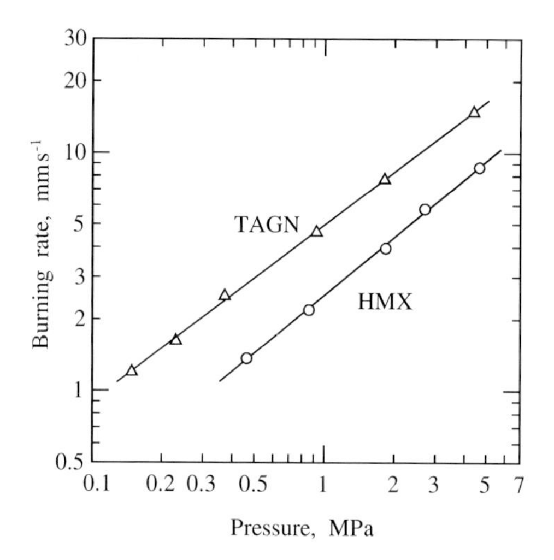

Fig. 5.3 Burning rates of HMX and TAGN showing that the burning rate of HMX is lower than that of TAGN.

sity and the thermal stress imparted to the crystal is absorbed by the voids between the HMX particles. The burning rate of an HMX sample pressed into pellet form is shown in Fig. 5.3. The pressed pellet is composed of a mixture of β-HMX particles of 20 μm in diameter (33 %) and 200 μm in diameter (67 %). The pellets have dimensions of diameter 8 mm and length 7 mm and their density is 1700 kg m^{-3}, which is 89 % of the theoretical maximum density (1900 kg m^{-3}). The burning rate of the HMX pellets is seen to increase linearly in an ln r versus ln p plot. The pressure exponent of the burning rate is 0.66 at an initial temperature of 293 K.

5.1.3.3 Gas-Phase Reaction

The overall initial decomposition reaction of HMX is represented by[18,19]

$$3(CH_2NNO_2)_4 \rightarrow 4NO_2 + 4N_2O + 6N_2 + 12CH_2O$$

Since nitrogen dioxide reacts quite rapidly with formaldehyde,[16,20,25] the gas-phase reaction represented by

$$7NO_2 + 5CH_2O \rightarrow 7NO + 3CO + 2CO_2 + 5H_2O$$

is probably the dominating reaction immediately after the initial decomposition reaction. The reaction between NO_2 and CH_2O is highly exothermic and the reaction rate is faster than the reaction rates of the other gaseous species.

The products generated by the above reactions react again at a later stage, i. e., NO and N_2O act as oxidizers, and H_2 and CO act as fuels. The reactions involving NO and N_2O are represented by[25]

$$2NO + 2H_2 \rightarrow N_2 + 2H_2O$$
$$2NO + 2CO \rightarrow N_2 + 2CO_2$$
$$2N_2O \rightarrow 2NO + N_2$$

The overall reactions involving NO and N_2O take place slowly and are trimolecular,[25] hence the reaction rate is very slow at low pressure and increases rapidly as the pressure is increased.

5.1.3.4 Combustion Wave Structure and Heat Transfer

Typical flame structures of HMX pellets as a function of pressure are shown in Fig. 5.4. A thin luminous flame sheet stands some distance from the burning surface and a reddish flame is produced above this luminous flame sheet. The flame sheet approaches the burning surface as the pressure is increased.[17] When the pressure is less than 0.18 MPa, the luminous flame sheet is blown off from the burning surface as shown in Fig. 5.4 (a). As the pressure is increased, the luminous flame sheet rapidly approaches the burning surface. However, the luminous flame sheet becomes very unstable above the burning surface and forms a wave-shaped flame sheet in the pressure range between 0.18 MPa and 0.3 MPa, as shown in Fig. 5.4 (b). On further increasing the pressure above 0.3 MPa, the luminous flame sheet becomes stable and one-dimensional in shape, just above the burning surface, as shown in Fig. 5.4 (c).

The combustion wave of HMX is divided into three zones: crystallized solid phase (zone I), solid and/or liquid condensed phase (zone II), and gas phase (zone III). A schematic representation of the heat transfer process in the combustion wave is shown in Fig. 5.5. In zone I, the temperature increases from the initial value T_0 to the decomposition temperature T_u without reaction. In zone II, the temperature increases from T_u to the burning surface temperature T_s (interface of the condensed phase and the gas phase). In zone III, the temperature increases rapidly from T_s to the luminous flame temperature (that of the flame sheet shown in Fig. 5.4). Since the condensed-phase reaction zone is very thin (~ 0.1 mm), T_s is approximately equal to T_u.

The heat flux transferred back from zone III to zone II, Λ_{III}, is given by

$$\Lambda_{III} = \lambda_{III}(dT/dx)_{III} \qquad (5.1)$$

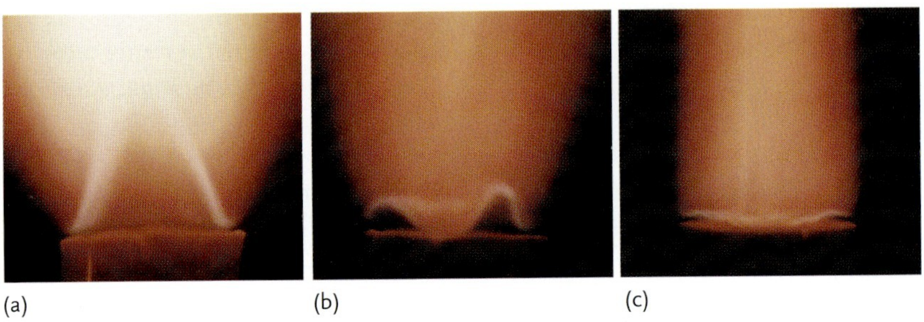

(a) (b) (c)

Fig. 5.4 Flame photographs of HMX at three different pressures: (a) 0.18 MPa, (b) 0.25 MPa, and (c) 0.30 MPa.

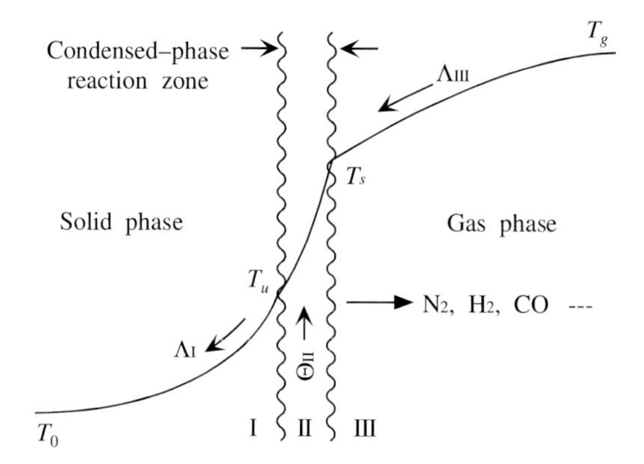

Fig. 5.5 Heat transfer model of the combustion wave of an energetic material.

and the heat flux produced in zone II, Θ_{II}, is given by

$$\Theta_{II} = \rho_I \, r \, Q_{II} \tag{5.2}$$

where Q_{II} is the heat of reaction in zone II. The heat balance equation at the burning surface is represented by

$$\Theta_{II} = \rho_I \, r \, c_I \, (T_s - T_0) - \Lambda_{III} \tag{5.3}$$

The temperature gradient in zone III, $(dT/dx)_{III}$, increases as the pressure is increased, according to $(dT/dx)_{III} \sim p^{0.7}$. However, T_s remains relatively constant (~ 700 K) in the pressure range between 0.1 MPa and 0.5 MPa. Using the physical parameter values of HMX, $\rho_I = 1700$ kg m^{-3}, $c_I = 1.30$ kJ kg^{-1} K^{-1}, and $\lambda_{III} = 8.4 \times 10^{-5}$ kW m^{-1} K^{-1}, Q_{II} is determined as 300 kJ kg^{-1}.

Fig. 5.6 shows the heat flux produced in zone II and the heat flux transferred back from zone III to zone II as a function of pressure. Θ_{II} is approximately equal to Λ_{III}, both of which increase with increasing pressure according to $\Lambda_{III} \sim p^{0.75}$ and $\Theta_{II} \sim p^{0.65}$. It is evident from Eq. (5.2) that the pressure sensitivity of Θ_{II} is approximately equal to that of the burning rate. The pressure sensitivity of the HMX burning rate ($\sim p^{0.66}$) is therefore dependent on Λ_{III}, i. e., the pressure sensitivity of the gas-phase reaction.[17]

5.1.4 Triaminoguanidine Nitrate (TAGN)

5.1.4.1 Thermal Decomposition

The oxidizer fragment (HNO_3) of TAGN is attached by an ionic bond in the molecular structure and the physicochemical processes of TAGN combustion are different from those of HMX and RDX, the oxidizer fragment of which (-N–NO_2) is attached by a covalent bond in their molecular structures. Though the flame temperature of TAGN is lower than that of HMX by 1200 K, the value of the thermodynamic parameter $(T_f/M_g)^{1/2}$ appears to be approximately the same for both materials. The

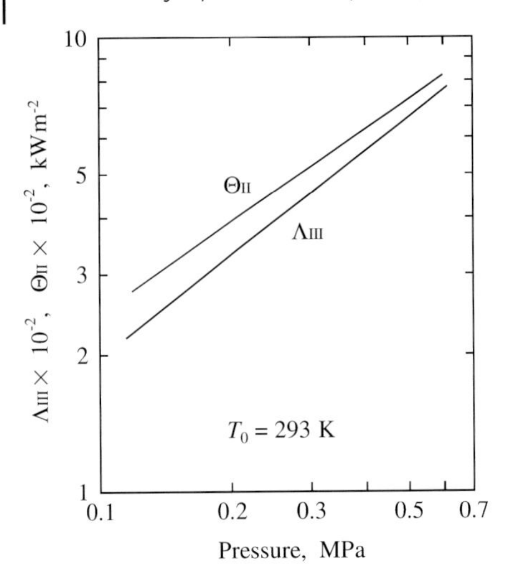

Fig. 5.6 Heat flux transferred back from the gas phase to the burning surface and heat flux produced at the burning surface.

major combustion products of TAGN are N_2, H_2, and H_2O, and those of HMX are N_2, CO, and H_2O. The molecular mass, M_g, of TAGN is 18.76 kg kmol^{-1} and that of HMX is 24.24 kg kmol^{-1}. TAGN produces a high concentration of hydrogen gas, which increases the value of the thermodynamic parameter even though the flame temperature is low.

Fig. 5.7 shows scanning electron microphotographs of a TAGN surface before combustion (a) and after quenching (b). The quenched surface is prepared by a rapid pressure decay in the strand burner shown in Appendix B. The quenched sur-

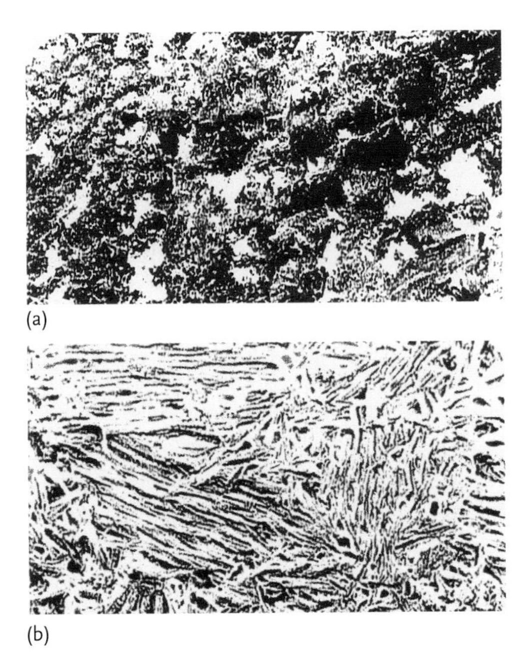

Fig. 5.7 Scanning electron microphotographs of a TAGN surface before combustion (a) and after quenching (b) from the burning pressure of 1.0 MPa.

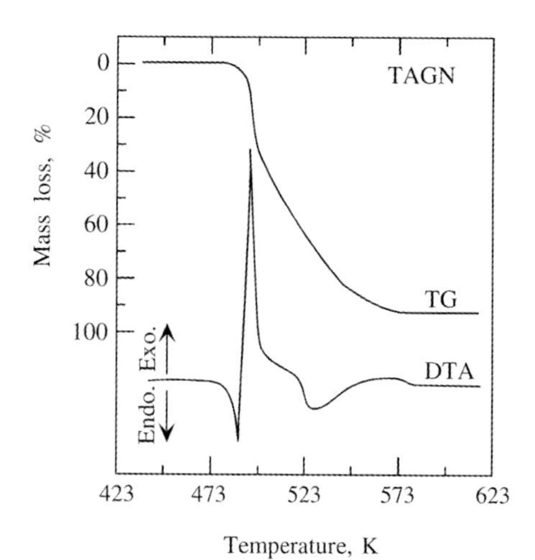

Fig. 5.8 DTA and TG results for TAGN showing a rapid exothermic reaction in the first stage of the decomposition process.

face shows finely divided recrystallized material dispersed homogeneously over it. It is evident that TAGN forms a melted layer and then decomposes to produce reactive gaseous species at the burning surface during burning.

The thermal decomposition of TAGN consists of a three-stage heat generation and mass loss process when measured by means of differential thermal analysis(DTA) and thermogravimetry (TG), as shown in Fig. 5.8.[26] The first stage corresponds to the rapid exothermic reaction between mass losses of 0% (488 K) and 27% (498 K), the second stage corresponds to the relatively slow endothermic reaction between mass losses of 27% (498 K) and 92% (573 K), and the third stage corresponds to the very slow endothermic reaction between mass losses of 92% (573 K) and 100% (623 K). The endothermic peak at 488 K corresponds to the phase change from solid to liquid. The exothermic rapid reaction in the first stage is the process representing the nature of the energetics of TAGN. This exothermic reaction occurs immediately after the endothermic phase change.

Thermally treated TAGN, as obtained by interrupted decomposition at 27% mass loss, decomposes without the exothermic peak observed as the first stage for TAGN. As shown in Fig. 5.9 (dotted lines), the gasification reaction starts at 498 K and the main decomposition reaction is complete at about 553 K. The exothermic peak observed as the first stage of the TAGN decomposition process shown in Fig. 5.8 is completely absent. This indicates that the main energetic fragment of TAGN is used up in the first stage of the decomposition reaction, accompanied by the 27% mass loss.

The molecular structure of guanidine nitrate (GN: $CH_6N_4O_3$) is similar to that of TAGN, except that the latter has three additional amino groups:

Molecular structure of GN	Molecular structure of TAGN
$NH_2 \cdot HNO_3$	$NHNH_2 \cdot HNO_3$
|	|
HN=C	$H_2N\text{–}N\text{=}C$
|	|
NH_2	$NHNH_2$

Molecular structure of GN   Molecular structure of TAGN

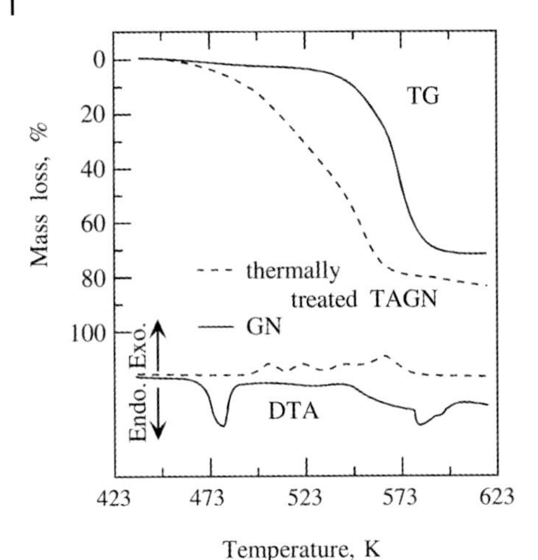

Fig. 5.9 DTA and TG results for a thermally treated TAGN and for GN, showing the lower energetic nature of both materials.

The results of DTA and TG of GN are shown in Fig. 5.9 (solid lines). The endothermic peak at 483 K corresponds to the phase change from solid to liquid and a very slow reaction occurs after this phase change. The main decomposition of GN begins at about 530 K and is complete at 539 K (70% mass loss). The remaining 30% mass fragment decomposes endothermically at higher temperatures. Although GN is also an energetic material (T_f= 1370 K and ΔH_f= −3.19 MJ kg^{-1}) consisting of an oxidizer fragment (HNO_3) attached by an ionic bond and fuel fragments, neither a rapid gasification reaction nor an exothermic reaction occurs. This is a significant contrast in comparison with the decomposition process of TAGN. Thus, one can conclude that the HNO_3 moiety attached to the molecular structure of TAGN is not the fragment responsible for the exothermic rapid reaction in the first stage of the decomposition process.[26]

Fig. 5.10 shows the infrared spectra of TAGN, thermally treated TAGN, and GN. It is evident that the -NH_2 and -C–N bonds of TAGN disappear almost completely when it is thermally treated until the exothermic peak occurs (27% mass loss). The spectrum of the thermally treated TAGN is similar to that of GN. Through the C–N bond breakage, liquefied gaseous fragments are formed; the –NH_2 bond of TAGN is no longer present in thermally treated TAGN. This implies that N–NH_2 bond breakage occurs, liberating NH_2 fragments in the first stage of the decomposition. Since the weakest chemical bond in the TAGN molecule is the N–N bond (159 kJ mol^{-1}), the initial bond breakage cleaves the amino groups. The NH_2 radicals attached to the TAGN molecule are split off. The mass fraction of 3(NH_2) within the TAGN molecule is 0.288, which is approximately equal to the observed mass loss fraction (0.27) in the first stage of the decomposition process. The chemical enthalpy difference between TAGN and GN is 344 kJ mol^{-1}, and the energy released by the reaction of NH_2 radicals to produce N_2 and H_2 is 168 kJ mol^{-1}.[26] This energy is the heat produced in the first stage of the decomposition process of TAGN.

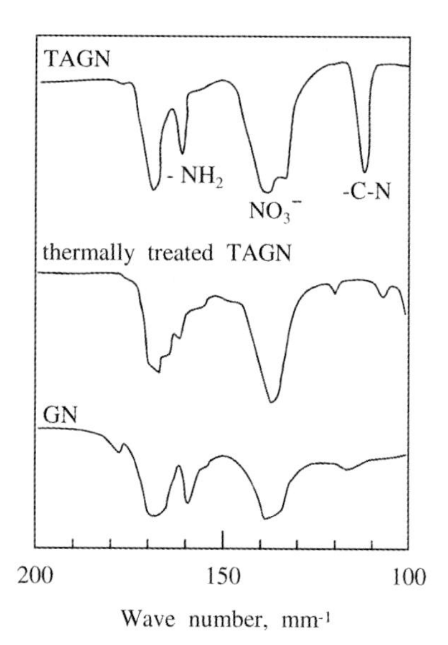

Fig. 5.10 Infrared spectra of TAGN, thermally treated TAGN, and GN.

5.1.4.2 **Burning Rate**

Since TAGN is composed of crystalline particles, its burning rate is measured using pressed pellets. The burning rate of TAGN as a function of pressure is shown in Fig. 5.3. Each pressed pellet is made from TAGN particles 5 μm in diameter and has a density of 1470 kg m^{-3}, which is 98 % of the theoretical maximum density. The burning rate is seen to increase linearly in an ln r versus ln p plot and the pressure exponent n is 0.78, which is equivalent to the pressure exponent of HMX, as is also shown in Fig. 5.3 as a reference.[21] The burning rate of TAGN is almost double that of HMX at the same pressure, even though its energy density is lower than that of HMX, as shown in Table 2.7.

5.1.4.3 **Combustion Wave Structure and Heat Transfer**

Fig. 5.11 shows a flame photograph of TAGN burning at 0.2 MPa. The luminous flame of TAGN stands some distance from the burning surface, but the luminous flame front approaches the burning surface when the pressure is increased, similarly to the luminous flame of HMX described in Section 5.1.3. As for double-base propellants and nitramines, the flame stand-off distance is represented by

$$L_g = ap^d \tag{5.4}$$

where d is −1.00. In an analysis based on Eq. (3.70), the overall reaction rate in the gas phase, $[\omega_g]$, can be expressed as:

$$[\omega_g] = \rho pr / L_g \tag{5.5}$$

$$\sim p^{m\text{-}d} \sim p^m \tag{5.6}$$

Fig. 5.11 Flame structure of TAGN showing a luminous flame standing above the burning surface: the flame front approaches the burning surface as the pressure is increased (not shown).

The reaction rate is seen to increase linearly in an ln $[\omega_g]$ versus ln p plot, and the overall order of the reaction in the gas phase is determined to be $m = 1.78$ based on the relationship $m = n - d$. This indicates that the reaction rate of TAGN in the gas phase is less pressure-sensitive than that of nitropolymer propellants; for example, $m = 2.5$ for double-base propellants.[26]

The heat transfer process in the combustion wave of TAGN consists of three zones, similar to what was illustrated for HMX in Fig. 5.5. Zone I is the solid phase, the temperature of which increases exponentially from the initial temperature, T_0, to the decomposition temperature, T_u, without chemical reaction. Zone II is the condensed phase, the temperature of which increases from T_u to the burning surface temperature, T_s, in an exothermic reaction. Zone III is the gas phase, the temperature of which increases rapidly from T_s to the final combustion temperature, T_g, in an exothermic reaction.

Fig. 5.12 shows the result of a measurement of the temperature profile in the combustion wave of TAGN. The melt layer temperature T_u is about 750 K and T_s is about 950 K, both of which remain relatively unchanged when the pressure is increased. Both the thickness of the condensed-phase reaction zone II and the heat flux feedback from zone III to zone II, Λ_{III}, increase as the pressure is increased, as shown in Fig. 5.13. At 0.3 MPa, the heat flux in zone II, Θ_{II}, is approximately 13 times higher than that in zone III, Λ_{III}, as shown in Fig. 5.14. The heat of reaction in zone II, Q_{II}, is determined to be 525 kJ kg^{-1}. It is evident that Q_{II} of TAGN is approximately 75 % higher than Q_{II} of HMX. Thus, the higher burning rate of TAGN compared to that of HMX shown in Fig. 5.3 is caused by the higher Q_{II} of TAGN, even though the adiabatic flame temperature of TAGN is about 1200 K lower than that of HMX. The decomposition and combustion of TAGN are discussed in detail in a symposium report.[26]

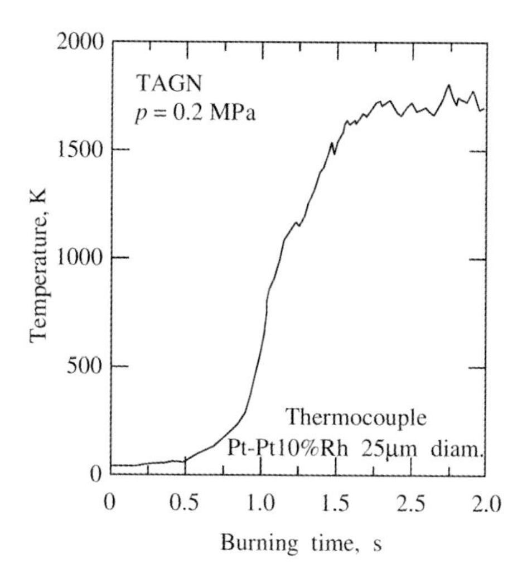

Fig. 5.12 Temperature profile in the combustion wave of TAGN.

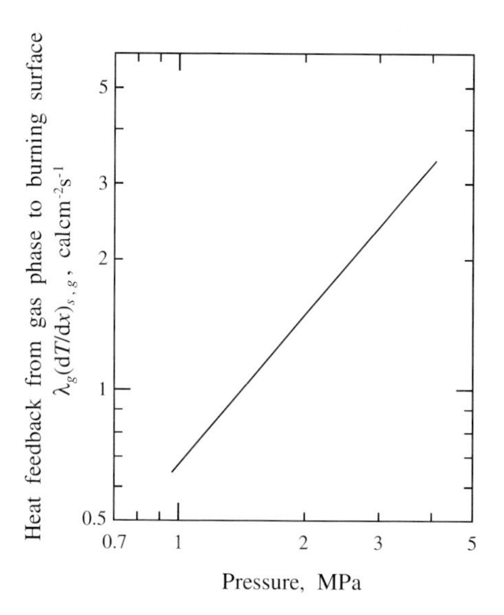

Fig. 5.13 Heat flux transferred back from the gas phase to the burning surface of TAGN.

5.1.5
ADN (Ammonium Dinitramide)

DSC and DTA measurements show melting of ADN, $NH_4N(NO_2)_2$, at 328 K, the onset of decomposition at 421 K, and an exothermic peak at 457 K.[27] Gasification of 30% of the mass of ADN occurs below the exothermic peak temperature, and the remaining 70% decomposes after the peak temperature. The decomposition is initiated by dissociation into ammonia and hydrogen dinitramide. The hydrogen dinitramide further decomposes to ammonium nitrate and N_2O. The final decomposition products in the temperature range 400–500 K are NH_3, H_2O, NO,

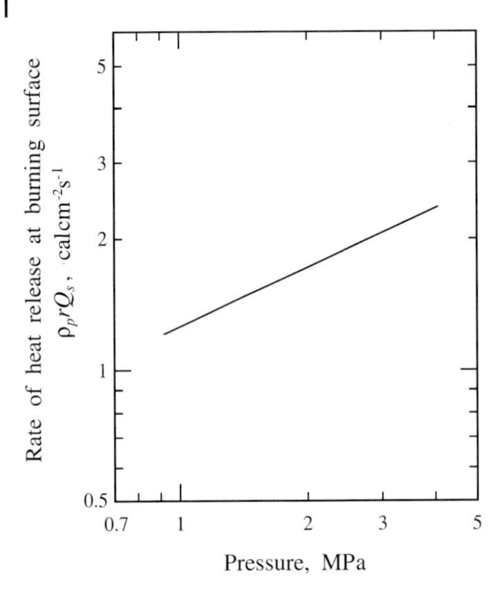

Fig. 5.14 Rate of heat release at the burning surface of TAGN as a function of pressure.

N_2O, NO_2, HONO, and HNO_3, with a total heat release of approximately 240 kJ mol^{-1}.[28] These molecules react in the gas phase to form O_2, H_2O, and N_2 as the final combustion products and the adiabatic flame temperature reaches 3640 K. The excess oxygen molecules act as an oxidizer when ADN is mixed with fuel components.

The combustion wave structure of ADN consists of three zones: the melt layer zone, the preparation zone, and the flame zone. The temperature remains relatively unchanged in the melt layer zone, then increases rapidly just above the melt layer zone to form the preparation zone, in which it rises from about 1300 to 1400 K. At some distance above the melt layer zone, the temperature increases rapidly to form the flame zone, in which the final combustion products are formed.

Similar to the situation with nitrate esters, the two-stage gas-phase reaction resulting from the combustion of ADN occurs due to the reduction of NO to N_2, which is reported to be a termolecular reaction. The heat flux transferred back from the preparation zone to the melt layer zone dominates the gasification process occurring in the melt layer zone.

5.1.6
HNF (Hydrazinium Nitroformate)

HNF, $N_2H_5C(NO_2)_3$, melts at 397 K and completely decomposes at 439 K, accompanied by an energy release of 113 kJ mol^{-1}. DTA and TG analyses reveal that the thermal decomposition of HNF occurs in two steps. The first step is an exothermic reaction accompanied by 60% mass loss in the temperature range 389–409 K. The second step is another exothermic reaction accompanied by 30% mass loss in the temperature range 409–439 K. These two steps occur successively and the decomposition mechanism seems to switch at 409 K.

As in the case of AN, the thermal decomposition process of HNF varies with the temperature of decomposition. Deflagatory decomposition of HNF produces ammonium nitroformate, which decomposes to hydrazine, nitroform, and ammonia.[28] These products react to generate heat in the gas phase and the final combustion products are formed according to:

$$N_2H_5C(NO_2)_3 \rightarrow 2NO + CO_2 + 2H_2O + 3/2N_2 + 1/2H_2$$

The combustion wave structure for HNF consists of two gas-phase zones, similar to that for ADN. However, the melt layer zone observed for ADN is not seen for HNF. The temperature increases rapidly in the gas phase just above the decomposing surface of HNF. It then increases relatively slowly in the first gas-phase zone. It increases rapidly once more at the beginning of the second gas-phase zone, with formation of the final combustion products. Thus, the second reaction zone stands some distance above the decomposing surface of the HNF. The second reaction zone is considered to involve the reaction between the 2NO and $1/2H_2$ produced in the first reaction zone, as represented by

$$2NO + 1/2H_2 \rightarrow N_2 + 1/2H_2O + 3/4O_2$$

This reduction of NO is highly exothermic but relatively slow at low pressure, because it appears to be a third-order reaction, similar to the dark-zone reaction of nitropolymer combustion. The overall reaction of HNF is represented by

$$N_2H_5C(NO_2)_3 \rightarrow CO_2 + 5/2H_2O + 5/2N_2 + 3/4O_2$$

and the adiabatic flame temperature reaches 3120 K. The large number of oxygen molecules formed by the combustion of HNF serve as oxidizers when HNF is mixed with fuel components.

5.2 Combustion of Polymeric Materials

5.2.1 Nitrate Esters

Nitrate esters are either liquid or solid and are characterized by an O–NO_2 chemical bond. Typical nitrate esters are nitrocellulose (NC) and nitroglycerin (NG). NC is known as gun cotton, a single-base propellant used for guns. Since NC is a fibrous material, grains are formed by treating it with a solvent. On the other hand, NG is a liquid at room temperature. It is mixed together with NC to form a rubber-like energetic material. This NC/NG mixture is used in explosives and propellants. When this mixture is used for guns and rockets, it is referred to as a double-base propellant.

Extensive experimental and theoretical studies have been performed in an effort to determine the decomposition and combustion processes of nitrate esters. This subsection provides a summary of the decomposition and combustion mechanisms of various types of nitrate esters presented previously. It is important to acquire sufficient understanding of the governing factors of the combustion of nitrate esters so that the combustion properties encountered during applications can be anticipated. An understanding of the basic steady-state combustion mechanism is a necessary prerequisite for models describing the burning rate as a function of pressure and initial temperature.

5.2.1.1 Decomposition of Methyl Nitrate

The simplest nitrate ester is methyl nitrate, which has the chemical structure CH_3ONO_2. The decomposition process is given by:[20,29]

$$\begin{aligned} CH_3ONO_2 &\rightarrow CH_3O^{\cdot} + NO_2 & Q_p &= -147\ \mathrm{kJ\ mol^{-1}} \\ &\rightarrow CH_2O + 1/2H_2 + NO_2 & Q_p &= -34\ \mathrm{kJ\ mol^{-1}} \\ &\rightarrow CO,\ H_2O,\ NO,\ H_2,\ N_2,\ N_2O,\ CH_2O,\ CO_2 & & \end{aligned}$$

The first two reaction steps are endothermic; however, the overall reaction is exothermic and the final flame temperature is 1800 K. The observed pressure dependence of the burning rate follows a second-order rate law; the overall activation energy is consistent with the oxidation reaction by NO_2 being the slowest and hence the rate-controlling step.

5.2.1.2 Decomposition of Ethyl Nitrate

The primary step in the decomposition of ethyl nitrate ($C_2H_5ONO_2$) is again the breaking of the $C_2H_5O–NO_2$ bond,[30] and the decomposition rate obeys a first-order law. The decomposition process of ethylene glycol dinitrate can be written as:[29]

$$\begin{matrix} CH_2ONO_2 & & CH_2O^{\cdot} & & & \\ | & \rightarrow & | & + & NO_2 & Q_p = -147\ \mathrm{kJ\ mol^{-1}} \\ CH_2ONO_2 & & CH_2ONO_2 & & & \end{matrix}$$

$$\rightarrow 2CH_2O + NO_2 \qquad Q_p = -113\ \mathrm{kJ\ mol^{-1}}$$

The breaking of one $O–NO_2$ bond gives a free radical, which decomposes to formaldehyde and nitrogen dioxide. When butane-2,3-diol dinitrate is decomposed at atmospheric pressure, nitrogen dioxide and acetaldehyde are formed.[31] However, these react rapidly to form nitric oxide within a distance of 2 mm from the decomposing surface. The decomposition of butane-1,4-diol dinitrate produces nitrogen dioxide, formaldehyde, and ethene. A steep temperature increase and a rapid concentration decrease of nitrogen dioxide and formaldehyde are observed within 1 mm of the decomposing surface. It is proposed[31] that the dinitrate decomposes to produce equal amounts of aldehyde and nitrogen dioxide according to:

$$\begin{array}{l} R\text{–}CHONO_2 \\ | \qquad\qquad \rightarrow 2RCHO + 2NO_2 \\ R\text{–}CHONO_2 \end{array}$$

NO_2 is then converted into NO by the oxidation reaction with RCHO.

5.2.1.3 Overall Decomposition Process of Nitrate Esters

As is evident from experimental measurements, most kinds of nitrate esters appear to decompose to NO_2 and C,H,O species with the breaking of the O–NO_2 bond as the initial step. A strong heat release occurs in the gas phase near the decomposing surface due to the reduction of NO_2 to NO accompanied by the oxidation of C,H,O species to H_2O, CO, and CO_2. NO reduction, however, is slow and this reaction is not observed in the decomposition of some nitrate ester systems. Even when the reaction occurs, the heat release does not contribute to the heat feedback to the surface because the reaction occurs at a distance far from the surface.

The decomposition process can be essentially divided into three stages for simple nitrate esters:

Stage 1. $RNO_2 \rightarrow NO_2$ + organic molecules (mainly aldehydes)
Stage 2. NO_2 + intermediate organic products
$\rightarrow NO + H_2$, CO, CO_2, H_2O, etc., at low pressure
Stage 3. NO + H_2, CO, etc. $\rightarrow N_2$, CO_2, H_2O, etc. at high pressure
Stage 2 occurs at both high and low pressures.

The decomposition process of double-base propellants is autocatalytic, NO_2 being evolved first and then reacting to increase the rate of its evolution.[32] The first step of the decomposition is the breaking of the RO–NO_2 bond, which is followed by the production of complex organic gases.[33]

Mixtures of HCHO and NO_2 react very rapidly at temperatures above 430 K; NO_2 is reduced almost quantitatively to NO, and the aldehyde is oxidized to CO, CO_2, and H_2O.[34] This process is discussed in detail in Section 5.2.1.4.

5.2.1.4 Gas-Phase Reactions of NO_2 and NO

From the discussion in the previous section, it is clear that, above all, NO_2, and then NO, are the principal oxidizers produced in the flames of nitrate esters. The reaction of NO_2 with aldehydes plays an important role in the combustion of nitrate esters, since these molecules are the major decomposition products of these materials in the first stage of combustion. Pollard and Wyatt studied the combustion process of HCHO/NO_2 mixtures at sub-atmospheric pressures.[34] They found that the reaction occurs very rapidly at temperatures above 433 K, the NO_2 being reduced almost quantitatively to NO, and the aldehyde being oxidized to CO, CO_2, and H_2O. The order of reaction was found to be one with respect to both reactants. The same result has been reported by McDowell and Thomas.[35] The proposed reaction steps are:

$$CH_2O + NO_2 \rightarrow CH_2O_2 + NO$$

$$CH_2O_2 \rightarrow CO + H_2O$$

$$CH_2O_2 + NO_2 \rightarrow CO_2 + H_2O + NO$$

The flame velocity is independent of pressure and the maximum velocity is 1.40 m s^{-1} for a mixture containing 43.2 mol% HCHO. The velocity depends significantly on the mixture ratio; for example, it drops to about half of this value for a mixture containing 60 % HCHO. Powling and Smith measured the flame velocities of CH_3CHO/NO_2 mixtures.[31] The velocity proved to be very sensitive to the CH_3CHO/NO_2 ratio, as Pollard and Wyatt observed for the $HCHO/NO_2$ flame. The velocity is about 0.10 m s^{-1} at 37 % CH_3CHO and decreases to 0.04 m s^{-1} at 60 % CH_3CHO.

The combustion of H_2, CO, and hydrocarbons with NO is important in both the dark zone and the flame zone of nitropolymer propellants. It is well known that NO behaves in a complex way in combustion processes, in that at certain concentrations it may catalyze a reaction to promote a process, while at other concentrations it may inhibit the reaction. Sawyer and Glassman[36] attempted to establish a measurable reaction between H_2 and NO in a flow reactor at 0.1 MPa. Over a wide range of mixture ratios, they found that the reaction did not occur readily below the temperature of NO dissociation, except in the presence of some radicals. Mixtures of CO and NO are also difficult to ignite, and only mixtures rich in NO could be ignited at 1720 K.

Cummings[37] measured the burning velocity of a mixture of NO and H_2 in a burner flame over a wide pressure range, and found it to be independent of pressure (about 0.56 m s^{-1}) between 0.1 MPa and 4.0 MPa. However, Strauss and Edse[38] found that, at a mixture ratio of 1:1, the burning velocity increased from 0.56 m s^{-1} to 0.81 m s^{-1} at 5.2 MPa.

An extensive experimental study on reaction mechanisms involving NO_2 and NO was conducted by Sawyer.[39] He found the reaction of H_2 with NO_2 to be about three times faster than the reaction of H_2 with a 2:1 NO/O_2 mixture, and no reaction was observed between H_2 and NO.

Generally, in calculating the flame speed of a premixed gas, the net reaction is assumed to be second order in the gas phase. However, it is well known that oxidation reactions involving NO are usually termolecular, for example:[40–42]

$$2NO + Cl_2 \rightarrow 2NOCl$$

$$2NO + Br_2 \rightarrow 2NOBr$$

$$2NO + O_2 \rightarrow 2NO_2$$

The experimental results of Hinshelwood and Green[42] tend to support the following mechanism for the reaction between NO and H_2:

$$2NO + H_2 \rightarrow N_2 + H_2O_2 \text{ (slow)}$$

$$H_2O_2 + H_2 \rightarrow 2H_2O \text{ (fast)}$$

The measured order of the overall reaction varies between 2.60 and 2.89. However, Pannetier and Souchay[43] suggested that the above reaction could not be expected to occur owing to the improbable nature of the termolecular process. They proposed

$$2NO \rightleftharpoons N_2O_2 \text{ (fast)}$$

$$N_2O_2 + H_2 \rightarrow N_2 + H_2O_2 \text{ (slow)}$$

$$H_2O_2 + H_2 \rightarrow 2H_2O \text{ (fast)}$$

In the above series of reactions, the slow reaction involving N_2O_2 is the rate-controlling step. The reaction involving NO is fast enough to maintain equilibrium with the N_2O_2. Consequently, it can be seen that the rate of production of N_2 and H_2O is third order with respect to NO and H_2. The overall sum of these reaction steps is indeed third order, while the elementary reactions are all bimolecular, i. e., second order.

In summary, gas-phase reactions between aldehydes and NO_2 occur readily and with strong exothermicity. The rate of reaction is largely dependent on the aldehyde/NO_2 mixture ratio, and is increased with increasing NO_2 concentration for aldehyde-rich mixtures. On the other hand, no appreciable gas-phase reactions involving NO are likely to occur below 1200 K. The overall chemical reaction involving NO appears to be third order, which implies that it is sensitive to pressure. The reactions discussed above are important in understanding the gas-phase reaction mechanisms of nitropolymer propellants.

5.2.2 Glycidyl Azide Polymer (GAP)

5.2.2.1 Thermal Decomposition and Burning Rate

The terminal OH groups of GAP prepolymer are cured with the NCO groups of hexamethylene diisocyanate (HMDI) and crosslinked with trimethylolpropane(TMP) in order to formulate GAP copolymer consisting of 84.8% GAP prepolymer, 12.0% HMDI, and 3.2% TMP, as shown in Fig. 4.8. The thermochemical data for GAP copolymer obtained by differential thermal analysis (DTA) and thermal gravimetry (TG) are shown in Fig. 5.15. The exothermic peak accompanied by mass loss between 475 K and 537 K is attributed to the decomposition and gasification reactions. A two-stage gasification reaction occurs: the first stage occurs rapidly and is accompanied by the evolution of heat, whereas the second stage occurs relatively slowly without the evolution of heat.[44] The activation energy of the first-stage exothermic gasification is 174 kJ mol^{-1}.

A thermally degraded GAP copolymer is produced at 532 K, accompanied by β = 0.25, where β is the mass fraction loss obtained by thermal degradation.[44] The ex-

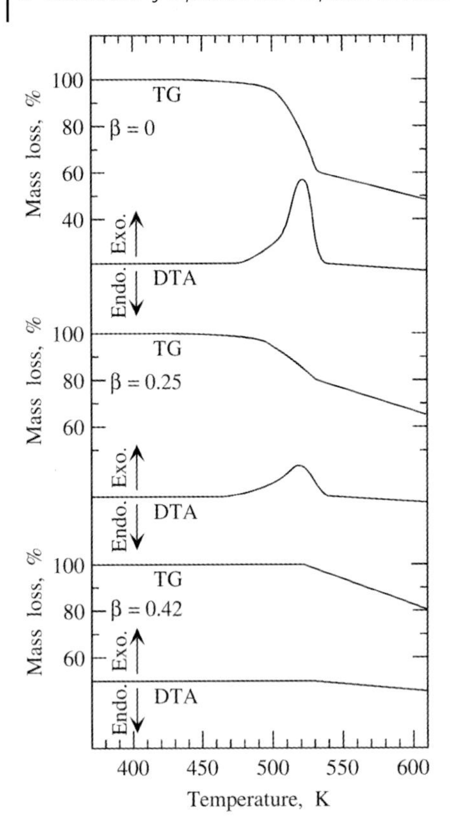

Fig. 5.15 Decomposition of thermally degraded GAP, showing that the exothermic peak decreases as the mass fraction of degradation is increased.

othermic peak is reduced and the first reaction stage, with $\beta = 0.25$, is complete at 529 K, as shown in Fig. 5.15. A thermally degraded GAP copolymer obtained by interruption of the heating at the end of the first reaction stage (537 K, accompanied by $\beta = 0.42$) shows no exothermic peak. The exothermic reaction of GAP copolymer occurs only in the early stages of the decomposition, and no exothermic reaction occurs upon gasification of the remaining mass.

The fraction of nitrogen atoms contained within the GAP copolymer decreases linearly as β increases, as shown in Fig. 5.16. A mass fraction of 0.68 nitrogen atom is present at the first-stage reaction process ($\beta < 0.41$) and the remaining mass fraction of 0.32 nitrogen atoms is gasified in the second-stage reaction process ($\beta > 0.41$). Similar to the loss of the nitrogen atoms, the fractions of hydrogen, carbon, and oxygen atoms within GAP copolymer also decrease linearly as β increases in the region $\beta < 0.41$.

Infrared analysis of GAP copolymer before and after thermal degradation monitored by TG shows that the absorption of the azide bond of the starting GAP copolymer ($\beta = 0.0$) is seen at about $\nu = 2150\ cm^{-1}$.[44] This azide bond absorption is completely lost following thermal degradation ($\beta = 0.41$). The -N_3 bonds within the GAP copolymer decompose thermally above 537 K to produce N_2. Thus, the gasification of the GAP copolymer observed as the first reaction stage occurs due to split-

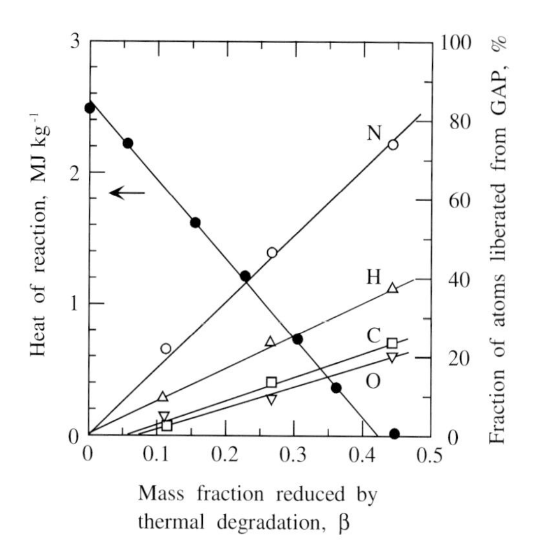

Fig. 5.16 Heat of reaction and fractions of atoms lost from GAP upon thermal degradation.

ting-off of the two nitrogen atoms accompanied by the liberation of heat. The remaining C,H,O molecular fragments decompose at the second reaction stage without the liberation of heat.

The burning rate of GAP copolymer increases linearly with increasing pressure in an ln r versus ln p plot, as shown in Fig. 5.17. The pressure exponent of burning rateat a constant initial temperature, as defined in Eq. (3.71), is 0.44. The temperature sensitivity of burning rate at constant pressure, as defined in Eq. (3.73), is 0.010 K^{-1}.

5.2.2.2 **Combustion Wave Structure**

The combustion wave of GAP copolymer is divided into three zones: zone I is a non-reactive heat-conduction zone, zone II is a condensed-phase reaction zone,

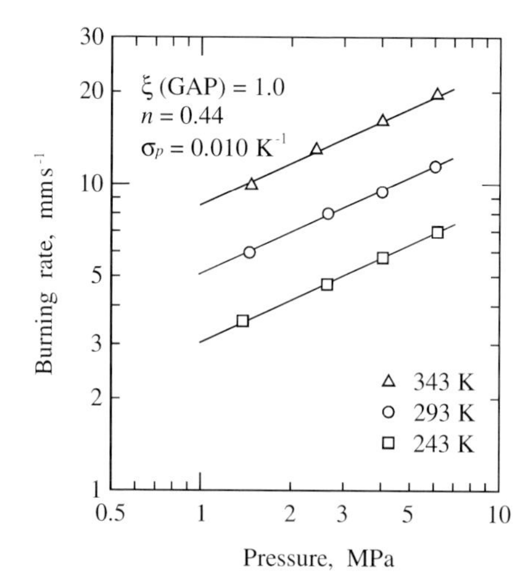

Fig. 5.17 Burning rates of GAP copolymer at three different initial temperatures.

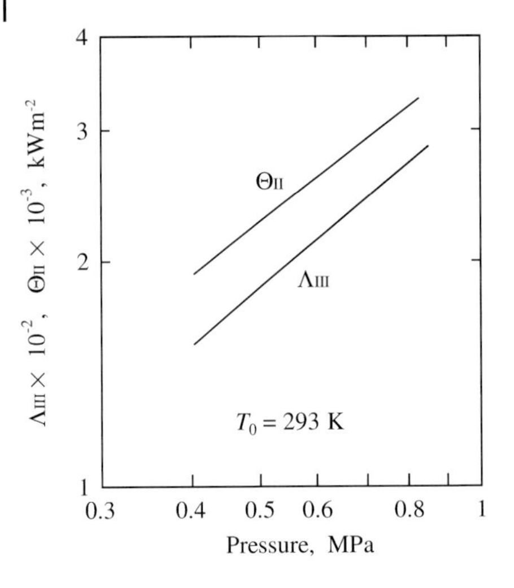

Fig. 5.18 Heat flux transferred back from the gas phase to the burning surface and heat flux produced at the burning surface of GAP copolymer as a function of pressure.

and zone III is a gas-phase reaction zone in which the final combustion products are formed. Decomposition occurs at T_u in zone II, and gasification is complete at T_s in zone II. This reaction scheme is similar to that of HMX or TAGN, as shown in Fig. 5.5.

Using Eqs. (5.1) and (5.2), the heat flux in zone II, Θ_{II}, and the heat flux in zone III (Λ_{III}) are determined from temperature profile data in the combustion wave. As shown in Fig. 5.18, both Θ_{II} and Λ_{III} increase linearly with increasing pressure in a log-log plot: $\Theta_{II} \sim p^{0.75}$ and $\Lambda_{III} \sim p^{0.80}$. The heat of reaction in zone II, Q_{II}, is determined as 624 kJ kg^{-1}.[44] It is noteworthy that the heat of reaction of HMX in zone II is 300 kJ kg^{-1}, even though the adiabatic flame temperature of HMX is 1900 K higher than that of GAP copolymer. Furthermore, Λ_{III} of GAP is of the same order of magnitude as Λ_{III} of HMX, despite the fact that Θ_{II} of GAP is approximately ten times larger than the Θ_{II} of HMX shown in Fig. 5.6.

5.2.3
Bis-azide methyl oxetane (BAMO)

5.2.3.1 Thermal Decomposition and Burning Rate

When BAMO is heated, an exothermic gasification reaction occurs, which is complete when the mass fraction loss reaches 0.35, as shown in Fig. 5.19. During this gasification process, scission of the two azide bonds within BAMO occurs, producing nitrogen gas accompanied by the release of heat.[45] The remaining part continues to gasify without exothermic reaction at higher temperatures. For reference, the thermal decomposition process of BCMO is also shown in Fig. 5.19. No exothermic reaction occurs during the gasification process. Fig. 5.20 shows the decomposition temperature, T_d, and the exothermic peak temperature, T_p, as a function of the heating rate of BAMO. Both T_d and T_p are seen to shift to higher values in an ln (θ/T^2) versus $1/T$ plot when the heating rate (θ) is increased. The activation energy associated with the decomposition is 158 kJ mol^{-1}.[30]

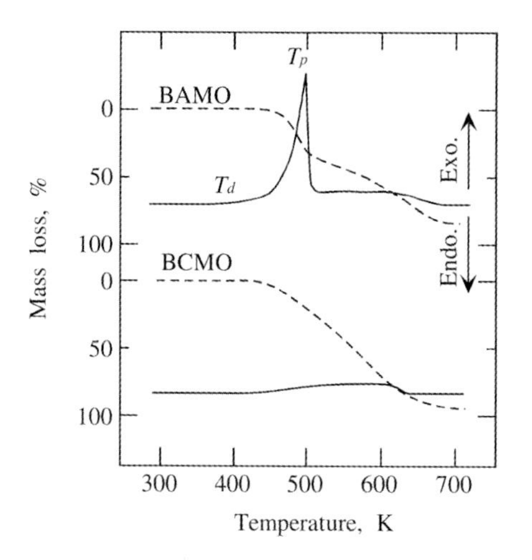

Fig. 5.19 BAMO, containing two C–N_3 bonds, decomposes with rapid heat release accompanied by a mass fraction loss of 0.3. On the other hand, BMCO, containing C–Cl bonds, decomposes relatively smoothly without heat release.

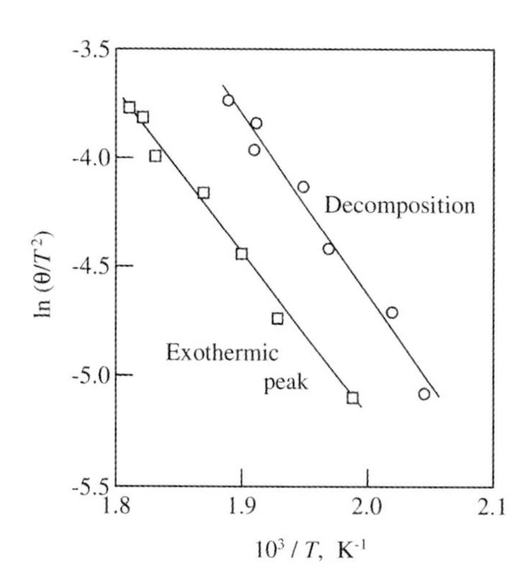

Fig. 5.20 The exothermic peak temperature and decomposition temperature of BAMO shift to higher values as the heating rate is increased.

The heat of decomposition, Q_d, of BAMO copolymer containing different levels of N_3 bond density, $\xi(N_3)$, is shown as a function of $\xi(N_3)$ in Fig. 5.21. BAMO prepolymer is copolymerized with THF. The N_3 bond density is varied by adjusting the mass fraction ratio of BAMO prepolymer and THF.

The relationship between Q_d (MJ kg^{-1}) and $\xi(N_3)$ (mol mg^{-1}) measured by differential scanning calorimetry (DSC) is represented by[46]

$$Q_d = 0.6\xi(N_3) - 2.7 \tag{5.7}$$

The heat of decomposition increases linearly with increasing N_3 bond density.

The condensed-phase reaction zone of a burning-interrupted BAMO copolymer is identified by infrared (IR) spectral analysis. In the non-heated zone, the absorption of the N_3 bond, along with the absorptions of the C–O, C–H, and N–H bonds,

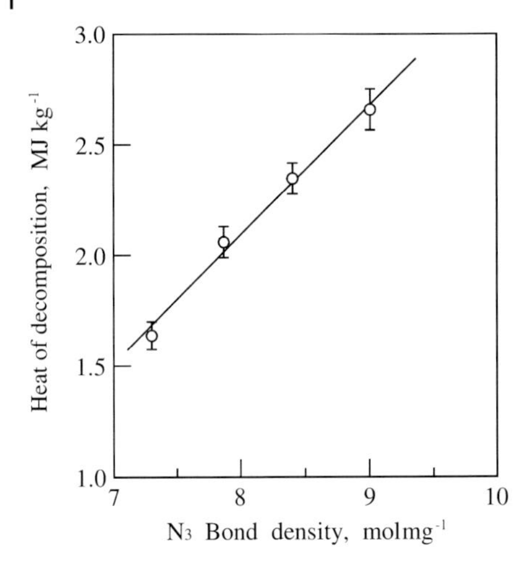

Fig. 5.21 Heat of decomposition increases with increasing N_3 bond density in BAMO copolymer.

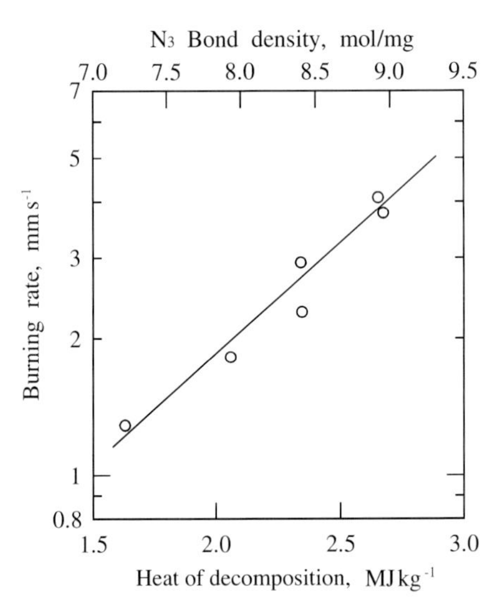

Fig. 5.22 The burning rate of BAMO copolymer increases with increasing heat of decomposition and also with increasing N_3 bond energy.

are seen. In the surface reaction zone (0–0.5 mm below the burning surface), the absorption of the N_3 bond is lost. However, the absorptions of the C–O, C–H, and N–H bonds remain as observed in the non-heated zone. This suggests that an exothermic reaction occurs due to decomposition of the N_3 bonds in the sub-surface and surface reaction zones.[45]

The burning rate of BAMO copolymer at $T_0 = 293$ K is shown as a function of Q_d and $\xi(N_3)$ in Fig. 5.22. The linear dependence of the burning rate, r (mm s^{-1}), in the semi-log plot at 3 MPa is represented by

$$r = 1.84 \times 10^{-4} \exp\,(1.14 Q_d) \qquad (5.8)$$

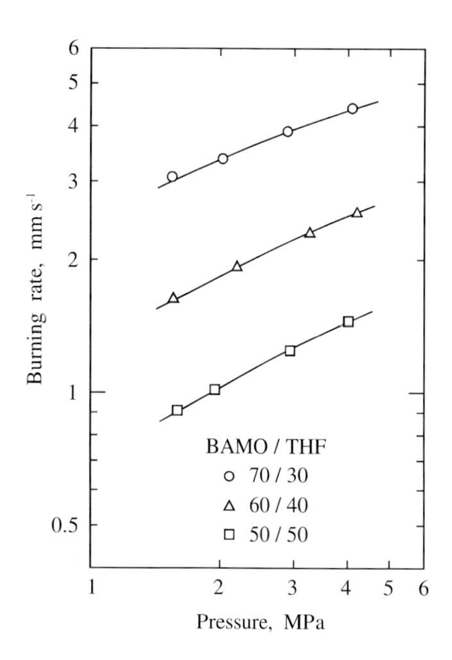

Fig. 5.23 The burning rate of BAMO/THF copolymer increases as the mass fraction of BAMO is increased at constant pressure.

The burning rates of BAMO copolymer samples with the compositions BAMO/THF = 70:30, 60:40, and 50:50 are shown as a function of pressure in Fig. 5.23.[31] The burning rate is seen to increase linearly in an ln r versus ln p plot with increasing pressure at constant T_0. The burning rate is also seen to be highly sensitive to the mixture ratio of BAMO and THF. Furthermore, the burning rate of BAMO copolymer is also sensitive to T_0. For example, the burning rate of BAMO/THF = 60:40 increases drastically when T_0 is increased at constant p, as shown in Fig. 5.24. The burning rate is represented by

$$r = 0.55 \times 10^{-3}\, p^{0.82} \text{ at } T_0 = 243 \text{ K}$$

$$r = 2.20 \times 10^{-3}\, p^{0.61} \text{ at } T_0 = 343 \text{ K}$$

The temperature sensitivity of the burning rate, as defined in Eq. (3.73), is 0.0112 K^{-1} at 3 MPa.

5.2.3.2 **Combustion Wave Structure and Heat Transfer**

During BAMO copolymer burning in a pressurized inert atmosphere, gaseous and carbonaceous solid fragments are formed exothermically at the burning surface. The temperature in the combustion wave of BAMO copolymer increases from the initial value, T_0, to the burning surface temperature, T_s, and then to the flame temperature, T_f. The burning surface temperature of BAMO copolymer with the composition BAMO/THF = 60:40 increases as T_0 increases at constant pressure. For example, at p = 3 MPa, T_s = 700 K at T_0 = 243 K and T_s = 750 K at T_0 = 343 K. The effect of T_0 on T_s as expressed by $(dT_s/dT_0)_p$ is determined as 0.50.[46]

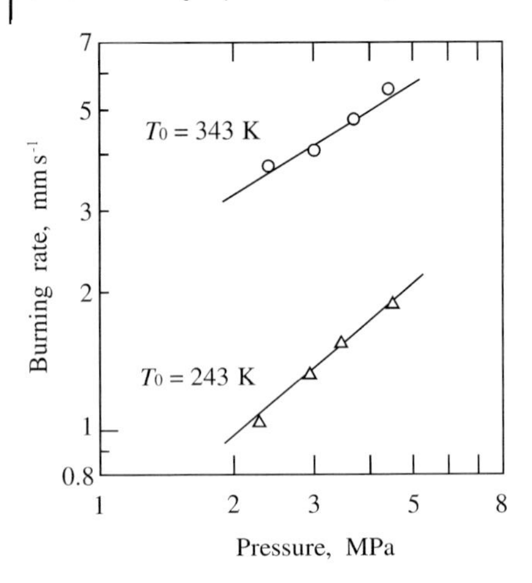

Fig. 5.24 Effect of initial temperature on the burning rate of BAMO/THF = 60:40 copolymer.

As shown in Fig. 5.25, the temperature gradient in the gas phase at $T_0 = 243$ K and $p = 3$ MPa reaches a maximum at the burning surface, decreases with distance from the burning surface, and becomes zero at a distance of about 1.1 mm. When the initial temperature, T_0, is increased from 243 K to 343 K, the temperature gradient is increased and becomes zero at about 0.7 mm from the burning surface.

The energy conservation equation in the gas phase for steady-state burning is given by Eq. (3.41). If one assumes that the physical parameters λ_g and c_g are constant in the gas phase, Eq. (3.14) can be represented by:[46,47]

$$q_d(x) + q_v(x) + q_c(x) = 0 \tag{5.9}$$

$$q_d(x) = \lambda_g\, d^2T/dx^2: \quad \text{heat flux by conduction} \tag{5.10}$$

$$q_v(x) = -mc_g\, dT/dx: \quad \text{heat flux by convection} \tag{5.11}$$

$$q_c(x) = Q_g\omega_g(x): \quad \text{heat flux by chemical reaction} \tag{5.12}$$

The overall reaction rate in the gas phase $[\omega_g]$ can be represented by

$$[\omega_g] = \int_0^\infty \omega_g(x)dx = m \tag{5.13}$$

Using the burning rate data shown in Fig. 5.24 and the temperature gradient data shown in Fig. 5.25, the heat fluxes given by Eqs. (5.10)–(5.12) can be determined as a function of burning distance. As shown in Fig. 5.25, $q_c(x)$ is maximal at the burning surface and decreases with increasing distance for both low and high initial temperatures.[46] The convective heat flux, $q_c(0)$, at 343 K is 3.3 times higher than that at 243 K, and the reaction distance to complete the gas-phase reaction is 1.1 mm at $T_0 = 243$ K and 0.7 mm at $T_0 = 343$ K. Using the data in Figs. 5.24 and 5.25 and Eq. (5.13), $[\omega_g]$ at 3 MPa is determined as 1.58×10^3 kg m^{-3}s^{-1} at $T_0 = 243$ K and 7.62×10^3 kg m^{-3}s^{-1} at $T_0 = 343$ K.

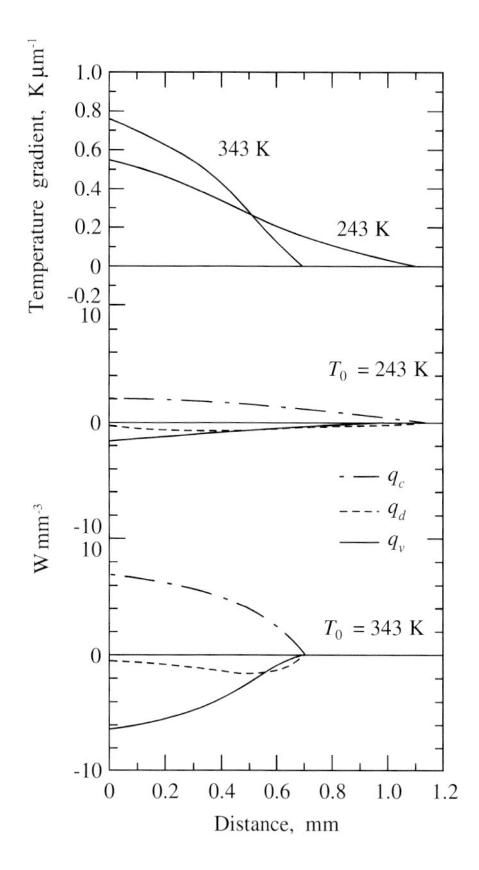

Fig. 5.25 Temperature gradient, conductive heat flux, convective heat flux, and heat flux by chemical reaction as a function of distance from the burning surface at 3 MPa (initial temperatures: 243 K and 343 K) for BAMO/THF = 60:40 copolymer.

Substituting the experimentally derived T_s data into Eqs. (3.75) and (3.76), one obtains:[46]

$$Q_s = 457 \text{ kJ kg}^{-1} \text{ at } T_0 = 243 \text{ K}$$

$$Q_s = 537 \text{ kJ kg}^{-1} \text{ at } T_0 = 343 \text{ K}$$

Substituting the T_s and Q_s data into Eqs. (3.75) and (3.76), the temperature sensitivity of the gas phase, Φ, as defined in Eq. (3.79), and of the solid phase, Ψ, as defined in Eq. (3.80), are determined as 0.0028 K^{-1} and 0.0110 K^{-1}, respectively; Ψ is approximately four times greater than Φ. The computed σ_p represented by the sum of Φ and Ψ is therefore 0.014 K^{-1}, which is approximately equal to the σ_p derived from burning rate experiments. The heat of reaction at the burning surface is the dominant factor on the temperature sensitivity of the burning rate of the BAMO copolymer.

References

1 Arden, E. A., Powling, J., and Smith, W. A. W., Observations on the Burning Rate of Ammonium Perchlorate, *Combustion and Flame*, Vol. 6, No. 1, 1962, pp. 21–33.

2 Levy, J. B., and Friedman, R., Further Studies of Ammonium Perchlorate Deflagration, 8th Symposium (International) on Combustion, The Williams & Wilkins, Baltimore (1962), pp. 663–672.

3 Bircumshaw, L. L., and Newman, B. H., The Thermal Decomposition of Ammonium Perchlorate I, *Proceedings of the Royal Society*, Vol. A227, No. 1168, 1954, pp. 115–132; see also Bircumshaw, L. L., and Newman, B. H., The Thermal Decomposition of Ammonium Perchlorate II, *Proceedings of the Royal Society*, Vol. A227, No. 1169, 1955, pp. 228–241.

4 Jacobs, P. W. M., and Pearson, G. S., Mechanism of the Decomposition of Ammonium Perchlorate, *Combustion and Flame*, Vol. 13, 1969, pp. 419–429.

5 Jacobs, P. W. M., and Whitehead, H. M., Decomposition and Combustion of Ammonium Perchlorate, *Chemical Reviews*, Vol. 69, 1969, pp. 551–590.

6 Jacobs, P. W. M., and Powling, J., The Role of Sublimation in the Combustion of Ammonium Perchlorate Propellants, *Combustion and Flame*, Vol. 13, 1969, pp. 71–81.

7 Hightower, J. D., and Price, E. W., Combustion of Ammonium Perchlorate, 11th Symposium (International) on Combustion, The Combustion Institute, Pittsburgh, PA, 1967, pp. 463–470.

8 Steinz, J. A., Stang, P. L., and Summerfield, M., The Burning Mechanism of Ammonium Perchlorate-Based Composite Solid Propellants, Aerospace and Mechanical Sciences Report No. 830, Princeton University, 1969.

9 Beckstead, M. W., Derr, R. L., and Price, C. F., The Combustion of Solid Monopropellants and Composite Propellants, 13th Symposium (International) on Combustion, The Combustion Institute, Pittsburgh, PA, 1971, pp. 1047–1056.

10 Manelis, G. B., and Strunin, V. A., The Mechanism of Ammonium Perchlorate Burning, *Combustion and Flame*, Vol. 17, 1971, pp. 69–77.

11 Brill, T. B., Brush, P. J., and Patil, D. G., Thermal Decomposition of Energetic Materials 60. Major Reaction Stages of a Simulated Burning Surface of NH_4ClO_4, *Combustion and Flame*, Vol. 94, 1993, pp. 70–76.

12 Mitani, T., and Niioka, T., Double-Flame Structure in AP Combustion, 20th Symposium (International) on Combustion, The Combustion Institute, Pittsburgh, PA (1984), pp. 2043–2049.

13 Guirao, C., and Williams, F. A., A Model for Ammonium Perchlorate Deflagration Between 20 and 100 atm, *AIAA Journal*, Vol. 9, 1971, pp. 1345–1356.

14 Tanaka, M., and Beckstead, M. W., A Three-Phase Combustion Model of Ammonium Perchlorate, AIAA 96-2888, 32nd AIAA Joint Propulsion Conference, AIAA, Reston, VA, 1996.

15 Sarner, S. F., Propellant Chemistry, Reinhold Publishing Corporation, New York (1966).

16 Boggs, T. L., The Thermal Behavior of Cyclotrimethylenetrinitramine (RDX) and Cyclotetramethylenetetranitramine (HMX), Fundamentals of Solid-Propellant Combustion (Eds.: Kuo, K. K., and Summerfield, M.), Progress in Astronautics and Aeronautics, Vol. 90, Chapter 3, AIAA, New York, 1984.

17 Kubota, N., Combustion Mechanism of HMX, *Propellants, Explosives, Pyrotechnics*, Vol. 14, 1989, pp. 6–11.

18 Suryanarayana, B., Graybush, R. J., and Autera, J. R., Thermal Degradation of Secondary Nitramines: A Nitrogen Tracer Study of HMX, *Chemistry and Industry*, Vol. 52, 1967, p. 2177.

19 Kimura, J., and Kubota, N., Thermal Decomposition Process of HMX, *Propellants and Explosives*, Vol. 5, 1980, pp. 1–8.

20 Fifer, R. L., Chemistry of Nitrate Ester and Nitramine Propellants, Fundamentals of Solid-Propellant Combustion (Eds.: Kuo, K. K., and Summerfield, M.), *Progress in Astronautics and Aeronautics*, Vol. 90, Chapter 4, AIAA, New York, 1984.

21 Beal, R. W., and Brill, T. B., Thermal Decomposition of Energetic Materials 77. Behavior of N–N Bridged Bifurazan

Compounds on Slow and Fast Heating, *Propellants, Explosives, Pyrotechnics*, Vol. 25, 2000, pp. 241–246.

22 Beal, R. W., and Brill, T. B., Thermal Decomposition of Energetic Materials 78. Vibrational and Heat of Formation Analysis of Furazans by DFT, *Propellants, Explosives, Pyrotechnics*, Vol. 25, 2000, pp. 247–254.

23 Nedelko, V. V., Chukanov, N. V., Raevskii, A. V., Korsounskii, B. L., Larikova, T. S., Kolesova, O. I., and Volk, F., Comparative Investigation of Thermal Decomposition of Various Modifications of Hexanitrohexaazaisowurtzitane (CL-20), *Propellants, Explosives, Pyrotechnics*, Vol. 25, 2000, pp. 255–259.

24 Häußler, A., Klapötke, T. M., Holl, G., and Kaiser, M., A Combined Experimental and Theoretical Study of HMX (Octogen, Octahydro-1,3,5,7-tetranitro-1,3,5,7-tetrazocine) in the Gas Phase, *Propellants, Explosives, Pyrotechnics*, Vol. 27, 2002, pp. 12–15.

25 Hinshelwood, C. N., The Kinetics of Chemical Change, Oxford University Press, Oxford, 1950.

26 Kubota, N., Hirata, N., and Sakamoto, S., Combustion Mechanism of TAGN, 21 st Symposium (International) on Combustion, The Combustion Institute, Pittsburgh, PA, 1986, pp. 1925–1931.

27 Santhosh, G., Venkatachalam, S., Krishnan, K., Catherine, B. K., and Ninan, K. N., Thermal Decomposition Studies on Advanced Oxidiser: Ammonium Dinitramide, Proceedings of the 3 rd International Conference on High-Energy Materials, Thiruvananthapuram, India, 2000.

28 Varma, M., Chatterjee, A. K., and Pandey, M., Ecofriendly Propellants and Their Combustion Characteristics, Advances in Solid Propellant Technology, 1 st International HEMSI Workshop (Eds.: Varma, M., and Chatterjee, A. K.), Birla Institute of Technology, India, 2002, pp. 144–179.

29 Adams, G. K., and Wiseman, L. A., The Combustion of Double-Base Propellants, Selected Combustion Problems, Butterworth's Scientific Publications, London, 1954, pp. 277–288.

30 Adams, G. K., The Chemistry of Solid Propellant Combustion: Nitrate Esters of Double-Base Systems, Proceedings of the 4 th Symposium on Naval Structural Mechanics, Purdue University, Lafayette, IN (1965), pp. 117–147.

31 Powling, J., and Smith, W. A. W., The Combustion of the Butane-2,3- and 4-Diol Dinitrates and Some Aldehyde-Nitrogen Dioxide Mixtures, *Combustion and Flame*, Vol. 2, No.2, 1958, pp. 157–170.

32 Hewkin, D. J., Hicks, J. A., Powling, J., and Watts, H., The Combustion of Nitric Ester-Based Propellants: Ballistic Modification by Lead Compounds, *Combustion Science and Technology*, Vol. 2, 1971, pp. 307–327.

33 Robertson, A. D., and Napper, S. S., The Evolution of Nitrogen Peroxide in the Decomposition of Gun Cotton, *Journal of the Chemical Society*, Vol. 91, 1907, pp. 764–786.

34 Pollard, F. H., and Wyatt, P. M. H., Reactions Between Formaldehyde and Nitrogen Dioxide; Part III, The Determination of Flame Speeds, *Transactions of the Faraday Society*, Vol. 46, No. 328, 1950, pp. 281–289.

35 McDowell, C. A., and Thomas, J. H., Oxidation of Aldehydes in the Gaseous Phase; Part IV. The Mechanism of the Inhibition of the Gaseous Phase Oxidation of Acetaldehyde by Nitrogen Peroxide, *Transactions of the Faraday Society*, Vol. 46, No. 336, 1950, pp. 1030–1039.

36 Sawyer, R. F., and Glassman, I., The Reactions of Hydrogen with Nitrogen Dioxide, Oxygen, and Mixtures of Oxygen and Nitric Oxide, 12 th Symposium (International) on Combustion, The Combustion Institute, Pittsburgh, PA, 1969, pp. 469–479.

37 Cummings, G. A. McD., Effect of Pressure on Burning Velocity of Nitric Oxide Flames, *Nature*, No. 4619, May 1958, p. 1327.

38 Strauss, W. A., and Edse, R., Burning Velocity Measurements by the Constant Pressure Bomb Method, 7 th Symposium (International) on Combustion, The Combustion Institute, Pittsburgh, 1958, pp. 377–385.

39 Sawyer, R. F., The Homogeneous Gas-Phase Kinetics of Reactions in Hydrazine-Nitrogen Tetraoxide Propellant System, Ph.D. Thesis, Princeton University, 1965.

40 Heath, G. A., and Hirst, R., Some Characteristics of the High-Pressure Combustion of Double-Base Propellants, 8th Symposium (International) on Combustion, Williams & Wilkins Co., Baltimore, 1962, pp. 711–720.
41 Penner, S. S., Chemistry Problems in Jet Propulsion, Pergamon Press, New York, 1957.
42 Hinshelwood, C. N., and Green, T. E., The Interaction of Nitric Oxide and Hydrogen and the Molecular Statistics of Termolecular Gaseous Reactions, *J. Chem. Soc.*, 1926, pp. 730–739.
43 Pannetier, G., and Souchay, P., Chemical Kinetics, Elsevier Publishing Co., New York, 1967.
44 Kubota, N., and Sonobe, T., Combustion Mechanism of Azide Polymer, *Propellants, Explosives, Pyrotechnics*, Vol. 13, 1988, pp. 172–177.
45 Miyazaki, T., and Kubota, N., Energetics of BAMO, *Propellants, Explosives, Pyrotechnics*, Vol. 17, 1992, pp. 5–9.
46 Kubota, N., Combustion of Energetic Azide Polymers, *Journal of Propulsion and Power*, Vol. 11, No. 4, 1995, pp. 677–682.
47 Kubota, N., Propellant Chemistry, *Journal of Pyrotechnics*, 11, 2000, pp. 25–45.

6
Combustion of Double-Base Propellants

6.1
Combustion of NC-NG Propellants

6.1.1
Burning Rate Characteristics

Since the discovery of double-base propellants, also referred to as smokeless powders, numerous investigators have attempted to improve or control their burning characteristics. In the early twentieth century, it was established that the burning rates of many double-base propellants obey a relationship known as Vieille's Law, represented by Eq. (3.68), in which the burning rate is proportional to the pressure raised to a power n, known as the pressure exponent of burning rate. Accordingly, extensive experimental and theoretical studies have been carried out on the combustion of double-base propellants and a number of combustion models have been proposed which describe the combustion wave structure and burning rate characteristics. Some of these results are summarized in Refs. [1–6]. The combustion model was first presented by Crawford, Huggett, and McBrady,[7] and theoretical models were proposed by Rice and Ginell[8] and by Parr and Crawford.[9] The models describe the fundamental process of burning and the rate-determining domains. Photographic observations of flame structure were first made by Heller and Gordon,[10] and then by Kubota,[11] Eisenreich,[12] and Aoki and Kubota.[13]

In the burning process, some oxidizer from the nitrate group, released by thermal decomposition, reacts with the other molecular decomposition products to produce heat. In examining the details of this process, one seeks to understand how they translate into the global characteristics of double-base propellants, such as the pressure dependence of the burning rate. Typical burning rate versus pressure plots of double-base propellants are shown in Fig. 6.1, while in Fig. 6.2 burning rate is plotted against heat of explosion, H_{exp}. The burning rate of each propellant increases linearly with increasing pressure in a log burning rate versus log pressure plot. Though the burning rate increases as the energy density (H_{exp}) contained within the propellant increases at the same pressure, the pressure exponent of burning rate, n, is 0.64 and appears to be independent of H_{exp} (in the range 3.47–4.95 MJ kg^{-1}).

Propellants and Explosives. Naminosuke Kubota

ISBN: 978-3-527-31424-9

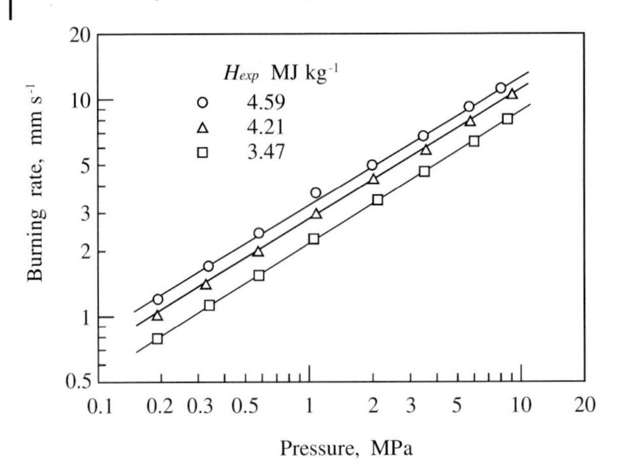

Fig. 6.1 Burning rate increases with increasing energy density of NC-NG double-base propellants at constant pressure; the pressure exponent remains unchanged when the energy density is changed.

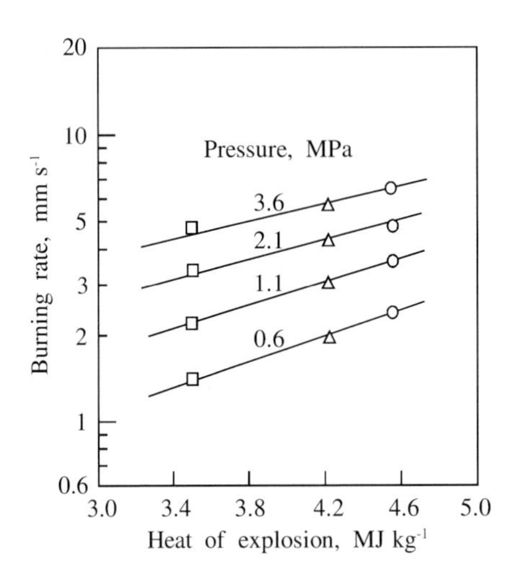

Fig. 6.2 The burning rates of NC-NG double-base propellants increase as the heat of explosion increases at constant pressure.

6.1.2
Combustion Wave Structure

The combustion wave of a double-base propellant consists of the following five successive zones, as shown in Fig. 6.3: (I) heat conduction zone, (II) solid-phase reaction zone, (III) fizz zone, (IV) dark zone, and (V) flame zone.[7,10,13–15]

(I) Heat conduction zone: Though there is a thermal effect due to heat conduction from the burning surface, no chemical changes occur. The temperature in-

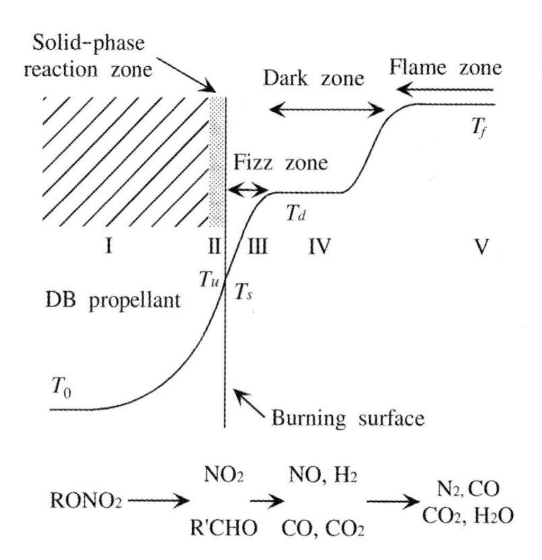

Fig. 6.3 Combustion wave structure of a double-base propellant.

creases from the initial propellant temperature, T_0, to the onset temperature of the solid-phase reaction, T_u.

(II) Solid-phase reaction zone: Nitrogen dioxide and aldehydes are produced in the thermal degradation process. This reaction process occurs endothermically in the solid phase and/or at the burning surface. The interface between the solid phase and the burning surface is composed of a solid/gas and/or solid/liquid/gas thin layer. The nitrogen dioxide fraction exothermically oxidizes the aldehydes at the interface layer. Thus, the overall reaction in the solid-phase reaction zone appears to be exothermic. The thickness of the solid-phase reaction zone is very small, and so the temperature is approximately equal to the burning surface temperature, T_s.

(III) Fizz zone: The major fractions of nitrogen dioxide and the aldehydes and other C,H,O and HC species react to produce nitric oxide, carbon monoxide, water, hydrogen, and carbonaceous materials. This reaction process occurs very rapidly in the early stages of the gas-phase reaction zone, just above the burning surface.

(IV) Dark zone: In this zone, oxidation reactions of the products formed in the fizz-zone reaction take place. Nitric oxide, carbon monoxide, hydrogen, and carbonaceous fragments react to produce nitrogen, carbon dioxide, water, etc. These exothermic reactions occur only very slowly unless the temperature and/or pressure is sufficiently high.

(V) Flame zone: When the dark-zone reactions occur rapidly after an induction period, they produce a flame zone in which the final combustion products are formed and attain a state of thermal equilibrium. When the pressure is low, below about 1 MPa, no flame zone is produced because the reduction of nitric oxide is too slow to produce nitrogen.

The solid-phase reaction zone is also termed the "subsurface reaction zone" or "condensed-phase reaction zone". As the dark zone reaction represents an induction zone ahead of the flame zone, the dark zone is also termed the "preparation zone" when it produces a luminous flame. Since the flame zone is luminous, it is also termed the "luminous flame zone".

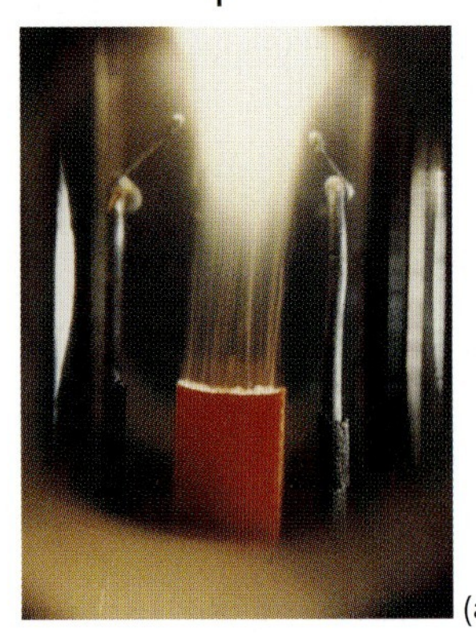
(a)

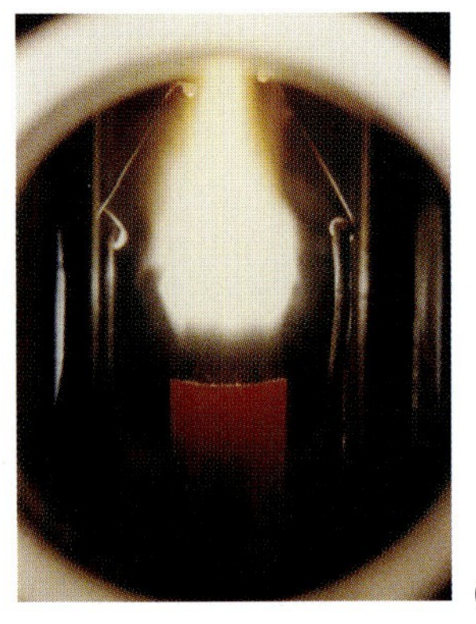
(b)

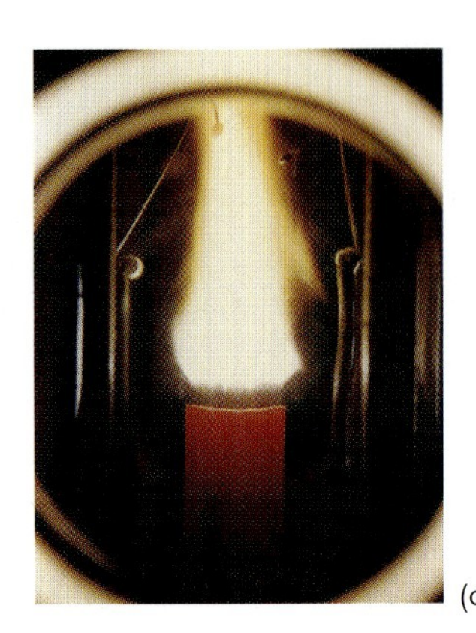
(c)

Fig. 6.4 Typical flame photographs of an NC-NG double-base propellant.

	p (MPa)	*r* (mm s^{-1})
(a)	1.0	2.2
(b)	2.0	3.1
(c)	3.0	4.0

Photographic observations of flame structures are useful to understand the overall characteristics of double-base propellant combustion. Fig. 6.4 shows flame photographs of a double-base propellant at three different pressures. The propellant strand is burned in a chimney-type strand burner pressurized with nitrogen gas. The dark zone is clearly discernible between the burning surface and the luminous flame zone. The dark zone length, L_d, defined as the distance between the burning surface and the luminous flame front (the fizz zone length, L_z, is included in L_d since $L_d >> L_z$), decreases with increasing pressure, i. e., the luminous flame approaches the burning surface as the pressure is increased.

The thermal structure of the combustion wave of a double-base propellant is revealed by its temperature profile trace. In the solid-phase reaction zone, the temperature increases rapidly from the initial temperature in the heat conduction zone, T_0, to the onset temperature of the solid-phase reaction, T_u, which is just below the burning surface temperature, T_s. The temperature continues to increase rapidly from T_s to the temperature at the end of the fizz zone, T_d, which is equal to the temperature at the beginning of the dark zone. In the dark zone, the temperature increases relatively slowly and the thickness of the dark zone is much greater than that of the solid-phase reaction zone or the fizz zone. Between the dark zone and the flame zone, the temperature increases rapidly once more and reaches the maximum flame temperature in the flame zone, i. e., the adiabatic flame temperature, T_g.

Thus, the combustion wave structure of double-base propellants appears to show a two-stage gas-phase reaction, taking place in the fizz zone and the dark zone. The thickness of the fizz zone is actually dependent on the chemical kinetics of the gaseous species evolved at the burning surface, which, in turn, is dependent on the pressure. The thickness decreases with increased pressure, resulting in an increased temperature gradient. Therefore, the rate of heat input by conduction from the gas phase to the solid phase increases with increasing pressure. The length of the dark zone, i. e., the flame stand-off distance, also decreases as pressure increases. Consequently, the luminous flame front approaches the burning surface, as shown in Fig. 6.4. As the lengths of both the fizz zone and the dark zone are decreased by an increase in pressure, there is a concomitant increase in burning rate.[10,13]

According to the results of numerous gas-composition analyses[7,10] of the combustion products of double-base propellants, only trace amounts of nitric oxide (NO) (~ 1 %) are found in the final combustion products at high pressures, almost all being consumed in the final luminous flame. In the dark zone, relatively large amounts of NO are found at both low pressures (~ 30 %) and high pressures (~ 20 %).[7,10] The main oxidation reactions producing heat in the dark zone involve NO.

The probable set of chemical reactions taking place in the dark zone has been analyzed theoretically by Sotter.[16] Sixteen reversible and four irreversible chemical reactions involving twelve chemical species were considered. The following reactions were taken to be the most important:

$$H + H_2O \rightarrow H_2 + OH$$

$$CO + OH \rightarrow CO_2 + H$$

$$2NO + H_2 \rightarrow 2HNO$$

$$2HNO + H_2 \rightarrow 2H_2O + N_2 \text{ (fast)}$$

The third chemical equation, involving nitric oxide, represents a termolecular reaction. Therefore, the overall order of the reaction is expected to exceed that of the second-order reaction generally assumed in the pre-mixed gas burning model. The high exothermicity accompanying the reduction of NO to N_2 is responsible for the appearance of the luminous flame in the combustion of a double-base propellant, and hence the flame disappears when insufficient heat is produced in this way, i. e., during fizz burning.

In the dark zone, the temperature increases relatively slowly and so for the most part the temperature gradient is much less steep than that in the fizz zone. However, the temperature increases rapidly at about 50 μm from where the flame reaction starts to produce the luminous flame zone. The gas flow velocity increases with increasing distance due to the increase in temperature. The mole fractions of NO, CO, and H_2 decrease and those of N_2, CO_2, and H_2O increase with increasing distance in the dark zone. The results imply that the overall reaction in the dark zone is highly exothermic and that the order of reaction is higher than second order because of the reduction reaction involving NO. The derivative of temperature with respect to time t in the dark zone is expressed empirically by the formula[16]

$$dT/dt = c\rho_g^{1.9} \exp(-E_g/RT) \quad (6.1)$$

where c is a constant, ρ_g is the density, T is the temperature, and E_g is the activation energy, which is in the range 100–150 kJ mol^{-1}.

If the luminous flame spreading speed is lower than the efflux velocity of the product gases (at the edge of the dark zone), blow-off of the luminous flame occurs and no such flame is observed. The efflux speed of the product gases is proportional to the burning rate, which is essentially independent of the luminous flame. According to laminar flame-speed theory, this flame speed, S_L, is proportional to $p^{(m/2)-1}$, where m is the order of the chemical reaction. On the other hand, the efflux velocity, u_g, of the product gases in the dark zone is proportional to p^{n-1}, where n is the pressure exponent of the burning rate. The stand-off distance of the luminous flame, i. e., the dark zone length, is represented by Eq. (3.7). Based on the empirically obtained Eq. (6.1) and the theoretical analysis described in Eqs. (3.65)–(3.70 a), the temperature gradient in the dark zone is represented by $dT/dx \propto \rho_g^{2.9}/r$. The dark zone length is then given by $L_d \propto p^{n-2.9}$. This result indicates that the overall order of chemical reaction, m, as defined in Eq. (3.70), is given by $m = n - d = 2.9$ for the gas-phase reaction. The reaction in the dark zone is more sensitive to pressure than other gas-phase reactions observed in conventional pre-mixed flames.

6.1.3
Burning Rate Model

6.1.3.1 Model for Heat Feedback from the Gas Phase to the Condensed Phase

The heat flux transferred back from the gas phase to the condensed phase is obtained by integrating Eq. (3.64) with the boundary condition that heat flux at infinity is zero. Using the results of the temperature measurements shown in Table 6.1,[11] the term in the exponent in Eq. (3.47), $\exp(-x_g/\delta_g) = \exp(-\rho_p r c_g x_g/\lambda_g)$, is evaluated as follows: $c_g = 1.55$ kJ kg^{-1} K^{-1}, $\rho_p = 1.54 \times 10^3$ kg m^{-3}, $\lambda_g = 0.084$ W m^{-1} K^{-1}.

Table 6.1 Measured parameter values of the flame stand-off distance in the fizz zone.

***p* (MPa)**		**0.12**	**0.8**	**2.0**	**10.0**
r	$\times 10^{-3}$ m s^{-1}	0.46	1.58	8.21	10.5
x_g	$\times 10^{-6}$ m	200	170	120	80
δ_g	$\times 10^{-6}$ m	76	22	4.3	3.3
x_g/δ_g		2.6	7.6	28	24

The term in the exponent is quite large above 0.1 MPa so that Eq. (3.47) is represented by Eq. (3.48) for double-base propellants.

6.1.3.2 **Burning Rate Calculated by a Simplified Gas-Phase Model**

The burning rate of a double-base propellant can be calculated by means of Eq. (3.59), assuming the radiation from the gas phase to the burning surface to be negligible. Since the burning rate of a double-base propellant is dominated by the heat flux transferred back from the fizz zone to the burning surface, the reaction rate parameters in Eq. (3.59) are the physicochemical parameters of choice for describing the fizz zone. The gas-phase temperature, T_g, is the temperature at the end of the fizz zone, i. e., the dark-zone temperature, as obtained by means of Eq. (3.60). Since the burning surface temperature, T_s, is dependent on the regression rate of the propellant, it should be determined by the decomposition mechanism of the double-base propellant. An Arrhenius-type pyrolysis law represented by Eq. (3.61) is used to determine the relationship between the burning rate and the burning surface temperature.

A calculated example of burning rate versus pressure for a double-base propellant is shown in Fig. 6.5. The physical and chemical parameters used in this calculation are shown in Table 6.2. The burning rate is seen to increase linearly with increasing pressure in an ln r versus ln p plot. The pressure exponent of burning rate defined in Eq. (3.68) is approximately 0.85, and the temperature sensitivity of burning rate defined in Eq. (3.73) decreases with increasing pressure. The burning surface temperature increases with increasing pressure. These calculated results have been confirmed by experimentally determined values of burning rate, pressure exponent, temperature sensitivity, and burning surface temperature. Though the burning rate model is a very simplified one, it provides a fundamental basis for understanding the combustion mechanisms of double-base propellants.

Table 6.2 Physicochemical parameters used in a burning rate model for a double-base propellant.

c_g	1.55 kJ kg^{-1} K^{-1}
c_p	1.55 kJ kg^{-1} K^{-1}
λ_g	0.084 W m^{-1} K^{-1}
ϱ_p	1540 kg m^{-3}
E_g	71 MJ $kmol^{-1}$
$\varepsilon_g^2 Z_g$	2.3×10^{12} m^3 kg^{-1} s^{-1}
Q_g	1.40 MJ kg^{-1}
R_g	0.29 kJ $kmol^{-1}$ K^{-1}
Q_s	340 kJ kg^{-1}
Z_s	5000 m s^{-1}
E_s	71 MJ $kmol^{-1}$

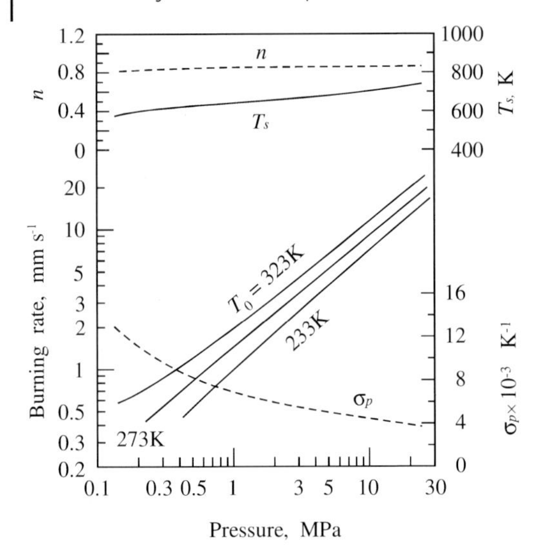

Fig. 6.5 Calculated burning rate characteristics of a double-base propellant.

6.1.4
Energetics of the Gas Phase and Burning Rate

Since the energy contained in unit mass of a double-base propellant is altered by changing the mass fraction of NO_2, propellants with varying energy content can be formulated by way of three methods:[13] (1) changing the concentration of plasticizer at a constant NC/NG mixture ratio, (2) changing the NC/NG mixture ratio, and (3) changing the concentration of nitrate groups in the NC used. Table 6–3 shows the chemical compositions and properties of NC-NG propellants made by method (1). The NC/NG mixture ratio is fixed at 1.307 for the propellants listed. The energy content is changed by the addition of DEP. The mass fraction of NO_2, $\xi(NO_2)$, and the mass fraction of NO, $\xi(NO)$, contained within the propellants range from 0.466 to 0.403 and from 0.304 to 0.263, respectively. The adiabatic flame temperature, T_g, ranges from 2760 K to 1880 K.

Table 6.3 Propellant compositions (% by mass) and physico-chemical properties.

NC	NG	DEP	2NDPA	$\xi(NO_2)$	$\xi(NO)$	T_g (K)	H_{exp} (MJ kg^{-1})
53.0	40.5	4.0	2.5	0.466	0.304	2760	4.36
51.3	39.3	7.0	2.4	0.452	0.295	2560	4.22
50.2	38.4	9.0	2.4	0.442	0.288	2420	3.95
48.0	36.7	13.0	2.3	0.422	0.275	2150	3.49
45.8	35.0	17.0	2.2	0.403	0.263	1880	2.98

Fig. 6.6 shows burning rate characteristics as a function of $\xi(NO_2)$. The burning rates are seen to increase linearly with increasing pressure in ln r versus ln p plots

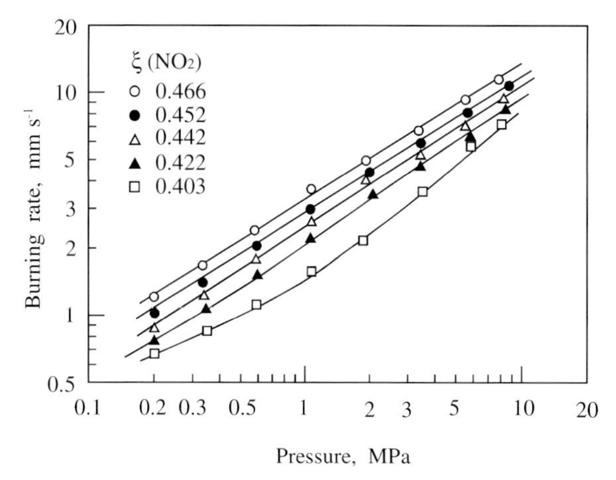

Fig. 6.6 Burning rates of double-base propellants increase with increasing pressure at constant mass fraction of NO_2, $\xi(NO_2)$.

and also increase with increasing $\xi(NO_2)$. The effect on the burning rate of the addition of the plasticizer, DEP, can clearly be seen; the burning rate decreases with decreasing energy content (energy density) at constant pressure. The pressure exponent remains unchanged at n = 0.62, even when the plasticizer content is changed, except in the case of the propellant designated by H_{exp} = 2.98 MJ kg^{-1}, the pressure exponent for which is 0.45 in the low-pressure region below 1.8 MPa and 0.78 in the high-pressure region above 1.8 MPa.

Since the energy density of double-base propellants is directly correlated with $\xi(NO_2)$, the burning rates shown in Fig. 6.6 have been plotted as a function of $\xi(NO_2)$ at different pressures.[13] As shown in Fig. 6.7, the burning rate increases

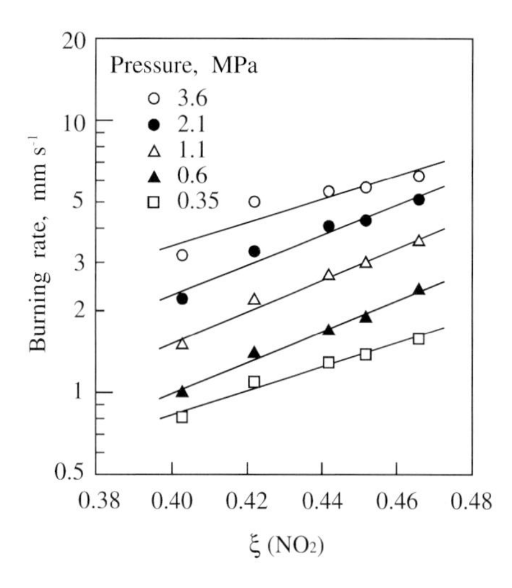

Fig. 6.7 Burning rates of double-base propellants increase with increasing $\xi(NO_2)$ at constant pressure.

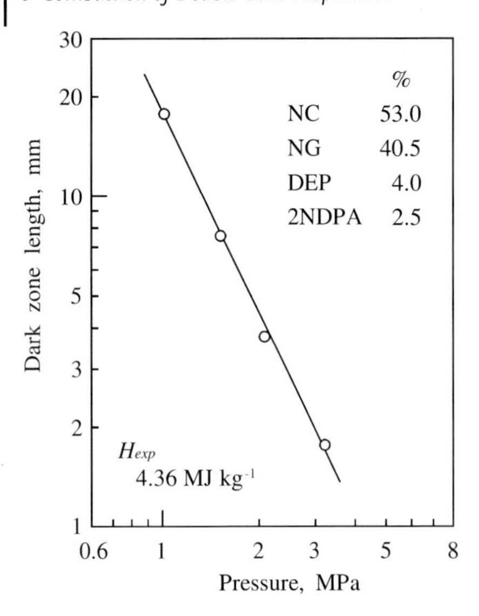

Fig. 6.8 Dark zone length (flame stand-off distance) decreases with increasing pressure.

linearly as $\xi(NO_2)$ increases in an ln r versus $\xi(NO_2)$ plot. Combining the results shown in Figs. 6.5 and 6.6, one obtains a burning rate expression represented by

$$r = 0.62 \exp\{10.0\xi(NO_2)\}\, p^{0.62} \tag{6.2}$$

for the propellants listed in Table 6.3.

Since the final gas-phase reaction to produce a luminous flame zone is initiated by the reaction in the dark zone, the reaction time is determined by the dark zone length, L_d, i. e., the flame stand-off distance. Fig. 6.8 and 6.9 show data for the dark zone length and the dark zone temperature, T_d, respectively, for the propellants listed in Table 6.3. The luminous flame front approaches the burning surface and

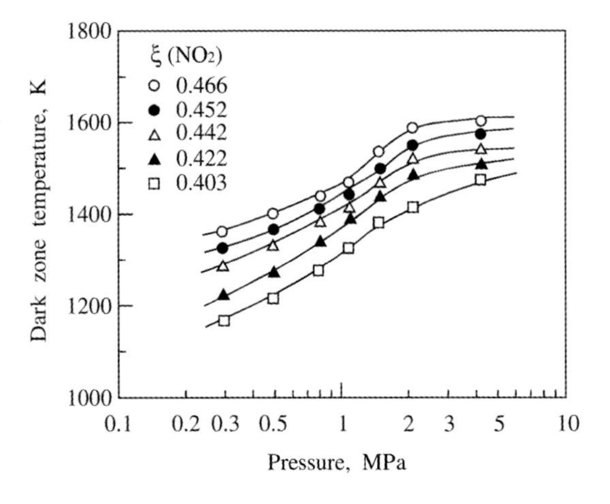

Fig. 6.9 Dark zone temperature (temperature at the end of the fizz zone) increases with increasing pressure as well as with increasing $\xi(NO_2)$ at constant pressure.

the dark zone length decreases with increasing pressure. No clear difference in dark zone length is seen between these propellants and the pressure exponent of the dark zone, $d = n - m$, as defined in Eq. (5.4), is determined as approximately –2.0. The overall order of the reaction in the dark zone is determined as $m = 2.6$ for all of the propellants. However, the dark zone temperature increases with increasing pressure at constant $\xi(NO_2)$ and also increases with increasing $\xi(NO_2)$ at constant pressure.

The reaction time to produce the luminous flame, τ_d, is given by

$$\tau_d = L_d / u_d \tag{6.3}$$

where L_d is the dark zone length, and u_d is the reactive gas velocity in the dark zone. The dark zone gas velocity is obtained from the mass continuity relationship

$$u_d = (\rho_p/\rho_d)\ r \tag{6.4}$$

where ρ_d is the density of the dark zone. Using the equation of state, Eq. (1.5), and Eqs. (6.3) and (6.4), τ_d is obtained according to

$$\tau_d = pL_d / rR_d\, T_d \tag{6.5}$$

where R_d is the gas constant in the dark zone. Since the dark zone reaction producing the luminous flame is dependent on the reaction involving NO, not NO_2, the reaction time in the dark zone is plotted as a function of $\xi(NO)$ at different pressures. As is clearly shown in Fig. 6.10, τ_d decreases rapidly with increasing $\xi(NO)$ at constant pressure, and also decreases with increasing pressure at constant $\xi(NO)$. Fig. 6.11 shows the relationship between the dark zone temperature and the reaction time in the dark zone, plotted as $\ln(1/\tau_d)$ versus $1/T_d$ at different values of $\xi(NO_2)$. The activation energy of the dark zone reaction is determined as 34 ± 2 kJ mol^{-1} from the slope of the plotted lines.

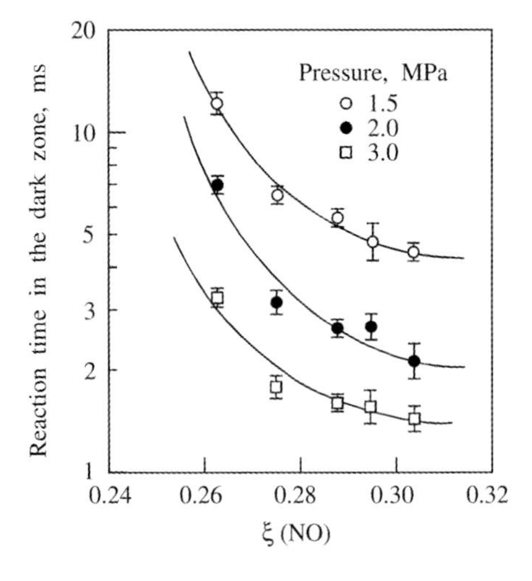

Fig. 6.10 Reaction time in the dark zone decreases with increasing $\xi(NO)$ at constant pressure.

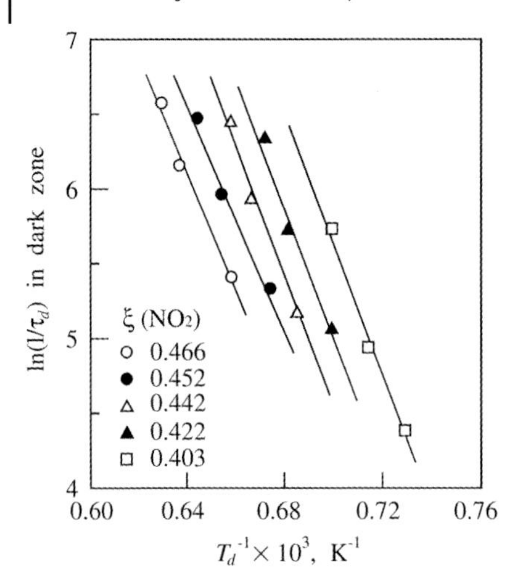

Fig. 6.11 Determination of the activation energy in the dark zone at different energy densities of double-base propellants.

The rate of temperature increase in the fizz zone, $(dT/dt)_{f,s}$, indicates the heating rate due to the exothermic reaction in the fizz zone. As shown in Fig. 6.12, the heating rate increases linearly in a plot of ln $(dT/dt)_{f,s}$ versus $\xi(NO_2)$ at 2.0 MPa. The reaction time in the fizz zone, τ_f, can be obtained by means of a similar relationship to Eq. (6.5), adapted to the fizz zone reaction. Fig. 6.13 shows τ_f versus $\xi(NO_2)$ at 2.0 MPa. The reaction time decreases linearly with increasing $\xi(NO_2)$ in a plot of ln τ_f versus $\xi(NO_2)$ and is represented by

$$\tau_f = 2.36 \exp \{-22.0\, \xi(NO_2)\} \tag{6.6}$$

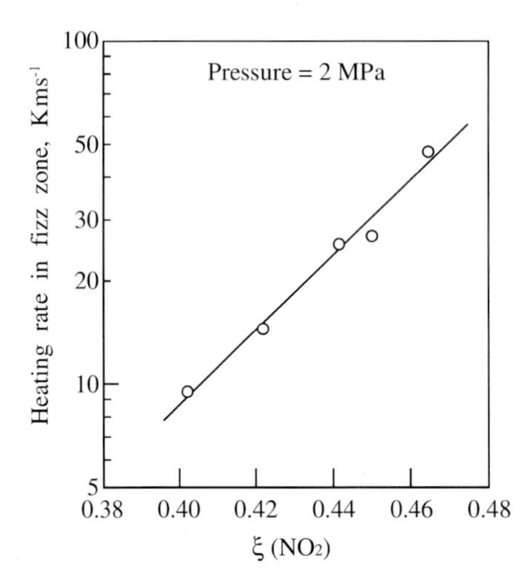

Fig. 6.12 Heating rate in the fizz zone increases with increasing $\xi(NO_2)$.

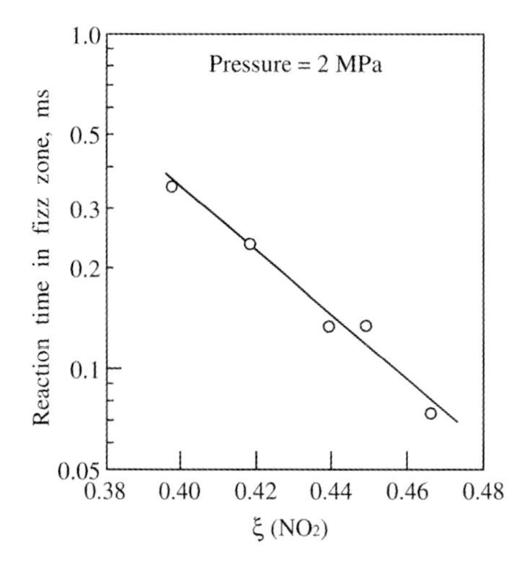

Fig. 6.13 Reaction time in the fizz zone decreases with increasing $\xi(NO_2)$.

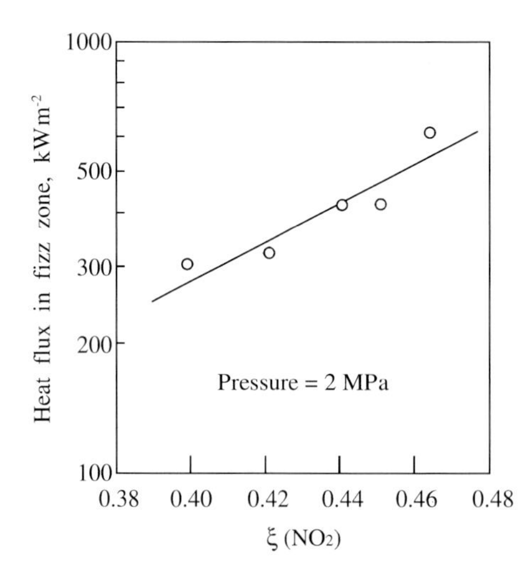

Fig. 6.14 Heat flux in the fizz zone increases with increasing $\xi(NO_2)$.

The heat flux feedback from the fizz zone to the burning surface, $(\lambda_g dT/dx)_{f,s}$, can also be computed from the temperature data in the fizz zone. Fig. 6.13 shows $(\lambda_g dT/dx)_{f,s}$ (kW m^{-2}) as a function of $\xi(NO_2)$ at 2.0 MPa, as represented by

$$(\lambda_g dT/dx)_{f,s} = 4.83 \exp\{10.2\, \xi(NO_2)\} \tag{6.7}$$

The heat flux feedback from the gas phase to the condensed phase increases with increasing $\xi(NO_2)$, and hence the burning rate increases with increasing energy density of double-base propellants, as shown in Figs. 6.1 and 6.2.

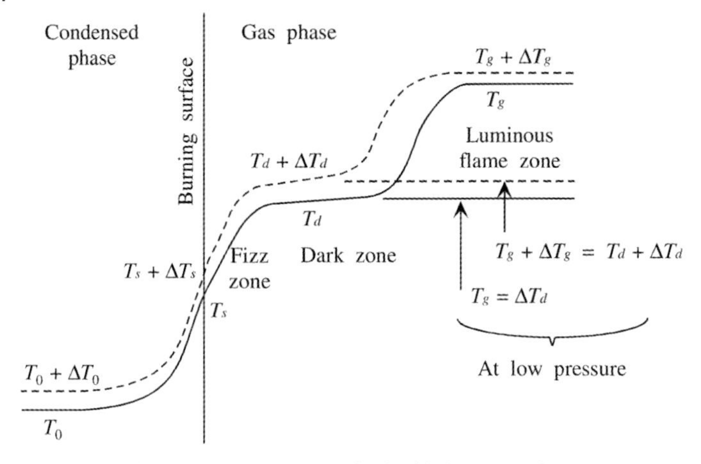

Fig. 6.15 Combustion wave structure of a double-base propellant at different initial propellant temperatures and at high and low pressures.

6.1.5
Temperature Sensitivity of Burning Rate

The temperature profile in the combustion wave of a double-base propellant is altered when the initial propellant temperature T_0 is increased to $T_0 + \Delta T_0$, as shown in Fig. 6.15. The burning surface temperature T_s is increased to $T_s + \Delta T_s$, and the temperatures of the succeeding gas-phase zones are likewise increased, that of the dark zone from T_g to $T_g + \Delta T_g$, and the final flame temperature from T_f to $T_f + \Delta T_f$. If the burning pressure is low, below about 1 MPa, no luminous flame is formed above the dark zone. The final flame temperature is T_g at low pressures. The burning rate is determined by the heat flux transferred back from the fizz zone to the burning surface and the heat flux produced at the burning surface. The analysis of the temperature sensitivity of double-base propellants described in Section 3.5.4 applies here.

Fig. 6.16 shows the burning rates and temperature sensitivities of two types of double-base propellants:[17] high-energy and low-energy propellants. Their chemical compositions are shown in Table 6.4. Since DEP is a low-energy material, it is added to the high-energy propellant to formulate the low-energy propellant. The

Table 6.4 Chemical compositions (% by mass) and adiabatic flame temperatures of high- and low-energy propellants.

Propellant	NC (12.2% N)	NG	DEP	T_f (K)
High energy	55.6	40.4	4.0	2720
Low energy	50.4	36.6	13.0	2110

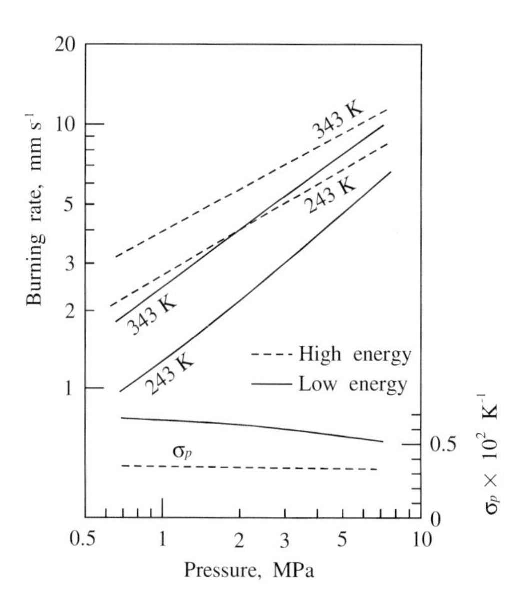

Fig. 6.16 Burning rates and temperature sensitivities of high- and low-energy double-base propellants.

burning rates of the high- and low-energy propellants yield approximately straight lines in an ln r versus ln p plot. The pressure exponents are 0.58 and 0.87 for the high- and low-energy propellants, respectively. When the initial propellant temperature T_0 is increased from 243 K to 343 K, the burning rate is increased for both propellants. The temperature sensitivities σp are 0.0034 K^{-1} for the high-energy and 0.0062 K^{-1} for the low-energy propellant at 2 MPa.

The dark zone temperature, T_d, increases with increasing pressure at constant T_0 and also increases with increasing T_0 at constant pressure for both propellants, as shown in Fig. 6.17. Though the burning surface temperature, T_s, increases with increasing T_0, the heat of reaction at the burning surface, Q_s, remains constant for both propellants. Furthermore, the heat of reaction is approximately the same for the high- and low-energy propellants. The temperature gradient in the fizz zone increases with increasing T_0 at constant pressure and also increases for both propellants. However, the temperature gradient for the high-energy propellant is approximately 50 % higher than that for the low-energy propellant. The higher temperature gradient implies that the heat flux transferred back from the fizz zone to the burning surface is higher and hence the burning rate becomes higher for the high-energy propellant than for the low-energy propellant. The activation energy in the fizz zone is 109 kJ mol^{-1} for the high-energy propellant and 193 kJ mol^{-1} for the low-energy propellant.

Substituting the data for temperature, activation energy in the fizz zone, and burning rate shown in Fig. 6.16 into Eqs. (3.86), (3.88), and (3.80), the temperature sensitivity of the gas phase, Φ, defined in Eq. (3.79), and the temperature sensitivity of the condensed phase, Ψ, defined in Eq. (3.80), are obtained as[17]

$$\sigma_p = \Phi + \Psi$$

$3.7 = 2.1 + 1.6 \times 10^{-3}\ K^{-1}$ for the high-energy propellant

$6.4 = 4.1 + 2.3 \times 10^{-3}\ K^{-1}$ for the low-energy propellant

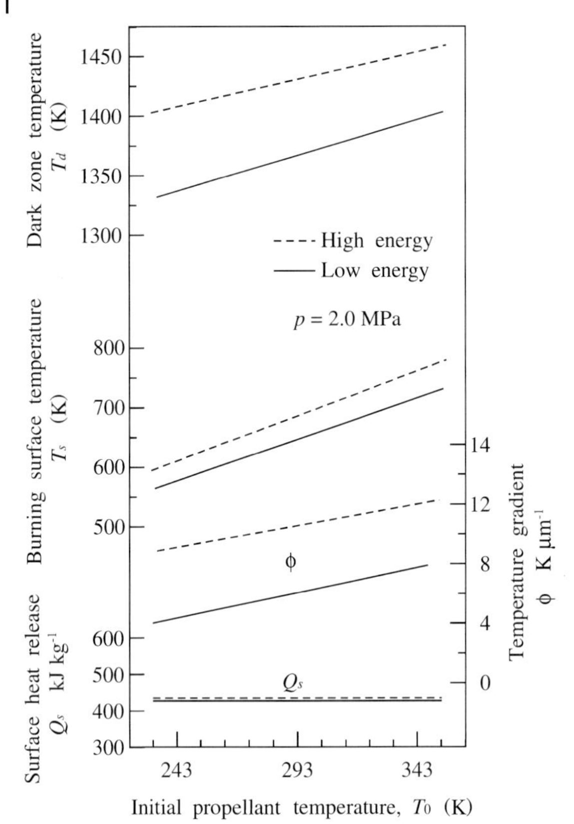

Fig. 6.17 Dark zone temperature, burning surface temperature, surface heat release, and temperature gradient in the fizz zone for high- and low-energy double-base propellants as a function of initial propellant temperature.

The temperature sensitivity of burning rate thus comprises 60% Φ and 40% Ψ. The lower σ_p of the high-energy propellant is due to the lower activation energy and the higher temperature in the fizz zone as compared to the low-energy propellant.

6.2 Combustion of NC-TMETN Propellants

6.2.1 Burning Rate Characteristics

Fig. 6.18 shows a typical comparative example of the burning rates of two propellants composed of NC-TMETN and NC-NG. The chemical compositions (% by mass) and thermochemical properties are shown in Table 6.5. The energy densities of these two propellants are approximately equivalent.

The burning rate of the NC-NG propellant is higher than that of the NC-TMETN propellant in the pressure range between 0.1 MPa and 10 MPa. However, the pres-

Table 6.5 Chemical compositions (% by mass) and thermochemical properties of NC-NG and NC-TMETN propellants.

	NC-NG	NC-TMETN
NC	39.6	53.8
NG	49.4	
TMETN		39.1
DBP	10.0	
TEGDN		7.0
DPA	1.0	
EC		0.1
T_g (K)	2690	2570
M_g (kg kmol^{-1})	24.6	23.5
I_{sp} (s)	242	240
Combustion products (mole fraction)		
CO	0.397	0.398
CO_2	0.124	0.104
H_2	0.115	0.143
H_2O	0.238	0.236
N_2	0.124	0.118
H	0.002	0.001

sure exponent appears to be $n = 0.74$ for both propellants. The fundamental burning rate characteristics of NC-TMETN propellants are equivalent to those of NC-NG propellants.[11]

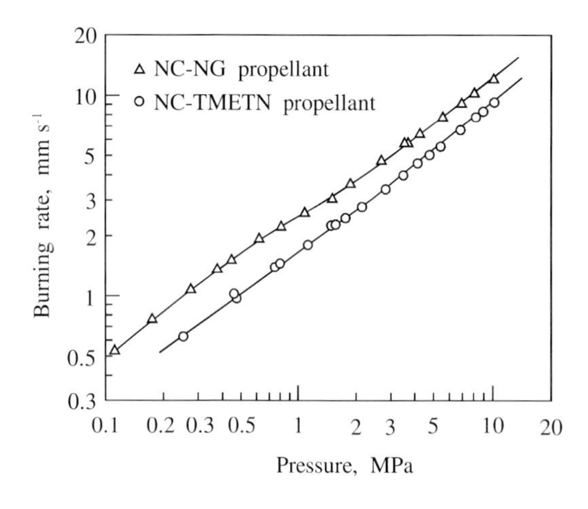

Fig. 6.18 Burning rates of NC-NG and NC-TMETN double-base propellants as a function of pressure.

6.2.2 Combustion Wave Structure

Since TMETN and TEGDN are both nitrate esters and liquid at room temperature, their fundamental thermochemical properties are equivalent to those of NG. The burning rates and the pressure exponents of NC-NG and NC-TMETN propellants appear to be approximately equivalent. The burning process and the combustion wave structure for NC-TMETM also appear to be equivalent to those for NC-NG. The flame structure of NC-TMETN consists of a two-stage reaction zone: fizz zone, dark zone, and luminous flame zone, as shown in Fig. 6.3. The major oxidizer fragment in the fizz zone is NO_2 and there is a rapid temperature rise due to the exothermic reduction of NO_2. In the dark zone, there is a slow oxidation by NO, and this generates the flame zone at some distance from the burning surface. Accordingly, the burning rate is dependent on the heat flux transferred back from the fizz zone to the burning surface and the heat flux generated at the burning surface. The temperature sensitivity of burning rate and the action of catalysts on the combustion zone are equivalent to those for NC-NG propellants.[11]

6.3 Combustion of Nitro-Azide Propellants

6.3.1 Burning Rate Characteristics

The burning rates of a nitro-azide propellant composed of NC, NG, and GAP are shown in Fig. 6.19. For comparison, the burning rates of a double-base propellant composed of NC, NG, and DEP are shown in Fig. 6.20. The chemical compositions of both propellants are shown in Table 6.6. The adiabatic flame temperature is increased from 2560 K to 2960 K and the specific impulse is increased from 237 s to 253 s when 12.5 % of DEP is replaced with the same amount of GAP.

Though both propellants contain equal amounts of NC and NG, the burning rate of NC-NG-GAP is approximately 70 % higher than that of NC-NG-DEP at T_0 = 293 K. The pressure exponent of burning rate remains relatively unchanged at n = 0.7 by the replacement of DEP with GAP. However, the temperature sensitivity of burning rate defined in Eq. (3.73) is increased significantly from 0.0038 K^{-1} to 0.0083 K^{-1}.

6.3.2 Combustion Wave Structure

The combustion wave of an NC-NG-GAP propellant consists of successive two-stage reaction zones.[18] The first gas-phase reaction occurs at the burning surface and the temperature increases rapidly in the fizz zone. The second zone is the dark zone, which separates the luminous flame zone from the burning surface. Thus, the luminous flame stands some distance above the burning surface. This structure

Table 6.6 Chemical compositions (% by mass) and physico-chemical properties (at 10 MPa) of NC-NG-DEP and NC-NG-GAP propellants.

	NC-NG-DEP	NC-NG-GAP
NC	48.5	48.5
NG	39.0	39.0
GAP		12.5
DEP	12.5	
T_g (K)	2560	2960
M_g (kg kmol^{-1})	24.0	25.0
I_{sp} (s)	237	253
Combustion products (mole fraction)		
CO	0.414	0.337
CO_2	0.110	0.134
H_2	0.134	0.091
H_2O	0.222	0.259
N_2	0.119	0.169

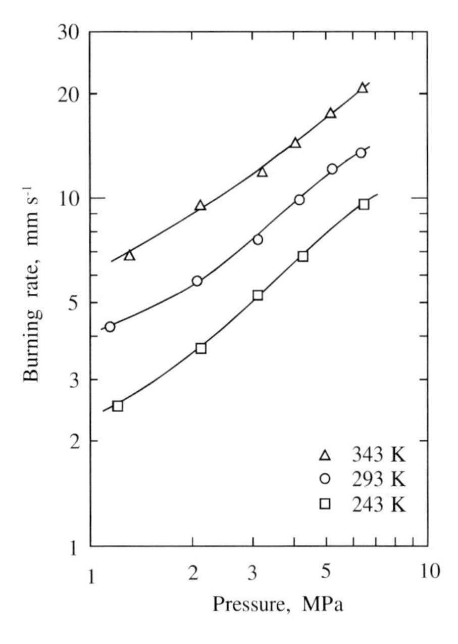

Fig. 6.19 Burning rates of a nitro-azide (NC-NG-GAP) propellant at different initial propellant temperatures, showing high temperature sensitivity.

is equivalent to that described for NC-NG double-base propellants in Section 6.1.2. The temperature in the dark zone is increased from 1400 K to 1550 K at 3 MPa when DEP is replaced with GAP. Though the reaction time in the dark zone is

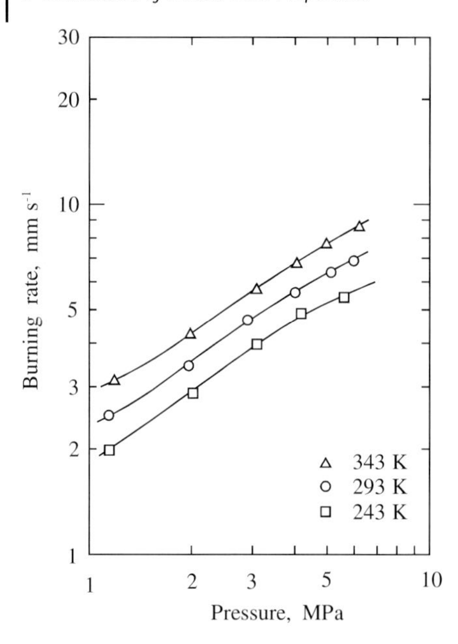

Fig. 6.20 Burning rates of a double-base (NC-NG-DEP) propellant at different initial propellant temperatures, showing low temperature sensitivity.

shortened by this replacement with GAP, the flame stand-off distance is increased because of the increased flow velocity in the dark zone. The overall order of the hemical reaction determined by the relationship $d = n - m$ is determined as $d = -1.7$ and $m = 2.4$ for the NC-NG-DEP propellant and $d = -1.7$ and $m = 2.5$ for the NC-NG-GAP propellant.[18] The results indicate that the basic chemical reaction mechanism in the gas phase, which involves the reduction of NO to N_2, remains unchanged by the replacement of DEP with GAP. However, the increased burning rate and temperature sensitivity of burning rate when DEP is replaced with GAP indicate that the process controlling the burning rate is changed by the increased surface heat release, Q_s, of GAP, cf. the high burning rate and high temperature sensitivity of burning rate of GAP[18] described in Section 5.2.2.

6.4
Catalyzed Double-Base Propellants

6.4.1
Super-Rate, Plateau, and Mesa Burning

During the Second World War, the accidental discovery was made that the use of lead compounds as lubricants in the propellant extrusion process resulted in a greatly increased pressure exponent of burning rate at low pressures, as manifested in an increased burning rate in this range. Investigation of this phenomenon brought to light the fact that the presence of small quantities of a variety of lead compounds leads to similar increases in burning rate at low pressures. Further exploration of this "super-rate burning" phenomenon led to the discovery of the "plateau-burning" region and the "mesa-burning" region.

It was soon realized that platonized propellants, with their reduced temperature sensitivity in the plateau- and mesa-burning range, could be effectively used to minimize the sensitivity of the performance of a rocket to the temperature of the environment. Much work has been devoted to understanding the mechanism of plateau and mesa burning, with a view to optimizing the performance characteristics of rocket motors.

The super-rate, plateau, and mesa-burning characteristics of catalyzed double-base propellants are defined in Fig. 6.21. The first published reference to the latter was from the Allegheny Ballistics Laboratory (ABL) in 1948.[19] Since then, extensive work has been carried out, largely with metal compounds, in developing super-rate, plateau, and mesa-burning propellants for practical use. For a time it appeared that the addition of a wide variety of metal compounds increased the burning rate. However, the increases in burning rate so obtained were insignificant compared with the increases obtained when lead compounds were added to the propellants. Furthermore, it was recognized that metal compounds other than lead compounds did not give plateau and mesa burning in the pressure range of rocket fuel combustion. Thereafter, the search for metal compounds giving plateau burning was focused largely on lead compounds; it was soon discovered that most lead compounds in adequate amounts give plateau-type burning.[11]

The pressure exponent of burning rate is commonly used to evaluate the effectiveness of catalysts in producing plateau and mesa burning.[19–23] The pressure exponent is approximately zero for plateau burning and is negative for mesa burning, as shown in Fig. 6.21. The pressure exponent and the domains of super-rate, plateau, and mesa burning are greatly dependent upon the physical and chemical properties of the lead compounds used, such as quantity, particle size, and chemical

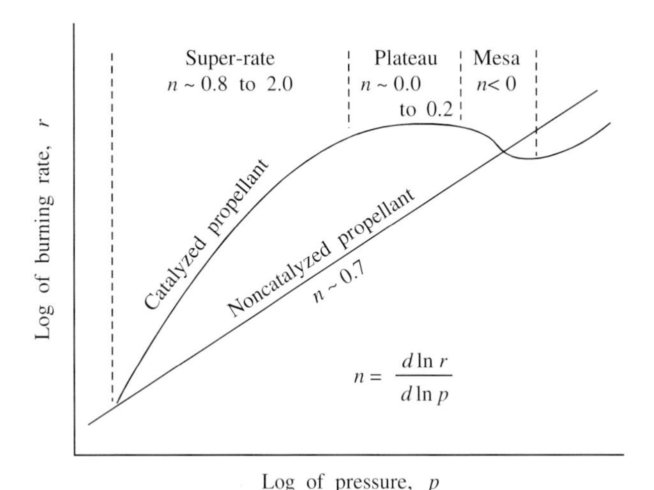

Fig. 6.21 Definition of super-rate, plateau, and mesa burnings of a catalyzed double-base propellant.

structure. A number of lead compounds, such as $PbBr_2$, PbI_2, and $PbCl_2$, do not yield a reduced pressure exponent. Aliphatic lead salts give plateau burning at low pressures with low burning rates, while aromatic lead salts give plateau burning at high pressures with high burning rates.[11]

6.4.2 Effects of Lead Catalysts

6.4.2.1 Burning Rate Behavior of Catalyzed Liquid Nitrate Esters

In an attempt to understand the combustion mechanism of catalyzed double-base propellants, several investigators have conducted experiments to measure the burning rates of strands of liquid nitrate esters. The various measurement techniques were very similar to that employed in a conventional solid propellant strand burner. The liquid esters were placed in a tubular container, and the liquid surface regression speeds were measured by optical methods or by the fuse-wire method used in solid-propellant strand burners. The only important difference between the solid and the liquid strand burning-rate measurements is that the liquid strand burning speed is very much dependent on the diameter of the container.

Steinberger and Carder[24,25] measured the burning rates of liquid strands composed of 63 % nitroglycerin and 37 % diethylene glycol. When 5 % lead aspirate was added to the basic liquid strand, they observed an increased burning rate between 4.0 MPa and 13.5 MPa. This effect is qualitatively the same as in the case of a double-base propellant. The burning rate of the liquid strand was increased by 70 % at 6.8 MPa by the addition of the lead aspirate.

Powling and coworkers[20] measured the burning rates of ethyl nitrate, butane-2,3-diol dinitrate, glycol dinitrate, and glycol dinitrate/triacetin mixtures. They used lead acetylsalicylate as a catalyst, which is slightly soluble in the liquid nitrate esters. They concluded from their experimental results that catalysis by the lead salt occurs in all cases, but that the effect is never pronounced. However, a mixture of glycol dinitrate and 3 % lead acetylsalicylate yielded a substantially higher burning rate, 47 % higher than that of the basic glycol dinitrate at 3.4 MPa. From the above discussion, it is concluded that it is possible to produce super-rate burning with liquid nitrate esters, although the effect is not as pronounced or as common as it is in the case of leaded double-base propellants.

6.4.2.2 Effect of Lead Compounds on Gas-Phase Reactions

It has been shown by many investigators that gas-phase reactions of nitrate esters are affected by the addition of lead compounds. Adams and coworkers studied the gaseous combustion zone of ethyl nitrate.[26] The flame temperature of the ethyl nitrate was found to be reduced by the addition of 0.1–1.0 % by mass of lead tetramethyl. The results are shown in Table 6.7.

The results indicate that the reduction of NO to N_2 is inhibited by the reduced temperature in the gas phase on the addition of lead tetramethyl. In addition, the same authors examined the effect of alkyl radicals on the flame speed by using different kinds of metal alkyls. They found the flame speed not to be affected by

Table 6.7 Flame temperatures and combustion products of ethyl nitrate without and with lead tetramethyl (LTM) at 0.1 MPa.

	Flame temperature	NO	N_2	N_2O	H_2	CO	CO_2	CH_4
without LTM	1470 K	28.9	3.7	2.5	7.0	30	2.5	25.4
with LTM	1070–1170 K	47.5	< 0.1	1.4	4.0	21.7	6.7	17.9

alkyl radicals; rather, the inhibition effect is due to the high degree of dispersion of the lead oxide formed during the flame reaction.

Ellis and coworkers studied the effect of lead oxide on the thermal decomposition of ethyl nitrate vapor.[27] They proposed that the surface provided by the presence of a small amount of PbO particles could retard the burning rate due to the quenching of radicals. However, the presence of a copper surface accelerates the thermal decomposition of ethyl nitrate, and the rate of the decomposition process is controlled by a reaction step involving the NO_2 molecule. Hoare and coworkers studied the inhibitory effect of lead oxide on hydrocarbon oxidation in a vessel coated with a thin film of PbO.[28] They suggested that the process of aldehyde oxidation by the PbO played an important role. A similar result was found in that lead oxide acts as a powerful inhibitor in suppressing cool flames and low-temperature ignitions.[29]

Hoare and coworkers proposed the following reaction scheme for formaldehydeon the surface of the lead oxide:[28]

$$CH_2O + PbO \rightarrow CO_2 + H_2 + Pb$$

The re-oxidation of lead was expected to occur rapidly at 600 K; the surface of the lead oxide would then remain unchanged in the presence of oxygen. The authors concluded that, as a consequence, general hydrocarbon combustion in which formaldehyde is a degenerate branching intermediate is inhibited in the presence of PbO by the rapid removal of formaldehyde.

It is known that a small amount of lead tetraethyl is effective as an anti-knock agent in gasoline engines. This effect is due to the inhibition of the gas-phase reaction by the lead or lead oxide derived from the lead tetraethyl.[30,31] The inhibitory action of lead compounds on hydrocarbon combustion is largely exerted in the gaseous phase of the combustion zone.

6.4.3 Combustion of Catalyzed Double-Base Propellants

6.4.3.1 Burning Rate Characteristics

Fig. 6.22 shows a typical set of burning rates for catalyzed and non-catalyzed NC-NG double-base propellants. The burning rate of the non-catalyzed propellant composed of 53 % NC, 40 % NG, and 7 % DEP is seen to increase linearly with increasing pressure in an $\ln r$ versus $\ln p$ plot. When the propellant is catalyzed with a com-

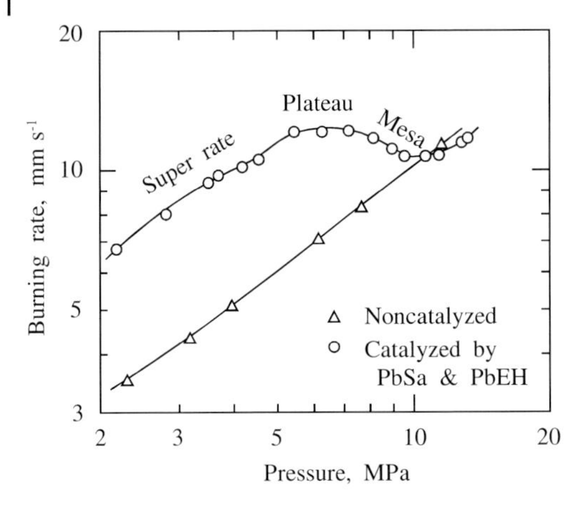

Fig. 6.22 A typical burning rate profile of a platonized double-base propellant.

bination of 1.5% lead salicylate (PbSa), 1.5%, lead 2-ethylhexoate (Pb2EH), and 0.2% graphite, the burning rate increases drastically (super-rate burning) in the low-pressure region below 5 MPa, becomes pressure-insensitive (plateau burning) between 5 MPa and 7 MPa, and then decreases with a further increase in pressure (mesa burning) between 7 MPa and 11 MPa. The burning rate returns to that of the non-catalyzed propellant above 11 MPa, whereupon the effect of the addition of the catalysts diminishes.[11]

Fig. 6.23 shows a comparison of the burning rates of catalyzed NC-NG and NC-TMETN propellants. As shown in Table 6.8, the chemical compositions of both propellants contain equal quantities of the same catalysts. The burning rates of the non-catalyzed NC-NG and NC-TMETN propellants are shown in Fig. 6.18. The energy densities of the two catalyzed propellants are approximately equal.

Table 6.8 Chemical compositions of catalyzed NC-TMETN and NC-NG double-base propellants (% by mass).

Propellant	NC	TMETN	TEGDN	NG	DEP	EC	PbSa	Pb2EH
NC-TMETN	50.0	40.4	7.1			0.1	1.2	1.2
NC-NG	50.0			37.1	10.4	0.1	1.2	1.2

The burning rate curves show approximately similar characteristics; at low pressures, no super-rate burning is seen for either propellant. Since two kinds of lead catalyst are added to each propellant, two super-rate and two plateau burning zones are seen in the same pressure regions in each case. The maximum super-rate burning occurs at around 0.5 MPa, and further relatively high super-rate burning is seen at about 4 MPa. Above 10 MPa, super-rate burning is no longer in evidence and the burning rates of both catalyzed propellants are almost the same as for the non-catalyzed propellants. Though the burning rate versus pressure relationships of NC-

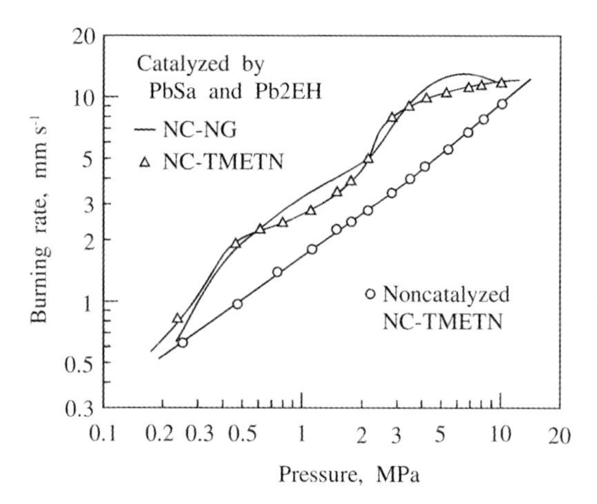

Fig. 6.23 Burning rates of catalyzed NC-NG and NC-TMETN double-base propellants, showing two platonized regions for both propellants resulting from the addition of a combination of two lead compounds.

NG and NC-TMETN are not quite the same due to small differences in chemical structure and in the energy levels of the propellants, the burning characteristics of NC-NG and NC-TMETN propellants are broadly similar and the action of the catalysts in terms of producing super-rate, plateau, and mesa burning is the same for both propellants.

In the super-rate and plateau burning regimes, the temperature sensitivity-decreases with increasing pressure; moreover, the lowest temperature sensitivity always appears at the upper end of the plateau burning region, i. e., in the mesa burning region.[11] At pressures above that of mesa burning, greatly increased temperature sensitivity is observed. A further observation indicates that between certain initial propellant temperatures there is negative temperature sensitivity in the mesa burning region of some propellants. No such negative temperature sensitivity has been reported in the super-rate region.

Plateau burning characteristics are dependent on the chemical components and the nature of the catalysts. The effects of aromatic lead and copper salts on burning rate behavior are shown in Fig. 6.24. The addition of PbSa (1 %) increases the burn-

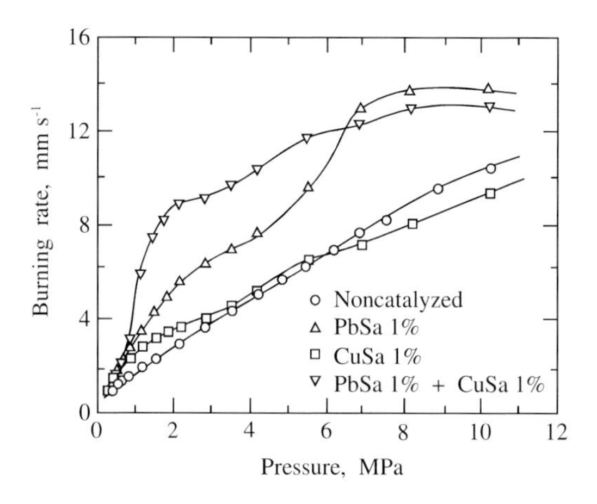

Fig. 6.24 Various types of plateau burning obtained by the addition of different types of catalysts.

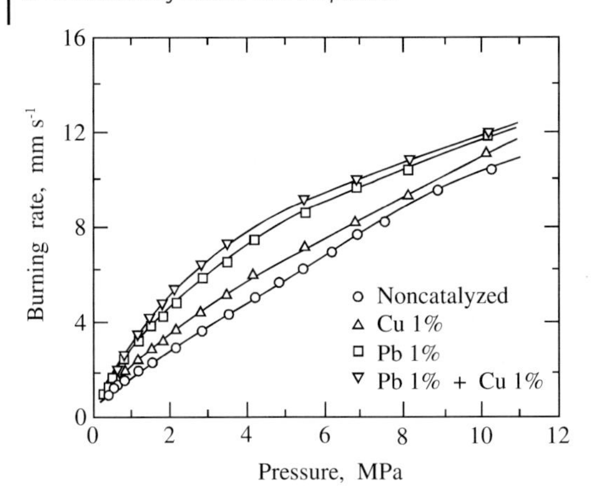

Fig. 6.25 Increased burning rates of a double-base propellant resulting from the addition of metallic copper and/or lead powders.

ing rate in the range 0.1 MPa to 7 MPa and produces plateau and mesa burning above 7 MPa. CuSa (1 %) increases the burning rate below 3 MPa and decreases the burning rate above 6 MPa. When PbSa (1 %) and CuSa (1 %) are mixed together, enhanced super-rate burning is seen in the region between 0.7 MPa and 6 MPa, and plateau and mesa burning are seen above 7 MPa. The effect on the burning rate of synergistic coupling between the PbSa and CuSa catalysts is clearly evident. No additional burning rate increase is seen when further PbSa or CuSa is incorporated; the rate-accelerating effect is apparently saturated.

The effect of the addition of metallic lead and copper powders on burning rate is shown in Fig. 6.25. Pb powder is more effective than Cu powder in increasing the burning rate. When both powders are mixed together within the non-catalyzed propellant, no drastic burning rate increase, i. e., no super-rate burning, is observed. However, when PbO powder (1 %) is added, the burning rate is increased by approximately 250 %, i. e., super-rate burning occurs.

Higher lead salt concentrations move the region of plateau burning to lower pressures, and shifting of the plateau also occurs when carbon powder is added to plateau-burning propellants. Above 0.5 % carbon, the plateau slope increases with increasing carbon content at a less-than-proportional rate, and the relative increase of plateau pressure falls off. Above 1 % carbon, the plateau disappears. The use of small-particle carbon powder apparently results in higher burning rates, but with particles greater than 0.1 μm in size, little or no rate increase is observed. Preckel observed a similar effect on plateau burning when hydrated alumina or acetylene black was added to plateau propellants.[19] However, titanium oxide, magnesium oxide, and levigated alumina are not effective in this regard. Carbon powder is not the only catalyst to enhance super-rate burning; there are others such as copper salicylate. The effectiveness of carbon powder in enhancing super-rate burning increases with increasing specific area of the carbon particles (0.01–1.0 μm).

Powling and coworkers[20] examined the effects of metal oxides on burning rate, and found that the burning rates of propellants containing Fe_2O_3, Co_2O_3, CuO, ZnO, SnO_2, and Al_2O_3 increased linearly with pressure, i. e., no plateau or mesa burning was observed. PbO was the only metal oxide to produce super-rate, plateau, and mesa burning. The increased burning rate at low pressure seen with PbO was much greater than that seen with the other metal oxides. The burning rate of the basic propellant was reduced by the addition of MgO and NiO over the entire pressure range in which the tests were conducted. Finally, it was reported that nickel powder decreases the burning rate at low pressures and causes a slight increase at high pressures.

6.4.3.2 **Reaction Mechanism in the Dark Zone**

When the burning rate of a double-base propellant is increased by means of lead catalysts, the luminous flame zone, where the greatest heat release occurs in the burning process, is displaced a greater distance from the burning surface than during non-catalyzed combustion. It is important to note that a higher burning-rate propellant has a longer dark-zone length at the same pressure. In the following example, the dark zone index, d, of the non-catalyzed propellant is –1.69, while that of the catalyzed propellant varies between –1.96 and –2.27 in the range 1.4–5.0 MPa. The chemical compositions of the non-catalyzed and catalyzed propellants are shown in Table 6.9.

Table 6.9 Chemical compositions of non-catalyzed and catalyzed propellants (% by mass).

Propellant	Basic composition					Additives	
	NC	TMETN	TEGDN	EC	C	PbSa	CuSa
Non-catalyzed	53.70	39.10	7.02	0.08	0.10		
Catalyzed L	52.69	38.33	6.88	0.08	0.10	0.98	0.98
Catalyzed S	52.69	38.33	6.88	0.08	0.10	0.98	0.98

Note: "catalyzed L" additives are particles of PbSa and CuSa both 10 μm in diameter; "catalyzed S" additives are particles of PbSa and CuSa both 3 μm in diameter.

The chemical reaction process in the dark zone is actually a complex sequence of reactions. With reference to the mathematical model represented by Eq. (3.70), the reaction rate and the overall order of the reaction, m, can be determined. The calculated results are tabulated in Table 6.10, together with the available data of other investigators. It is observed that the pressure exponent of burning rate, n, varies between 0.28 and 0.80 for non-catalyzed and catalyzed propellants, and the dark zone index, $d = n - m$, varies between –1.69 and –2.27, these parameters being dependent upon the propellant composition and the pressure range. It is also observed that the propellant with the lowest pressure exponent has the largest negative dark zone index, i. e., the pressure-dependence of the dark zone is larger. The dark zone length decreases rapidly with increasing pressure in the mesa-burning region. Note, however, that the calculated overall order of reaction, m, is approximately 2.5 for all of the propellants, irrespective of whether or not they are catalyzed. The results indicate that the overall reaction rate in the dark zone of the catalyzed propellant is fundamentally the same as that in the dark zone of the non-catalyzed propellant.

Table 6.10 Overall reaction order, m, in the dark zone determined from the pressure exponent of burning rate, n, and the dark zone index, d.

Propellant	Catalyst	Pressure range, MPa	n	d	m	Reference
NC/TMETN	non-catalyzed	1.5–6.0	0.80	−1.69	2.49	[11]
NC/TMETN	catalyzed L	1.5–6.0	0.45	−1.96	2.41	[11]
NC/TMETN	catalyzed S	1.5–6.0	0.28	−2.27	2.55	[11]
NC/NG	non-catalyzed	2.0–10.0	0.56	−1.95	2.51	[32]
NC/NG	non-catalyzed	1.1–3.5	0.60	−2.20	2.80	[10]
NC/NG	non-catalyzed	1.7–4.1	0.45	−2.00	2.45	[15]

In attempting to understand the mechanism of mesa burning, it is also important to note that the flame zone is not distended as the burning rate decreases, but in fact moves closer to the burning surface. This again indicates that the conductive energy transfer from the luminous flame to the burning surface is not important, and that there must instead be some inhibiting reaction near or at the burning surface, i. e., in the fizz zone, which produces the mesa burning.

6.4.3.3 Reaction Mechanism in the Fizz Zone Structure

The thickness of the fizz zone (i. e., zone III) is actually dependent on the chemical kinetics of the gaseous species evolved at the burning surface, which, in turn, is dependent on pressure. The thickness decreases with increased pressure, resulting in an increased temperature gradient. Therefore, the rate of heat input by conduction from the gas phase to the condensed phase increases with increasing pressure. The pressure-dependence of the temperature gradient in the gas phase just above the burning surface, $(dT/dx)_{s,g}$, is similar for both non-catalyzed and catalyzed propellants. However, the temperature gradient for the catalyzed propellant is less than that for the non-catalyzed propellant. Therefore, the heat feedback from the gas phase to the condensed phase for the catalyzed propellant is clearly larger than that for the non-catalyzed propellant; it is evident that the reaction in the fizz zone of the catalyzed propellant is accelerated by the effect of the catalysts in the regime of super-rate burning. This implies that heat flux transferred back from the gas phase to the burning surface is increased by the addition of lead compounds in the region of super-rate burning, even though the luminous flame is distended from the burning surface.

6.4.4 Combustion Models of Super-Rate, Plateau, and Mesa Burning

The regions directly affected by lead compounds are the condensed phase just below the burning surface and the burning surface itself (a distance of less than 100 μm at 0.1 MPa or 20 μm at 0.2 MPa), where the lead compounds ultimately decompose into finely divided metallic lead or lead oxide particles. The decomposition products of the lead catalyst react with the nitrate esters in this surface reaction layer, where the chemical degradation leads to NO_2, aldehydes, etc., altering their normal thermal decomposition paths so as to produce an increased amount of solid carbon at the burning surface of the catalyzed propellant. Despite this chemical pathway alteration, the net exothermicity of the surface reaction layer is not significantly changed.

Most importantly, the presence of lead compounds results in a strong acceleration of the fizz zone reactions, i. e., those in the gas phase close to the burning surface. Acceleration of the reactions in the subsequent dark zone or in the luminous flame zone is not significant. The net result of the fizz zone reaction rate acceleration is an increased heat feedback to the surface (e. g., by as much as 100 %), which produces super-rate burning.

Consequently, this action of the lead compounds to produce increased carbon and the acceleration of the fizz zone reactions are directly coupled as follows. The portion of decomposed organic molecules which appears at the surface in the form of carbon rather than readily oxidizable aldehydes could reduce the effective fuel/oxidizer (aldehydes/NO_2) ratio. This increased NO_2 proportion constitutes a shift in equivalence ratio toward the stoichiometric value. Such a shift for aldehydes/NO_2 mixtures results in a greatly accelerated reaction rate. The excess carbon is oxidized downstream of the fizz zone. Furthermore, the carbon formed at the burning surface that protrudes into the fizz zone increases the overall value of the thermal conductivity in the fizz zone, λ_g, shown in Eq. (3.59). The thermal conductivity of carbon is more than 1000 times higher than that of the gaseous products in the fizz zone. In other words, the averaged value of the thermal conductivity in the fizz zone is considered to be increased during super-rate burning due to the formation of carbonaceous materials therein. This also indicates that the heat flux transferred back from the fizz zone to the burning surface is increased significantly when carbon formation occurs at the burning surface. In fact, the effectiveness of lead compounds in producing super-rate burning decreases as the propellant's heat of explosion increases.[11] This implies that propellants with a lower NG content or those made with NC with a lower degree of nitration are more strongly influenced by lead compounds.

Though it is impossible to formulate a complete mathematical representation of the super-rate burning, it is possible to introduce a simplified description based on a dual-pathway representation of the effects of a shift in stoichiometry. Generalized chemical pathways for both non-catalyzed and catalyzed propellants are shown in Fig. 6.26. The shift toward the stoichiometric ratio causes a substantial increase in the reaction rate in the fizz zone and increases the dark zone temperature, a consequence of which is that the heat flux transferred back from the gas phase to the burning surface increases.

Reaction zone	Subsurface	→	Fizz zone	→	Dark zone	→	Flame zone
Non-catalyzed propellant	Normal reaction path	→	Normal gas-phase reaction NO_2 + RCHO species	→	Normal gas-phase reaction NO + C,H,O species	→	Normal flame reaction H_2O, CO, CO_2, etc.
Catalyzed propellant	Pb salts in association with nitrate ester	→	Increased reaction rate [RCHO]/[NO_2] decreases	→	Increased reaction rate by increased temperature	→	Normal flame reaction H_2O, CO, CO_2, etc.
	Solid carbon formation	→	Toward stoichiometric ratio	→	Increased temperature	→	Increased temperature

Fig. 6.26 Reaction pathways for super-rate burning.

The simplified burning-rate model given by Eq. (3.59) only represents the increased burning rate, i. e., super-rate burning, within the pressure region in which carbonaceous materials are formed on the burning surface.

As the burning rate increases (resulting both from increased pressure and the addition of lead compounds), the thermal wave thickness of the condensed phase, δ_p, defined according to $\delta_p = (\lambda_p \rho_p / c_p)/r$, decreases, and the time available for the initial catalytic action, τ_p, defined according to $\tau_p = \delta_p/r$, in the surface reaction layer decreases, where ρ_p is the density and c_p is the specific heat of the propellant. For example, $\tau_p = 0.2$ s at 0.1 MPa and $\tau_p = 0.002$ s at 2 MPa. Thus, the fraction of the reactants affected by the lead compounds (and a higher concentration of NO_2) decreases in the fizz zone with increasing burning rate. Consequently, the reaction rate in the fizz zone approaches that of the normal reaction pathway, and hence the increased burning rate, i. e., super-rate burning, diminishes with increasing pressure. The rate of disappearance of super-rate burning with increased pressure diminishes the slope of the burning rate plateau.

A negative slope for the burning rate, i. e. mesa burning, occurs as a result of an inhibition reaction in the fizz zone caused by the lead or lead oxide, which are derived from the lead compounds incorporated into the propellant. Since the reaction rate in the fizz zone increases with increasing pressure, the decreased reaction rate due to the inhibition reaction is more than compensated by the increased reaction rate due to the increased pressure. Accordingly, the burning rate of plateau mesa burning returns to the normal burning rate in the high-pressure region.

Experimental results indicate that various phenomena occur on platonized propellants.[11,33] When lead catalysts are added to high-energy double-base propellants, no super-rate burning occurs. Lead catalysts only act effectively on low-energy double-base propellants. On the other hand, lead catalysts are known to retard the rate of aldehyde oxidation. Carbonaceous materials are formed on the burning surface during super-rate burning, which diminish as the super-rate burning diminishes. The following combustion mechanisms are proposed to rationalize the observed super-rate, plateau, and mesa burning: The initial action of the catalysts is in the condensed phase, and the reaction process at the burning surface and the ensuing reaction pathways in the gas phase are also altered. Due to the formation of carbonaceous materials, the equivalence ratio of the fuel and oxidizer components shifts towards the optimal stoichiometry. The reaction rate in the gas phase is accelerated by this change in stoichiometry. The heat flux transferred back from the fizz zone to the burning surface increases and super-rate burning occurs. Furthermore, the conductive heat is also increased as a result of the increased average heat conductivity in the fizz zone due to the formation of carbonaceous materials therein. As the burning rate increases in the high-pressure region, the formation of carbonaceous materials diminishes and hence the super-rate burning also diminishes and becomes plateau burning. This negative catalytic effect of lead compounds is considered to produce mesa burning.

6.4.5 LiF-Catalyzed Double-Base Propellants

Super-rate burning occurs when lithium fluoride (LiF) is incorporated into NC-NG or NC-TMETN double-base propellants. As shown in Fig. 6.27, the burning rate of a propellant catalyzed with 2.4% LiF and 0.1% C increases drastically in the pressure region between 0.3 MPa and 0.5 MPa. This super-rate burning effect diminishes gradually as the pressure is increased above 0.5 MPa. The non-catalyzed propellant is a conventional NC-NG double-base propellant composed of 55% NC, 35% NG, and 10% DEP. The maximum burning rate increase is about 230% at 0.5 MPa.

The dark zone length of LiF-catalyzed propellants is increased by the addition of LiF in the region of super-rate burning, similar to the case of Pb-catalyzed propellants, as shown in Fig. 6.28. Table 6–11 shows the dark zone lengths and reaction times τ_g in the dark zone producing the luminous flame at two different pressures, 1.5 MPa and 3.0 MPa. The reaction time defined in Eq. (6.3) remains relatively unchanged and no significant effect on the dark zone reaction is seen by the addition of LiF and C.

Table 6.11 Dark zone lengths and reaction times in the dark zone.

Propellant	p MPa	r mm s^{-1}	L_g mm	T_d K	τ_g ms
Non-catalyzed	1.5	2.6	6.1	1100	8.7
	3.0	3.9	1.8	1100	3.0
Catalyzed with LiF	1.5	4.9	9.1	1200	6.4
	3.0	5.9	2.7	1200	2.5

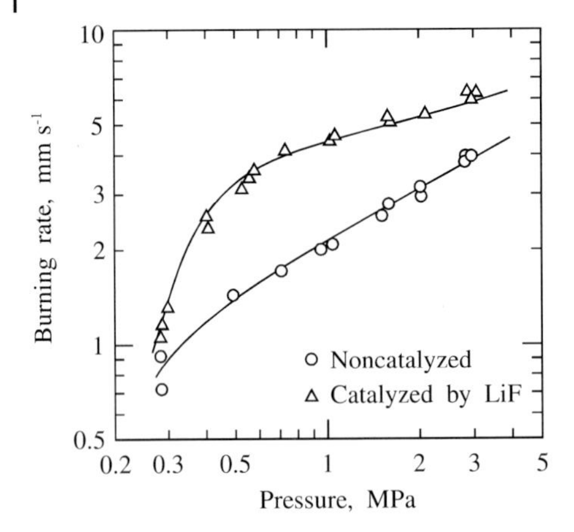

Fig. 6.27 Super-rate burning of an LiF-catalyzed double-base propellant.

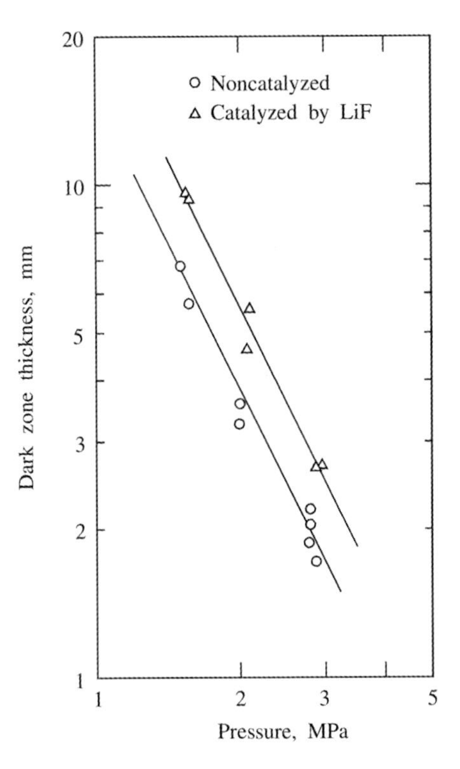

Fig. 6.28 Flame stand-off distance is increased by the addition of LiF in the super-rate burning region.

The temperature gradient in the fizz zone just above the burning surface increases from 1.9 K μm^{-1} to 2.5 K μm^{-1} at 0.7 MPa when 2.4% LiF and 0.1% C are added. The gas-phase reaction rate in the fizz zone is increased and the heat flux feedback from the fizz zone to the burning surface is increased by the addition of the catalysts. However, the effect of the addition of the catalysts is not seen in the dark zone.

The gas-phase and burning-surface structures of the LiF-catalyzed propellant are very similar to those of Pb-catalyzed propellants showing super-rate burning. When super-rate burning is induced by the addition of LiF, large amounts of carbonaceous materials are formed on the burning surface. These carbonaceous materials disappear at high pressures, with a simultaneous loss of the super-rate burning. The site and mode of action of LiF on double-base propellants are the same as those for Pb compounds. Since LiF is a lead-free and eco-friendly material, lead compounds are being replaced with LiF as a super-rate and plateau catalyst of double-base propellants.

6.4.6 Ni-Catalyzed Double-Base Propellants

Metallic nickel is known to catalyze the reduction of NO in its reactions with aldehydes or hydrocarbon gases,[33] and the primary reaction in the dark zone of a double-base propellant is the reduction of NO to N_2. Small amounts of fine Ni particles or organic nickel compounds are incorporated into double-base propellants to increase the reaction rate in the dark zone.

Fig. 6.29 shows the effect, or lack thereof, of the addition of Ni particles on the burning rate of a double-base propellant. The double-base propellant is composed of $\xi_{NC}(0.44)$, $\xi_{NG}(0.43)$, $\xi_{DEP}(0.11)$, and $\xi_{EC}(0.02)$ as a reference propellant. This propellant is catalyzed with 1.0% Ni particles (2 μm in diameter). No burning rate change is seen upon the addition of Ni particles.[28] However, the flame structure is altered significantly by the addition of Ni. The flame stand-off distance between the burning surface and the luminous flame front is shortened, as shown in Fig. 6.30. Though the flame stand-off distance of the reference propellant is about 8 mm at 1.5 MPa and decreases rapidly with increasing pressure (1 mm at 4.0 MPa), the flame stand-off distance of the Ni-catalyzed propellant remains unchanged (0.3 mm) when the pressure is increased.

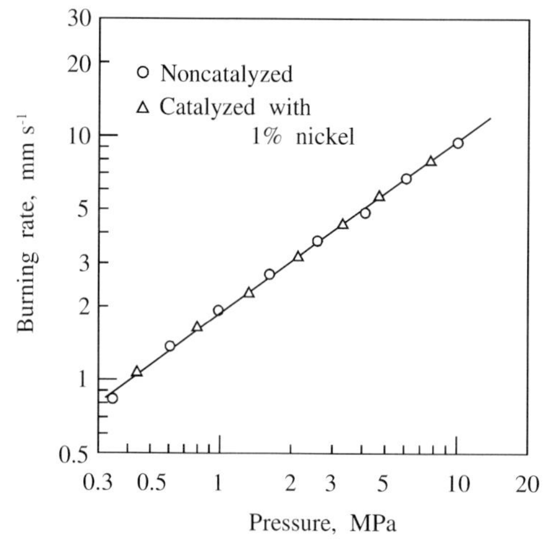

Fig. 6.29 Burning rate remains unchanged upon the addition of nickel powder.

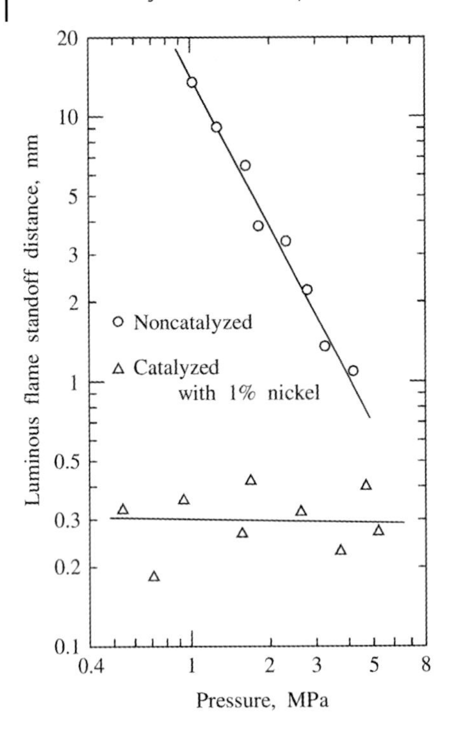

Fig. 6.30 Flame stand-off distance is decreased significantly by the addition of nickel powder even though the burning rate remains unchanged (see Fig. 6–29).

The temperature in the dark zone of the non-catalyzed propellant is approximately 1300 K at 0.3 MPa and increases with increasing pressure (1500 K at 2.0 MPa). A significant increase in this temperature (to above 2500 K at 0.3 MPa) is seen for the propellant catalyzed with 1% Ni powder. The dark zone is almost completely eliminated and a high-temperature luminous flame approaches the burning surface. However, the temperature gradient in the fizz zone just above the burning surface (0.2–0.3 mm from the surface) remains unchanged when Ni particles are added. This indicates that the heat flux feedback from the fizz zone to the burning surface is also unchanged. The Ni particles act only on the reaction in the dark zone, not on that in the fizz zone or on the condensed-phase reaction. Thus, the burning rate remains unchanged by the addition of the Ni particles.[33]

As described in Section 6.4.4, lead catalysts act on the condensed-phase reaction and on the reaction in the fizz zone, not on that in the dark zone, and they increase the burning rate. On the contrary, Ni catalysts act on the dark zone reaction, not on the condensed-phase reaction or on the fizz zone reaction, and they do not increase the burning rate. No luminous flame is produced when double-base propellants burn at low pressure, for example, at 0.5 MPa. However, when propellants catalyzed with metallic Ni or organic Ni compounds burn, a luminous flame is produced just above the burning surface at the same pressure. Ni catalysts act to accelerate the dark zone reaction of NO + gaseous hydrocarbon $\rightarrow$ N_2, H_2O, CO_2, CO, etc., not to accelerate the fizz zone reaction of NO_2 + aldehydes $\rightarrow$ NO, H_2, CO, etc. shown in Fig. 6.3.

6.4.7
Suppression of Super-Rate and Plateau Burning

Fig. 6.31 shows the effect of the addition of potassium salts on platonized propellants. The chemical compositions of the propellants are shown in Table 6.12. Propellants KN-0 and KS-0 are platonized by 3.2% and 4.4% PbSt, respectively. When 4.0% KNO_3 particles is incorporated into the platonized propellant KN-0, the super-rate, plateau, and mesa burning characteristics are almost completely suppressed, as shown in Fig. 6.31. Observation of the surface structure during burning indicates that the carbonaceous materials formed on the burning surface of the propellant KN-0 are no longer seen on the burning surface of the propellant KN-4 containing KNO_3 particles in the region of the super-rate burning pressure. It is considered that the carbonaceous materials are oxidized by the KNO_3 particles on the burning surface, and that the heat flux transferred back from the gas phase to the burning surface is reduced. The super-rate burning arising from the PbSt is thereby reduced to the normal burning of the propellant without PbSt catalyst.

Table 6.12 Chemical compositions of nitropolymer propellants catalyzed with KNO_3 and K_2SO_4.

Propellant	ξ_{NC}	ξ_{NG}	ξ_{DEP}	ξ_{EC}	ξ_{PbSt}	ξ_{Al}	ξ_{KNO3}	ξ_{K2SO4}
KN-0	0.466	0.369	0.104	0.029	0.032			
KN-4	0.466	0.369	0.104	0.029	0.032		0.040	
KS-0	0.468	0.381	0.073		0.044	0.025		
KS-8	0.468	0.381	0.073		0.044	0.025		0.080

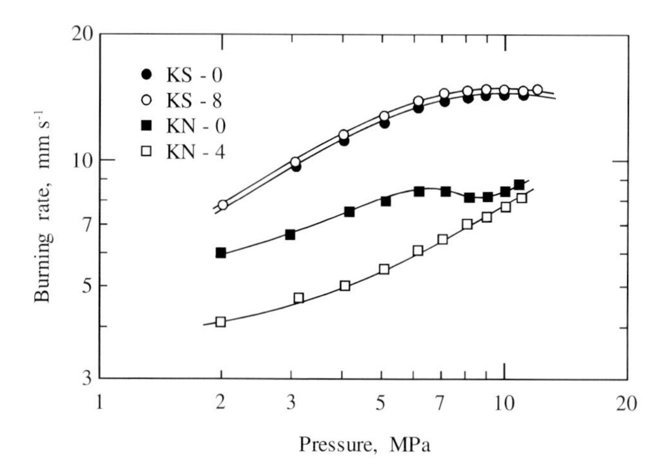

Fig. 6.31 Super-rate and plateau burning are suppressed by the addition of KNO_3, but not by the addition of K_2SO_4.

The propellant KS-0 platonized by 4.4% PbSt shows super-rate burning in the low-pressure region below 9 MPa and plateau burning in the high-pressure region between 9 MPa and 12 MPa with a higher burning rate, as shown in Fig. 6.31. The addition of 8.0% K_2SO_4 particles to KS-0 is seen to have no effect on the super-rate and plateau burning. The burning rate characteristics of the propellant KS-8 remain the same as those of KS-0.

The difference between KNO_3 and K_2SO_4 as additives to double-base propellants is evident from Table 6.13. The melting point and decomposition temperatures of KNO_3 are much lower than those of K_2SO_4. When KNO_3 particles are incorporated into a catalyzed double-base propellant, the particles melt at the burning surface, the temperature of which is approximately 600 K, and are decomposed on and above the burning surface in the fizz zone. The oxidizer fragments formed by the decomposition of the KNO_3 particles react with carbonaceous materials at the burning surface. In fact, the carbonaceous materials formed on the burning surface of the catalyzed propellant KN-0 are eliminated from the burning surface of the propellant KN-4 incorporating KNO_3.

On the other hand, when K_2SO_4 particles are incorporated into the propellant KS-0, the particles are ejected from the burning surface into the fizz zone. No melting or decomposition occur on or above the burning surface due to the high melting point and decomposition temperatures of K_2SO_4. Thus, the K_2SO_4 particles exert no appreciable effect on or above the burning surface. This implies that K_2SO_4 particles act as inert particles in the combustion wave. The plateau burning effect remains unchanged when K_2SO_4 particles are incorporated into platonized double-base propellants.

Table 6.13 Melting points and decomposition temperatures of potassium nitrate and potassium sulfate.

	KNO_3	K_2SO_4
Melting point temperature, K	606	1342
Decomposition temperature, K	673	1962

It should be noted that both KNO_3 and K_2SO_4 are useful additives for eliminating the luminous flame generated at a rocket nozzle and also for suppressing the formation of muzzle flash generated at the exit of a gun barrel. The potassium atoms generated in the gun barrel by the decomposition of these potassium salts are believed to act as a flame retardant.

References

1 Fifer, R. L., Chemistry of Nitrate Ester and Nitramine Propellants, Fundamentals of Solid-Propellant Combustion (Eds.: Kuo, K. K., and Summerfield, M.), *Progress in Astronautics and Aeronautics*, Vol. 90, Chapter 4, AIAA, New York, 1984.

2 Kubota, N., Flame Structure of Modern Solid Propellants, Nonsteady Burning and Combustion Stability of Solid Propellants (Eds.: De Luca, L., Price, E. W., and Summerfield, M.), *Progress in Astronautics and Aeronautics*, Vol. 43, Chapter 4, AIAA, Washington DC, 1990.

3 Lengelle, G., Bizot, A., Duterque, J., and Trubert, J. F., Steady-State Burning of Homogeneous Propellants, Fundamentals of Solid-Propellant Combustion (Eds.: Kuo, K. K., and Summerfield, M.), *Progress in Astronautics and Aeronautics*, Vol. 90, Chapter 7, AIAA, New York (1984).

4 Huggett, C., Combustion of Solid Propellants, Combustion Processes, High-Speed Aerodynamics and Jet Propulsion Series, Vol. 2, Princeton University Press, Princeton, NJ, 1956, pp. 514–574.

5 Timnat, Y. M., Advanced Chemical Rocket Propulsion, Academic Press, New York, 1987.

6 Lengelle, G., Duterque, J., and Trubert, J. F., Physico-Chemical Mechanisms of Solid Propellant Combustion, Solid Propellant Chemistry, Combustion, and Motor Interior Ballistics (Eds.: Yang, V., Brill, T. B., and Ren, W.-Z.), *Progress in Astronautics and Aeronautics*, Vol. 185, Chapter 2.2, AIAA, Virginia, 2000.

7 Crawford, B. L., Huggett, C., and McBrady, J. J., The Mechanism of the Double-Base Propellants, *Journal of Physical and Colloid Chemistry*, Vol. 54, No. 6, 1950, pp. 854–862.

8 Rice, O. K., and Ginell, R., Theory of Burning of Double-Base Rocket Propellants, *Journal of Physical and Colloid Chemistry*, Vol. 54, No. 6, 1950, pp. 885–917.

9 Parr, R. G., and Crawford, B. L., A Physical Theory of Burning of Double-Base Rocket Propellants, *Journal of Physical and Colloid Chemistry*, Vol. 54, No. 6, 1950, pp. 929–954.

10 Heller, C. A., and Gordon, A. S., Structure of the Gas-Phase Combustion Region of a Solid Double-Base Propellant, *Journal of Physical Chemistry*, Vol. 59, 1955, pp. 773–777.

11 Kubota, N., The Mechanism of Super-Rate Burning of Catalyzed Double-Base Propellants, AMS Report No. 1087, Aerospace and Mechanical Sciences, Princeton University, Princeton, NJ, 1973.

12 Eisenreich, N., A Photographic Study of the Combustion Zones of Burning Double-Base Propellant Strands, *Propellants and Explosives*, Vol. 3, 1978, pp. 141–146.

13 Aoki, I., and Kubota, N., Combustion Wave Structures of High- and Low-Energy Double-Base Propellants, *AIAA Journal*, Vol. 20, No. 1, 1982, pp. 100–105.

14 Beckstead, M. W., Model for Double-Base Propellant Combustion, *AIAA Journal*, Vol. 18, No. 8, 1980, pp. 980–985.

15 Crawford, B. L., Huggett, C., and McBrady, J. J., Observations on the Burning of Double-Base Powders, National Defense Research Committee Armor and Ordnance Report, No. A-268 (OSRD No. 3544), April 1944.

16 Sotter, J. G., Chemical Kinetics of the Cordite Explosion Zone, 10th Symposium (International) on Combustion, The Combustion Institute, Pittsburgh, PA, 1965, pp. 1405–1411.

17 Kubota, N., and Ishihara, A., Analysis of the Temperature Sensitivity of Double-Base Propellants, 20th Symposium (International) on Combustion, The Combustion Institute, Pittsburgh, PA (1984), pp. 2035–2041.

18 Nakashita, G., and Kubota, N., Energetics of Nitro/Azide Propellants, *Propellants, Explosives, Pyrotechnics*, Vol. 16, 1991, pp. 171–181.

19 Preckel, R. F., Ballistics of Catalyst-Modified Propellants, Bulletin of Fourth Meeting of the Army-Navy Solid Propellant Group, Armour Research Foundation, Illinois Institute of Technology, Chicago (1948), pp. 67–71.

20 Hewkin, D. J., Hicks, J. A., Powling, J., and Watts, H., The Combustion of Nitric

Ester-Based Propellants: Ballistic Modification by Lead Compounds, *Combustion Science and Technology*, Vol. 2, 1971, pp. 307–327.

21 Eisenreich, N., and Pfeil, A., The Influence of Copper and Lead Compounds on the Thermal Decomposition of Nitrocellulose in Solid Propellants, *Thermochimica Acta*, Vol. 27, 1978, pp. 339–346.

22 Kubota, N., Determination of Plateau Burning Effect of Catalyzed Double-Base Propellant, 17 th Symposium (International) on Combustion, The Combustion Institute, Pittsburgh, PA, 1979, pp. 1435–1441.

23 Kubota, N., Ohlemiller, T. J., Caveny, L. H., and Summerfield, M., Site and Mode of Action of Platonizers in Double-Base Propellants, *AIAA Journal*, Vol. 12, No. 12, 1974, pp. 1709–1714; see also Kubota, N., Ohlemiller, T. J., Caveny, L. H., and Summerfield, M., The Mechanism of Super-Rate Burning of Catalyzed Double-Base Propellants, 15 th Symposium (International) on Combustion, The Combustion Institute, Pittsburgh, PA, 1973, pp. 529–537.

24 Steinberger, R., and Carder, K. E., Mechanism of Burning of Nitrate Esters, Bulletin of the 10 th Meeting of the JANAF Solid Propellant Center, Dayton, June 1954, pp. 173–187.

25 Steinberger, R., and Carder, K. E., Mechanism of Burning of Nitrate Esters, 5 th Symposium (International) on Combustion, Reinhold, New York, 1955, pp. 205–211.

26 Adams, G. K., Parker, W. G., and Wolfhard, H. G., Radical Reaction of Nitric Oxide in Flames, *Discussions Faraday Soc.*, Vol. 14, 1953, pp. 97–103.

27 Ellis, W. R., Smythe, B. M., and Theharne, E. D., The Effect of Lead Oxide and Copper Surfaces on the Thermal Decomposition of Ethyl Nitrate Vapor, 5 th Symposium (International) on Combustion, Reinhold, New York, 1955, pp. 641–647.

28 Hoare, D. E., Walsh, A. D., and Li, T. M., The Oxidation of Tetramethyl Lead and Related Reactions, 13 th Symposium (International) on Combustion, 1971, pp. 461–469.

29 Bardwell, J., Inhibition of Combustion Reactions by Inorganic Lead Compounds, *Combustion and Flame*, Vol. 5, No. 1, 1961, pp. 71–75.

30 Lewis, B., and von Elbe, G., Combustion Flames and Explosions of Gases, Academic Press Inc., New York, 1961.

31 Ashmore, P. G., Catalysis and Inhibition of Chemical Reactions, Butterworths, London, 1963.

32 Heath, G. A., and Hirst, R., Some Characteristics of the High-Pressure Combustion of Double-Base Propellants, 8 th Symposium (International) on Combustion, The Williams & Wilkins Co., Baltimore, 1962, pp. 711–720

33 Kubota, N., Role of Additives in Combustion Waves and Effect on Stable Combustion Limit of Double-Base Propellants, *Propellants and Explosives*, Vol. 3, 1978, pp. 163–168.

7
Combustion of Composite Propellants

7.1
AP Composite Propellants

7.1.1
Combustion Wave Structure

Ammonium perchlorate (AP) is the most widely used crystalline oxidizer to formulate composite propellants for rockets. Unlike double-base propellants, AP composite propellants are not used as gun propellants because their combustion products include a high mass fraction of hydrogen chloride (HCl), which acts to oxidize and erode the interior surface of gun barrels. AP composite propellants are composed of fine AP particles as oxidizers with polymeric hydrocarbons as fuel components. The polymeric hydrocarbons also act as binders (BDR) of the fine AP particles, allowing the formulation of propellant grains of the requisite physical shape. Since AP composite propellants are highly heterogeneous in their physical structure, the combustion wave structure is also heterogeneous because of the diffusional process between the oxidizer gases produced by the AP particles and the fuel gases produced by the binder above the burning surface. Typical polymeric hydrocarbons are HTPB (hydroxy-terminated polybutadiene), CTPB (carboxy-terminated polybutadiene), and HTPE (hydroxy-terminated polyether).

7.1.1.1 Premixed Flame of AP Particles and Diffusion Flame

Extensive experimental and theoretical studies have been carried out on the decomposition and combustion of AP composite propellants with a view to obtaining a wide spectrum of burning rates.[1–13] The combustion mode of AP composite propellants is controlled by the diffusion process of the gaseous decomposition products of the AP particles and the surrounding polymeric hydrocarbons at the burning surface.[1,2,4–10,12] The AP particles decompose to produce perchloric acid, $HClO_4$, and the fuel binder decomposes to produce hydrocarbon fragments and hydrogen.[9] These gaseous decomposition products react to produce heat on and above the burning surface. A close-up view of the combustion process of the AP particles and the decomposing fuel binder (HTPB) is shown in Fig. 7.1. An endothermic reaction involving pyrolysis of the binder to form gaseous fuel products,

Propellants and Explosives. Naminosuke Kubota

ISBN: 978-3-527-31424-9

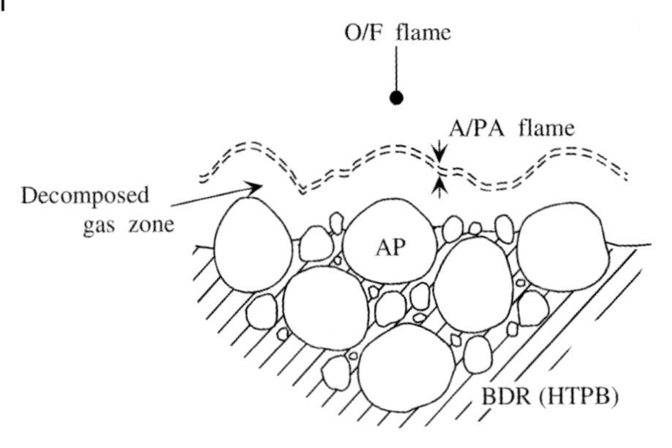

Fig. 7.1 A close-up view of the combustion process of AP particles and fuel binder.

coupled with dissociative sublimation and/or decomposition of the AP particles to form ammonia and perchloric acid, occurs at the burning surface. The ammonia and perchloric acid molecules react exothermically to form premixed flamelets just above each AP particle. Since the reaction products of these premixed flamelets contain excess oxidizer fragments, these react with the fuel gases produced by the decomposition of the polymeric binder to produce diffusional flamelets. Thus, the combustion process of AP composite propellants consists of a two-step reaction.

The temperature in the condensed phase increases from the initial propellant temperature, T_0, to the burning surface temperature, T_s, through conductive heat feedback from the burning surface. Then, the temperature increases in the gas phase because of the exothermic reaction above the burning surface and reaches the final combustion temperature, T_g. Since the physical structure of AP composite propellants is highly heterogeneous, the temperature fluctuates from time to time and also from location to location. The temperature profile shown in Fig. 7.2 thus illustrates a time-averaged profile. This is in a clear contrast to the combustion wave

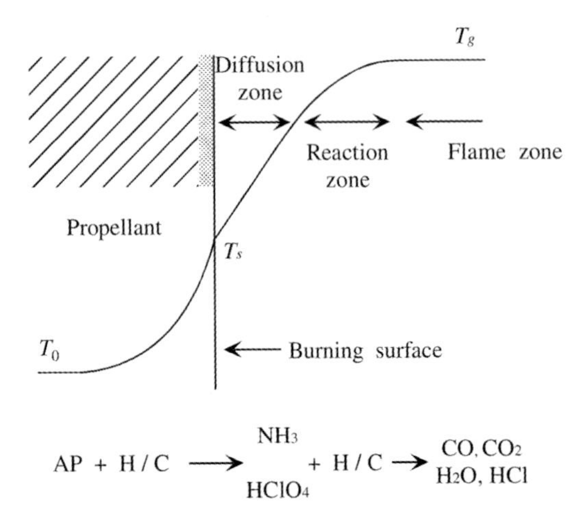

Fig. 7.2 An overall view of the reaction process and the temperature profile in the combustion wave of an AP composite propellant.

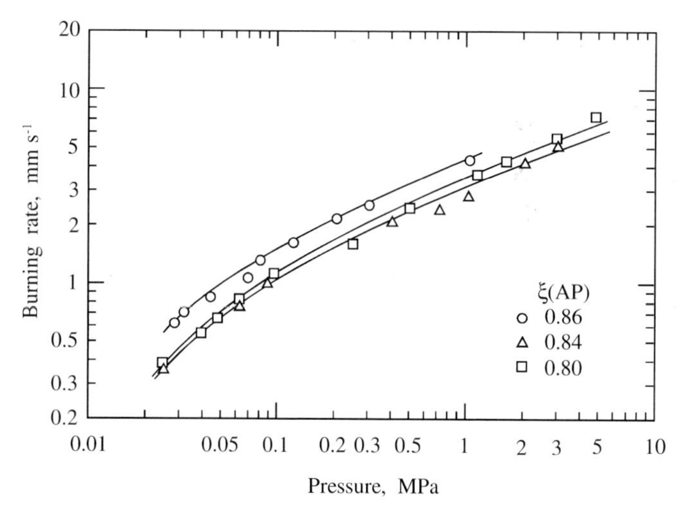

Fig. 7.3 Burning rates of three AP-HTPB composite propellants at low pressures below 1 MPa.

structure for double-base propellants shown in Fig. 6.3. Thus, the burning rate of AP composite propellants depends largely on the particle size of AP,[3,11] the mass fraction of AP, and the type of binder used.[3,12]

In order to clarify the combustion wave structure of AP composite propellants, photographic observations of the gas phase at low pressure are very informative. The reaction rate is lowered and the thickness of the reaction zone is increased at low pressure. Fig. 7.3 shows the reduced burning rates of three AP-HTPB composite propellants at low pressures below 0.1 MPa.[3] The chemical compositions of the propellants are shown in Table 7.1. The burning rate of the propellant with the composition $\xi_{AP}(0.86)$ is higher than that of the one with $\xi_{AP}(0.80)$ at constant pressure. However, the pressure exponents are 0.62 and 0.65 for the $\xi_{AP}(0.86)$ and $\xi_{AP}(0.80)$ propellants, respectively; that is, the burning rate is represented by $r \sim p^{0.62}$ for the $\xi_{AP}(0.86)$ propellant and by $r \sim p^{0.65}$ for the $\xi_{AP}(0.80)$ propellant.

Fig. 7.4 shows flame photographs of the $\xi_{AP}(0.86)$ propellant shown in Table 7.1 at pressures of 0.07 MPa (a) and 0.1 MPa (b). A bluish flame is formed on the burning

Table 7.1 Chemical compositions of AP-HTPB propellants.

Propellant	ξ_{AP}/ξ_{HTPB}	AP (c)	AP (f)	HTPB	T_f (K)
A	0.86/0.14	0.43	0.43	0.14	2680
B	0.84/0.16	0.42	0.42	0.16	2480
C	0.80/0.20	0.40	0.40	0.20	2310

HTPB: $C_{7.057}H_{10.647}O_{0.223}N_{0.063}$
AP (c): d_0 = 200 μm; AP (f): d_0 = 20 μm
T_f: adiabatic flame temperature at 0.1 MPa

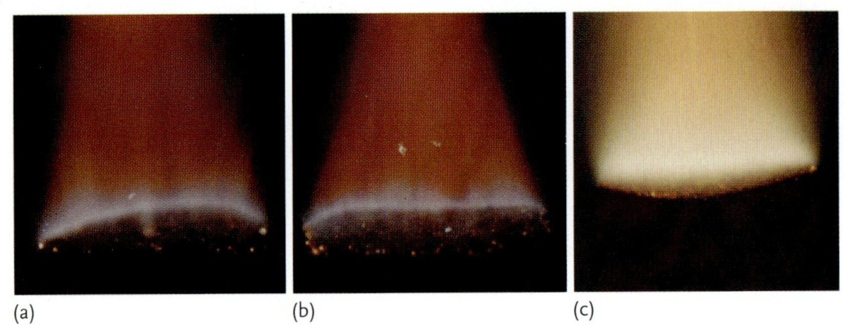

Fig. 7.4 Flame photographs of AP-HTPB composite propellants at low pressures:

	Mass fraction AP	HTPB	p MPa	r mm s^{-1}
(a)	0.86	0.14	0.07	1.2
(b)	0.86	0.14	0.10	1.5
(c)	0.80	0.20	0.10	1.0

surface. The thickness of the bluish flame is decreased with increasing pressure, as shown in Fig. 7.5. The relationship between the thickness of the bluish flame, δ_g, and pressure, p, is represented by $\delta_g \sim p^{-0.60}$: 0.5 mm at 0.02 MPa and 0.2 mm at 0.1 MPa. A reddish flame is seen above the bluish flame. On the other hand, no bluish and reddish flames are seen when the $\xi_{AP}(0.80)$ propellant burns at a pressure of 0.1 MPa, as shown in Fig. 7.4 (c). The flame is entirely yellowish, which can be ascribed to the fuel-rich nature of the diffusion flame generated by the gaseous decomposition products of the binder and the AP particles.[8] Indeed, the flame

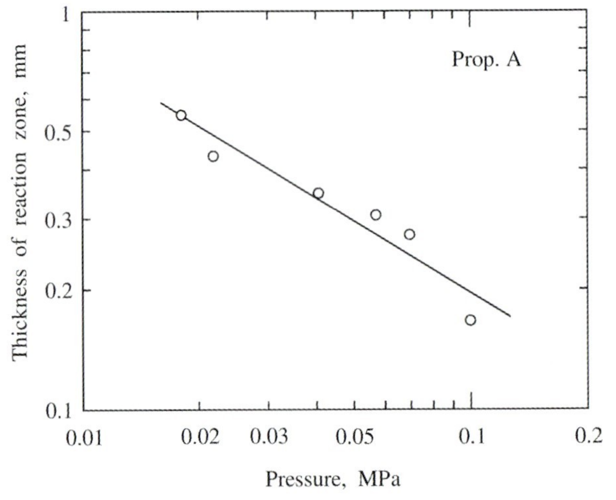

Fig. 7.5 Thickness of reaction zone as a function of pressure for AP-HTPB composite propellants at low pressures.

temperature of the propellant with the composition $\xi_{AP}(0.86)$ is higher than that of the one with $\xi_{AP}(0.80)$, as shown in Table 7.1. Since the flame photographs are taken with a relatively long exposure time (1/60 s), the thickness of the bluish flame appears as a time-averaged value of the numerous diffusion-type flamelets generated by the gaseous decomposition products of the binder and the AP particles.

The overall order of the chemical reaction in the reaction zone is determined from the δ_g data. This overall reaction order, m, is correlated with the pressure exponent of burning rate, n, as described in Section 3.5:

$$\delta_g \propto p^{n-m} = p^d \tag{3.70}$$

where d is the pressure exponent of the gas-phase reaction. From the data shown in Fig. 7.5, d is determined to be –0.60, and from the data shown in Fig. 7.3, n is determined to be 0.62. We then obtain $m = n - d = 1.22$. This overall order of reaction is approximately equal to that of hydrocarbon/air flame reactions. The overall order of the chemical reaction in the dark zone of double-base propellants is approximately 2.5.

The heat flux transferred back from the gas phase to the burning surface is dependent on the temperature gradient in the gas phase, which is inversely proportional to the thickness of the reaction zone in the gas phase. Since the reaction in the gas phase is complete at the upper end of the bluish flame, the heat flux defined by Λ_{III} conforms to the proportionality relationship $\Lambda_{III} \sim 1/\delta_g \sim p^{0.60}$. The observed pressure dependence of the burning rate, $r \sim p^{0.62}$, is caused by the pressure dependence of the reaction rate in the bluish flame.[8]

7.1.1.2 Combustion Wave Structure of Oxidizer-Rich AP Propellants

If the binder (BDR) concentration of AP-HTPB composite propellants is less than the stoichiometric ratio, the burning rate increases as ξ(BDR) increases, as shown in Fig. 7.6.[8] The burning rate of a propellant with the composition $\xi_{BDR}(0.08)$ is

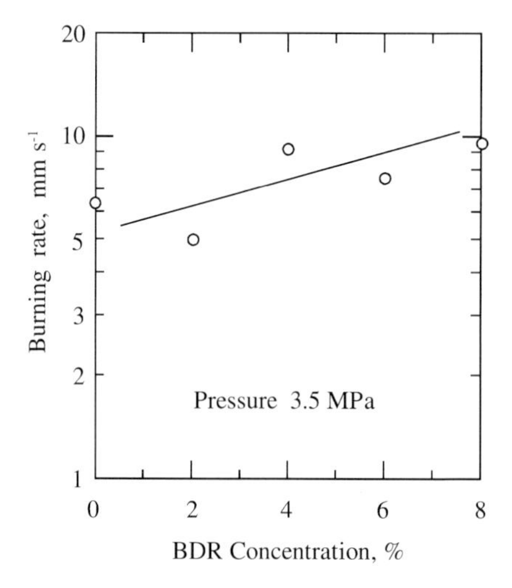

Fig. 7.6 Burning rate versus ξ(BDR) of oxidizer-rich AP-HTPB propellants at 3.5 MPa.

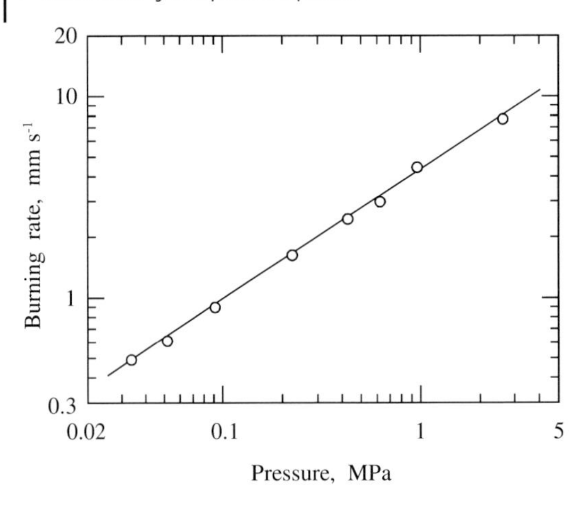

Fig. 7.7 Burning rate of an oxidizer-rich AP-HTPB propellant with the composition $\xi_{HTPB}(0.08)$ in the low-pressure region.

shown in Fig. 7.7. The AP particles mixed within the binder are trimodal: large-sized (400 μm in diam.) 60%, medium-sized (200 μm in diam.) 20%, and small-sized (10 μm in diam.) 20%. The burning rate is expressed by

$$r = 4.30\ p^{0.66} \tag{3.71}$$

Fig. 7.8 shows the temperature profiles [temperature (K) versus burning time (s)] in the combustion waves of this propellant at 0.0355 MPa and at 0.0862 MPa. The temperature is seen to increase relatively smoothly in the condensed phase, but then increases with large fluctuations in the gas phase at both pressures. However, the rate of temperature increase is clearly much higher at 0.0862 MPa than at 0.0355 MPa.

From the data of the temperature profile measurements, the heat transfer process in the combustion wave is analyzed on the basis of the simplified model

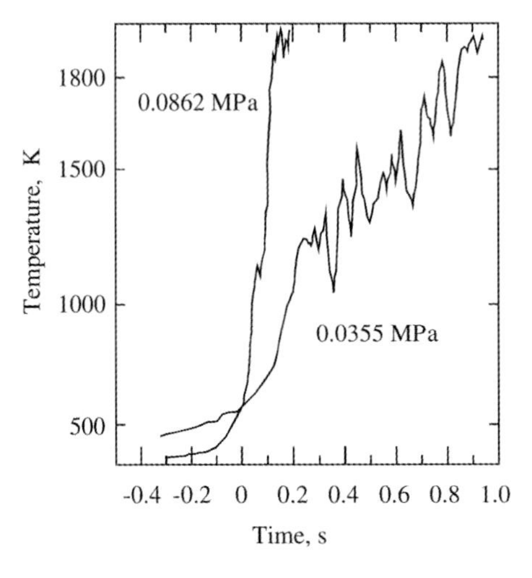

Fig. 7.8 Temperature profiles in the combustion waves at 0.0862 MPa and 0.0355 MPa.

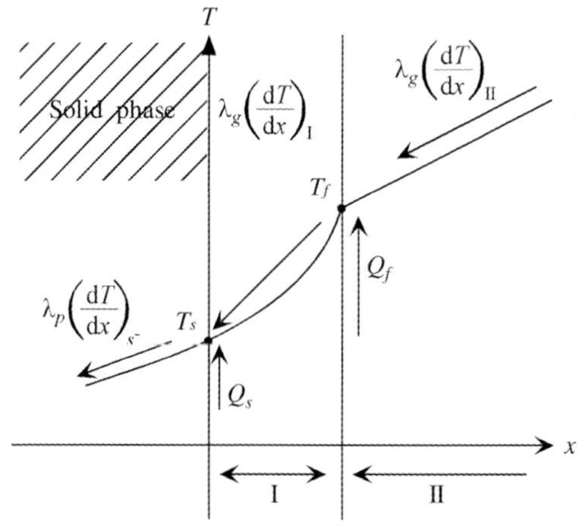

Fig. 7.9 Heat flux and heat of reaction in the gas phase and in the condensed phase for an oxidizer-rich AP-HTPB propellant at low pressures below 0.1 MPa.

shown in Fig. 7.9. The combustion wave is divided into three zones: the heat conduction zone in the solid phase, the preparation zone in the gas phase just above the burning surface (I), and the exothermic reaction zone (II). The burning surface temperature, T_s, and the onset temperature of the gas-phase reaction between zones I and II, T_f, are plotted against pressure in Fig. 7.10.

The heat balance at the burning surface, i. e., at the interface between the solid phase and zone I, is represented by

$$\lambda_p(dT/dx)_{s-} = \lambda_g(dT/dx)_{s+} + \rho_p r Q_s \tag{3.72}$$

$$\lambda_g(dT/dx)_{I,II} = \lambda_g(dT/dx)_{II+} + \rho_g r Q_f \tag{3.73}$$

where λ is thermal conductivity, ρ is density, Q is heat of reaction, and the meanings of the subscripts are as follows: g = gas phase, p = solid phase, s = burning surface, $s-$ = burning surface in the solid phase, $s+$ = burning surface in the gas phase, and f = the interface between zones I and II. Since the zone thickness is too thin to measure ac-

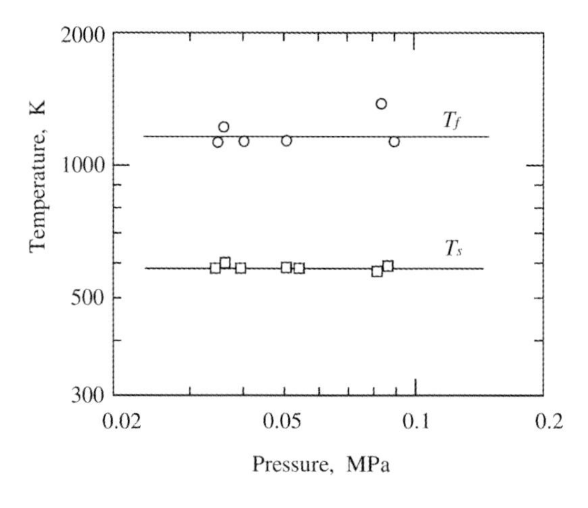

Fig. 7.10 Burning surface temperature (T_s) and the temperature of the A/PA flame (T_f) as a function of pressure below 0.09 MPa.

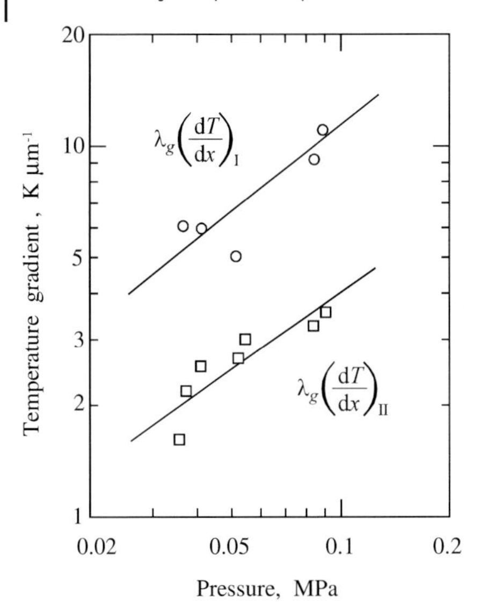

Fig. 7.11 Heat fluxes in zones I and II as a function of pressure in the low-pressure region.

curately, it is assumed that $(dT/dx)_{s+} = (dT/dx)_{I,II} = (dT/dx)_I$. The temperature gradients $(dT/dx)_I$ and $(dT/dx)_{II}$ are shown in Fig. 7.11. Based on the data for r, T_s, T_f, $(dT/dx)_I$, and $(dT/dx)_{II}$, Q_s and Q_f are determined by using Eqs. (3.72) and (3.73), respectively. The heat of reaction at the burning surface, Q_s, is a negative value, representing the heat consumed to decompose AP and HTPB, including the reaction $NH_4ClO_4 \rightarrow NH_3 + HClO_4$. The heat of reaction for the gas-phase reaction at the interface between zones I and II, Q_f, is a positive value, as shown in Fig. 7.12, which is considered to be the heat of reaction of $NH_3 + HClO_4 \rightarrow NO + 3/2H_2O + HCl + 3/4O_2$. The heat transferred back from zone II to zone I is represented by

$$\lambda_g(dT/dx)_{II} = 1.71 \times 10^4\ p^{0.67} \tag{3.74}$$

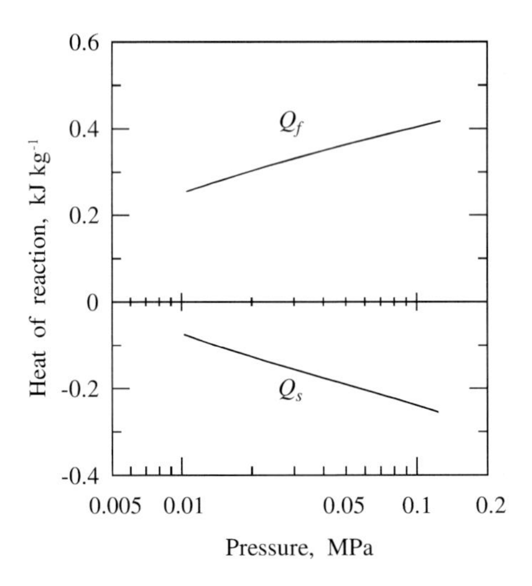

Fig. 7.12 Heats of reaction at the burning surface (Q_s) and in the gas phase at the interface between zone I and zone II (Q_f).

This heat flux is responsible for the burning rate given by Eq. (3.71) in the low-pressure zone, as shown graphically in Fig. 7.7.

7.1.2 Burning Rate Characteristics

7.1.2.1 Effect of AP Particle Size

Fig. 7.13 shows the effect of the particle size of AP on burning rate.[3] The propellants have the composition $\xi_{AP}(0.80)$ and $\xi_{HTPB}(0.20)$. The AP particles are bimodal large-sized, with a 350 μm/200 μm mixture ratio of 4:3, and bimodal small-sized, with a 15 μm/3 μm mixture ratio of 4:3. The burning rate of the small-sized AP propellant is more than double that of the large-sized AP propellant. The pressure exponent of burning rate is 0.47 for the large-sized AP propellant and 0.59 for the small-sized AP propellant.

Fig. 7.14 shows the effect of the AP particle size on the burning rate of monomodal AP composite propellants.[3] The burning rate increases as the AP particle size, d_0, decreases. However, the effect of the particle size on the burning rate diminishes with increasing pressure. Though the propellants shown in Fig. 7.14 are all fuel-rich, with $\xi_{AP}(0.65)$, the effect of AP particle size on burning rate is clearly evident.

7.1.2.2 Effect of the Binder

The burning rates of AP composite propellants are not only dependent on the AP particles, but also on the binder used as a fuel component. There are many types of binders, with varying physicochemical properties, as described in Section 4.2. The

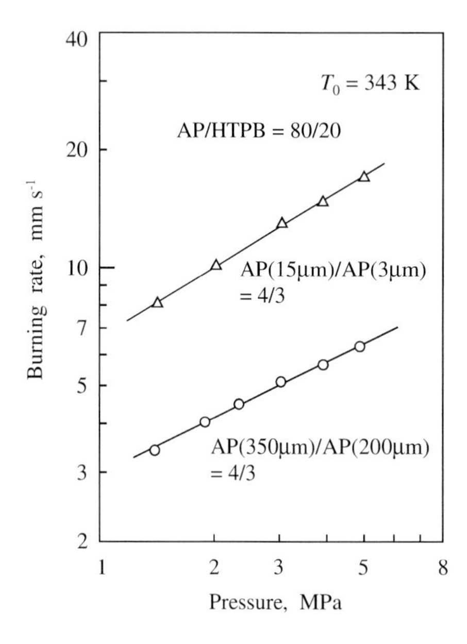

Fig. 7.13 Effect of AP particle size (bimodal) on burning rate.

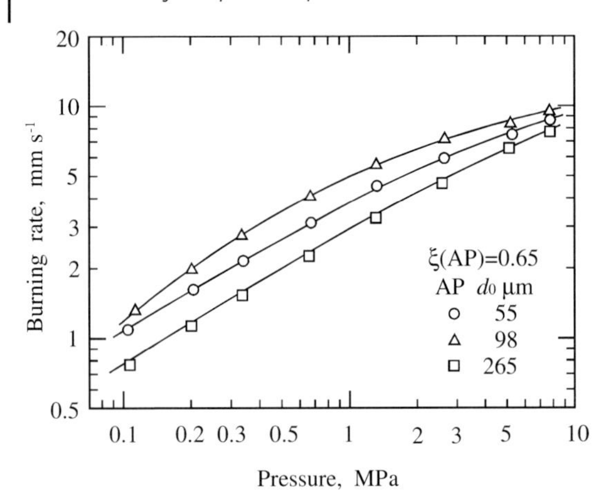

Fig. 7.14 Effect of AP particle size (monomodal) on burning rate.

specific impulse is also dependent on the type of binder and the mass fraction thereof, ξ(BDR), that is mixed with the AP particles. Fig. 7.15 shows the effect of ξ(BDR) on I_{sp} for the binders PB, PU, and PA. It can be seen that a relatively high I_{sp} is maintained in the region of higher mass fraction in the case of PU. This can be ascribed to the oxygen content of PU as opposed to the absence of oxygen in PB or PA. Fig. 7.16 shows the effect of ξ_{HTPB} on the burning rate of AP-HTPB propellants. The burning rate increases with decreasing ξ_{HTPB} at 1 MPa. The adiabatic flame temperature increases from 2300 K to 2700 K when ξ_{HTPB} is decreased from 0.20 to 0.14.

Various types of binders are used to formulate AP composite propellants. Binders such as HTPB and HTPE decompose endothermically or exothermically at the burning surface. The burning rates of AP composite propellants thus appear to be dependent on the thermochemical properties of the binders used. Figs. 7.17 and 7.18 show ln r versus ln p plots for AP composite propellants made with five differ-

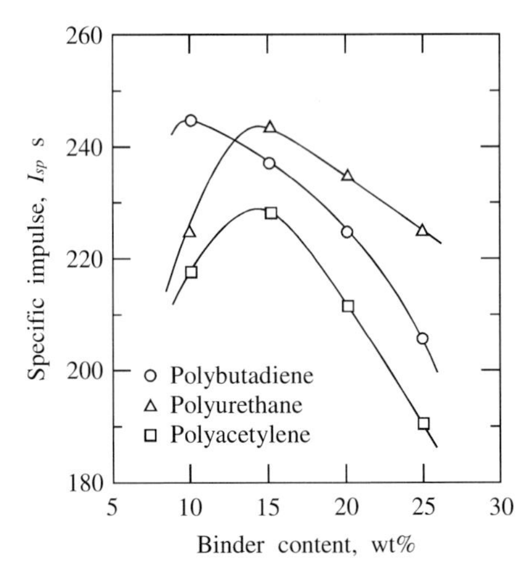

Fig. 7.15 Effect of type of binder on *Isp* as a function of ξ(BDR).

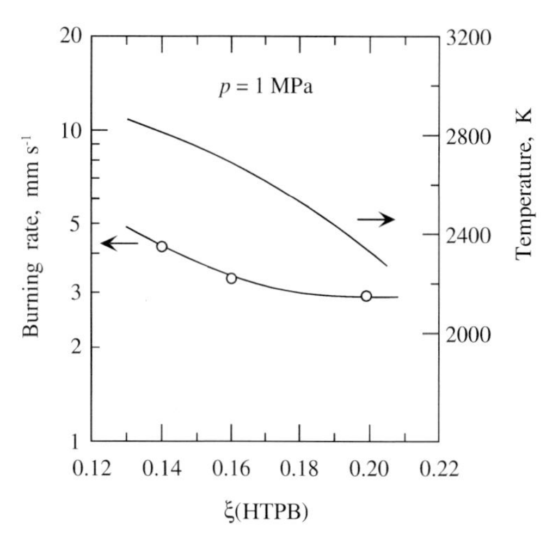

Fig. 7.16 Effect of ξ(HTPB) on the burning rate and adiabatic flame temperature of AP-HTPB propellants.

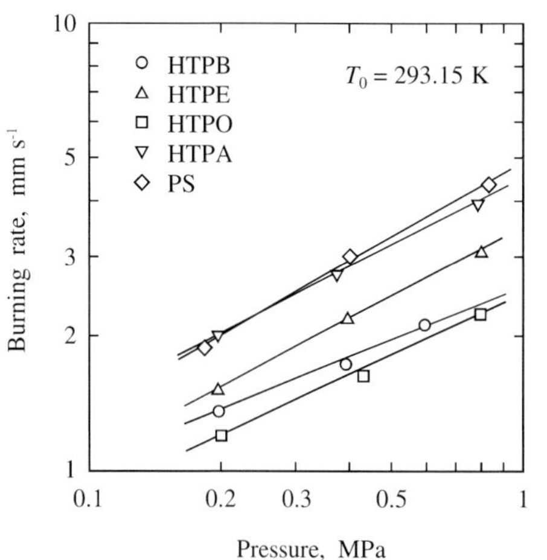

Fig. 7.17 Effect of binder on burning rate in the low-pressure region.

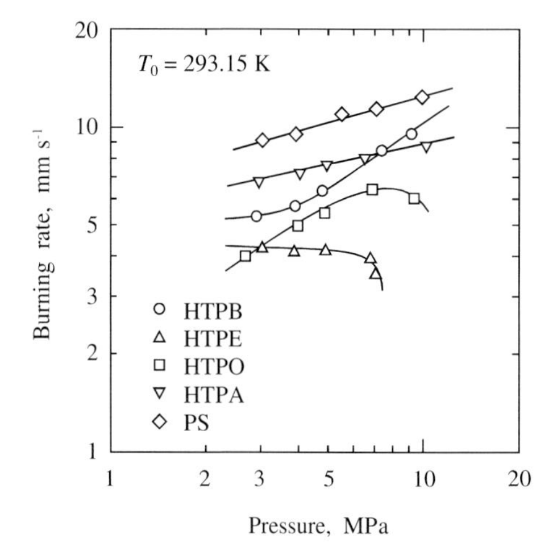

Fig. 7.18 Effect of binder on burning rate in the high-pressure region.

ent types of binders: HTPB, HTPE, HTPO, HTPA, and PS. These AP composite propellants are composed of $\xi_{AP}(0.80)$ and $\xi_{BDR}(0.20)$, and the AP particle diameters are bimodal at 200 μm and 20 μm with a mixture ratio of 56:24. The burning rate is seen to be highly dependent on the binder used. A high burning rate in the low-pressure zone below 1 MPa is obtained when HTPA or PS is used as the binder. The burning rate of HTPO propellant is doubled when this binder is replaced with HTPA. However, the pressure exponent of the burning rate is between 0.3 and 0.5 for all of the propellants. When the burning pressure is raised beyond 3 MPa, the burning rate of the HTPE propellant first becomes pressure-insensitive in a plateau region, and then further pressure has a negative effect on the burning rate, that is, mesa burning is seen at about 8 MPa. The burning is interrupted above this pressure. Similar burning rate characteristics are observed in the case of the HTPO binder. Thus, the binder not only affects the burning rate, but also the pressure exponent of the burning rate in the high-pressure zone.

7.1.2.3 **Temperature Sensitivity**

Fig. 7.19 shows the burning rates of AP-HTPB composite propellants at 243 K and 343 K. The propellants are composed of bimodal fine or coarse AP particles. The chemical compositions of the propellants are shown in Table 7.2. The burning rates of both propellants are seen to increase linearly in an ln r versus ln p plot in the pressure range 1.5–5 MPa, and also increase with increasing initial propellant temperature at constant pressure.[13] The burning rate increases and the temperature sensitivity decreases with decreasing AP particle size.

Table 7.2 Chemical compositions of AP composite propellants (% by mass).

Propellant	Binder	AP particle size (μm)				Catalyst
	HTPB	400	200	20	3	BEFP
AP(fn)	20			40	40	
AP(cn)	20	40	40			
AP(fc)	20			40	40	1.0
AP(cc)	20	40	40			1.0

Fig. 7.20 shows the effect of an added catalyst on the burning rates of propellants composed of fine or coarse AP particles. The added catalyst is 2,2-bis(ethylferrocenyl)propane (BEFP). The burning rates of both propellants are seen to be increased significantly by the addition of 1.0% BEFP. BEFP has a more pronounced effect on the burning rate of the propellant composed of fine AP particles than on that of the propellant composed of coarse AP particles. The temperature sensitivity of the propellant composed of fine AP particles with 1.0% BEFP is lower than that of the propellant composed of coarse AP particles with 1.0% BEFP.

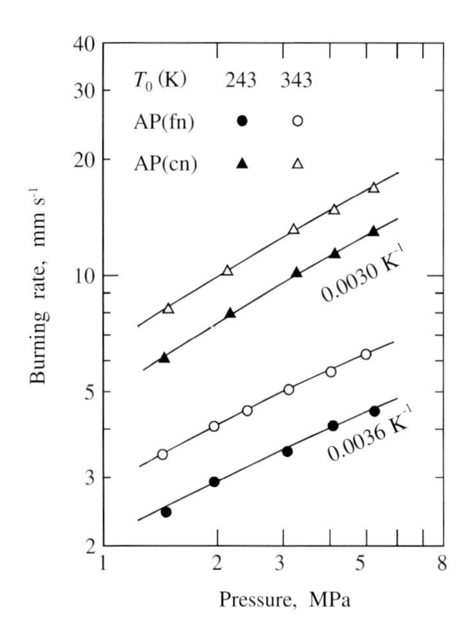

Fig. 7.19 Burning rates and their temperature sensitivity for AP-HTPB composite propellants composed of fine or coarse AP particles.

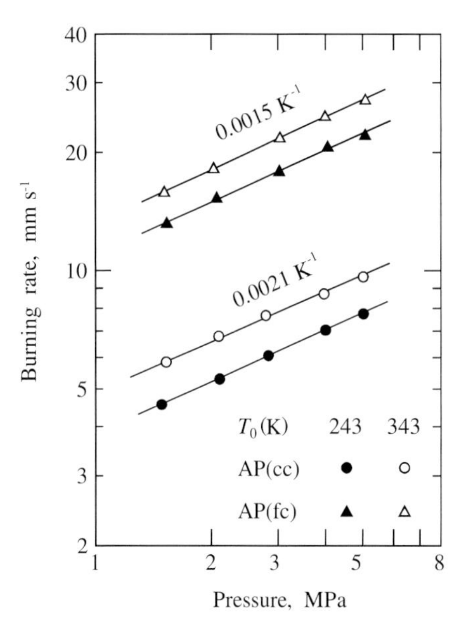

Fig. 7.20 Burning rates and their temperature sensitivity for BEFP-catalyzed AP-HTPB composite propellants composed of fine or coarse AP particles.

The relationship between temperature sensitivity and burning rate is shown in Fig. 7.21 as a function of AP particle size and burning rate catalyst (BEFP).[13] The temperature sensitivity decreases when the burning rate is increased, either by the addition of fine AP particles or by the addition of BEFP. The results of the temperature sensitivity analysis shown in Fig. 7.22 indicate that the temperature sensitivity of the condensed phase, Ψ, defined in Eq. (3.80), is higher than that of the gas phase, Φ, defined in Eq. (3.79). In addition, Φ becomes very small when the propel-

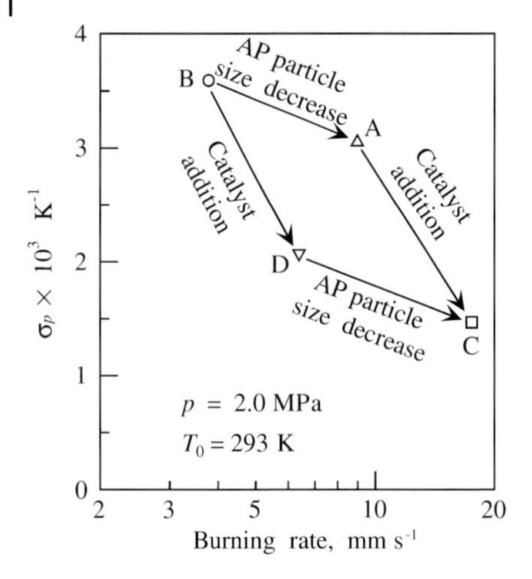

Fig. 7.21 Temperature sensitivity characteristics of AP-HTPB composite propellants.

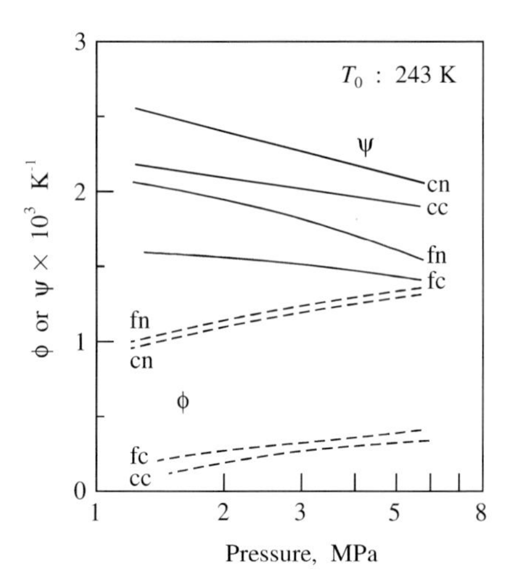

Fig. 7.22 Temperature sensitivities of the gas phase and the condensed phase of AP-HTPB composite propellants.

lant is catalyzed. The temperature sensitivity of AP-HTPB propellants depends largely on the reaction process in the condensed phase.

7.1.3
Catalyzed AP Composite Propellants

The burning rate of propellants is one of the important parameters for rocket motordesign. As described in Section 7.1.2, the burning rate of AP composite propellants is altered by changing the particle size of the AP used. The diffusional mixing process between the gaseous decomposition products of the AP particles and of the polymeric binder used as a fuel component determines the heat flux feedback from the gas phase to the condensed phase at the burning surface.[6,9] This process is a dominant factor in determining the burning rate, and it is for this reason that the burning rate is altered by changing the size of the AP particles.

The heat flux feedback from the gas phase to the burning surface is also determined by the chemical reaction rate in the gas phase. The reaction rate in the gas phase is altered by the addition of catalysts. The catalysts act either on the decomposition reaction of the condensed phase or on the reaction in the gas phase of the gaseous decomposition products. There are two types of catalysts: positive catalysts that increase the burning rate and negative catalysts that decrease the burning rate. Iron oxides and organic iron compounds act as positive catalysts, whereas LiF, $CaCO_3$, and $SrCO_3$ act as negative catalysts.

7.1.3.1 **Positive Catalysts**

The burning rates of AP composite propellants are increased by the addition of positive catalysts, which act to accelerate the decomposition reaction and/or the gas-phase reaction of the AP particles within the propellants.[14] Since catalysts act on the surface of the AP particles, the total surface area of the catalyst particles at a fixed concentration is an important factor for obtaining high catalyst efficiency. Though very finely divided iron oxides are used to increase the burning rate of AP composite propellants, organic iron compounds are more effective because of the formation of discrete iron oxide molecules during their decomposition. Typical iron compounds used are ferric oxides (Fe_2O_3 and Fe_3O_4), hydrated ferric oxide(FeO(OH)), *n*-butyl ferrocene (*n*BF), di-*n*-butyl ferrocene (D*n*BF), BEFP, and iron acetate. Organic iron compounds are also chemically bonded to polymers such as polybutadienes and polyesters.

As shown in Fig. 7.20, the burning rate of the AP composite propellant is increased approximately twofold by the addition of 1% BEFP. In general, the degree of the burning rate increase is proportional to the amount of catalyst added when the catalyst constitutes less than about 3% of the total mass, and the effect of the catalyst addition shows saturation behavior at about 5% by mass. Fig. 7.23 shows the burning rates of AP-HTPB composite propellants composed of $\xi_{AP}(0.80)$ and

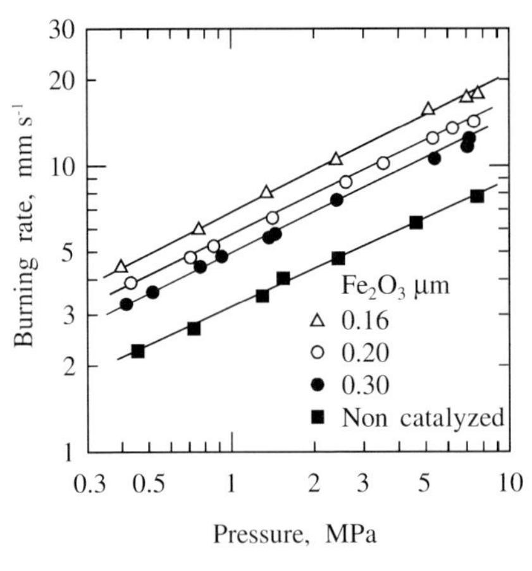

Fig. 7.23 Effect of the addition of a positive catalyst (Fe_2O_3) on burning rate and pressure exponent as a function of the particle size of Fe_2O_3.

ξ_{HTPB}(0.20) with and without the positive catalyst Fe_2O_3. The burning rate in the absence of Fe_2O_3 is seen to increase linearly in an ln burning rate versus ln pressure plot. When Fe_2O_3 particles are incorporated into the propellant, the burning rate increases at constant pressure, but the pressure exponent of the burning rate remains unchanged. The burning rate increases with decreasing d_0 at constant ξ_{AP} at constant pressure. For example, the burning rate at 5 MPa is increased from 6.2 mm s^{-1} to 10.3 mm s^{-1} by the addition of particles with d_0 = 0.30 μm and from 6.2 mm s^{-1} to 16.5 mm s^{-1} by the addition of particles with d_0 = 0.16 μm at constant ξ_{Fe2O3}(0.016). The surface area of each Fe_2O_3 particle plays a crucial role in determining the catalyst effect. Since the pressure exponent remains unchanged upon the addition of Fe_2O_3, the catalyst is considered to increase the rate of the gas-phase reaction, but not to have an effect on the reaction in the condensed phase.

Though copper oxides and $CuCrO_4$ are effective catalysts for increasing the burning rates of AP composite propellants, the thermal stability of the propellants is lowered and they become prone to spontaneous ignition. Organoboron compoundssuch as *n*-butyl carborane (*n*-BC), iso-butyl carborane (*i*-BC), and *n*-hexyl carborane(*n*-HC) are also effective catalysts. The burning rate of AP-HTPB propellants increases from 1 mm s^{-1} to 9 mm s^{-1} at 7 MPa by the addition of 13 % *n*-HC, as shown in Fig. 7.24. The pressure exponent remains relatively unchanged by the addition of *n*-HC in the wide pressure range from 2 MPa to 10 MPa. Since the amount of *n*-HC incorporated into the propellant is 13 % by mass, the boron atoms of the *n*-HC are considered to serve as a fuel component in the combustion wave. The thermal decomposition of *n*-HC generates boron atoms, which are oxidized by the AP decomposition fragments to produce heat at the burning surface. The increased heat flux enhances the decomposition of the AP particles and of the binder on the burning surface.

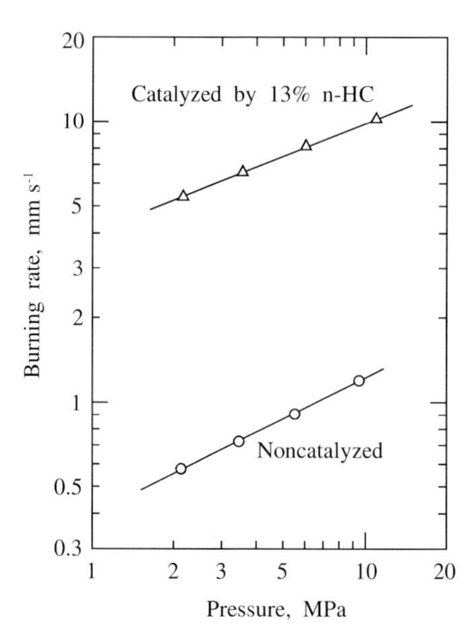

Fig. 7.24 Burning rate of an *n*-HC-catalyzed AP composite propellant, showing that the burning rate is increased drastically but that the pressure exponent remains unchanged by the addition of the catalyst.

The friction sensitivity of AP composite propellants increases significantly when organic iron and organic boron compounds are added. The ignition of AP composite propellants depends on their friction sensitivity, and this, in turn, relates to their burning rates.[14] Effective catalysts such as *n*-HC, ferrocene, and copper oxides render the composites very sensitive to mechanical friction. Propellants containing these catalysts are easily ignited during the formulation process.

7.1.3.2 **LiF Negative Catalyst**

7.1.3.2.1 Reaction of AP with LiF

The thermal decomposition process of AP particles is altered significantly when 10% LiF is added, as shown in Fig. 7.25. The decomposition of AP particles without LiF commences at about 570 K and 50% mass loss occurs at 667 K, which corresponds to the exothermic peak. The TG curve consists of a two-stage mass-loss process. The first stage corresponds to the first exothermic reaction at 635 K, and the second stage corresponds to the second exothermic reaction in the high-temperature region between 723 K and 786 K observed in the DTA experiments.[15]

When 10% LiF is added to the AP particles, an endothermic peak is observed at 516 K. The exothermic peak originally at 725 K now appears at 635 K. An endothermic reaction occurs between 520 K and 532 K. An additional exothermic peak is observed in the higher temperature region between 720 K and 790 K.

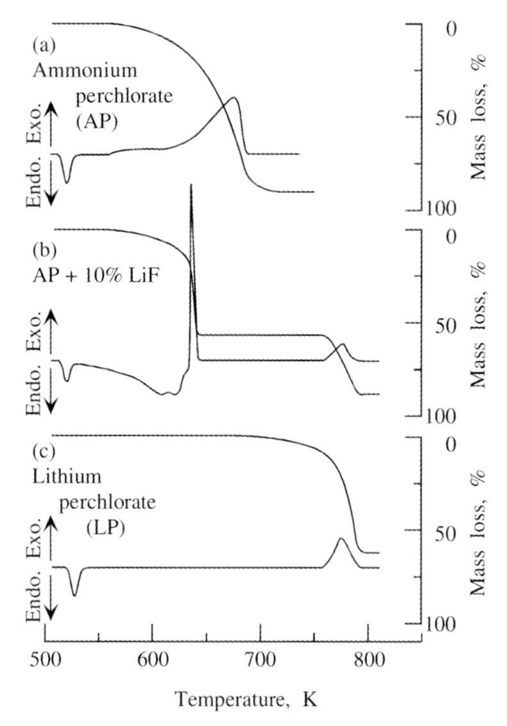

Fig. 7.25 Thermal decomposition processes of AP, AP + 10% LiF, and lithium perchlorate.

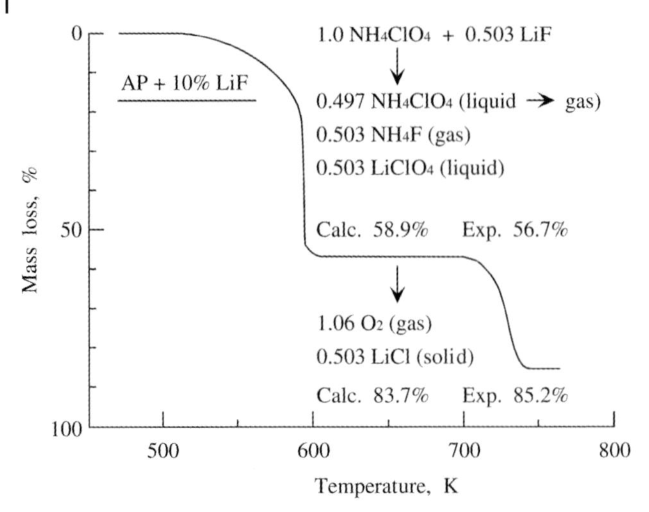

Fig. 7.26 Two-stage decomposition process of AP + 10% LiF measured by thermogravimetry.

An endothermic peak at 525 K is observed when lithium perchlorate ($LiClO_4$: LP) is thermally decomposed, caused by the phase change from solid to liquid. As the temperature is increased further, the molten $LiClO_4$ begins to decompose at about 680 K and rapid mass-loss decomposition occurs in the temperature range between 720 K and 790 K. This decomposition is similar to that of the AP particles with 10% LiF added.

The mass loss of the first-stage decomposition of the AP particles with 10% LiF is 56.7%, and that of the second-stage decomposition is 28.5%. The residue remaining above 790 K is 14.8%. As shown in Fig. 7.26, the two-stage decomposition process can be summarized as follows:[15]

(1) primary endothermic reaction

$$1.0\ NH_4ClO_{4\ (s)} + 0.503\ LiF_{\ (s)} \rightarrow 0.497\ NH_4ClO_{4\ (l)} + 0.503\ NH_4F_{\ (g)} + 0.503\ LiClO_{4\ (l)}$$

(2) first-stage decomposition

$$\rightarrow 0.497\ NH_{3\ (g)} + 0.497\ HClO_{4\ (g)} + 0.503\ LiClO_{4\ (l)}$$

(3) second-stage decomposition

$$\rightarrow 1.06\ O_{2\ (g)} + 0.503\ LiCl_{\ (s)}$$

where (s), (l), and (g) denote solid, liquid, and gas, respectively.

The decomposition process of the AP particles is greatly altered by the addition of LiF. After the rapid decomposition and gasification reactions at about 630 K, an undecomposed liquefied residue remains. This residue is the same material as observed in the temperature range between 640 K and 720 K. When the temperature is increased further, the liquefied residue again decomposes at about 750 K and produces a solidified residue at 790 K. This solidified residue is determined by chemical analysis to be LiCl.

7.1.3.2.2 Combustion Wave Structures with and without LiF

The burning rate characteristics of AP composite propellants with and without LiF are shown in Fig. 7.27. The burning rate of the AP composite propellant without LiF increases with increasing pressure. At constant pressure, the burning rates of the propellants with LiF decrease somewhat with increasing concentration of LiF. A further increase of the LiF concentration at a given pressure eventually results in self-extinction of the propellant.[15] Thus, one can state that LiF not only decreases the burning rate but also inhibits steady-state combustion at or below a certain pressure.

Temperature gradients just above the burning surface for non-catalyzed and 0.5 % LiF-catalyzed AP composite propellants are shown in Fig. 7.28 as a function of burning rate. The conductive heat flux feedback from the gas phase to the burning surface is relatively unchanged by the addition of 0.5 % LiF. The burning rate reduction on the addition of LiF is due to modification of the condensed-phase reaction of the AP particles. When both propellants are quenched during burning by rapid pressure decay, no AP particles are seen on the quenched catalyzed propellantwith 0.5 % LiF, as shown in Fig. 7.29. All AP particles on the burning surface decompose in the course of the pressure decay. LiCl particles remain on the quenched burning surface, which is consistent with the results obtained in the DTA and TG experiments.[15]

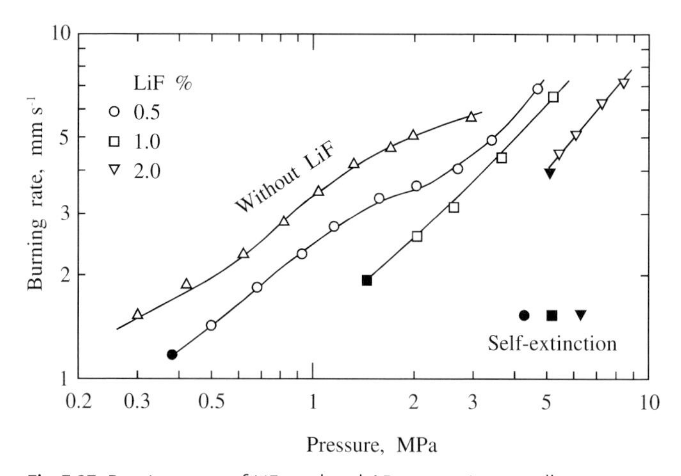

Fig. 7.27 Burning rates of LiF-catalyzed AP composite propellants, showing that the burning rate decreases and the pressure of self-interruption increases with increasing concentration of LiF.

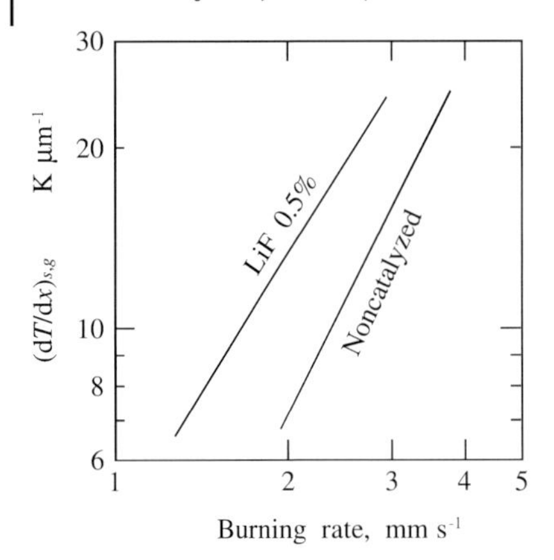

Fig. 7.28 Temperature gradients in the gas phase just above the burning surfaces of non-catalyzed and 0.5 % LiF-catalyzed AP composite propellants.

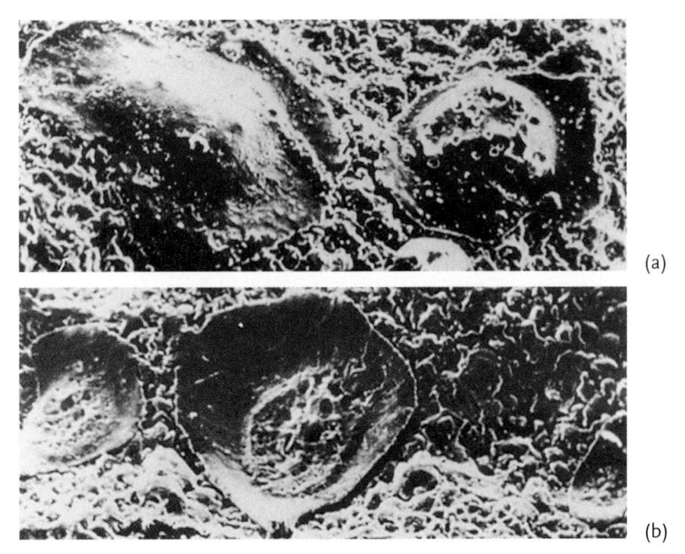

Fig. 7.29 Scanning electron microphotographs of quenched AP composite propellant burning surfaces without LiF (a) and with 0.5 % LiF (b), obtained by a pressure decay from 2 MPa to 0.1 MPa; the width of each photograph is 0.60 mm.

7.1.3.3 **$SrCO_3$ Negative Catalyst**

7.1.3.3.1 Burning Rate Characteristics

Similar to LiF, $SrCO_3$ acts as a negative catalysteduce the burning rate of AP composite propellants. As shown in Fig. 7.30, the burning rate is decreased in the high-pressure region by the addition of $SrCO_3$. The burning rate of a propellant with the composition $\xi_{AP}(0.88)$ and $\xi_{HTPB}(0.12)$ is decreased from 7.5 mm s^{-1} to 4.4 mm s^{-1} at 1.8 MPa by the addition of 2.0 % $SrCO_3$, which represents a decrease of 41 %. The burning rate remains unchanged on adding this catalyst in the low-pressure region

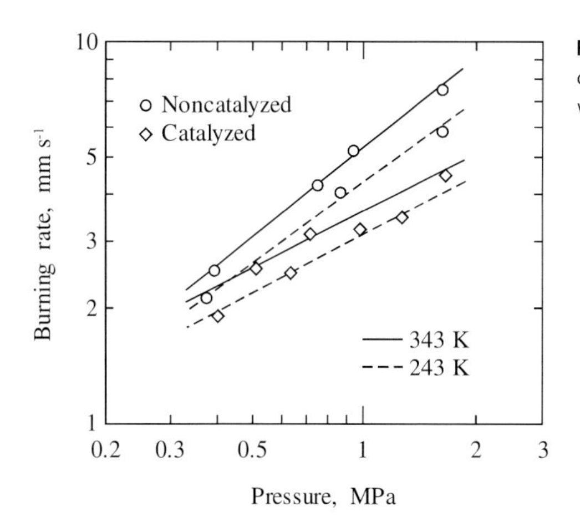

Fig. 7.30 Burning rates of AP-HTPB composite propellants with and without the negative catalyst $SrCO_3$.

at about 0.3 MPa. The pressure exponent of the burning rate is decreased from 0.70 to 0.50 at 343 K, and from 0.70 to 0.55 at 243 K. The temperature sensitivity of the burning rate is also decreased from 2.61×10^{-3} K^{-1} to 1.33×10^{-3} K^{-1} at 1.0 MPa when 2.0% $SrCO_3$ is added.[16,17]

7.1.3.3.2 Combustion Wave Structure

Studies on the combustion waves of AP composite propellants with and without $SrCO_3$ by micro-thermocouple techniques have revealed that the burning surface temperature is increased from 700 K to 970 K and that the heat of reaction is increased by approximately 40% in the pressure range between 0.3 MPa and 1.5 MPa. However, the reaction rate in the gas phase remains unchanged by the addition of the catalyst. The heat flux transferred back to the condensed phase of the AP propellant is reduced and hence the burning rate is reduced when $SrCO_3$ is added. The results suggest that $SrCO_3$ acts on the decomposition process of AP particles at the burning surface and has no effect on the gas-phase reaction.

7.1.3.3.3 Reaction of AP with $SrCO_3$

The thermal decomposition processes of mixtures of AP particles and $SrCO_3$ have been measured by DTA and TG. As shown in Fig. 7.31, the activation energy is 220 kJ mol^{-1} for the gasification of AP particles, but this is reduced to 101 kJ mol^{-1} by the addition of 2.0% $SrCO_3$.[16] Fig. 7.32 shows the results of a TG analysis of a mixture of 2 moles of AP and 1 mole of $SrCO_3$. The decomposition process is a two-stage reaction, similar to that for AP + LiF shown in Fig. 7.26. The mixture starts to gasify at about 630 K and produces CO_2, NH_3, H_2O, and liquid $Sr(ClO_4)_2$. This liquid $Sr(ClO_4)_2$ starts to decompose at about 710 K to produce O_2 and solid $SrCl_2$, and the reaction is complete at about 740 K. The reaction process is represented by:

(1) first-stage decomposition

$$2NH_4ClO_4 + SrCO_3 \rightarrow CO_{2\,(g)} + 2NH_{3\,(g)} + H_2O_{\,(g)} + Sr(ClO_4)_{2\,(l)}$$

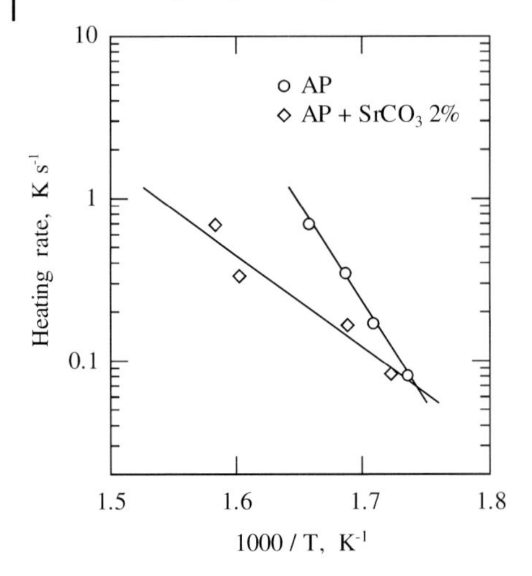

Fig. 7.31 Arrhenius plots for the thermal decomposition of AP with and without $SrCO_3$.

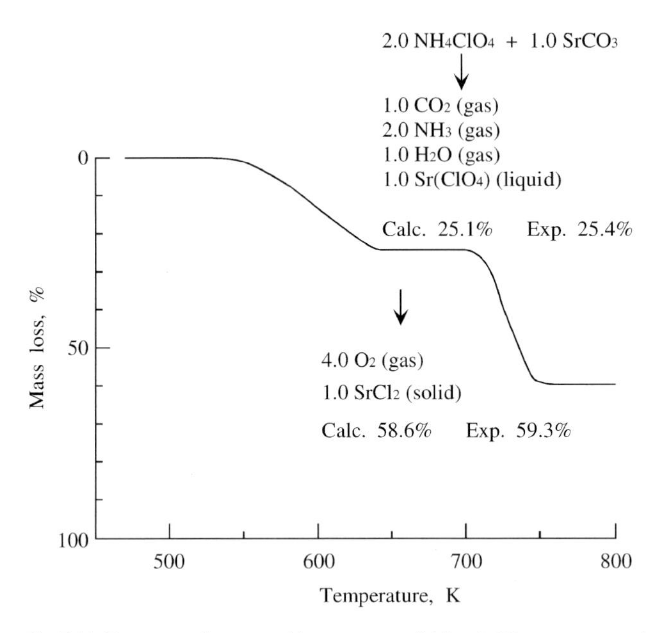

Fig. 7.32 Two-stage decomposition process of AP + $SrCO_3$ as measured by thermogravimetry.

(2) second-stage decomposition

$$Sr(ClO_4)_{2\ (l)} \rightarrow 4O_{2\ (g)} + SrCl_{2\ (s)}$$

Though the detailed mechanism of the action of $SrCO_3$ on AP propellants has yet to be fully understood, the results of the thermograms indicate that the site and mode of the action of $SrCO_3$ are in the condensed phase rather than in the gas phase. The decomposition process of AP particles is substantially altered by the addition of $SrCO_3$.

7.2 Nitramine Composite Propellants

Nitramines such as HMX and RDX are high energy density materials that produce high-temperature gaseous products. When nitramine particles are mixed with a polymeric material, a nitramine composite propellant is formed. The nitramine particles are decomposed and gasified at the burning surface of the propellant. The heat generated from the nitramine particles is partly consumed in decomposing the polymeric material that surrounds each nitramine particle. This decomposition is endothermic and so the average combustion temperature of the propellant is reduced by the incorporation of the polymeric material. The nitramines are stoichiometrically balanced materials and do not generate excess oxidizer fragments. Accordingly, the gaseous fuel fragments of the decomposition products of the polymeric material serve neither as fuel components nor as oxidizer components in the combustion of nitramine composite propellants. The gaseous fuel fragments serve to increase the volume of gas produced and to increase the specific impulse of the nitramine composite propellants.

7.2.1 Burning Rate Characteristics

7.2.1.1 Effect of Nitramine Particle Size

When particles of nitramines such as RDX and HMX are mixed with polyurethane binder, nitramine composite propellants are formed.[1,18–24] Plotting burning rate against pressure in an ln r versus ln p plot results in a straight line for both HMX and RDX composite propellants. The burning rate of an RDX propellant is higher than that of an HMX propellant for equal mass fractions, e. g. $\xi_{HMX}(0.80) = \xi_{RDX}(0.80)$, at the same pressure, and the pressure exponents of HMX and RDX propellants are 0.64 and 0.55, respectively.[19] When ξ_{HMX} or ξ_{RDX} is increased from 0.80 to 0.85, the burning rate is increased in both cases, and the pressure exponent is increased from 0.55 to 0.60 for the RDX propellant. However, the burning rate is less dependent on the particle size of the RDX or HMX used, e. g., when the mixture ratio of bimodal large and small particles is reversed: 120 μm/2 μm = 7:3 and 3:7 for RDX propellants and 225 μm/20 μm = 7:3 and 3:7 for HMX propellants.

7.2.1.2 Effect of Binder

The effects of four types of binders, HTPS, HTPE, HTPA, and HTPB, on the burning rates of HMX composite propellants are shown in Fig. 7.33. The physicochemi-

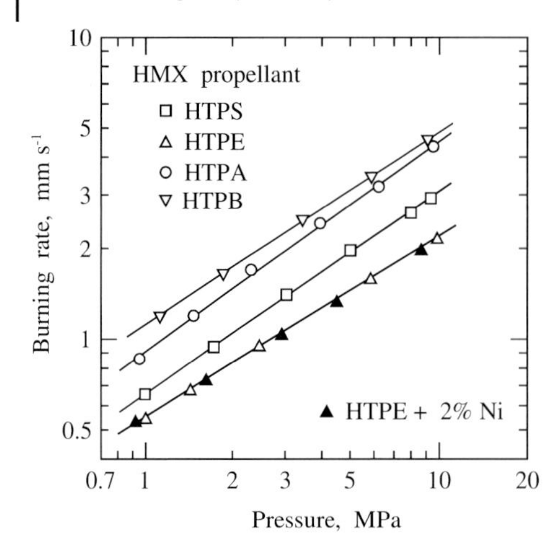

Fig. 7.33 Burning rates of HMX composite propellants composed of HTPS, HTPE, HTPA, and HTPB binders.

cal properties of these binders are shown in Table 4.3. These propellants each have the composition $\xi_{HMX}(0.80)$ and $\xi_{BDR}(0.20)$. The HMX particles are a bimodal mixture consisting of 70% large (220 µm in diameter) and 30% small (20 µm in diameter). Of these four binders, HTPS has the highest oxygen content and HTPB the lowest.[21] The adiabatic flame temperatures of the HTPS, HTPE, HTPA, and HTPB propellants are 1940, 1910, 2040, and 1800 K, respectively.

The burning rates of all four composite propellants give approximately straight lines when ln *r* is plotted against ln *p*, and the pressure exponents range from 0.62 to 0.73. For three of the propellants, the order of their burning rates correlates with the order of their adiabatic flame temperatures, the exception being the HTPB propellant, which exhibits the highest burning rate even though its adiabatic flame temperature is the lowest. The temperature sensitivities of burning rate, as defined in Eq. (3.73), are 0.0022 K^{-1} for the RDX-HTPA propellant and 0.0039 K^{-1} for the RDX-HTPS propellant.[21]

7.2.2
Combustion Wave Structure

Since the physical structures of nitramine composite propellants are heterogeneous, their combustion wave structures are also considered to be heterogeneous, as in the case of AP composite propellants. However, a difference between AP propellants and RDX propellants is evident in Figs. 7.34 (a) and (b). The luminous flame appears almost directly above the burning surface in the case of the AP propellant (a). On the other hand, the luminous flame stands some distance above the burning surface in the case of the RDX propellant (b). The flame stand-off distance between the burning surface and the luminous flame front is similar to the dark zone of double-base propellants.[19] However, the burning surface of the RDX propellant composed of HTPB binder is covered with carbonaceous fragments and so the gas phase is heterogeneous. Thus, the luminous flame stems directly from these carbonaceous fragments as well as from the burning surface.

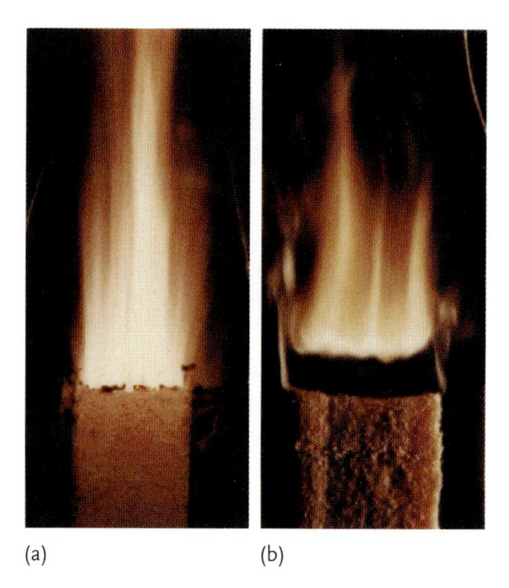

Fig. 7.34 Flame photographs of an AP composite propellant (a) and an RDX composite propellant (b) showing that the luminous flame front of the RDX composite propellant is distended from the burning surface:

	Mass fraction			p	r
	AP	RDX	PU	MPa	mm s^{-1}
(a)	0.80	–	0.20	2.0	5.3
(b)	–	0.80	0.20	2.0	1.1

The presence of carbonaceous fragments can be attributed to the low oxygen content in the HTPB binder, 3.6 %, compared to oxygen contents in the HTPS, HTPE, and HTPA binders of more than 25 %, as shown in Table 4.3.

The combustion wave structure of RDX composite propellants is homogeneous and the temperature in the solid phase and in the gas phase increases relatively smoothly as compared with AP composite propellants. The temperature increases rapidly on and just above the burning surface (in the dark zone near the burning surface) and so the temperature gradient at the burning surface is high. The temperature in the dark zone increases slowly. However, the temperature increases rapidly once more at the luminous flame front. The combustion wave structure of RDX and HMX composite propellants composed of nitramines and hydrocarbon polymers is thus very similar to that of double-base propellants composed of nitrate esters.[19]

The flame stand-off distance, L_d, defined in Eq. (3.70), decreases with increasing pressure, and the pressure exponent of the flame stand-off distance, d, ranges from –1.9 to –2.3 for RDX and HMX propellants. The overall order of the reaction in the dark zone is determined to be $m = 2.5$–2.8. This is approximately equal to the overall order of the reaction in the dark zone in the case of double-base propellants, $m = 2.5$, which would suggest close similarity of the reaction pathways in the dark zone for nitramine composite propellants and double-base propellants.

The initial decomposition products of RDX and HMX contain relatively high concentrations of NO_2. Exothermic reaction between the NO_2 and other decomposition fragments from the nitramine occurs at the burning surface. The gaseous decomposition products from the binder diffuse into the exothermic reaction zone and hence the temperature increases rapidly at the burning surface. The reduction of NO_2 produces NO, which also reacts exothermically to produce the luminous flame zone. Since the reaction involving NO is known to be slow and highly pressure-dependent, the luminous flame stands some distance above the burning surface.

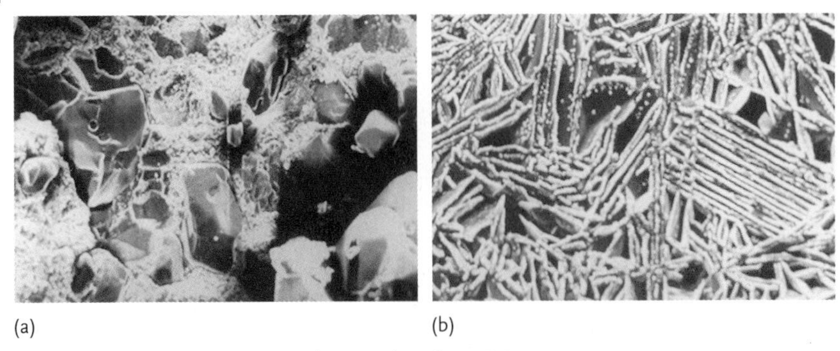

(a) (b)

Fig. 7.35 Scanning electron microphotographs of an RDX composite propellant surface before combustion (a) and after quenching (b) by rapid pressure decay from burning at 2 MPa.

However, the luminous flame front rapidly approaches the burning surface when the pressure is increased. This reaction is also caused by the reduction of NO to N_2, as in the reaction process of double-base propellants.

RDX and HMX are stoichiometrically balanced materials and no excess oxidizer fragments are produced in their combustion products. The binders surrounding the nitramine particles do not generate fuel components, unlike in the case of the binders of AP composite propellants. Fig. 7.35 shows a pair of microphotographs of an RDX-PU propellant before burning (a) and after quenching (b). The quenched burning surface shows finely divided recrystallized RDX homogeneously dispersed on the surface. The RDX particles and the binder melt and diffuse into each other and form an energetic mixture on the burning surface during combustion.[19] This energetic mixture acts as a homogeneous propellant, much like a double-base propellant. The gaseous decomposition products ejected from this energetic mixture form a homogeneous gas-phase structure and produce the successive reaction zones, i. e. the dark zone and the luminous flame zone.

As discussed in Section 6.4.6, metallic nickel is known to catalyze the reaction between NO and hydrocarbon gases. Similar to the situation with double-base propellants, nickel powder acts on the gas-phase reaction of nitramine composite propellants. When 2.0% nickel powder (0.1 μm in diameter) is added to HMX-HTPE propellant, the burning rate remains unchanged as shown in Fig. 7.33. However, the dark zone between the burning surface and the luminous flame front is eliminated completely, and so the luminous flame approaches the burning surface. The NO molecules produced by the reduction of NO_2 at the burning surface react with hydrocarbon gases produced by the decomposition of the HTPE binder. This reaction is enhanced by the addition of the nickel powder. However, the heat flux feedback from the gas phase to the burning surface remains unchanged because the enhanced NO reduction does not have an effect on the NO_2 reduction at the burning surface, as is the case for NO_2 reduction in the fizz zone of double-base propellants.

7.2.3 HMX-GAP Propellants

7.2.3.1 Physicochemical Properties of Propellants

Though no excess oxidizer fragments are formed when HMX burns, high-temperature combustion products are formed. As shown in Fig. 5.3, the burning rate of HMX is very low even though the flame temperature is high.[25] On the other hand, the burning rate of GAP is very high even though the flame temperature is low, as shown in Fig. 5.17. Mixing HMX particles with GAP formulates HMX-GAP propellants, the burning rate characteristics of which differ from those of other types of propellants such as double-base propellants and AP composite propellants.

The physicochemical properties of propellants with the compositions $\xi_{HMX}(0.4)$, $\xi_{HMX}(0.6)$, and $\xi_{HMX}(0.8)$ are shown in Table 7.3. Since the energy density of HMX is higher than that of GAP, the adiabatic flame temperatures of HMX-GAP propellants increase with increasing ξ_{HMX}.

Table 7.3 Physicochemical properties of HMX-GAP composite propellants.

ξ_{HMX}	0.4	0.6	0.8
Flame temperature at 5 MPa, T_f (K)	1628	1836	2574
Molecular mass, M_g (kg kmol^{-1})	19.2	18.9	21.1
Density, ρ_p (kg m^{-3})	1460	1580	1770

7.2.3.2 Burning Rate and Combustion Wave Structure

The burning rate of a GAP-HMX propellant with the composition $\xi_{HMX}(0.80)$ is seen to increase linearly with increasing pressure in an ln *r* versus ln *p* plot, as shown in Fig. 7.36. The pressure exponent increases and the temperature sensitivity decreases with increasing ξ_{HMX}, as shown in Fig. 7.37. The burning rate of GAP binder shown in Fig. 5.17 is higher than that of HMX shown in Fig. 5.3, even though the flame temperature of GAP binder is 1890 K lower than that of HMX (see Tables 4.6 and 4.8). In Fig. 7.38, the burning rates of GAP-HMX propellants are shown as a function of ξ_{HMX} at different pressures. The burning rate decreases with increasing ξ_{HMX} in the range $\xi_{HMX} < 0.6$ and increases with increasing ξ_{HMX} in the range $0.6 < \xi_{HMX}$.

The gas-phase reaction of HMX-GAP propellants is a two-stage process occurring in two zones:[25] in the first-stage reaction zone, the temperature increases rapidly on and just above the burning surface. In the second-stage reaction zone, the temperature again increases rapidly at some distance from the burning surface. In the preparation zone between the first and second stages, the temperature increases very slowly. In the second-stage reaction zone, the luminous flame is produced. The flame stand-off distance, L_g, of a propellant with $\xi_{HMX}(0.80)$ is seen to decrease linearly with increasing pressure in an ln L_g versus ln *p* plot, as shown in Fig. 7.39 (the line marked "non-catalyzed").

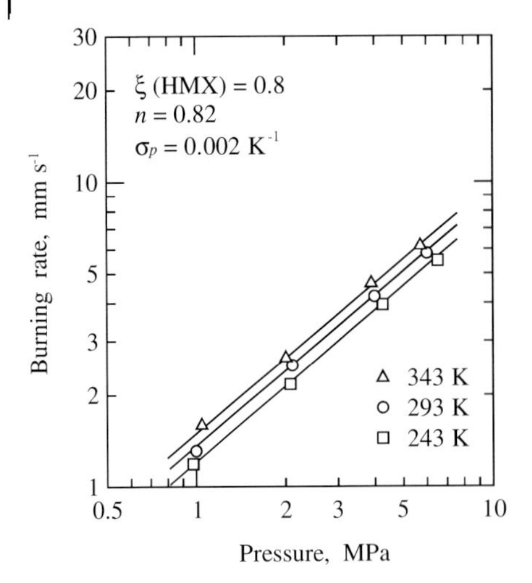

Fig. 7.36 Dependence of burning rate on pressure and temperature sensitivity thereof for an HMX composite propellant.

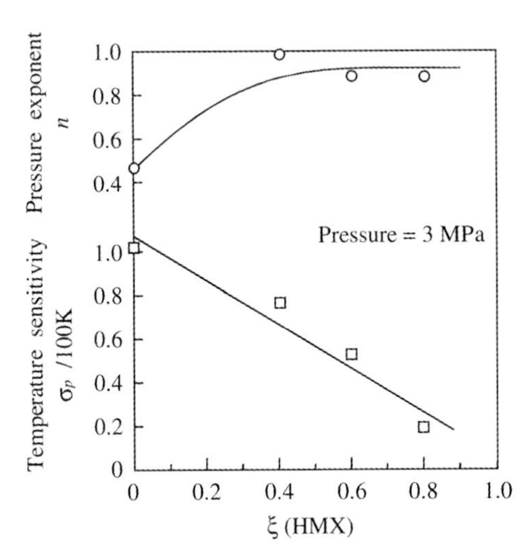

Fig. 7.37 Pressure exponent and temperature sensitivity of an HMX composite propellant as a function of ξ_{HMX}.

The overall reaction rate in the second-stage reaction zone (preparation zone), ω_g, is determined according to Eq. (5.5). The computed results show that the reaction rate increases linearly with increasing pressure in an ln ω_g versus ln p plot. Since the burning rate is represented by Eq. (3.68), ϕ and φ defined in Eqs. (3.74) and (3.75) can be obtained. As is clear from Eq. (3.73), the burning rate increases as ϕ increases and as φ decreases. The burning surface temperature and the temperature gradient at the burning surface are approximately 695 K and 2.3×10^6 K m^{-1} at 0.5 MPa, respectively. The heat flux transferred back from the gas phase to the burning surface is 190 kW m^{-2}. In the computations of ϕ and Q_s, the physical parameter values used are: $\lambda_g = 8.4 \times 10^{-5}$ kW m^{-1} K^{-1}, ρ_p = 1770 kg m^{-3}, and c_p = 1.30 kJ kg^{-1} K^{-1}. Substituting T_0 = 293 K, T_s = 695 K,

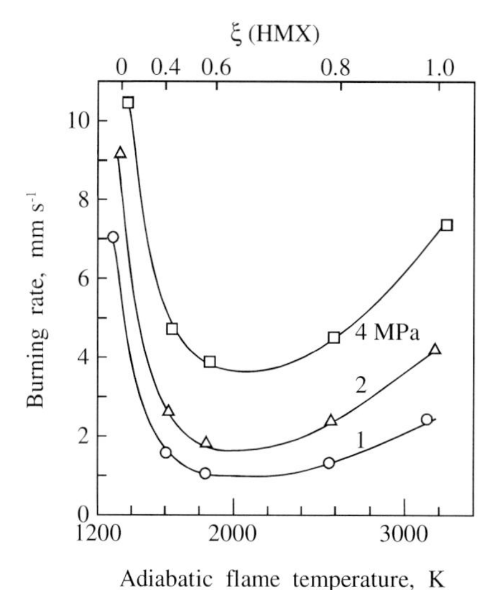

Fig. 7.38 Burning rates of HMX composite propellants as a function of adiabatic flame temperature or ξ_{HMX}, showing the existence of a minimum in the burning rate at about ξ_{HMX} (0.6).

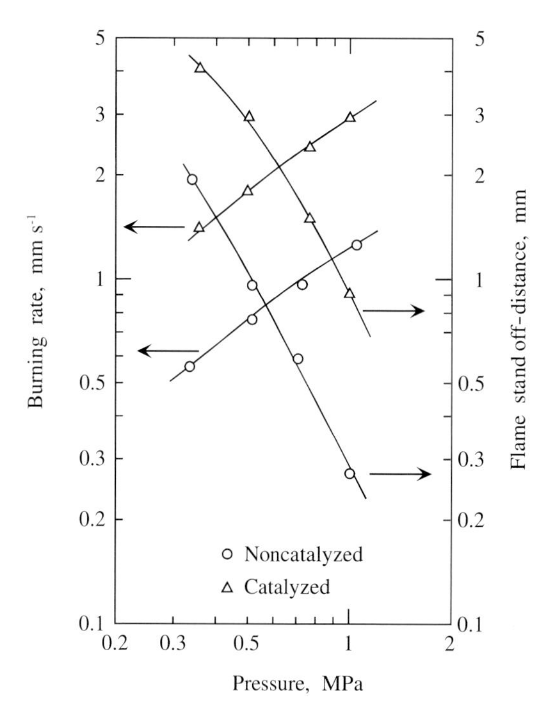

Fig. 7.39 Burning rates and flame stand-off distances of non-catalyzed and catalyzed HMX-GAP composite propellants.

and $\phi = 2.3 \times 10^6$ K m^{-1} into Eqs. (3.74) and (3.75), Q_s of the propellant with the composition $\xi_{HMX}(0.8)$ is determined as 369 kJ kg^{-1}. The results indicate that ϕ and Q_s play dominant roles in determining the burning rates of GAP-HMX composite propellants.[22]

7.2.4 Catalyzed Nitramine Composite Propellants

Lead compounds are effective in increasing the burning rate of double-base propellants, leading to so-called super-rate burning. A similar super-rate burning effect is observed when lead compounds are incorporated into composite propellants consisting of nitramine particles and a polymeric binder. Thus, although the chemical structures and properties of nitramines differ from those of nitrate esters, super-rate burning effects are obtained by the use of the same lead compounds.

7.2.4.1 Super-Rate Burning of HMX Composite Propellants

The super-rate burning rates of HMX-HTPE and HMX-HTPS propellants are shown in Figs. 7.40 and 7.41, respectively. The basic chemical compositions and

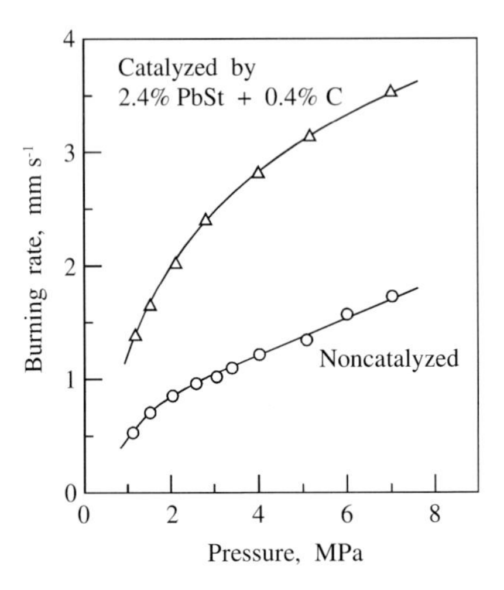

Fig. 7.40 Super-rate burning of catalyzed HMX-HTPE composite propellant.

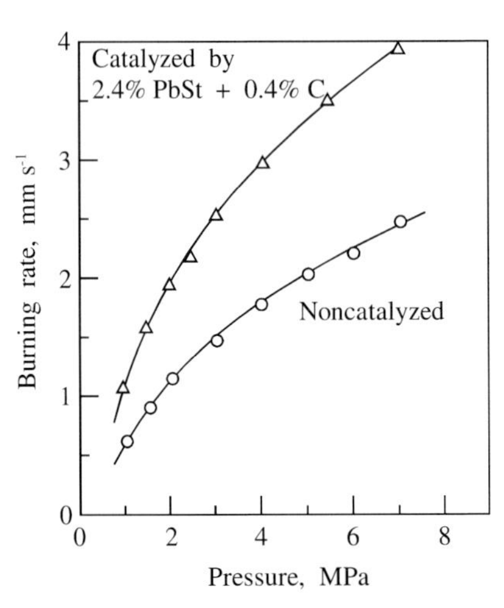

Fig. 7.41 Super-rate burning of catalyzed HMX-HTPS composite propellant.

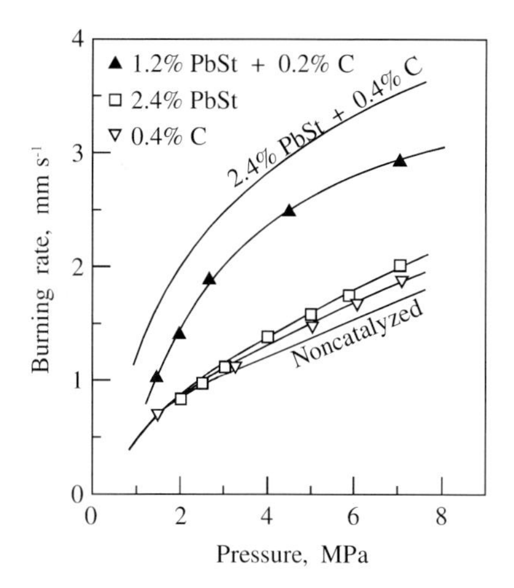

Fig. 7.42 Effect of catalyst on super-rate burning of HMX-HTPE composite propellant.

physicochemical properties of the non-catalyzed HMX-HTPE and HMX-HTPS propellants correspond to those of the respective propellants that yielded the data in Fig. 7.33. The catalysts, 2.4% PbSt and 0.4% carbon, were incorporated into the non-catalyzed propellants. The nature of the super-rate burning observed for HMX-inert polymer propellants is similar to that of the super-rate burning of double-base propellants.[23]

With a view to delineating the roles of PbSt and C in promoting super-rate burning, the burning rates of propellants to which 2.4% PbSt and 0.4% C had been added individually have been measured. As shown in Fig. 7.42, only very small increases in burning rate are seen for both propellants when used alone. However, when a mixture of 1.2% PbSt and 0.2% C is incorporated, super-rate burning occurs.[26] The degree of super-rate burning is increased when the amounts of both additives are doubled. Lead stearate acts as a super-rate burning catalyst and C acts as a catalyst modifier. The two additives act synergistically to produce a high degree of super-rate burning.

Though the physicochemical properties of HTPE and HTPS are different, both are subject to a similar super-rate burning effect. However, the magnitude of the effect is dependent on the type of binder used. As in the case of double-base propellants, the combustion wave structures of the respective propellants are homogeneous, even though the propellant structures are heterogeneous and the luminous flames are produced above the burning surfaces.

7.2.4.2 **Super-Rate Burning of HMX-GAP Propellants**

A typical super-rate burning of an HMX-GAP composite propellant is shown in Fig. 7.43. The lead catalyst is a mixture of lead citrate (LC: PbCi), $Pb_3(C_6H_5O_7)_2 \cdot xH_2O$, and carbon black (CB). The composition of the catalyzed HMX-GAP propellant in terms of mass fractions is as follows: $\xi_{GAP}(0.194)$, $\xi_{HMX}(0.780)$, $\xi_{LC}(0.020)$, and $\xi_{C}(0.006)$. GAP is cured with 12.0% hexamethylene diisocyanate (HMDI) and then crosslinked with 3.2% trimethylolpropane (TMP) to

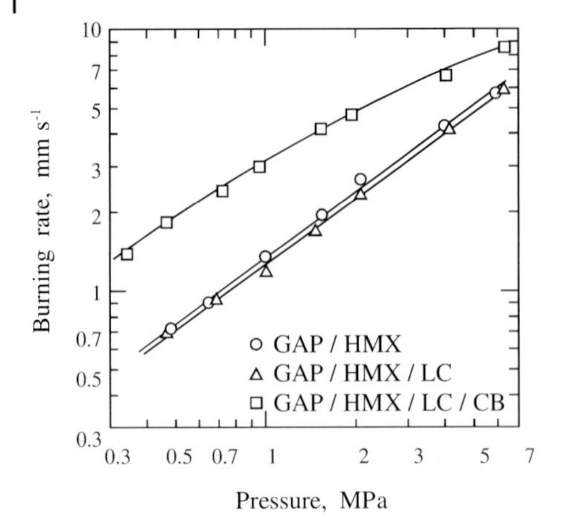

Fig. 7.43 Effect of catalyst on super-rate burning of HMX-GAP composite propellant.

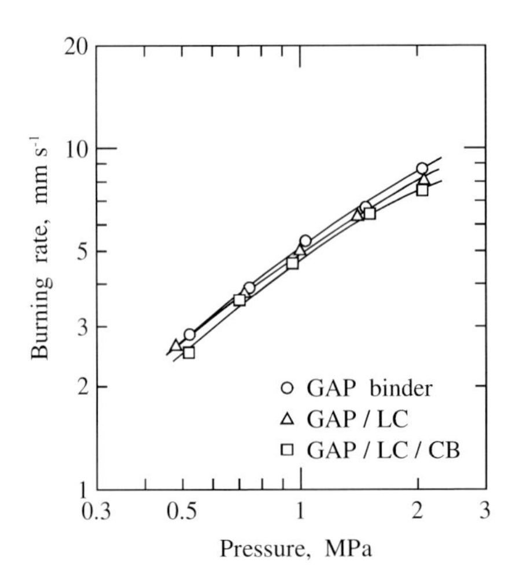

Fig. 7.44 Effect of catalyst on the burning rate of GAP copolymer, showing no burning rate increase by the addition of lead citrate and/or carbon black.

formulate the GAP binder. The HMX particles are finely divided crystalline β-HMXwith a bimodal size distribution (70% 2 μm and 30% 20 μm in diameter).

Super-rate burning only occurs when a combination of LC and CB is incorporated into an HMX-GAP propellant.[25] Fig. 7.44 shows that the addition of LC and/or CB to GAP binder has little or no effect on the burning rate. Likewise, no effect on burning rate is seen when these additives are incorporated into HMX pressed pellets. These experimental observations indicate that super-rate burning of HMX-GAP propellants occurs only when HMX, GAP, a lead compound, and carbon are mixed together.

Fig. 7.45 shows a set of flame photographs of HMX-GAP propellants with and without catalysts. The luminous flame front of the non-catalyzed propellant is almost attached the burning surface at 0.5 MPa (a). When the propellant is catalyzed, the luminous flame is distended from the burning surface at the same pressure (b). Since the heat flux transferred back from the gas phase and the heat of reaction at

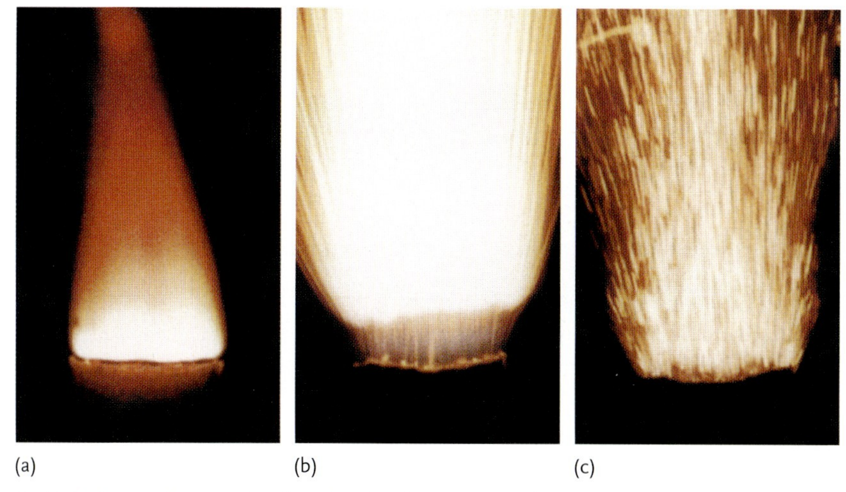

(a) (b) (c)

Fig. 7.45 Flame photographs of catalyzed and non-catalyzed HMX-GAP composite propellants:

	Mass fraction			p	r
	ξ_{HMX}	ξ_{GAP}	$\xi_{Cat.}$	MPa	mm s^{-1}
(a)	0.80	0.20	0	0.5	0.8
(b)	0.80	0.20	0.03	0.5	1.9
(c)	0.80	0.20	0.03	1.5	4.1

the burning surface are increased, the flow velocity in the gas phase is increased,[26] and so too is the reaction distance to produce the luminous flame. It is evident that the heat flux transferred back from the luminous flame front to the burning surface has a negligible effect on the burning rate. Carbonaceous materials are formed in the super-rate burning regime, as in the case of catalyzed double-base propellants. When the pressure is increased to 1.5 MPa, the luminous flame front is blown downstream and numerous carbonaceous particles are ejected from the burning surface (c).

7.2.4.3 **LiF Catalysts for Super-Rate Burning**

In order to avoid the use of lead compounds on environmental grounds, lithium fluoride (LiF) has been chosen to obtain super-rate burning of nitramine composite propellants.[27,28] Typical chemical compositions of HMX composite propellants-with and without LiF are shown in Table 7.4. The non-catalyzed HMX propellant is used as a reference pyrolant to evaluate the effect of super-rate burning. The HMX particles are of finely divided, crystalline β-HMX with a bimodal size distribution. Hydroxy-terminated polyether (HTPE) is used as a binder, the OH groups of which are cured with isophorone diisocyanate. The chemical properties of the HTPE binder are summarized in Table 7.5.

Table 7.4 Typical compositions of HMX composite propellants.

Pyrolants	HMX	HTPE	LiF	C
Non-catalyzed	0.80	0.20		
Catalyzed	0.80	0.20	0.01	
Catalyzed	0.80	0.20	0.01	0.01

Table 7.5 Chemical properties of HTPE polymer.

Chemical formula	$C_{5.194}H_{9.840}O_{1.608}N_{0.149}$
Oxygen content	25.7% by mass
Heat of formation	–302 MJ kmol^{-1} at 298 K

Fig. 7.46 shows the burning rates of the catalyzed HMX propellants and demonstrates a drastically increased burning rate, i. e., super-rate burning. However, LiF or C alone are seen to have little or no effect on burning rate. The super-rate burning occurs only when a combination of LiF and C is incorporated into the HMX propellant. The results indicate that LiF acts as a catalyst to produce super-rate burning of the HMX propellant only when used in tandem with a small amount of C. The C (carbon black) is considered to act as a catalyst promoter. A similar super-rate burning effect is observed when the same catalysts are added to nitropolymer propellants.

HMX composite propellants are composed of crystalline HMX particles and polymeric materials, and so their physical structures are heterogeneous. On the other hand, nitropolymer propellants are composed of mixtures of nitrate esters such as NC and NG, and their physical structures are homogeneous. Moreover, HMX pro-

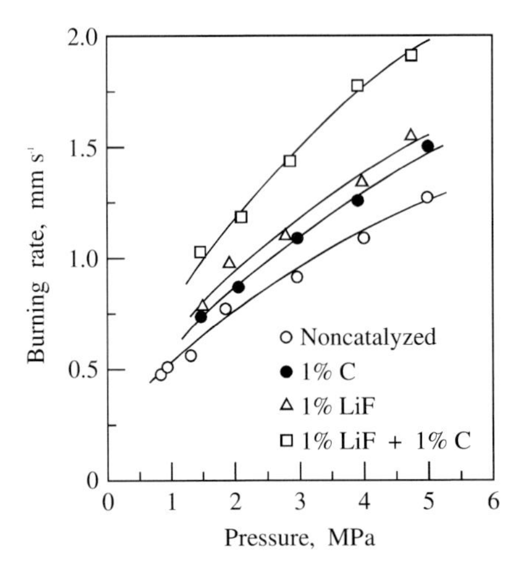

Fig. 7.46 Burning rate characteristics of non-catalyzed and catalyzed HMX composite propellants.

pellants are characterized by -N–NO_2 bonds, whereas nitropolymer propellants are characterized by -O–NO_2 bonds.

It is well known that the super-rate burning of nitropolymer propellants diminishes with increasing pressure in the region 5–100 MPa and that the pressure exponent of burning rate decreases.[6–9] This burning rate mode is called plateau burning. As for these nitropolymer propellants catalyzed with LiF and C, HMX propellants catalyzed with LiF and C also show plateau burning.

7.2.4.4 **Catalyst Action of LiF on Combustion Wave**

The combustion wave structure of HMX propellants catalyzed with LiF and C is similar to that of catalyzed nitropolymer propellants: the luminous flame stands some distance above the burning surface at low pressures and approaches the burning surface with increasing pressure. The flame stand-off distance from the burning surface to the luminous flame front is increased at constant pressure when the propellant is catalyzed. The flame stand-off distance decreases with increasing pressure for both non-catalyzed and catalyzed propellants.

The burning surface of an HMX propellant only becomes covered with carbonaceous materials when the propellant is catalyzed with both LiF and C. This surface structure is similar to the burning surface of an HMX propellant catalyzed with a lead compound and C. The results indicate that the combustion mode and the action of LiF are the same as those resulting from the use of lead compounds to produce super-rate and plateau burning of nitramine propellants.

The combustion wave of an HMX composite propellant consists of successive reaction zones: the condensed-phase reaction zone, a first-stage reaction zone, a second-stage reaction zone, and the luminous flame zone. The combustion wave structure and temperature distribution for an HMX propellant are shown in Fig. 7.47. In the condensed-phase reaction zone, HMX particles melt together with the polymeric binder HTPE and form an energetic liquid mixture that covers the burning surface of the propellant. In the first-stage reaction zone, a rapid exother-

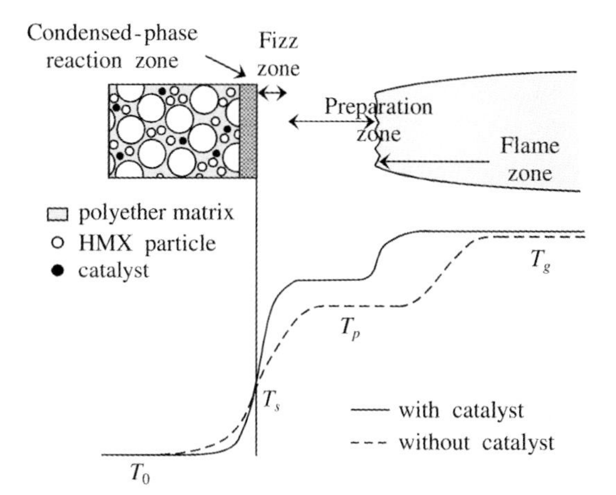

Fig. 7.47 Combustion wave structures of non-catalyzed and catalyzed HMX composite propellants.

mic reaction between NO_2 and fuel fragments occurs and increases temperature in the gas phase just above the burning surface. In the second-stage reaction zone, the NO and N_2O produced in the first-stage reaction zone slowly react with fuel fragments and so the temperature rise in this zone is only moderate. In the luminous flame zone, the gas-phase reaction between NO and N_2O and the remaining fuel fragments occurs and produces the final reaction products of N_2, H_2O, CO, and $C_{(s)}$, and the temperature reaches its maximum. The major light emission from the luminous flame zone is caused by the carbon soot. These reaction processes and the gas-phase structure are very similar to those resulting from the combustion of nitropolymer propellants.

The length of the second-stage reaction zone, i. e., the flame stand-off distance, decreases with increasing pressure, which is represented by $L_g = ap^d$, where L_g is the length of the second-stage reaction zone, p is pressure, d is the pressure exponent of the preparation zone, and a is a constant. It should be noted that the length of the first-stage reaction zone is of the order of 0.1 mm at 1 MPa. The pressure exponent, d, of the catalyzed HMX propellant is determined to be –2.0, and this remains unchanged by the addition of LiF and C. The reaction rate in the gas phase, ω_g, is given by Eq. (3.64). If one assumes that the reaction rate is constant throughout the second-stage reaction zone, the effective overall reaction rate in the gas phase, ω_g, is obtained by using the mass continuity equation at the interface of the gas phase and the solid phase, that is, $\omega_g \sim p^{n-d} \sim p^k$, where k is the reaction order in the preparation zone. Fig. 7.48 shows the reaction rate as a function of pressure. The overall order of the reaction in the preparation zone is given by the slope, $k = n - d$, and is determined to be 2.6 for HMX propellants both with and without LiF and C.

The results indicate that the basic chemical reaction mechanism in the gas phase, which involves the reduction of NO to N_2, remains unchanged by the addi-

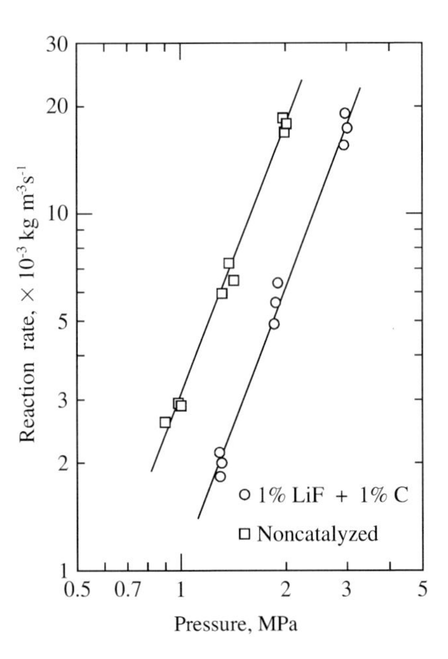

Fig. 7.48 Variation of reaction rate in the preparation zone with pressure for non-catalyzed and catalyzed HMX composite propellants.

tion of the catalyst. The temperature gradient in the gas phase just above the burning surface is increased significantly by the addition of LiF and C. This is again similar to the fizz zone of nitropolymer propellants catalyzed with lead compounds and C. The increased heat flux feedback from the gas phase to the burning surface increases the burning rate of catalyzed HMX composite propellants.

Though detailed mechanisms of the super-rate burning of nitramine composite propellants have yet to be fully understood, it is clear that the initial action of the catalysts (lead compounds or LiF in combination with C) is in the condensed phase and that the reaction rate in the gas phase is increased. This increased reaction rate increases the heat flux feedback from the gas phase to the burning surface, which increases the burning rate of the nitramine composite propellant. Furthermore, carbonaceous materials are formed when nitramine composite propellants are catalyzed by lead compounds or LiF together with C. Since the thermal diffusivity of these carbonaceous materials is higher than that of the reactive gases in the gas phase, the heat flux transferred back from the gas phase to the burning surface through the carbonaceous materials is higher. It is these two effects that cause the observed super-rate burning of both nitropolymer propellants and nitramine composite propellants.

7.3 AP-Nitramine Composite Propellants

7.3.1 Theoretical Performance

Since rocket propellants are composed of oxidizers and fuels, the specific impulseis essentially determined by the stoichiometry of these chemical ingredients. Nitramines such as RDX and HMX are high-energy materials and no oxidizers or fuels are required to gain further increased specific impulse. AP composite propellants composed of AP particles and a polymeric binder are formulated so as to make the mixture ratio as close as possible to their stoichiometric ratio. As shown in Fig. 4.14, the maximum I_{sp} is obtained at about $\xi_{AP}(0.89)$, with the remaining fraction being HTPB used as a fuel component.

When some portion of the AP particles contained within an AP composite propellant is replaced with nitramine particles, an AP-nitramine composite propellantis formulated. However, the specific impulse is reduced because there is an insufficient supply of oxidizer to the fuel components, i. e., the composition becomes fuel-rich. The adiabatic flame temperature is also reduced as the mass fraction of nitramine is increased. Fig. 7.49 shows the results of theoretical calculations of I_{sp} and T_f for AP-RDX composite propellants as a function of ξ_{RDX}. The propellants are composed of $\xi_{HTPB}(0.13)$ and the chamber pressure is 7.0 MPa with an optimum expansion to 0.1 MPa. Both I_{sp} and T_f decrease with increasing ξ_{RDX}. The molecular mass of the combustion products also decreases with increasing ξ_{RDX} due to the production of H_2 by the decomposition of RDX. It is evident that no excess oxidizer fragments are available to oxidize this H_2.

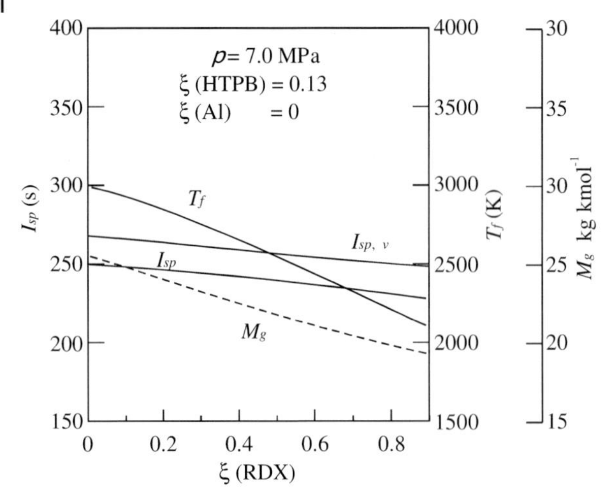

Fig. 7.49 Theoretical performance of AP-RDX composite propellants: $\xi_{AP} + \xi_{RDX} = 0.87$.

The addition of aluminum powder to AP-nitramine composite propellants increases the specific impulse, as in the case of AP composite propellants. Fig. 7.50 shows the theoretical I_{sp} and T_f values for AP-RDX composite propellants containing $\xi_{Al}(0.15)$ as a function of ξ_{RDX}. The propellants are composed of $\xi_{HTPB}(0.1105)$ and the chamber pressure is 7 MPa with an optimum expansion to 0.1 MPa. Though T_f decreases with increasing ξ_{RDX}, as shown in Fig. 7.50, I_{sp} increases with increasing ξ_{RDX} in the region $\xi_{RDX} < 0.40$. The maximum I_{sp} is obtained at about $\xi_{RDX}(0.40)$, and then it decreases with increasing ξ_{RDX} in the region $\xi_{RDX} > 0.4$.

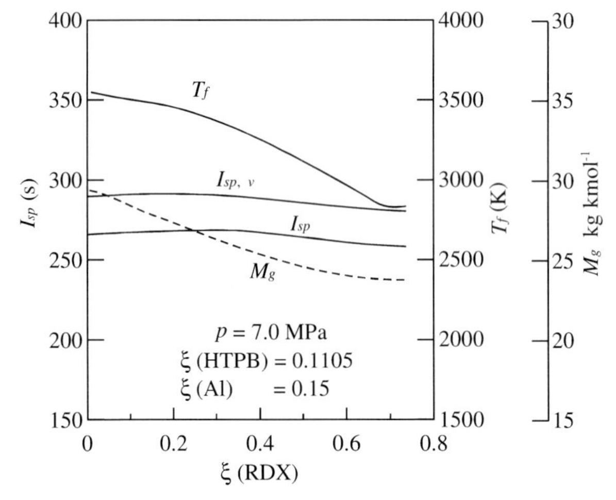

Fig. 7.50 Theoretical performance of AP-RDX-Al composite propellants: $\xi_{AP} + \xi_{RDX} = 0.7395$.

7.3.2 Burning Rate

7.3.2.1 Effects of AP/RDX Mixture Ratio and Particle Size

The burning rates of AP-RDX composite propellants are dependent on the physicochemical properties of the AP, RDX, and fuel used, such as particle size, as well as on mixture ratio and the type of binder. The results of burning rate measurements are reported in AIAA Paper No. 81–1582.[29] Various combinations of AP and RDX particles are used to formulate AP-RDX composite propellants, as shown in Table 7.6.[29] The particles incorporated into the propellants have bimodal combinations of sizes, where large RDX particles (RDX-*L*), small RDX particles (RDX-*S*), large AP particles (AP-*L*), and small AP particles (AP-*S*) are designated by d_R, d_r, d_A, and d_a, respectively. HTPB binder is used in all of the propellants shown in Table 7.6.

Table 7.6 Compositions of AP, RDX, and AP-RDX composite propellants.

Propellant no.	Rr– 73	Rr– 55	Rr– 37	Ra– 73	Ra– 55	Ra– 37	Ar– 73	Ar– 55	Ar– 37	Aa– 73	Aa– 55	Aa– 37
HTPB	20	20	20	20	20	20	20	20	20	20	20	20
RDX-*L*	56	40	24	56	40	24	–	–	–	–	–	–
RDX-*S*	24	40	56	–	–	–	24	40	56	–	–	–
AP-*L*	–	–	–	–	–	–	56	40	24	56	24	24
AP-*S*	–	–	–	24	40	24	–	–	–	24	40	56
d_L/d_S	7/3	5/5	3/7	7/3	5/5	3/7	7/3	5/5	3/7	7/3	5/5	3/7
	d_R = 140 μm			d_r = 5 μm			d_A = 200 μm			d_r = 20 μm		

As shown in Figs. 7.51 to 7.54, the burning rates are dependent on the combinations of AP and RDX. The burning rates of AP composite propellants designated by Aa-37, Aa-55, and Aa-73 shown in Fig. 7.51 are higher than those of RDX composite propellants designated by Rr-37, Rr-55, and Rr-73 shown in Fig. 7.52. The burning rates of Ra propellants containing small AP particles, d_a, shown in Fig. 7.53, are higher than those of Ar propellants containing small RDX particles, d_r, as shown in Fig. 7.54. The burning rate is increased by the use of the smaller AP particles. The pressure exponent of burning rate, n, remains unchanged at $n = 0.38$ on varying the d_R/d_a mixture ratio in Ra propellants. The effects of the mixture ratio of the different sizes of AP and RDX particles on the burning rate at 4 MPa are shown in Fig. 7.55. Similar burning rate characteristics to those observed for AP-RDX composite propellants are also observed for AP-HMX composite propellants.[29]

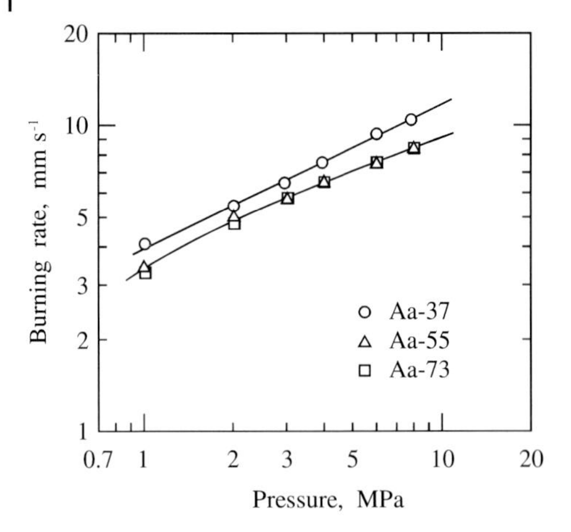

Fig. 7.51 Burning rate characteristics of AP composite propellants composed of coarse AP and fine AP particles.

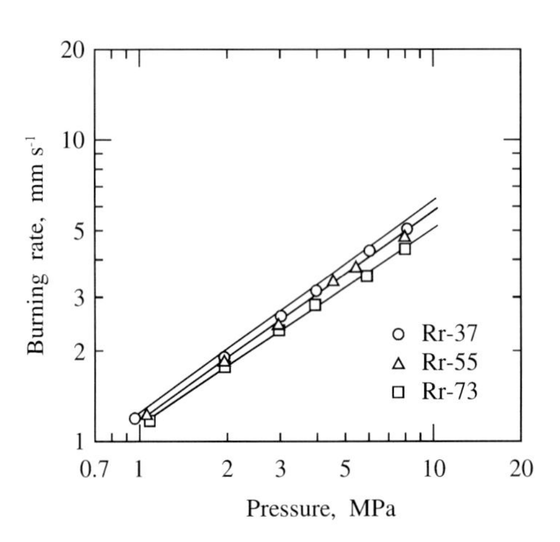

Fig. 7.52 Burning rate characteristics of RDX composite propellants composed of coarse RDX and fine RDX particles.

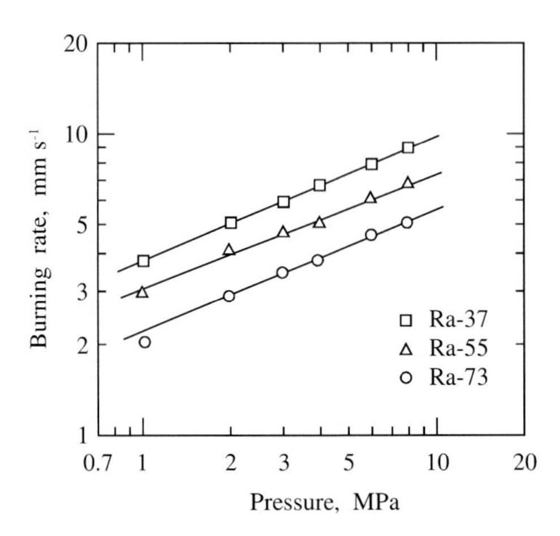

Fig. 7.53 Burning rate characteristics of AP-RDX composite propellants composed of coarse RDX and fine AP particles.

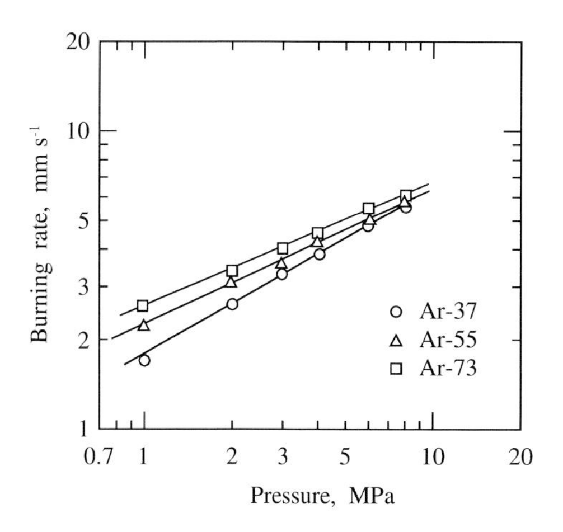

Fig. 7.54 Burning rate characteristics of AP-RDX composite propellants composed of coarse AP and fine RDX particles.

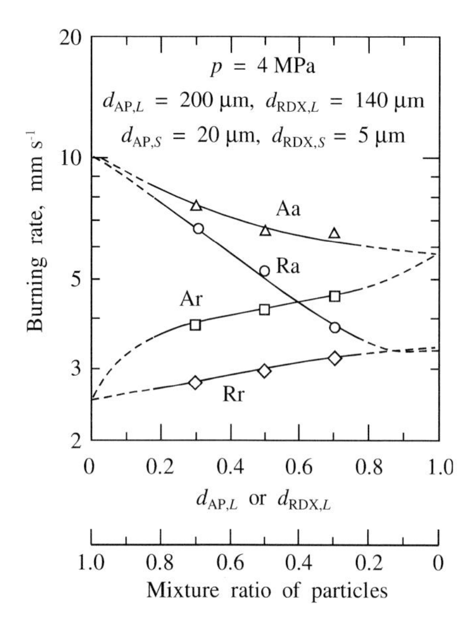

Fig. 7.55 Effects of particle size and mixture ratio of AP and RDX on the burning rates of AP, RDX, and AP-RDX composite propellants.

7.3.2.2 **Effect of Binder**

When HTPB binder is replaced with HTPE binder in AP-RDX composite propellants, the burning rate characteristics are significantly affected.[29] As shown in Fig. 7.56, the burning rate of Aa-55 (HTPE) propellant composed of AP and HTPE shows plateau burning characteristics, for which the pressure exponent, n, varies between 0.3 and –0.2 in the pressure range 1–10 MPa. The plateau effect is caused by the use of HTPE, which melts on the burning surface of the propellant and forms a melt layer. The burning AP particles on the burning surface are partially covered by the melt layer, and so the heat flux transferred back from the gas phase to the burning surface is reduced, as is the burning rate. The burning is interrupted

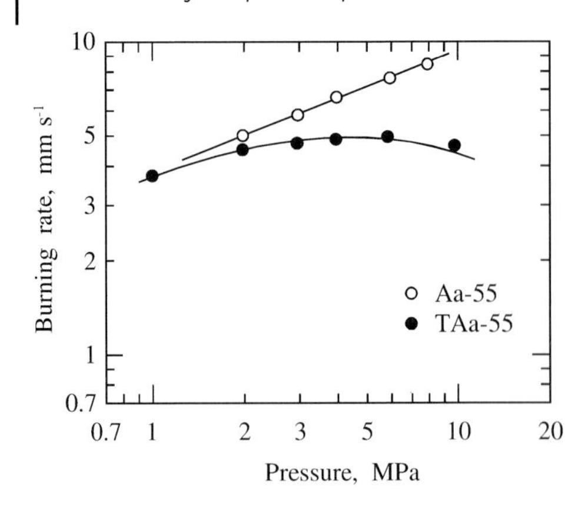

Fig. 7.56 Effect of binder on the burning characteristics of AP composite propellants (HTPB and HTPE).

in the high-pressure region above 15 MPa. Similar plateau burning and intermittent burning are observed when other fuels susceptible to melting are used. Typical fuels are diisocyanate-terminated polyester with polyol cure (Estane/Polyol), polyol with diisocyanate (Polyol/TDI), and polyisobutylene with imine cure and plasticizer(PIB/MAPO).[9]

On the other hand, RDX composite propellant formulated with HTPE binder, Rr-55 propellant, burns with a high pressure exponent, $n = 0.85$, as shown in Fig. 7.57. Since the burning rates of RDX composite propellants are low compared with those of AP composite propellants in the pressure region below about 10 MPa, the HTPE binder on the burning surface is given enough time to melt and decompose by gasification on the surface. Thus, the RDX particles protrude to allow for exothermic gasification. The burning rate increases and the pressure exponent decreases with increasing concentration of AP particles within RDX composite propellants. Fig. 7.58 also shows the burning rates of AP, AP-RDX, and RDX composite propellants formulated with HTPE binder. The effect of the mixture of coarse AP or RDX and fine RDX or AP is evident, in that plateau burning appears when AP particles are incorporated into the propellants.

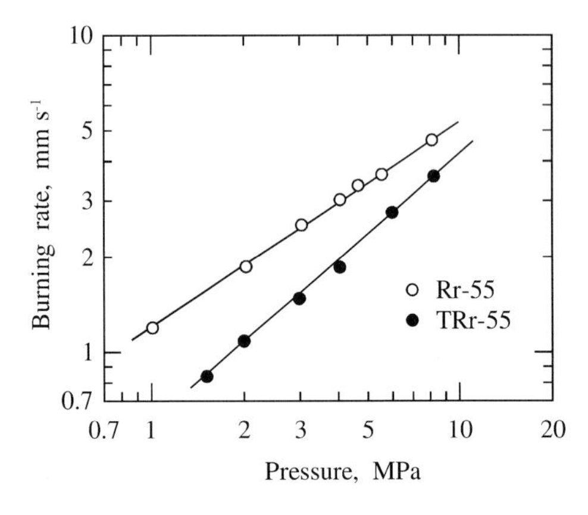

Fig. 7.57 Effect of binder on the burning rate characteristics of RDX composite propellants (HTPB and HTPE).

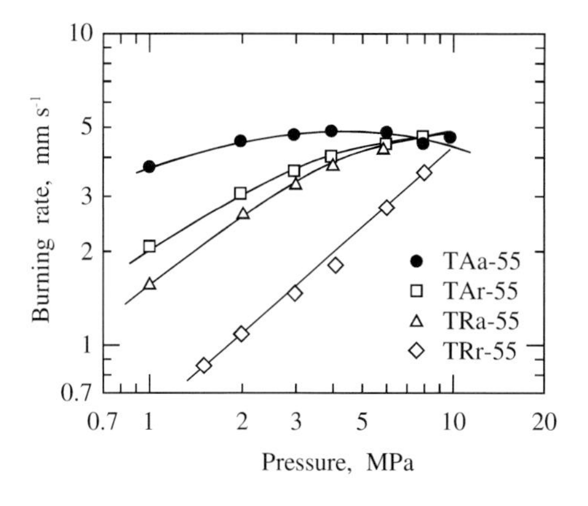

Fig. 7.58 Burning rate characteristics of AP, AP-RDX, and RDX composite propellants (HTPE).

7.4 TAGN-GAP Composite Propellants

7.4.1 Physicochemical Characteristics

As shown in Fig. 5.3, TAGN burns very rapidly even though the flame temperature is low. When TAGN is mixed with a polymeric material, a unique burning rate is observed. Unlike AP composite propellants, TAGN composite propellants are fuel-rich and the flame temperature is low. However, the energy density of composite propellants composed of TAGN and GAP is relatively high because of the high hydrogen content.[30] As shown in Table 7.7, the energy density of a propellant composed of $\xi_{TAGN}(0.20)$ mixed with GAP is approximately equivalent to that of a propellant composed of $\xi_{HMX}(0.20)$ mixed with GAP. The physicochemical properties of TAGN-GAP and HMX-GAP (as a reference) are shown in Table 7.7. The thermodynamic potential, Θ, defined by T_g/M_g, is approximately the same for both the TAGN-GAP and HMX-GAP composite propellants.

Table 7.7 Physicochemical properties of TAGN-GAP and HMX-GAP propellants composed of $\xi_{GAP}(0.80)$.

ξ(TAGN) or ξ(HMX)	TAGN-GAP ξ_{TAGN}(0.20)	HMX-GAP ξ_{HMX}(0.20)
Flame temperature at 5 MPa, T_f (K)	1380	1480
Molecular mass, M_g (kg kmol^{-1})	19.5	19.8
Density, ρ_p (kg m^{-3})	1320	1400
Θ (kmol K kg^{-1})	70.8	74.7

7.4.2
Burning Rate and Combustion Wave Structure

The burning rates at different temperatures of TAGN-GAP propellant composed of ξ_{TAGN}(0.20) are shown in Fig. 7.59. The pressure exponent is 0.95 and the temperature sensitivity is 0.010 K^{-1}. When TAGN and GAP are mixed together, the burning rate is lowered to less than that of each individual ingredient, i. e., GAP copolymer-shown in Fig. 5.17 and TAGN shown in Fig. 5.3.[30] However, when the mass fraction of TAGN is increased, the burning rate increases in the low-pressure region and the pressure exponent decreases, as shown in Fig. 7.60.

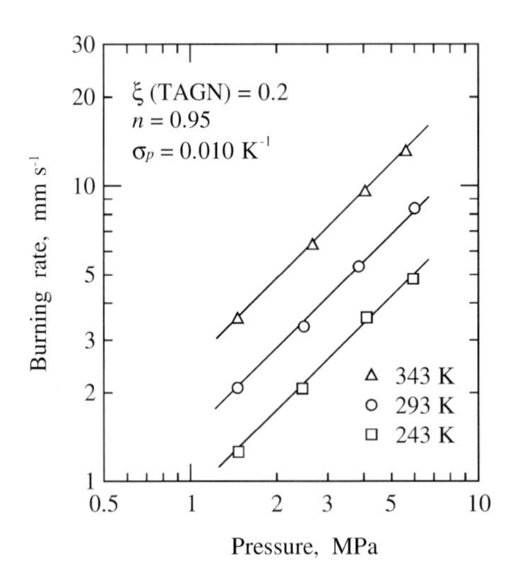

Fig. 7.59 Burning rates and temperature sensitivity thereof for TAGN-GAP composite propellant composed of ξ_{TAGN}(0.20).

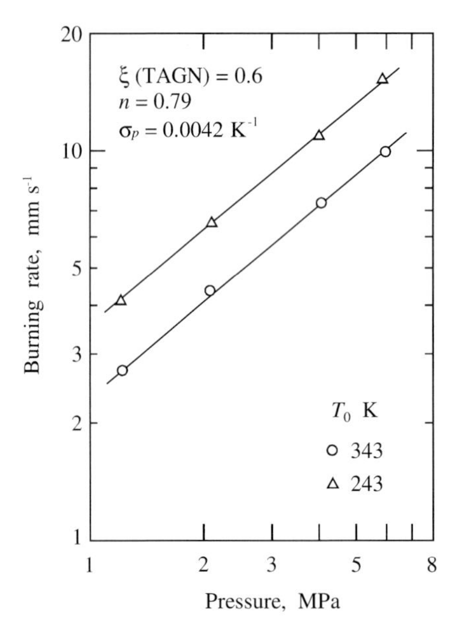

Fig. 7.60 Burning rates and temperature sensitivity thereof for TAGN-GAP composite propellant composed of ξ_{TAGN}(0.60).

The burning surface of a TAGN-GAP propellant is heterogeneous due to the decomposition of the TAGN particles and GAP. In the region of low ξ_{TAGN}, below about 0.3, the combustion wave is similar to that of GAP copolymer. In the region of high ξ_{TAGN}, above about 0.7, the luminous flame stands some distance above the burning surface, similar to the TAGN flame shown in Fig. 5.4. Since the heat flux generated at the burning surface is high for both TAGN and GAP, the burning rate of a TAGN-GAP propellant is controlled by the rate of heat release at the burning surface.

7.5 AN-Azide Polymer Composite Propellants

7.5.1 AN-GAP Composite Propellants

Similar to nitramine composite propellants and TAGN composite propellants, AN composite propellants produce halogen-free combustion products and thus represent smokeless propellants. However, their ballistic properties are inferior to those of other composite propellants: the burning rate is too low and the pressure exponent is too high to permit fabrication of rocket propellant grains. In addition, the mechanical properties of AN composite propellants vary with temperature due to the phase transitions of AN particles.

AN composite propellants composed of ξ_{AN} in the range 0.70 to 0.80 with PU binder have been extensively studied as smokeless rocket propellants and various ballistic properties have been obtained.[31] Since these propellants are made with halogen-free ingredients, in theory, no smoke should be produced from the exhaust nozzle of a rocket in which they are deployed. However, the specific impulse is too low to permit their use in advanced rocket propulsion. Furthermore, the phase transition of AN particles is detrimental to the mechanical properties of the propellant grains and the production capability is affected by humidity. The mixing of AN with the binder is limited due to the crystal shape of AN particles. Small-sized AN propellant particles are made as pressed grains, the compositions of which are $\xi_{AN}(0.85)$ or more, so as to obtain the smokeless nature of the combustion products. This class of propellant grains is used as gas-generating propellants.

The burning rates of AN propellants are rather limited, ranging from 0.8 mm s^{-1} to 10 mm s^{-1} at 7 MPa. The burning rate is less dependent on the AN particle size because the AN particles melt and mix with the molten PU binder on the burning surface.[31] The gaseous decomposition products produced at the burning surface are premixed and thus form a premixed-type luminous flame above the burning surface. The luminous flame stands some distance above the burning surface, as seen in the case of double-base propellants. In contrast, AP propellants produce a diffusion-type luminous flame on and just above the burning surface.

Combustion catalysts are needed to obtain complete combustion of AN composite propellants. Chromium compounds such as chromium trioxide (Cr_2O_3), ammonium dichromate ($(NH_4)_2Cr_2O_7$), and copper chromite ($CuCr_2O_4$) are known to be catalysts for the combustion of AN composite propellants. Carbon (C) is also added as a burning rate modifier in conjunction with these combustion catalysts. Very fine NiO particles are incorporated into AN propellants as a phase transition stabilizer of AN crystals.[21] The pressure exponent in the absence of a catalyst is approximately 0.8 at $\xi_{AN}(0.8)$ and is reduced to 0.5–0.6 by the addition of a catalyst.

If the production of a small amount of halogen molecules among the combustion products of AN composite propellants is permissible, AN propellants containing AP particles are formulated, i. e., AN-AP composite propellants. Though the addition of AP particles increases the concentration of HCl in the combustion products, the ballistic properties are improved significantly: the addition of AP particles to AN-PU propellants increases the burning rate and reduces the pressure exponentwithout the use of a catalyst. The specific impulse is also increased from 225 s to 235 s at 10 MPa by replacing $\xi_{AN}(0.30)$ with $\xi_{AP}(0.30)$. However, the smokeless nature of the propellant is lost due to the production of HCl and it is then categorized as a reduced-smoke propellant.

When AN particles are mixed with GAP, AN-GAP composite propellants are formulated. The specific impulse is increased by approximately 10 s by replacing PB or PU binder with GAP, as shown in Fig. 4.16. The burning rate is also increased due to the exothermic decomposition of GAP. Since GAP burns by itself, the burning of the AN particles is supported by the exothermic decomposition reaction of the GAP at the burning surface of the propellant. As shown in Fig. 7.61, the burning rate is drastically decreased by the addition of AN particles. When $\xi_{AN}(0.3)$ is incorporated into GAP, the burning rate in the low-pressure region is decreased and the pressure exponent is increased from 0.55 to 1.05. The burning rate at 8 MPa is decreased from 18 mm s^{-1} to 3.7 mm s^{-1} when $\xi_{AN}(0.7)$ is incorporated into GAP,

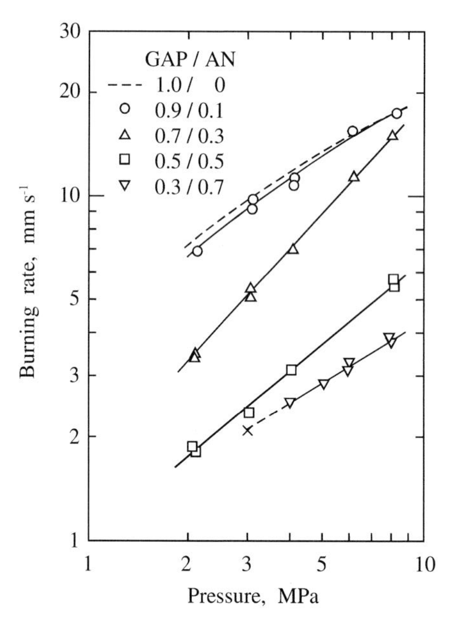

Fig. 7.61 Burning rate characteristics of AN-GAP propellants.

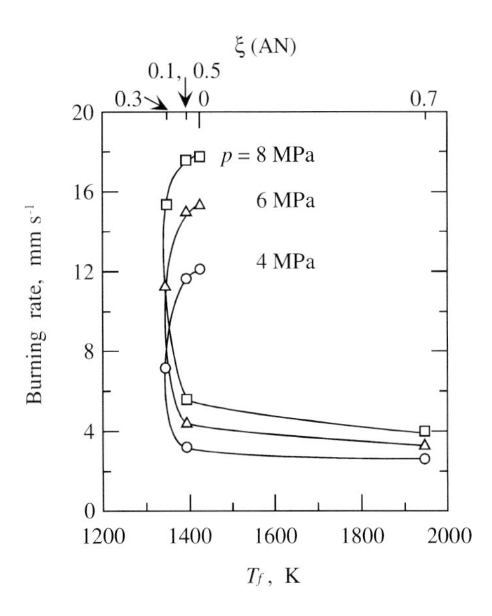

Fig. 7.62 Relationship between adiabatic flame temperature and burning rate for AN-GAP composite propellants, showing that the burning rate decreases even though the flame temperature is increased by the addition of AN particles.

and the burning is interrupted at about 3 MPa. Fig. 7.62 shows the relationship between burning rate and adiabatic flame temperature for AN-GAP composite propellants. The flame temperature ranges from 1350 K to 1410 K when mass fractions of AN particles in the range $\xi_{AN}(0.1)$–$\xi_{AN}(0.5)$ are added. However, when $\xi_{AN}(0.7)$ is added, the flame temperature is increased from 1400 K to 1950 K. Despite this considerable increase in the flame temperature, the burning rate is actually decreased by the addition of $\xi_{AN}(0.7)$.

7.5.2
AN-(BAMO-AMMO)-HMX Composite Propellants

Fig. 7.63 shows plots of ln burning rate versus ln pressure for AN-(BAMO-AMMO)-HMX propellants with and without $(NH_4)_2Cr_2O_7$ and $CuCr_2O_4$.[32] The chemical compositions of these propellants are shown in Table 7.8.

The burning rate of the propellant is clearly increased by the addition of these catalysts. The burning rate is approximately doubled by the addition of 1%

Table 7.8 Chemical compositions of AN-(BAMO-AMMO)-HMX composite propellants with and without burning rate catalysts.

Propellant	ξ_{AN}	$\xi_{BAMO\text{-}AMMO}$	ξ_{HMX}	$\xi_{(NH4)2Cr2O7}$	$\xi_{CuCr2O4}$
A	0.60	0.25	0.15	–	–
B	0.60	0.25	0.15	0.02	–
C	0.60	0.25	0.15	–	0.02
D	0.60	0.25	0.15	0.02	0.02

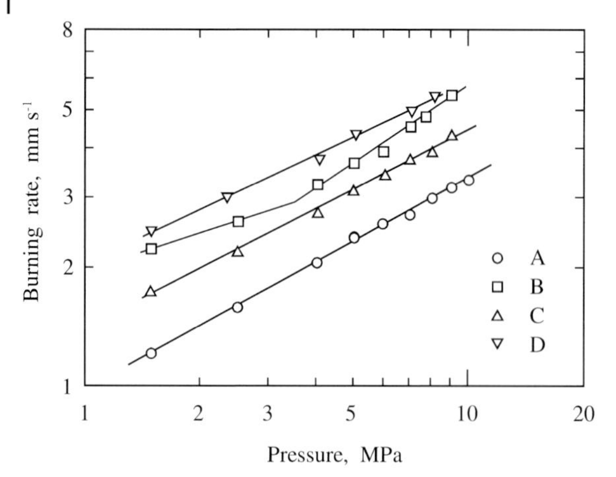

Fig. 7.63 Effect of catalysts on the burning rate of AN-(BAMO-AMMO)-HMX composite propellants.

$(NH_4)_2Cr_2O_7$ and 1% $CuCr_2O_4$. The pressure exponent remains almost unchanged by the addition of the catalysts. It can be seen that the burning rate is effectively increased when both catalysts are used simultaneously. Both catalysts act on the decomposition process of the AP particles. However, the sites and modes of action of these two catalysts are different, and so the overall catalytic effect is more than doubled when they are deployed simultaneously.

7.6 AP-GAP Composite Propellants

Though GAP burns by itself and produces high-temperature combustion products, large amounts of fuel fragments are produced simultaneously. When AP particles are mixed with GAP, an AP-GAP composite propellant is formulated. Similar to AP-HTPB composite propellants, AP-GAP composite propellants burn with diffusional flamelets produced by the mixing of the oxidizer gas and the fuel fragments in the gas phase. When $\xi_{AP}(0.20)$ is incorporated into GAP, the burning rates are seen to increase linearly in an ln burning rate versus ln pressure plot at different initial temperatures of 243 K, 293 K, and 343 K, as shown in Fig. 7.64. The pressure exponent is increased from 0.44 to 0.62 and the temperature sensitivity is decreased from 0.0100 K^{-1} to 0.0078 K^{-1} by the addition of $\xi_{AP}(0.20)$. In other words, the burning rate of GAP in the low-pressure region below about 5 MPa is decreased by the addition of $\xi_{AP}(0.2)$ when compared with the burning rate of GAP shown in Fig. 5.17.

Fig. 7.65 shows the relationship between ξ_{AP} and burning rate as a function of pressure. Though the burning rate is decreased by the addition of $\xi_{AP}(0.1)$, it tends to increase with increasing ξ_{AP} below 1 MPa. However, the burning rate decreases with increasing ξ_{AP} above 5 MPa. The flame temperature is increased with increas-

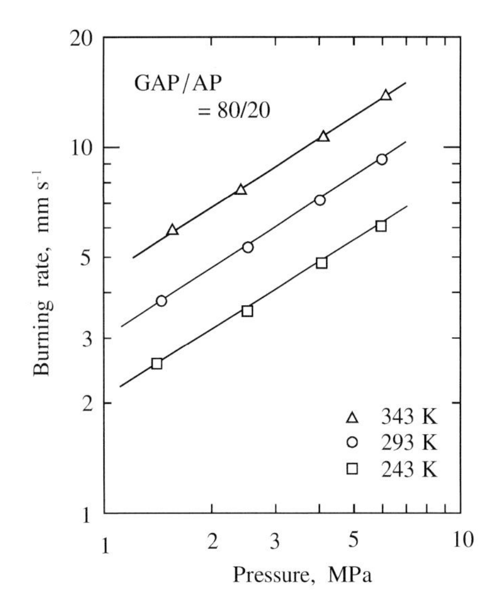

Fig. 7.64 Burning rate characteristics of AP-GAP propellant composed of $\xi_{AP}(0.20)$.

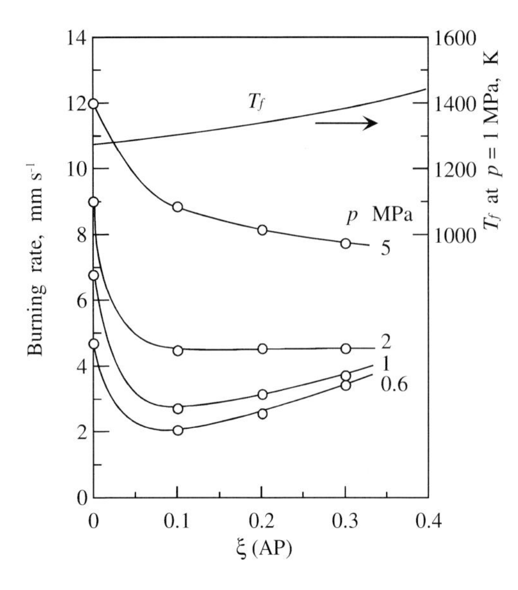

Fig. 7.65 Burning rate and adiabatic flame temperature of AP-GAP propellants as a function of ξ_{AP}.

ing ξ_{AP}. However, this effect is not significant when ξ_{AP} is less than about 0.6. It is shown in Fig. 4.15 that an I_{sp} of more than 260 s at 10 MPa and a flame temperature of 3000 K are obtained at $\xi_{AP}(0.75)$.

7.7
ADN , HNF, and HNIW Composite Propellants

The ballistic properties of ADN, HNF, and HNIW as monopropellants and as oxidizers in composite propellants have been extensively studied.[32–39] Since ADN, HNF, and HNIW particles produce excess oxygen among their combustion products, these particles are used as oxidizer crystals in composite propellants. The pressure exponents of crystalline ADN and HNIW particles are both approximately 0.7,[38] about the same value as those for HMX and RDX when they are burned as pressed pellets. However, the pressure exponent of HNF is 0.85–0.95,[39] higher than those of the other energetic crystalline oxidizers.

When these oxidizer particles are mixed with a binder such as HTPB, a nitropolymer, or GAP, the burning rate decreases with increasing mass fraction of ADN or HNF particles.[34] Though the temperature sensitivity of an ADN composite propellant is significantly high in the low-pressure region, 0.005 K^{-1} at 1 MPa, it decreases with increasing pressure, falling to 0.0023 K^{-1} at 10 MPa.[37] However, the temperature sensitivity of an HNF composite propellant is about 0.0018 K^{-1} in the pressure range 1–10 MPa.[38] The burning surfaces of ADN and HNF composite propellantsare covered with a molten layer when PU binder is used, and the luminous flames stand some distance above the burning surface for both propellants,[35] as in the case of HMX composite propellants.[24]

When HNF or ADN particles are mixed with a GAP copolymer containing aluminum particles, HNF-GAP and ADN-GAP composite propellants are formed, respectively. A higher theoretical specific impulse is obtained as compared to those of aluminized AP-HTPB composite propellants.[36] However, the ballistic properties of ADN, HNIW, and HNF composite propellants, such as pressure exponent, temperature sensitivity, combustion instability, and mechanical properties, still need to be improved if they are to be used as rocket propellants.

As shown in Fig. 7.66, the burning rate of a propellant composed of HNIW particles and BAMO/NIMO binder is higher than that of a propellant composed of HMX particles and BAMO/NIMO binder.[32] The burning rate is 7.3 mm s^{-1} at 3 MPa and 17 mm s^{-1} at 10 MPa, with a pressure exponent of 0.75, for a BAMO/NIMO–HNIW propellant composed of $\xi_{BAMO/NIMO}(0.40)$. The burning rate is less dependent on the particle size of the added HNIW. The burning rates of propellants composed of AP or HMX particles and BAMO/NIMO binder are also shown in Fig. 7.66 for comparison. When HMX is incorporated into BAMO/NIMO binder, the burning rate is low and the pressure exponent is 0.61. When AP is incorporated into this binder, the burning rate is high and the pressure exponent is 0.32, similar to the burning rate characteristics of conventional AP-HTPB composite propellants.

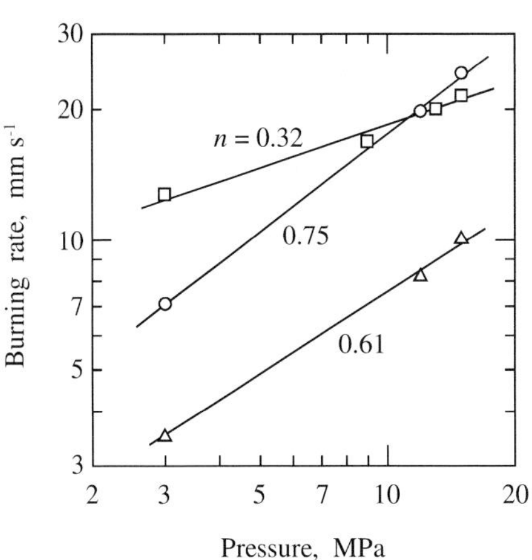

Fig. 7.66 Burning rate characteristics of composite propellants composed of HNIW, AP, or HMX, with BAMO/NIMO as a binder.

References

1 Lengelle, G., Duterque, J., and Trubert, J. F., Physico-Chemical Mechanisms of Solid Propellant Combustion (Eds.: Yang, V., Brill, T. B., and W.-Z. Ren), *Progress in Astronautics and Aeronautics*, Vol. 185, Chapter 2.2, AIAA, Washington DC (2000).

2 Ramohalli, K. N. R., Steady-State Burning of Composite Propellants, Fundamentals of Solid-Propellant Combustion (Eds.: Kuo, K. K., and Summerfield, M.), *Progress in Astronautics and Aeronautics*, Vol. 90, Chapter 8, AIAA, New York (1984).

3 Kubota, N., Temperature Sensitivity of Solid Propellants and Affecting Factors: Experimental Results, Nonsteady Burning and Combustion Stability of Solid Propellants (Eds.: De Luca, L., Price, E. W., and Summerfield, M.), *Progress in Astronautics and Aeronautics*, Vol. 143, Chapter 4, AIAA, Washington DC (1992).

4 Beckstead, M. W., Derr, R. L., and Price, C. F., Model of Composite Solid-Propellant Combustion Based on Multiple Flames, *AIAA Journal*, Vol. 8, No. 12, 1970, pp. 2200–2207.

5 Hermance, C. E., A Model of Composite Propellant Combustion Including Surface Heterogeneity and Heat Generation, *AIAA Journal*, Vol. 4, No. 9, 1966, pp. 1629–1637.

6 Summerfield, M., Sutherland, G. S., Webb, W. J., Taback, H. J., and Hall, K. P., The Burning Mechanism of Ammonium Perchlorate Propellants, *ARS Progress in Astronautics and Rocketry*, Vol. 1, Solid Propellant Rocket Research, Academic Press, New York (1960), pp. 141–182.

7 Boggs, T. L., Derr, R. L., and Beckstead, M. W., Surface Structure of Ammonium Perchlorate Composite Propellants, *AIAA Journal*, Vol. 8, No. 2, 1970, pp. 370–372.

8 Kuwahara, T., and Kubota, N., Low Pressure Burning of Ammonium Perchlorate Composite Propellants, *Combustion Science and Technology*, Vol. 47, 1986, pp. 81–91.

9 Steinz, J. A., Stang, P. L., and Summerfield, M., The Burning Mechanism of Ammonium Perchlorate-Based Composite Solid Propellants, AMS Report No. 830, Aerospace and Mechanical Sciences,

Princeton University, Princeton, NJ (1969).

10 Price, E. W., Handley, J. C., Panyam, R. R., Sigman, R. K., and Ghosh, A., Combustion of Ammonium Perchlorate-Polymer Sandwiches, *Combustion and Flame*, Vol. 7, No. 7, 1963.

11 Bastress, E. K., Modification of the Burning Rates of Ammonium Perchlorate Solid Propellants by Particle Size Control, Ph.D. Thesis, Department of Aeronautical Engineering, Princeton University, Princeton, NJ (1961).

12 Kubota, N., Kuwahara, T., Miyazaki, S., Uchiyama, K., and Hirata, N., Combustion Wave Structure of Ammonium Perchlorate Composite Propellants, *Journal of Propulsion and Power*, Vol. 2, No. 4, 1986, pp. 296–300.

13 Kubota, N., and Miyazaki, S., Temperature Sensitivity of Burning Rate of Ammonium Perchlorate Propellants, *Propellants, Explosives, Pyrotechnics*, Vol. 12, 1987, pp. 183–187.

14 Bazaki, H., and Kubota, N., Friction Sensitivity Mechanism of Ammonium Perchlorate Composite Propellants, *Propellants, Explosives, Pyrotechnics*, Vol. 16, 1991, pp. 43–47.

15 Kubota, N., and Hirata, N., Inhibition Reaction of LiF on the Combustion of Ammonium Perchlorate Propellants, 20th Symposium (International) on Combustion, The Combustion Institute, Pittsburgh, PA (1984), pp. 2051–2056.

16 Bazaki, H., Negative Catalyst Action of $SrCO_3$ on AP Composite Propellants, In-House Research Report 2005, Asahi Kasei Chemicals, 2005.

17 Miyata, K., and Kubota, N., Inhibition Reaction of $SrCO_3$ on the Burning Rate of Ammonium Perchlorate Propellants, *Propellants, Explosives, Pyrotechnics*, Vol. 15, 1990, pp. 127–131.

18 Beckstead, M. W., and McCarty, K. P., Modeling Calculations for HMX Composite Propellants, *AIAA Journal*, Vol. 20, No. 1, 1982, pp. 106–115.

19 Kubota, N., Combustion Mechanisms of Nitramine Composite Propellants, 18th Symposium (International) on Combustion, The Combustion Institute, Pittsburgh, PA (1981), pp. 187–194.

20 Cohen, N. S., and Price, C. F., Combustion of Nitramine Propellants, *AIAA Journal*, Vol. 12, No. 10, 1975, pp. 25–42.

21 Kubota, N., Physicochemical Processes of HMX Propellant Combustion, 19th Symposium (International) on Combustion, The Combustion Institute, Pittsburgh, PA (1982), pp. 777–785.

22 Beckstead, M. W., Overview of Combustion Mechanisms and Flame Structures for Advanced Solid Propellants (Eds.: Yang, V., Brill, T. B., and W.-Z. Ren), *Progress in Astronautics and Aeronautics*, Vol. 185, Chapter 2.1, AIAA, Washington DC, 2000.

23 Klager, K., and Zimmerman, G. A., "Steady Burning Rate and Affecting Factors: Experimental Results", Nonsteady Burning and Combustion Stability of Solid Propellants (Eds.: De Luca, L., Price, E. W., and Summerfield, M.), *Progress in Astronautics and Aeronautics*, Vol. 143, Chapter 3, AIAA, Washington DC, 1992.

24 Kubota, N., Sonobe, T., Yamamoto, A., and Shimizu, H., Burning Rate Characteristics of GAP Propellants, *Journal of Propulsion and Power*, Vol. 6, No. 6, 1990, pp. 686–689.

25 Kubota, N., and Sonobe, T., Burning Rate Catalysis of Azide/Nitramine Propellants, 23 rd Symposium (International) on Combustion, The Combustion Institute, Pittsburgh, PA (1990), pp. 1331–1337.

26 Kubota, N., and Hirata, N., Super-Rate Burning of Catalyzed HMX Propellants, 21 st Symposium (International) on Combustion, The Combustion Institute, Pittsburgh, PA (1986), pp. 1943–1951.

27 Kubota, N., "Nitramine Propellants for Rockets and Guns", Japanese Patent No. 60–177452, Aug. 12th, 1985.

28 Shibamoto, H., and Kubota, N., Super-Rate Burning of LiF-Catalyzed HMX Pyrolants, 29th International Pyrotechnics Seminar, Westminster, Colorado, July 14–19th, 2002, pp. 147–155.

29 Kubota, N., Takizuka, M., and Fukuda, T., Combustion of Nitramine Composite Propellants, AIAA/SAE/ASME 17th Joint Propulsion Conference, July 27–29th, Colorado Springs, Colorado, AIAA-81-1582, 1981.

30 Kubota, N., Hirata, N., and Sakamoto, S., Combustion Mechanism of TAGN, 21 st Symposium (International) on Combustion, The Combustion Institute, Pittsburgh, PA (1986), pp. 1925–1931.

31 In-house Report (unpublished), Daicel Chemical Industries, Tokyo (1966–1968).

32 Bazaki, H., In-house Research Report 2005, Asahi Kasei Chemicals, 2005.

33 Chan, M. L., Reed Jr., R., and Ciaramitaro, D. A., Advances in Solid Propellant Formulations, Solid Propellant Chemistry, Combustion, and Motor Interior Ballistics (Eds.: Yang, V., Brill, T. B., and W.-Z. Ren), *Progress in Astronautics and Aeronautics*, Vol. 185, Chapter 1.7, AIAA, Washington DC (2000).

34 Takishita, Y., and Shibamoto, H., Inhouse Report (unpublished), Hosoya Pyrotechnics Co., Tokyo (1999).

35 Price, E. W., Chakravarthy, S. R., Freeman, J. M., and Sigman, R. K., Combustion of Propellants with Ammonium Dinitramide, AIAA-98–3387, AIAA, Reston, Virginia (1998).

36 Korobeinichev, O. P., Kuibida, L. V., and Paletsky, A. A., Development and Application of Molecular Mass Spectrometry to the Study of ADN Combustion Chemistry, AIAA-98–0445, AIAA, Reston, Virginia (1998).

37 Gadiot, G. M. H. J. L., Mul, J. M., Meulenbrugge, J. J., Korting, P. A. O. G., Schnorhk, A. J., and Schoyer, H. F. R., New Solid Propellants Based on Energetic Binders and HNF, IAF-92–0633, 43 rd Congress of the International Astronautical Federation, Paris (1992).

38 Atwood, A. I., Boggs, T. L., Curran, P. O., Parr, T. P., and Hanson-Parr, D. M., Burn Rate of Solid Propellant Ingredients, Part 1: Pressure and Initial Temperature Effects, and Part 2: Determination of Burning Rate Temperature Sensitivity, *Journal of Propulsion and Power*, Vol. 15, No. 6, 1999, pp. 740–752.

39 Louwers, J., Gadiot, G. M. H. J. L., Breqster, M. Q., Son, S. F., Parr, T., and Hanson-Parr, D., Steady-State Hydrazinium Nitroformate (HNF) Combustion Modeling, *Journal of Propulsion and Power*, Vol. 15, No. 6, 1999, pp. 772–777.

8
Combustion of CMDB Propellants

8.1
Characteristics of CMDB Propellants

Since the energy contained within double-base propellants is limited because of the limited energies of nitrocellulose (NC) and nitroglycerin (NG), the addition of ammonium perchlorate or energetic nitramine particles such as HMX and RDX increases the combustion temperature and specific impulse. Extensive experimental studies have been carried out on the combustion characteristics of composite-modified double-base (CMDB) propellants containing AP, RDX or HMX particles[1–9] and several models have been proposed to describe the burning rates of these propellants.[1–3]

AP-CMDB propellants have a different burning mode compared to nitramine-CMDB propellants. The burning mode of AP-CMDB propellants is akin to that of AP composite propellants, that is, a diffusion-flame mode. On the other hand, the burning mode of nitramine-CMDB propellants is that of double-base propellants, that is, a premixed-flame mode. The added AP particles supply an oxidizer to the double-base propellant used as the base matrix. HMX and RDX are stoichiometrically balanced materials and produce high-temperature combustion gases, but unlike AP particles they yield no excess oxidizer fragments. The nitramine particles supply heat to the base matrix. Accordingly, the combustion wave structures and burning rate characteristics of nitramine-CMDB propellants are different from those of AP-CMDB propellants.[1]

8.2
AP-CMDB Propellants

8.2.1
Flame Structure and Combustion Mode

On addition of AP particles to a double-base matrix, very fine luminous flameletsare formed from each AP particle at the burning surface, which diffuse into the dark zone of the base matrix. As the number of AP particles added to the base matrix is increased, the number of flamelets also increases and the dark zone is replaced with a luminous flame.

Propellants and Explosives. Naminosuke Kubota

ISBN: 978-3-527-31424-9

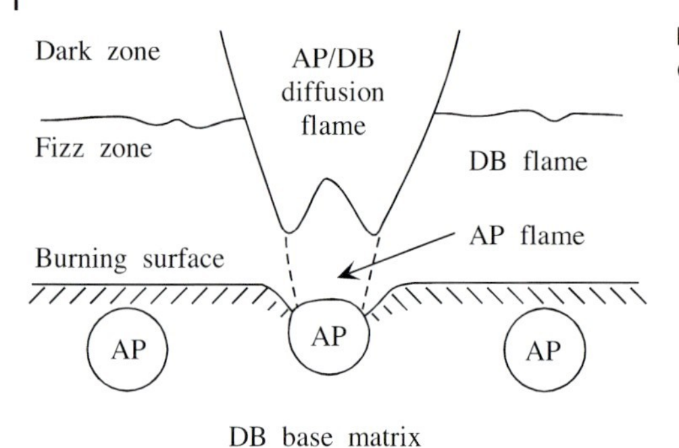

Fig. 8.1 Flame structure of an AP-CMDB propellant.

When large spherical AP particles (d_0 = 3 mm) are added, large flamelets are formed in the dark zone.[1] Close inspection of the AP particles at the burning surface reveals that a transparent bluish flame of low luminosity is formed above each AP particle. These are ammonia/perchloric acid flames, the products of which are oxidizer-rich, as are also observed for AP composite propellants at low pressures, as shown in Fig. 7.5. The bluish flame is generated a short distance from the AP particle and has a temperature of up to 1300 K. Surrounding the bluish flame, a yellowish luminous flame stream is formed. This yellowish flame is produced by interdiffusion of the gaseous decomposition products of the AP and the double-base matrix. Since the decomposition gas of the base matrix is fuel-rich and the temperature in the dark zone is about 1500 K, the interdiffusion of the products of the AP and the matrix shifts the relative amounts towards the stoichiometric ratio, resulting in increased reaction rate and flame temperature. The flame structure of an AP-CMDB propellant is illustrated in Fig. 8.1.

When the mass fraction of AP is increased, the dark zone of the base matrix is eliminated almost completely and the luminous flame approaches the burning surface as shown in Fig. 8.2(a). For reference, the flame structure of an RDX-CMDB propellant is also shown in Fig. 8.2(b). Since RDX is a stoichiometrically balanced

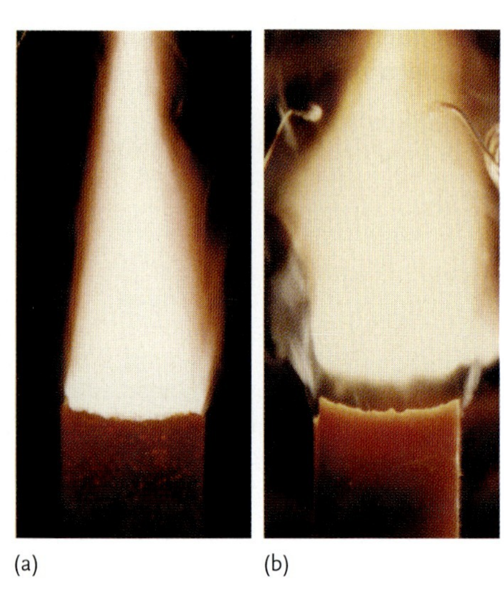

Fig. 8.2 Flame photographs of AP-CMDB propellant (a) and RDX-CMDB propellant (b) showing that the luminous flame front of the RDX-CMDB propellant is distended from the burning surface:

	Mass fraction			p	r
	AP	RDX	DB matrix	MPa	mm s^{-1}
(a)	0.30		0.70	2.0	9.0
(b)		0.30	0.70	2.0	2.2

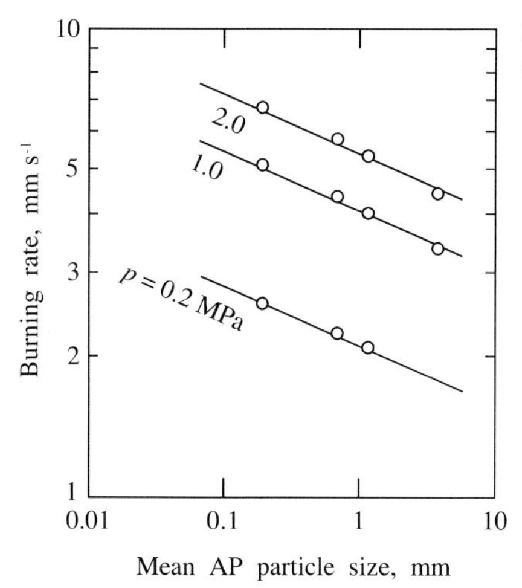

Fig. 8.3 Burning rates of AP particles in a DB matrix.

crystalline material, no diffusional flamelets are formed above the burning surface. The gaseous decomposition products of the RDX particles diffuse and mix with the gaseous decomposition products of the base matrix and form a reactive homogeneous gas, which reacts to produce a premixed flame above the burning surface. As in the case of double-base propellants, the luminous flame is distended from the burning surface for both RDX-CMDB and HMX-CMDB propellants.[1]

8.2.2 Burning Rate Models

Experimental observation indicates that the AP particles within the matrix of a double-base (DB) propellant appear to burn and regress in size spherically on the burning surface. Thus, the regression rate of the particles appears to depend on their instantaneous diameter. The average burning rate of each AP particle is determined as a function of pressure, as shown in Fig. 8.3. The burning rate is represented by $r_{AP} = \kappa_1 p^{0.45}$, where κ_1 is independent of pressure, r_{AP} is the burning rate of the AP particles, p is pressure, and d_0 is the initial AP particle size. The relationship between the burning rate and the particle diameter is determined to be approximately $r_{AP} = \kappa_2 / d_0{}^{0.15}$, where κ_2 is independent of the burning rate of the DB matrix. From these two expressions, the burning rate of an AP particle in a DB matrix is given by

$$r_{AP} = \kappa p^{0.45} / d_0{}^{0.15} \tag{8.1}$$

where κ is an experimentally determined constant equal to 0.38 for r_{AP} (mm s^{-1}), d_0 (mm), and p (MPa). Equation (8.1) is similar to the expression obtained by Barrere and Nadaud for the burning of AP spheres in gaseous fuel environments.[10]

In order to derive a simplified burning rate expression for the AP particles, it is assumed that for the area of fastest regression there is an effective region of thickness η; around each AP particle, which is independent of particle size. This effective region is also assumed to burn at the same rate as an AP particle in the DB matrix. This effective thickness is found experimentally to conform to:

$$\frac{\pi}{6} w(d_0 + 2\eta)^3 = \xi \tag{8.2}$$

$$w = \frac{6\rho_{DB}}{\pi d_0^3 (\rho_{AP} / \phi_{AP} + \rho_{DB} - \rho_{AP})} \tag{8.3}$$

where ξ is the volumetric fraction of the AP-CMDB propellant that burns at the same rate as an AP particle in the DB matrix, and w is the number of AP particles in unit volume of the propellant.

The critical mixture ratio at which excess AP particles begin to separate is approximately a 30% mixture of 18 μm AP particles within the DB matrix. Thus, the effective thickness is derived from Eqs. (8.2) and (8.3) as $\eta = 5$ μm with ξ assumed to be unity.

As the combustion zones of AP-CMDB propellants are highly heterogeneous, both physically and chemically, it is assumed that the burning surface of the DB matrix regresses at different rates for the two different regions: (1) the area distant from the AP particles, in which the heat feedback from the gas phase to the burning surface is due only to the DB flame, and (2) the area adjacent to the AP particles, which is influenced by heat feedback from the AP/DB diffusion flame. When the areas of fastest regression at the interface between the AP particles and the DB matrix begin to overlap, the base matrix encapsulating some of the AP particles is eroded and some of the burning AP particles are ejected into the gas phase. These AP particles are called "excess AP particles".

The regression rate in region (1) is regarded as being the same as that represented by Eqs. (3.57)–(3.60). The burning rate equations are obtained under the assumptions that the gas-phase reaction proceeds at a constant rate and is second-order over the fizz zone, and that the luminous flame is too far from the surface to influence the burning rate. The regression rate can also be expressed by a one-step, Arrhenius-type rate equation, as given by Eq. (3.61).

Region (2) is assumed to regress at approximately the same rate as the AP particles, as given by Eq. (8.1). Thus, the burning rate of an AP-CMDB propellant is expressed as the fractionally weighted sum of the two different regression rates:

$$r_A = \frac{1}{\xi / r_{AP} + (1 - \xi) / r_{DB}} \tag{8.4}$$

The results of the calculations are shown in Fig. 8.4. The assumed values for the physical constants and reaction kinetics are listed in Ref. [1]. The burning rate increases with increasing pressure, and also increases with increasing concentration and decreasing particle size of the AP particles. These calculated results compare favorably with the experimental results shown in Fig. 8.4. The calculated burning surface temperature of the DB matrix varies from 621 K at 1 MPa to 673 K at 8 MPa. The temperature sensitivity decreases with increasing pressure ($\sigma_p = 0.0056$ K^{-1} at 8 MPa).[1]

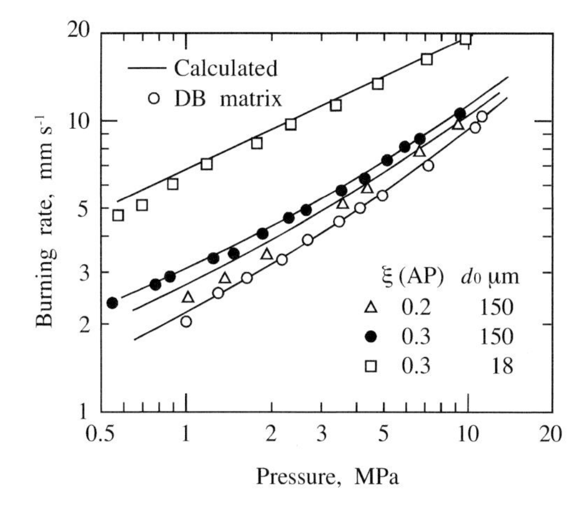

Fig. 8.4 Computed and experimental burning rates of AP-CMDB propellants.

8.3 Nitramine-CMDB Propellants

8.3.1 Flame Structure and Combustion Mode

The gaseous decomposition products produced by nitramine particles and the base matrix (double-base propellant) diffuse and mix together and produce a relatively homogeneous gas phase on and above the burning surface of the nitramine-CMDB propellant. Since the primary ingredients of double-base propellants are nitrate esters such as NC and NG, their energy densities are represented by the number of $O–NO_2$ chemical bonds contained within unit mass of propellant. The breaking of $O–NO_2$ bonds produces NO_2, aldehydes, and other C,H,O species. The NO_2 acts as an oxidizer component and reacts with the aldehydes and the other C,H,O species, which act as fuel components. This reaction is highly exothermic and determines the burning rate of double-base propellants. Thus, the burning rate of a double-base propellant correlates with $\xi(NO_2)$.[2,4–6]

The chemical structure of HMX is $(CH_2)_4(N–NO_2)_4$ and that of RDX is $(CH_2)_3(N–NO_2)_3$. The decomposed fragments of $N–NO_2$ bonds produce nitrogen oxides such as NO_2 and NO, as well as oxide radicals. These nitrogen oxides act as oxidizer components and the remaining C,H,O fragments act as fuel components. The oxidizer and fuel components react together to produce high-temperature combustion products in the gas phase. Since the mass fraction of NO_2, $\xi(NO_2)$, contained within unit mass of a nitramine is higher than that in a double-base propellant, the overall gas-phase reaction is enhanced by the addition of the nitramine. The physical structures of double-base propellants are homogeneous due to the gelatinized nature of the mixtures of NC, NG, and stabilizers. Though the addition of the nitramine crystalline particles renders the physical structure of nitramine-CMDB propellants heterogeneous, the gas-phase structure is homogeneous because the gaseous decomposition products from the HMX particles and the base matrix diffuse and mix together at or just above the burning surface and then react at some distance above

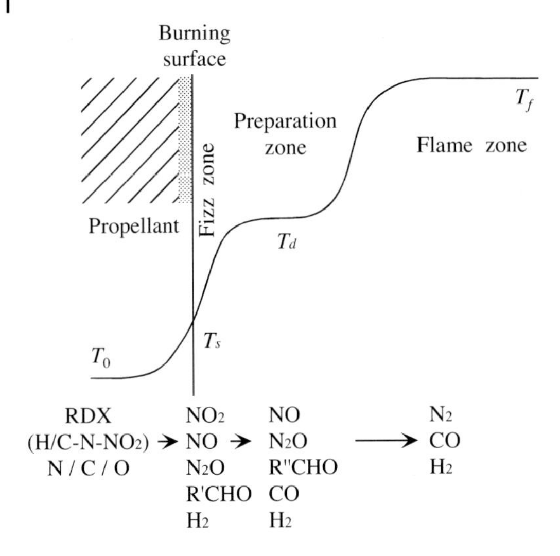

Fig. 8.5 Combustion wave structure of an RDX-CMDB propellant.

the surface to produce the luminous flame. The gas-phase structure of HMX-CMDB propellants is similar to that of RDX-CMDB propellants, as shown in Fig. 8.2. The combustion wave structure of an RDX-CMDB propellant is illustrated in Fig. 8.5. As in the case of double-base propellants, the luminous flame is distended from the burning surface and so there is a preparation zone between the burning surface and the luminous flame. The flame stand-off distance decreases with increasing pressure. An exothermic reaction zone is formed just above the burning surface, in which the temperature increases rapidly, similar to the fizz zone of double-base propellants. Fig. 8.6 shows the temperature profiles in combustion waves of an RDX-CMDB propellant at different pressures.

Fig. 8.7 shows the adiabatic flame temperatures and the heats of explosion of HMX-CMDB propellants as a function of $\xi(NO_2)$ under conditions of thermal equilibrium. The adiabatic flame temperatures, T_g, and the heats of explosion increase

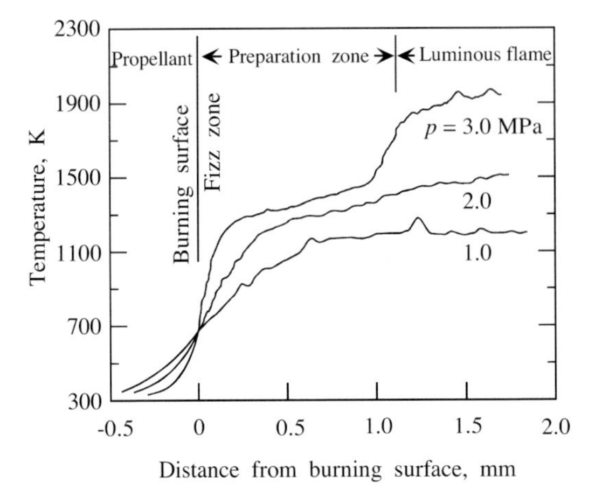

Fig. 8.6 Temperature profiles in the combustion wave structures of an RDX-CMDB propellant at different pressures.

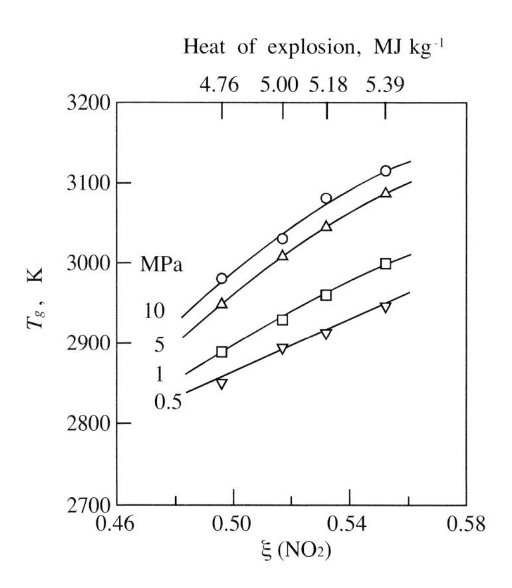

Fig. 8.7 Adiabatic flame temperatures of HMX-CMDB propellants as a function of $\xi(NO_2)$ or heat of explosion.

with increasing $\xi(NO_2)$, which increases with increasing ξ_{HMX}. The composition of the base matrix in terms of mass fractions is $\xi_{NC}(0.25)$, $\xi_{NG}(0.65)$, and $\xi_{DEP}(0.10)$. The nitrogen content of the nitrocellulose used is 12.2 %, the mass fraction of NO_2 contained within the base matrix, $\xi(NO_2)$, is 0.496, and the heat of explosion is 4.76 MJ kg^{-1}.

Fig. 8.8 shows the relationship between $\xi(NO_2)$ and the mole fractions of the final combustion products at 1 MPa. The mole fractions of CO, H_2O, and CO_2 are seen to decrease, while that of N_2 is seen to increase, with increasing $\xi(NO_2)$. The mole fraction of H_2 remains relatively unchanged in the range from $\xi_{HMX}(0.0)$ to $\xi_{HMX}(0.444)$.

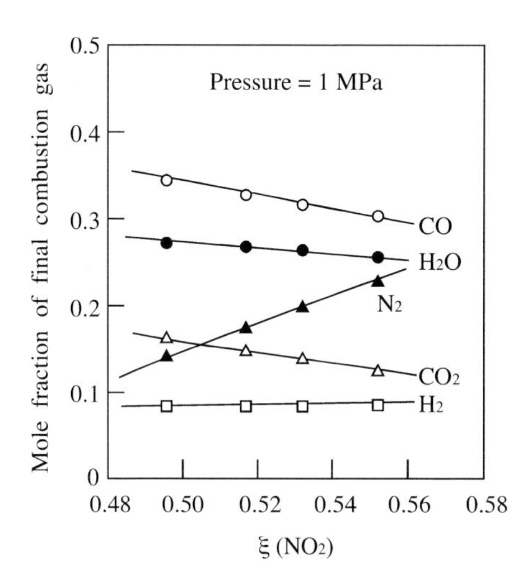

Fig. 8.8 Mole fractions of the final combustion gases of HMX-CMDB propellants as a function of $\xi(NO_2)$.

8.3.2
Burning Rate Characteristics

The burning rates of double-base propellants increase with increasing energy density of the propellant at constant pressure. On the contrary, the burning rates of HMX-CMDB propellants decrease with increasing ξ_{HMX}, as shown in Fig. 8.9, which implies that the burning rates of HMX-CMDB propellants decrease with increasing energy density of the propellant at constant pressure.[2,4–7] The HMX particles incorporated are of the β crystal form and the mean particle diameter is 20 μm. Though the burning rates of AP-CMDB propellants increase with increasing ξ_{AP} and are highly dependent on the size of the AP particles incorporated, the burning rate is less dependent on the size of the HMX particles used.[1]

Fig. 8.10 shows the relationship between burning rate and $\xi(NO_2)$ at different pressures.[2] The burning rate is seen to decrease linearly with increasing $\xi(NO_2)$

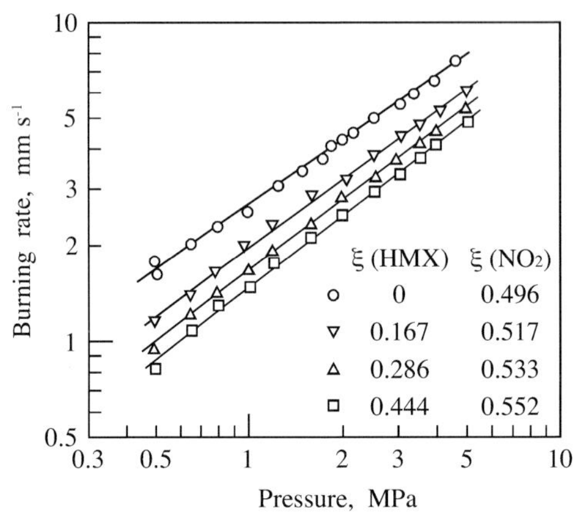

Fig. 8.9 Burning rates of HMX-CMDB propellants containing different mass fractions of HMX.

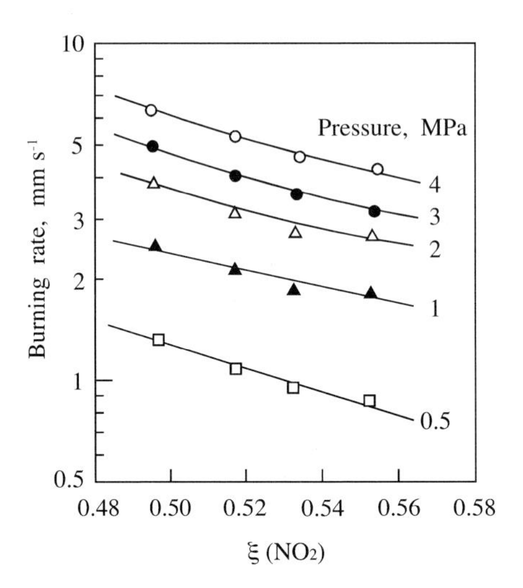

Fig. 8.10 Burning rates of HMX-CMDB propellants as a function of $\xi(NO_2)$.

at constant pressure in an ln r versus $\xi(NO_2)$ plot. The burning rate is represented by

$$r = c \exp \{a\xi(NO_2)\} p^n \tag{8.5}$$

where r (mm s^{-1}) is the burning rate, a is a constant, p (MPa) is the pressure, n is the pressure exponent, and c (mm s^{-1}) is a constant depending on the initial propellant temperature, T_0. The pressure exponent is less dependent on $\xi(NO_2)$ and pressure. The burning rate parameters given by Eq. (8.5) are $a = -5.62$, $c = 38.3$ mm s^{-1}, and $n = 0.70$ at $T_0 = 293$ K.

Since the reaction rate in the gas phase is dependent on the mole fractions of the reactive gases, the fundamental reaction pathway in the gas phase for double-base propellants remains relatively unchanged by the addition of HMX particles. On the other hand, the burning rate of double-base propellants increases with increasing $\xi(NO_2)$ at constant pressure, as discussed in Chapter 6. The gas-phase structure for HMX-CMDB propellants is similar to that for double-base propellants shown in Fig. 6.3. A luminous flame stands above the burning surface and the flame stand-off distance decreases with increasing pressure. The region between the burning surface and the flame front of the luminous flame zone is again referred to as the dark zone. The flame stand-off distance (the dark zone length) decreases with increasing ξ_{HMX} at constant pressure. The homogeneous nature of the gas phase of a double-base propellant remains unchanged even when crystalline HMX particles are incorporated into the propellant.[1,4,5]

8.3.3 Thermal Wave Structure

When an HMX-CMDB propellant burns steadily, the temperature in the condensed phase increases exponentially from the initial value, T_0, to the burning surface temperature, T_s, as shown in Fig. 6.3. A rapid temperature increase occurs above the burning surface, rising to the temperature in the dark zone, T_d. The temperature remains relatively unchanged over a certain distance from the burning surface, but then increases again further downstream in the gas phase. The gas-phase thermal structure for HMX-CMDB propellants appears to be similar to that for homogeneous double-base propellants, even though the HMX-CMDB propellants are heterogeneous.

Fig. 8.11 shows the relationship between the dark zone temperature, T_d, and the adiabatic flame temperature, T_g, at different burning pressures. T_d decreases with increasing T_g. The addition of HMX decreases T_d. On the other hand, T_d increases slightly with increasing pressure at constant $\xi(NO_2)$, as shown in Fig. 8.12. The burning rate is correlated with T_d, as manifested in a straight line in an ln r versus T_d plot, as shown in Fig. 8.13. The burning rate increases with increasing T_d at constant $\xi(NO_2)$, i. e., the burning rate increases with increasing pressure.[5] Fig. 8.14 shows the burning rate as a function of the adiabatic flame temperature, T_g. The burning rate decreases with increasing T_g, which implies that the burning rate decreases with increasing energy density of HMX-CMDB propellants.

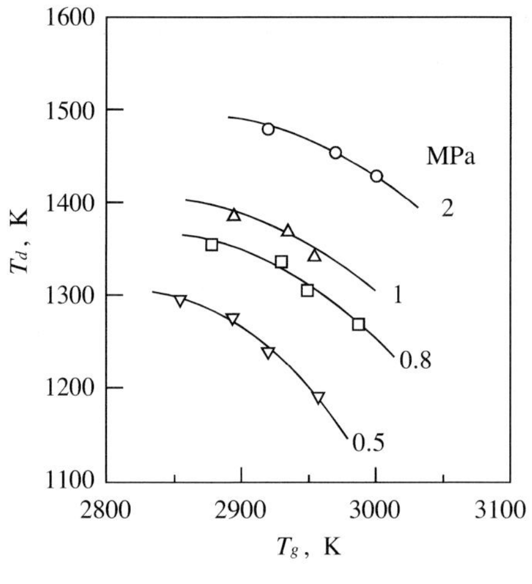

Fig. 8.11 Dark zone temperature versus adiabatic flame temperature at different pressures.

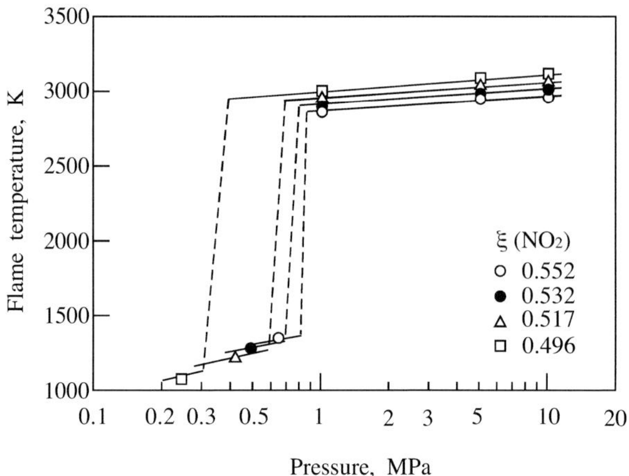

Fig. 8.12 Temperature distribution in the gas phase for HMX-CMDB propellants as a function of $\xi(NO_2)$.

As for double-base propellants, the burning rate of HMX-CMDB propellants is determined by the heat flux transferred back from the gas phase to the solid phase and the heat flux generated at the burning surface. In Fig. 8.15, the temperature gradient in the fizz zone of an HMX-CMDB propellant, $\phi_f = (dT/dx)_f$, is shown as a function of the dark zone temperature, where T is temperature, x is distance, and the subscript f denotes the fizz zone above the burning surface. As T_d increases, ϕ_f is seen to increase linearly in an ln ϕ_f versus T_d plot. The heat flux transferred back from the fizz zone to the burning surface is the dominant factor in determining the burning rate.[4,5] In fact, the relationship between ϕ_f and T_d shown in Fig. 8.15 is similar to that between the burning rate and T_d shown in Fig. 8.13.

An important result is that the dark zone temperature, T_d, decreases even though the flame temperature, T_g, increases with increasing $\xi(NO_2)$ at constant pressure,

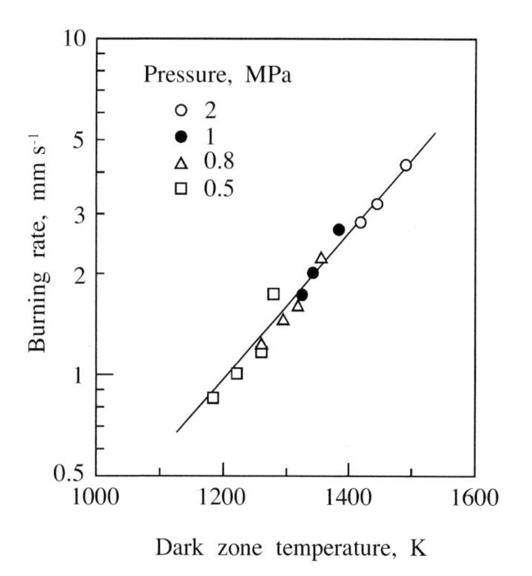

Fig. 8.13 Burning rate versus dark zone temperature for an HMX-CMDB propellant at different pressures.

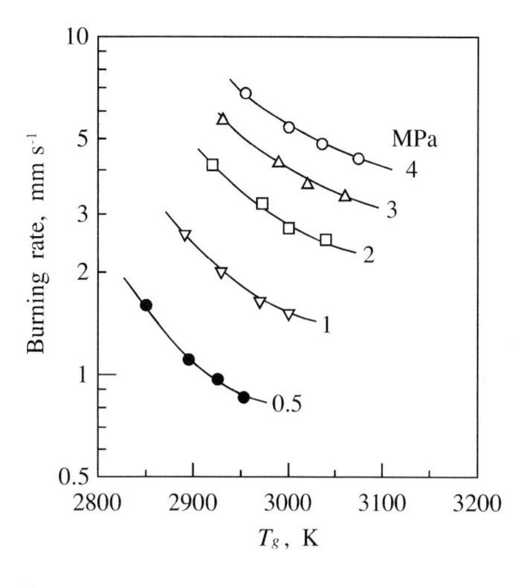

Fig. 8.14 Burning rate versus adiabatic flame temperature for an HMX-CMDB propellant at different pressures, showing that the burning rate decreases with increasing adiabatic flame temperature.

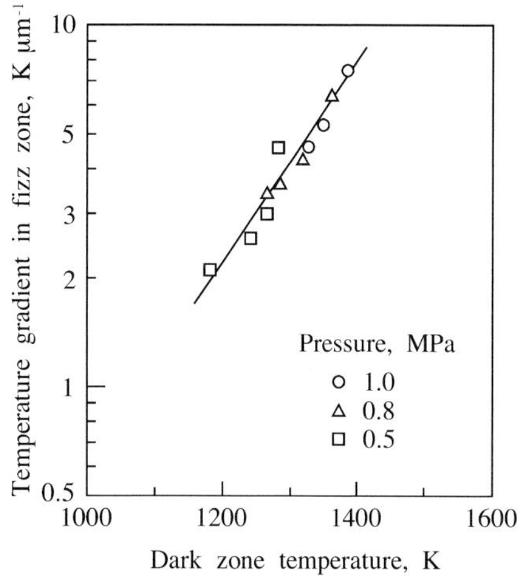

Fig. 8.15 Temperature gradient in fizz zone versus dark zone temperature for an HMX-CMDB propellant at different pressures, showing that the heat flux transferred back to the burning surface increases with increasing dark zone temperature.

as shown in Fig. 8.11. Furthermore, ϕ_f also decreases with increasing $\xi(NO_2)$, and so too does the burning rate, i. e., the burning rates of HMX-CMDB propellants decrease with increasing ξ_{HMX} at constant pressure. The observed burning rate characteristics of HMX-CMDB propellants are understood without consideration of the diffusional process and the chemical reaction between the gaseous decomposition products of the base matrix and the HMX particles. This is a significant difference compared to the burning rate characteristics of AP-CMDB propellants.

As established above, the burning rates of HMX-CMDB propellants decrease with increasing ξ_{HMX} in the region of ξ_{HMX} less than about 0.5. However, when ξ_{HMX} is increased beyond about 0.5, the burning rate tends to increase with increasing ξ_{HMX} at $T_0 = 243$ K, as shown in Fig. 8.16. The burning rate data for pure HMX, $\xi_{HMX}(1.0)$, shown in Fig. 8.16 are taken from the results of Boggs et al.,[11] and the dotted lines are extrapolated from the data for r versus ξ_{HMX}. As in the case of double-base propellants, the gas-phase structure remains homogeneous even when ξ_{HMX} is increased well above 0.5. Though a detailed analysis of the combustion wave structure of the HMX-CMDB propellants is not possible based on the experimental data, the decrease in burning rate with increasing ξ_{HMX} in the region with $\xi_{HMX} < 0.5$ is considered to be caused by a decreased reaction rate in the fizz zone, while the increase in burning rate in the region with $\xi_{HMX} > 0.5$ can be attributed to the increased heat of reaction at the burning surface. Furthermore, the results indicate that the reaction rate in the gas phase of pure HMX is lower and hence the heat feedback from the gas phase to the burning surface is also lower when compared with a conventional double-base propellant. In addition, the heat of reaction at the burning surface of pure HMX is higher than that in the case of a conventional double-base propellant.

The temperature sensitivity, σ_p, as defined in Eq. (3.73), of the HMX-CMDB propellants is shown as a function of ξ_{HMX} in Fig. 8.17. Though σ_p of the double-base propellant used as the base matrix decreases with increasing pressure, the σ_p values of the HMX-CMDB propellants remain relatively constant when ξ_{HMX} is kept con-

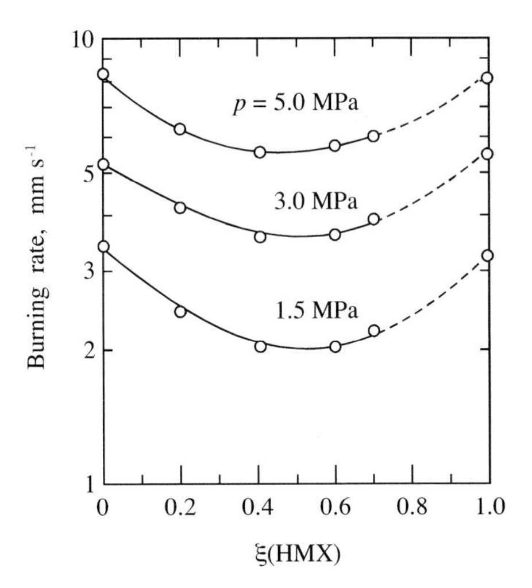

Fig. 8.16 Burning rate decreases with increasing ξ_{HMX} in the region below $\xi_{HMX}(0.5)$ and increases with increasing ξ_{HMX} in the region beyond $\xi_{HMX}(0.5)$ at $T_0 = 243$ K.

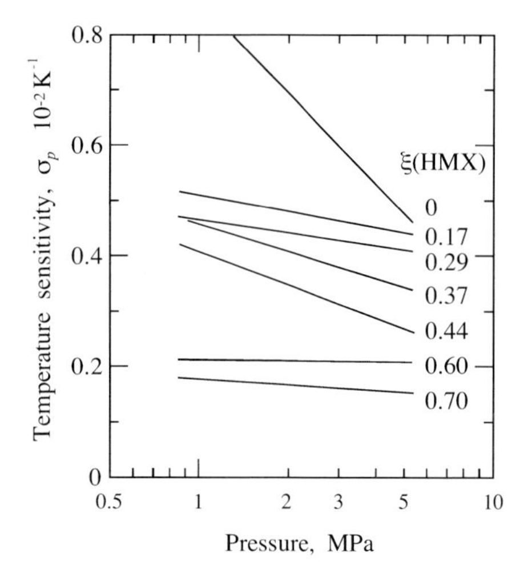

Fig. 8.17 Temperature sensitivity of HMX-CMDB propellants as a function of pressure.

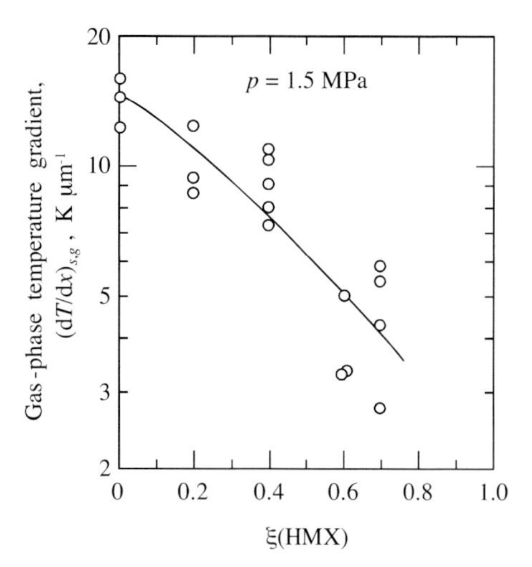

Fig. 8.18 Temperature gradient in the fizz zone decreases with increasing ξ_{HMX}.

stant. However, σ_p decreases when ξ_{HMX} is increased. The temperature just above the burning surface increases smoothly and reaches the dark zone temperature. The temperature gradient in the fizz zone, $(dT/dx)_f$, decreases with increasing ξ(HMX) at constant pressure, as shown in Fig. 8.18. Based on the burning rate data shown in Fig. 8.16 and the temperature gradient in the fizz zone data shown in Fig. 8.17, the gas-phase parameter ϕ and the condensed-phase parameter ψ can be determined as a function of ξ_{HMX}. The results indicate that ψ decreases with increasing ξ_{HMX} at constant pressure. The temperature sensitivity of the condensed phase, Ψ, as defined in Eq. (3.80), is determined from Eq. (3.78) using the data for σ_p and Φ. As shown in Fig. 8.19, σ_p decreases with increasing ξ_{HMX} at constant pressure. The relationship $\sigma_p = \Phi + \Psi$ indicates that σ_p of HMX-CMDB propellants consists of 70 % Φ and 30 % Ψ.[2] This result implies that the temperature sensitivity of burning rate is more dependent on the reaction in the gas phase than on that in the condensed phase.

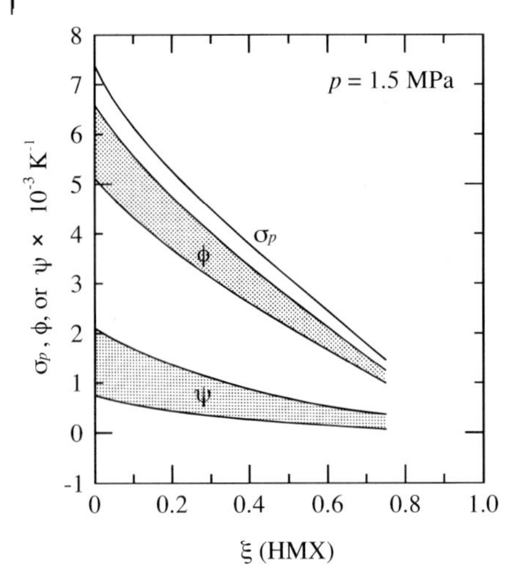

Fig. 8.19 Temperature sensitivity of burning rate decreases with increasing ξ_{HMX}, and Φ (gas phase) has a greater effect on σ_p than Ψ (condensed phase).

8.3.4
Burning Rate Model

The gas-phase structure of HMX-CMDB propellants is relatively homogeneous, even though the condensed phase is highly heterogeneous.[1,4,5] This is because the HMX particles within the double-base propellant used as a base matrix melt and diffuse into the double-base propellant at the burning surface. Consequently, a homogeneous gaseous mixture is formed on and above the burning surface. Thus, it is assumed that HMX-CMDB propellants burn in a similar manner as double-base propellants, and that the burning rate equation expressed by Eq. (3.75) is applicable. The effect of the addition of HMX particles on the burning rate of double-base propellants can be understood from the modified burning rate equation, Eq. (3.75), $r = \alpha_s\phi/\psi$. The effect of the addition of HMX on the burning rate of HMX-CMDB propellants is also dependent on two parameters: ϕ is the gas-phase reaction parameter in the fizz zone, and ψ is the condensed-phase reaction parameter at the burning surface.

Since the burning surface becomes highly heterogeneous with increasing ξ_{HMX}, the determination of T_s values is no longer possible, which precludes the determination of the heat of reaction at the burning surface, Q_s. However, it is assumed that the decomposition of the HMX particles within the base matrix occurs relatively independently of the decomposition of the base matrix, i. e., the double-base propellant. The overall heat release at the burning surface of an HMX-CMDB propellant is thus represented by[1]

$$Q_{s,CMDB} = \xi_{HMX}\, Q_{s,HMX} + \{(1 - \xi_{HMX}) Q_{s,DB}\} \tag{8.6}$$

where $Q_{s,CMDB}$, $Q_{s,HMX}$, and $Q_{s,DB}$ are the heat releases at the burning surface of unit mass of the HMX-CMDB propellant, the HMX particles, and the base matrix, respectively. Equation (8.6) indicates that the overall heat release of an HMX-CMDB propellant is the mass-averaged heat release of the HMX particles and the base matrix.

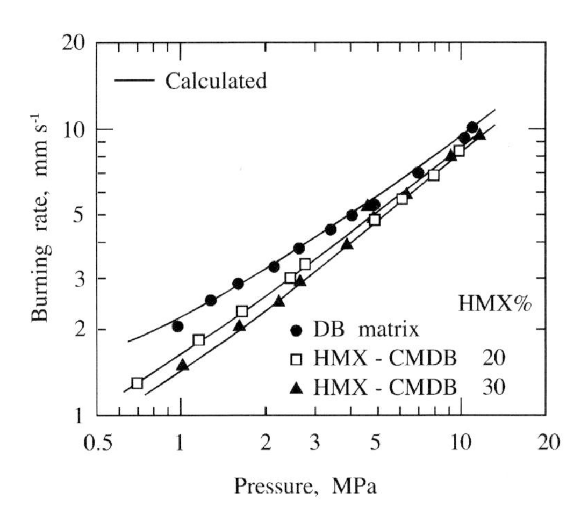

Fig. 8.20 Calculated and experimental burning rates for HMX-CMDB propellants, showing that the burning rate decreases with increasing HMX concentration (average HMX particle size = 200 μm).

According to the burning-surface and gas-phase observations, HMX particles in CMDB propellants are not active at the burning surface or in the fizz zone. The temperature profile in the fizz zone is not affected by the HMX particles. Thus, for an HMX-CMDB propellant, the gas-phase reaction can be assumed to be the same as that for a DB matrix in the region of $\xi_{HMX} < 0.5$, and the burning rate model can be applied. The heat release of HMX at the burning surface is not known, however, and it is assumed that the total heat release of the HMX-CMDB propellant at the surface is represented by Eq. (8.6). Thus, the burning rate of an HMX-CMDB propellant is obtained by simultaneously solving Eqs. (3.59)–(3.61).

Calculated HMX-CMDB propellant burning rate data with respect to both pressure and HMX concentration are consistent with experimental results, as shown in Fig. 8.20. The assumed values for the physical constants and reaction kinetics are listed in Ref. [1] and a further assumption is that $Q_{s,HMX} = 210$ kJ kg^{-1}. The burning rate increases with increasing pressure and decreases with increasing concentration of HMX. With an HMX content of 30%, the calculated surface temperature varies from 598 K at 1 MPa to 658 K at 8 MPa, and the calculated σp is 0.0048 K^{-1} at 8 MPa. These calculated values agree fairly well with actual measurements, as shown in Fig. 8.20. The burning rate model does not include the HMX particle size as a parameter since this has only a slight influence on the burning rate.

8.4 Plateau Burning of Catalyzed HMX-CMDB Propellants

8.4.1 Burning Rate Characteristics

Like double-base propellants, CMDB propellants show super-rate and plateau burning when they are catalyzed with small amounts of lead compounds. Fig. 8.21 shows a typical plateau burning for a propellant composed of NC-NG and HMX.[12] The chemical composition of the catalyzed propellant is shown in Table 8.1.

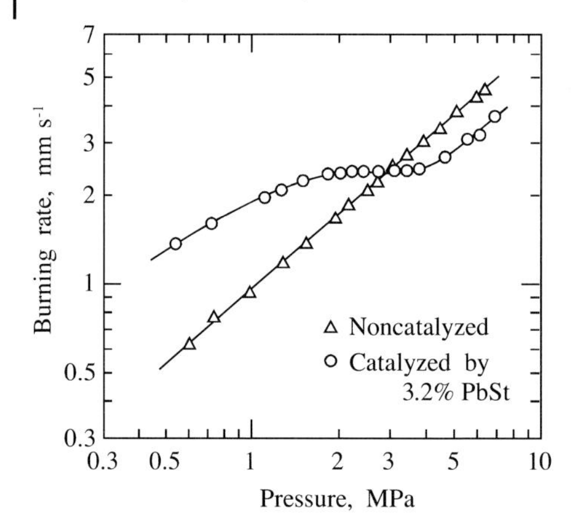

Fig. 8.21 Plateau burning over a wide pressure range (1.6–3.6 MPa) following the addition of 3.2% PbSt.

Table 8.1 Chemical composition of super-rate and plateau burning propellant.

ξ_{NC}	ξ_{NG}	ξ_{HMX}	ξ_{DEP}	ξ_{SOA}	ξ_{PbSt}
0.42	0.18	0.22	0.08	0.068	0.032

The burning rate of the non-catalyzed propellant (without PbSt) gives approximately a straight line in an ln r versus ln p plot. The pressure exponent n is 0.85 in the pressure range from 0.6 MPa to 7 MPa. When the non-catalyzed propellant is catalyzed with ξ_{PbSt}(0.032), super-rate burning is observed at low pressures below 1.6 MPa and the burning rate is significantly higher in this region than for the non-catalyzed propellant. A broad plateau burning of 2.4 mm s^{-1} is seen at pressures between 1.6 MPa and 3.6 MPa. Above the plateau burning region, the burning rate and pressure exponent increase once more with increasing pressure.[12]

8.4.2
Combustion Wave Structure

8.4.2.1 Flame Stand-off Distance

The flame structures of the non-catalyzed and catalyzed propellants used to obtain the data in Fig. 8.21 are similar, except with regard to the flame stand-off distances, i. e., the dark zone lengths, as shown in Fig. 8.22. The dark zone lengths of both propellants decrease with increasing pressure, in accordance with Eq. (5.4). The dark zone pressure exponent, d, of the non-catalyzed propellant is –1.8 throughout the pressure range from 1.2 MPa to 5 MPa. On the other hand, the pressure exponent of the catalyzed propellant is –2.0 in the super-rate region below 1.6 MPa and –2.6 in the plateau region between 1.6 MPa and 3.6 MPa.

The overall order of the reaction in the dark zone defined in Eq. (5.6) is 2.6 for both propellants, irrespective of whether or not the catalyst is incorporated. The

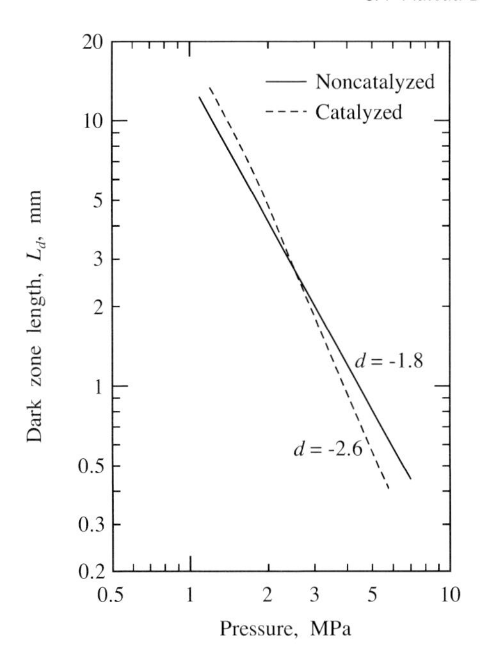

Fig. 8.22 The luminous flame front of the platonized propellant approaches the burning surface more rapidly than that of the non-catalyzed propellant when the pressure is increased in the plateau-burning pressure region.

value remains unchanged for the catalyzed propellant not only in the super-rate region ($n = 0.5$ and $d = -2.0$) but also in the plateau region ($n = 0.0$ and $d = -2.6$). This suggests that the reaction mechanism in the dark zone, which involves the reduction of NO to N_2, remains unchanged by the addition of the catalyst. It should be noted that while the burning rate is constant in the plateau region, the luminous flame of the catalyzed propellant stands somewhat further from the burning surface, and approaches the burning surface more rapidly than in the case of the non-catalyzed propellant when the pressure is increased. The luminous flame reactionoccurs too far away from the burning surface to cause the observed plateau

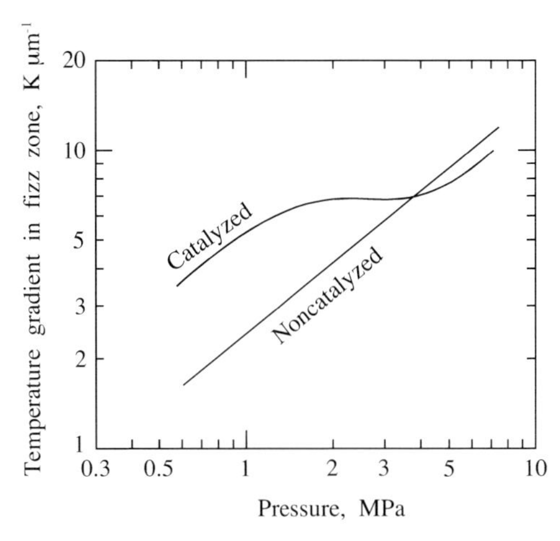

Fig. 8.23 Temperature gradient in the fizz zone increases in the super-rate burning region and then remains unchanged in the plateau-burning pressure region for the catalyzed propellant.

burning rate. Accordingly, the conductive heat flux from the luminous flame is not responsible for plateau burning, as it is in the case of normal burning.

8.4.2.2 **Catalyst Activity**

The temperature at the end of the fizz zone, i. e., at the beginning of the dark zone, is approximately 1400 K for the catalyzed propellant, which is slightly higher than that for the non-catalyzed propellant. The temperature gradient just above the burning surface, ϕ_f, i. e., in the fizz zone, is significantly different for the two propellants, as shown in Fig. 8.23. The slope increases linearly with pressure in an ln $\phi_{f,n}$ versus ln p plot for the non-catalyzed propellant. The catalyzed propellant exhibits a significantly different temperature gradient behavior compared to that of the non-catalyzed propellant. In the super-rate burning region, $\phi_{f,c}$ is more than doubled by the addition of the catalyst. In the plateau region, in which the burning rate is constant, $\phi_{f,c}$ becomes nearly pressure-independent. The gradient is represented by $\phi_{f,n} = b_n p^{0.85}$ for the non-catalyzed propellant, and by $\phi_{f,c} = b_c p^{0.0}$ for the catalyzed propellant,[12] where the subscripts n and c denote the non-catalyzed and the catalyzed propellants, respectively. These pressure exponents are approximately equal to the pressure exponent of burning rate, n, for each propellant. In other words, the burning rate behavior is governed by the conductive heat feedback from the fizz zone to the burning surface of the propellant.

The effect of catalyst activity of burning rate is defined by

$$r = (r_c - r_n)/r_n \tag{8.7}$$

while the catalyst activity in the fizz zone is defined by

$$\eta_f = (\tau_{f,n} - \tau_{f,c})/\tau_{f,n} \tag{8.8}$$

where τ_f is the reaction time in the fizz zone. If one assumes that $\phi_f = (T_d - T_s)/L_f$, where L_f is the fizz zone length, one gets

$$\eta_f = 1 - \phi_{f,n}\, r_n/\phi_{f,c}\, r_c \tag{8.9}$$

The catalyst activity in the dark zone is similarly defined by

$$\eta_d = (\tau_{d,n} - \tau_{d,c})/\tau_{d,n} \tag{8.10}$$

$$= 1 - r_n L_{d,c}/r_c L_{d,n} \tag{8.11}$$

where τ_d is the reaction time in the dark zone as given by Eq. (6.2).

If the catalyst acts in the fizz zone or in the dark zone, $\eta_{f,c}$ $\eta_{d,c}$ becomes smaller than $\eta_{f,n}$ or $\eta_{d,n}$, respectively. As shown in Fig. 8.24, the behavior of η_f corresponds to that of η_r. Both catalyst activities have positive values in the super-rate region, decrease with increasing pressure in the plateau region, and finally both become negative above 3 MPa. This indicates that the catalyst acts as a positive catalyst in

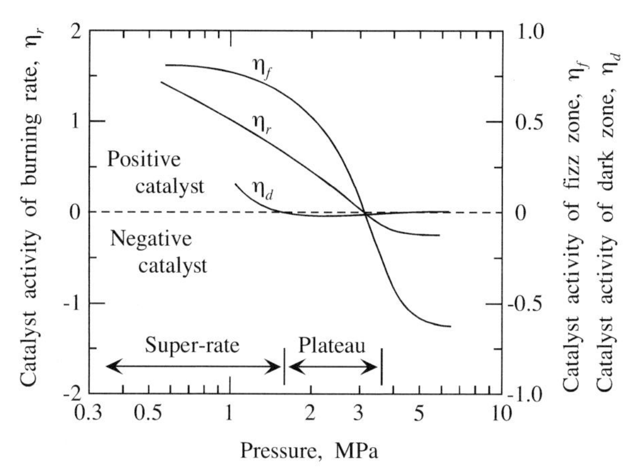

Fig. 8.24 Catalyst activity in the fizz zone decreases rapidly in the plateau region and becomes negative above the pressure region in which the burning rate of the catalyzed propellant is lower than that of the non-catalyzed propellant.

the fizz zone, increasing the burning rate below 3 MPa, but acts as a negative catalyst above 3 MPa, reducing the burning rate. On the other hand, η_d is approximately zero in all regions, i. e., super-rate, plateau, and mesa burning, indicating that the catalyst has no action in the dark zone and has no effect on burning rate.[12]

The effects of the catalyst on burning rate and flame reaction indicate that the super-rate burning phenomena observed in the combustion of HMX-CMDB propellants are fundamentally the same as the combustion phenomena of catalyzed double-base propellants. This implies that the lead catalysts act on the combustion of HMX to produce super-rate burning.

8.4.2.3 **Heat Transfer at the Burning Surface**

Referring to Eq. (3.49), the temperature gradient in the fizz zone, represented by ϕ_f, is given by the heat balance at the burning surface:

$$\phi_f = r\psi/\alpha_s \tag{3.75}$$

It can be seen that the temperature gradient could increase with increased heat release at the burning surface at a fixed burning rate. Since the temperature gradient in the fizz zone and the burning surface temperature are determined, the heat release at the burning surface can be computed using Eq. (3.75). The results indicate no clear difference between the non-catalyzed and catalyzed propellants. Although the heat released during plateau burning is apparently unaltered by the addition of the catalyst, as for double-base propellants catalyzed by lead compounds, neither super-rate burning nor plateau burning is associated with changes in the heat released in the subsurface and at the burning surface for catalyzed HMX-CMDB propellants.

Based on Eq. (3.75), the temperature gradient in the fizz zone is also represented by

$$\phi_f = \omega_f Q_f / \rho_p c_f \, r \tag{8.12}$$

Since ϕ_f and r of the catalyzed propellant are pressure-independent in the plateau region and Q_f is nearly constant, ω_f should also be pressure-independent. This is a significant difference when compared with commonly observed bimolecular gas-phase reactions. Referring to Eq. (3.33), the reaction rate in the gas phase, ω_g, is given as a function of pressure according to

$$\omega_g = p^k Z_g \exp(-E_g/RT) \tag{8.13}$$

where p is pressure, E_g is the activation energy, Z_g is a pre-exponential factor, and k is the order of the chemical reaction. Substituting Eq. (8.13) into Eq. (8.12) with $\omega_g = \omega_f$, the temperature gradient in the fizz zone can be related to pressure as follows:

$$\phi_f \propto p^{k-n} \tag{8.14}$$

The effective overall order of the fizz zone reaction, k, is determined to be zero for plateau burning, and approximately 1.4 for super-rate burning. The reaction order for the non-catalyzed propellant is also determined to be approximately 1.7, that is, nearly equal to the order of a conventional gas-phase reaction.

It is evident that the initial action of the lead compound is in the subsurface layer or at the burning surface.[13] The decomposition processes of the propellant are altered to form carbonaceous material. Thus, the gaseous products in the fizz zone are changed, and the reaction processes are changed further. These changes enhance the reaction rate in the fizz zone during super-rate burning. However, a pressure-independent reaction involving some inhibitory action of lead must occur in the fizz zone to keep ϕ_f constant during plateau burning. This inhibiting effect is due to the high degree of dispersion of the lead oxide formed at the burning surface, similar to the effect of lead catalysts during super-rate and plateau burning of catalyzed double-base propellants, as described in Section 6.4.2. It is important to note that super-rate and plateau burning of double-base propellants occur with propellants containing high concentrations of HMX particles, i. e., HMX-CMDB propellants. Similar super-rate and plateau burning are observed when the HMX particles are replaced with RDX particles. Double-base propellants are nitrate esters characterized by -O–NO_2 chemical bonds, while HMX and RDX are nitramines characterized by -N–NO_2 chemical bonds. NO_2 has been recognized as the dominant oxidizer molecule affecting the burning rates of both nitrate esters and nitramines. This implies that lead compounds act on gaseous NO_2 after the decomposition of nitrate esters and nitramines, both of which produce NO_2 in the initial stage of thermal decomposition, that is, in the fizz zone.

References

1 Kubota, N., and Masamoto, T., Flame Structures and Burning Rate Characteristics of CMDB Propellants, 16th Symposium (International) on Combustion, The Combustion Institute, Pittsburgh, PA, 1976, pp. 1201–1209.

2 Kubota, N., Energetics of HMX-Based Composite-Modified Double-Base Propellant Combustion, *J. of Propulsion and Power*, Vol. 15, No. 6, 1999, pp. 759–762.

3 Swaminathan, V., and Soosai, M., On the Burning Rate Characteristics of CMDB Propellants, *Propellants and Explosives*, Vol. 4, 1979, pp. 107–111.

4 Yano, Y., and Kubota, N., Combustion of HMX-CMDB Propellants (I), *Propellants, Explosives, Pyrotechnics*, Vol. 10, 1985, pp. 192–196.

5 Yano, Y., and Kubota, N., Combustion of HMX-CMDB Propellants (II), *Propellants, Explosives, Pyrotechnics*, Vol. 11, 1986, pp. 1–5.

6 Kubota, N., and Okuhara, H., Burning Rate Temperature Sensitivity of HMX Propellants, *J. of Propulsion and Power*, Vol. 5, No. 4, 1989, pp. 406–410.

7 Aoki, I., Burning Rate Characteristics of Double-Base and CMDB Propellants, Ph.D. Thesis, Department of Aeronautics, University of Tokyo, Tokyo, 1998.

8 Singh, H., Advances in Composite-Modified Double-Base Propellants (Eds.: Varma, M., and Chatterjee, A. K.), pp. 107–132, Tata McGraw-Hill Publishing Co., Ltd., India (2002).

9 Aoki, I., and Kubota, N., Combustion Wave Structure of HMX-CMDB Propellants (Eds. Varma, M., and Chatterjee, A. K.), pp. 133–143, Tata McGraw-Hill Publishing Co., Ltd., India (2002).

10 Barrere, N., and Nadaud, L., 10th Symposium (International) on Combustion, The Combustion Institute, Pittsburgh, PA, 1965, pp. 1381–1389.

11 Boggs, T. L., The Thermal Behavior of Cyclotrimethylenetrinitramine (RDX) and Cyclotetramethylenetetranitramine (HMX), Fundamentals of Solid-Propellant Combustion (Eds.: Kuo, K. K., and Summerfield, M.), *Progress in Astronautics and Aeronautics*, Vol. 90, Chapter 3, AIAA, New York, 1984.

12 Kubota, N., Determination of Plateau Burning Effect of Catalyzed Double-Base Propellant, 17th Symposium (International) on Combustion, The Combustion Institute, Pittsburgh, PA, 1979, pp. 1435–1441.

13 Kubota, N., The Mechanism of Super-Rate Burning of Catalyzed Double-Base Propellants, AMS Report No. 1087, Aerospace and Mechanical Sciences, Princeton University, Princeton, 1973.

9
Combustion of Explosives

9.1
Detonation Characteristics

9.1.1
Detonation Velocity and Pressure

In contrast to the detonation of gaseous materials, the detonation process of explosives composed of energetic solid materials involves phase changes from solid to liquid and to gas, which encompass thermal decomposition and diffusional processes of the oxidizer and fuel components in the gas phase. Thus, the precise details of a detonation process depend on the physicochemical properties of the explosive, such as its chemical structure and the particle sizes of the oxidizer and fuel components. The detonation phenomena are not thermal equilibrium processes and the thickness of the reaction zone of the detonation wave of an explosive is too thin to identify its detailed structure.[1–3] Therefore, the detonation processes of explosives are characterized through the details of gas-phase detonation phenomena.

The basic equations for describing the detonation characteristics of condensed materials are fundamentally the same as those for gaseous materials described in Sections 3.2 and 3.3. The Rankine–Hugoniot equations used to determine the detonation velocities and pressures of gaseous materials are also used to determine these parameters for explosives. Referring to Section 3.2.3, the derivative of the Hugoniot curve is equal to the derivative of the isentropic curve at point J. Then, Eq. (3.13) becomes

$$\left\{\frac{\partial p}{\partial(1/\rho)}\right\}_H = \left\{\frac{\partial p}{\partial(1/\rho)}\right\}_s = \frac{p_2 - p_1}{1/\rho_2 - 1/\rho_1} \tag{9.1}$$

The logarithmic form of Eq. (1.14) gives the specific heat ratio under isentropic change as

$$\gamma = -\left[\frac{\partial \ln p}{\partial \ln(1/\rho)}\right]_s = \frac{1 - p_1/p_2}{\rho_2/\rho_1 - 1} \tag{9.2}$$

The pressure p_J at the CJ point is obtained using Eqs. (3.12) and (9.2) as

$$p_J = (\rho_1 u_D^2 + p_1)/(\gamma + 1) \tag{9.3}$$

Propellants and Explosives. Naminosuke Kubota

ISBN: 978-3-527-31424-9

Since p_J is much larger than p_1 in the case of a detonation wave, Eq. (9.3) becomes

$$p_J = \rho_1 u_D^2 / (\gamma + 1) \tag{9.4}$$

The characteristic values at the Chapman–Jouguet point are also obtained from Eqs. (9.2) and (9.4) as[4]

$$\rho_J = (1 + 1/\gamma)\rho_1 \tag{9.5}$$

$$u_p = u_D / (\gamma + 1) \tag{9.6}$$

The specific heat ratio, γ, expressed by Eq. (1.28), is determined to be 2.85 for high explosives by statistical detonation experiments. Thus, Eqs. (9.4)–(9.6) become:[4]

$$p_J = 0.26\,\rho_1\,u_D^2 \tag{9.7}$$

$$\rho_J = 1.35\,\rho_1 \tag{9.8}$$

$$u_p = 0.26\,u_D \tag{9.9}$$

Though the theoretical detonation velocity and pressure at the *CJ* point are expressed by very simplified expressions, the computed results obtained by means of Eq. (9.7) are confirmed by measured data for RDX- and TNT-based explosives, as shown in Table 9.1[1] (Cp-B indicates "Composition B", with the two columns relating to different particle sizes).

Table 9.1 Density, detonation velocity, and pressure at the *CJ* point.

	RDX	TNT	Cp-B	Cp-B	Cyclotol
ρ_1 (kg m^{-3})	1767	1637	1670	1713	1743
u_D (m s^{-1})	8639	6942	7868	8018	8252
p_J by experiment (GPa)	33.79	18.91	27.2	29.22	31.25
p_J by Eq. (9.7) (GPa)	34.5	20.7	27.2	29.0	31.2

9.1.2
Estimation of Detonation Velocity of CHNO Explosives

A semi-empirical method for estimating the detonation velocities of explosives composed of C, H, N, and O atoms has been proposed.[5] The heat of detonation per mole, ΔH_D, defined by

$$\Delta H_D = \Delta H_{Df} - \Delta H_{ef} \tag{9.10}$$

is determined first, where ΔH_{Df} is the heat of formation of the detonation productsand ΔH_{ef} is the heat of formation of the explosive. The detonation temperature, T_D, is defined by

$$T_D = \left(298.15 \sum_i c_{p,i} - \Delta H_D \right) / \sum_i c_{p,i} \tag{9.11}$$

where $c_{p,i}$ is the heat capacity of the i-th component of the detonation products, which can be obtained from JANAF thermochemical data.[5] The parameters of the explosives, T_D and n, are assumed to be the values listed in Table 9.2.[6]

Table 9.2 Assumed reaction products and parameters of CHNO explosives.

Explosive	Reaction products	T_D (K)	n
NQ	$CO + 2N_2 + H_2O + H_2$	2716	0.0481
PA	$6CO + 1.5N_2 + H_2O + 0.5H_2$	2469	0.0393
DATB	$6CO + 2.5N_2 + 2.5H_2$	1962	0.0453
Tetryl	$7CO + 2.5N_2 + 1.5H_2 + H_2O$	3126	0.0418
TNT	$C_{(s)} + 6CO + 1.5N_2 + 2.5H_2$	2122	0.0441
NG	$3CO_2 + 1.5N_2 + 2.5H_2O + 0.25O_2$	4254	0.0319
NM	$CO + 0.5N_2 + 0.5H_2 + H_2O$	2553	0.0492
PETN	$3CO_2 + 2CO + 2N_2 + 4H_2O$	3808	0.0348
RDX	$3CO + 3N_2 + 3H_2O$	3750	0.0405

Based on the experimental data for various types of explosives shown in Table 9.2, the detonation velocity is represented by the following equation:

$$u_D = 0.314(nT_D)^{1/2}\rho_0 + 1.95 \tag{9.12}$$

where n is the number of moles of gaseous products per unit mass of explosive and ρ_0 is the loading density of the explosive. The measured detonation velocities increase linearly in a plot of u_D versus $(nT_D)^{1/2}\rho_0$ and these experimental data are well corroborated by Eq. (9.12) for the explosives shown in Table 9.2. Thus, this simplified equation predicts the detonation velocity for explosives composed of C, H, N, and O atoms.

9.1.3 Equation of State for Detonation of Explosives

Since the pressure generated by a gaseous detonation is of the order of 10 MPa, the equation of state for an ideal gas represented by Eq. (1.5) is considered to be valid for use in the computation of detonation characteristics. However, the pressure generated by the detonation of a solid explosive is of the order of 40 GPa and hence the equation of state for an ideal gas no longer applies because of the interaction between the molecules of the detonation products. The Rankine–Hugoniot equationsused for gaseous materials are also used for explosives, but not the equation of state given by Eq. (1.5).

Several types of equation of state have been proposed for detonation products. The equation of state based on the Van der Waals equation represented by

$$p(v - b) = RT \tag{9.13}$$

is a simple expression for high-pressure gases, where b is the effective volume of the molecules. The Kistiakowsky and Wilson equation represented by

$$pv = nRT\{1 + x \exp(x)\} \tag{9.14}$$

is commonly used for the detonation computation, where $x = K/vT^{1/2}$ and K is an adjustable constant.[2,3,7,8] Substituting Eq. (9.13) or Eq. (9.14) into the Rankine–Hugoniot equations given by Eqs. (3.9)–(3.12), the detonation velocities and pressures of explosives may be computed on the basis of their thermochemical properties.

9.2 Density and Detonation Velocity

9.2.1 Energetic Explosive Materials

Table 9.3 shows the measured detonation velocities and densities of various types of energetic explosive materials based on the data in Refs. [9–11]. The detonation velocity at the CJ point is computed by means of Eq. (9.7). The detonation velocity increases with increasing density, as does the heat of explosion. Ammonium nitrate(AN) is an oxidizer-rich material and its adiabatic flame temperature is low compared with that of other materials. Thus, the detonation velocity is low and hence the detonation pressure at the CJ point is low compared with that of other energetic materials. However, when AN particles are mixed with a fuel component, the detonation velocity increases. On the other hand, when HMX or RDX is mixed with a fuel component, the detonation velocity decreases because HMX and RDX are stoichiometrically balanced materials and the incorporation of fuel components decreases their adiabatic flame temperatures.

Table 9.3 Density, detonation velocity, particle velocity, and detonation pressure at the CJ point for various energetic materials.

	ρ_1 kg m^{-3}	u_D m s^{-1}	u_p m s^{-1}	p_J GPa
Ammonium picrate	1720	7150	1860	22.9
Diazodinitrophenol	1630	7000	1820	17.3
Diethylene glycol	1380	6600	1720	15.6
NG	1591	7600	1980	20.7
Nitroglycol	1480	7300	1900	20.5
NQ	1710	8200	2130	29.9
NIBGTN	1680	7600	1980	25.2
NM	1138	6290	1640	11.7
Hydrazine nitrate	1640	8690	2260	32.2
HMX	1900	9100	2370	40.9
RDX	1818	8750	2280	36.2
PETN	1760	8400	2180	32.3
TNT	1654	6900	1790	20.5
Trinanisol	1610	6800	1770	19.4
TNB	1760	7300	1900	24.4
TNChloroB	1797	7200	1880	24.2
Methyl nitrate	1217	6300	1640	12.6
Tetryl	1730	7570	1970	25.8
Picric acid	1767	7350	1910	24.8
AN	1720	2700	700	3.3
Lead azide	4600	5300	1380	33.6

9.2.2 Industrial Explosives

Though the detonation velocity of an explosive is an important property with regard to its use as an industrial explosive, safe handling and cost are further factors that need to be taken into account. Historically, nitroglycerin has been used as a major ingredient of industrial explosives and has been used to formulate various types of explosives with a view to obtaining high detonation velocity. However, nitroglycerin is highly sensitive to mechanical shock during formulation and handling, and this provided the impetus for the development of safer and more convenient industrial explosives.

Since ammonium nitrate (AN) is hygroscopic and very soluble in water and forms ammonium ions according to $NH_4NO_3 \rightleftharpoons NH_4^+ + NO_3^-$, it has long been used as a valuable ingredient of fertilizers. Thus, AN has been proved to be a safe material during handling and is also a very inexpensive material. On the other hand, the decomposition products of AN contain oxidizer-rich fragments that act as an oxidizer when mixed with fuel components. When AN is mixed with a polymeric material such as polyether or polyurethane, AN composite propellants for rockets are formulated, which are known as non-detonative propellants. However, when such AN propellants are subjected to a high-strength shock, they may detonate explosively. It must be noted that detonative propagation only occurs following initiation by a relatively high detonative strength. Otherwise, the AN propellants-burn very slowly through a deflagration phenomenon. AN explosives are formulated more simply than AN propellants as a mixture of AN and oil or of AN and water.

9.2.2.1 ANFO Explosives

A mixture of AN and light oil forms a low-strength explosive, which is used as a blasting compound in mining and industrial engineering. This class of explosives is named ammonium nitrate fuel oil explosives (ANFO explosives), which are typically composed of 94% AN and 6% fuel oil and have a density of 800–900 kg m^{-3}. The AN particles used for ANFO explosives are porous and spherical in shape, the so-called prill ammonium nitrate. Since the volumetric density is low and the flame temperature is also low, the detonation velocity falls in the range 2500–3500 m s^{-1}. The sensitivity of these materials to detonation is very low, hence the handling of ANFO explosives is easier than that of other industrial explosives.

9.2.2.2 Slurry and Emulsion Explosives

A mixture of AN and water forms a low-strength explosive referred to as a slurry or emulsion explosive. Since a mixture of AN and water cannot be detonated by initiation with a moderate detonation strength, to formulate practical slurry explosives nitrate esters such as monomethylamine nitrate, ethylene glycol mononitrate, or ethylamine mononitrate in conjunction with aluminum powder are added as sensitizers that facilitate the initiation of detonation.

The major chemical components of emulsion explosives are fundamentally the same as those of slurry explosives, as shown in Table 9.4.[10] Instead of the sensitizers used for slurry explosives, a large number of hollow microspheres made of glass or plastics are incorporated to formulate emulsion explosives in order to obtain successive detonation propagation after the initiation of detonation. During detonation propagation into the interior of the explosives, an adiabatic compression results

Table 9.4 Typical chemical ingredients of slurry and emulsion explosives (mass %).

Slurry explosives		Emulsion explosives	
Ammonium nitrate	45.0	Ammonium nitrate	76.0
Water	10.0	Water	10.0
Sodium nitrate	10.0	Fuel oil or wax	3.0
Monomethylamine nitrate	30.0	Hollow glass microspheres	5.0
Aluminum powder	2.0	Stabilizers	6.0
Stabilizers	3.0		

Table 9.5 Physicochemical characteristics of slurry and emulsion explosives.

	Slurry explosives	Emulsion explosives
Density, kg m^{-3}	1150–1350	1050–1230
Detonation velocity, m s^{-1}	5300–6000	5000–6500
Detonation completeness at 258 K	100%	100%

from the destruction of the hollow microspheres. Fuel oil or wax is also included as a fuel component. Typical chemical components and detonation characteristics of slurry and emulsion explosives are shown in Tabs. 9.4 and 9.5.[10]

9.2.3 Military Explosives

The explosives used for military purposes are different from those used in industry. Not only thermomechanical power for destruction, but also various other characteristics are required. Experimental tests, such as slow cook-off, fast cook-off, bullet impact, and sympathetic explosion tests, must be passed to meet the requirements for insensitive munitions (IM). The aerodynamic heating of warheads on flight projectiles is also an important factor in designing warheads.

9.2.3.1 TNT-Based Explosives

TNT is one of the most important materials used to formulate blasting charges for both industrial and military applications. Since TNT does not corrode metals, it can be cast directly into metal cases or pressed into warhead shells. In order to obtain high-explosive characteristics, TNT is mixed with other materials, such as AN, Tetryl, PETN, Al powder, or nitramine particles.[10] A mixture of TNT and AN is named Amatol, the composition of which ranges between ξ_{TNT}/ξ_{AN} mass fraction ratios of 0.5/0.5 and 0.2/0.8. These mixtures of TNT and AN are melted and then cast. A mixture of TNT and Al is named Tritonal and has a composition in terms of mass fraction ratio of $\xi_{TNT}/\xi_{Al} = 0.8/0.2$. A mixture of TNT and HMX is named Octol, the composition of which in terms of the ξ_{TNT}/ξ_{HMX} mass fraction ratio ranges from 0.3/0.7 to 0.25/0.75. The maximum detonation velocity of 8600 m s^{-1} is obtained when the density is 1800 kg m^{-3}.

A mixture TNT and RDX is referred to as "Composition A" when pressed with a small amount of wax. Its detonation velocity is approximately 8100 m s^{-1}. When TNT and RDX are cast together with a wax, the mixture is named "Composition B". The mass fraction ratio is $\xi_{TNT}/\xi_{RDX} = 0.4/0.6$ with 0.1% wax. The density ranges from 1600 kg m^{-3} to 1750 kg m^{-3} and the detonation velocity is about 8000 m s^{-1}. The melting point of TNT is 353.8 K, which is too low to permit its use in warhead charges intended for supersonic or hypersonic flight (Mach number > 5). Deformation of the charges would occur due to the aerodynamic heating.

9.2.3.2 **Plastic-Bonded Explosives**

In general, PBX materials are used for the warheads of rockets and guns. Thus, the detonation pressure p_J represented by Eq. (9.7) is the most important parameter above all others. Since the detonation velocity u_D can be measured more easily and more accurately than p_J, performance is evaluated by measuring u_D, which is converted into p_J by means of Eq. (9.7). Table 9.6 shows u_D and ρ data, along with computed detonation pressures at the *CJ* point, for various HMX-PBX and RDX-PBXmaterials.

As shown in Table 9.6, the detonation velocity is highly dependent on the density of the PBX, which, in turn, depends on the mass fraction of HMX or RDX.[11] When a mixture of nylon powder and HMX particles is pressed into an explosive of the desired shape, a high-density HMX-PBX is formed. However, during the formulation process, the material is sensitive to pressurization and to mechanical shock.

Though the density of aluminized PBX is high, the measured detonation velocity is low compared with that of non-aluminized PBX. Since HMX and RDX are stoichiometrically balanced materials, no extra oxygen is available to oxidize the aluminum particles. The aluminum particles are instead oxidized by CO molecules formed among the combustion products of HMX or RDX. Furthermore, the oxidation of the aluminum particles takes a much longer time compared to the decomposition of the crystalline HMX or RDX particles. The aluminum particles do not react in the detonation wave, but downstream of the *CJ* point shown in Fig. 3.5. Thus, there is no increase in the detonation velocity, even though the density of the PBX is increased by the addition of aluminum particles. However, when an aluminized PBX is used in water, the high-temperature aluminum particles react with the water to produce bubbles of hydrogen. These bubbles generate additional pressure and a shock wave in the water.

Table 9.6 Density, detonation velocity, particle velocity, and pressure at *CJ* point for various HMX-PBX and RDX-PBX materials.

Composition	Mass fraction ratio	ϱ_1 kg m^{-3}	u_D m s^{-1}	u_p m s^{-1}	p_J GPa
(a) HMX-PBX					
HMX/Nylon	0.95/0.05	1800	8670	2250	35.2
HMX/Nylon	0.86/0.14	1730	8390	2180	31.7
HMX/Polystyrene	0.82/0.18	1670	7980	2070	27.7
HMX/HGTPB	0.82/0.18	1640	8080	2100	27.8
HMX/GAP	0.82/0.18	1670	8010	2080	27.9
HMX/Al/Polystyrene	0.59/0.23/0.18	1800	7510	1950	26.4
(b) RDX-PBX					
RDX/HTPB	0.86/0.14	1650	8120	2110	28.3
RDX/Al/HTPB	0.71/0/17/0.12	1750	7700	2000	27.0

9.3 Critical Diameter

The detonation velocity propagated within an explosive charge, u_D, decreases with decreasing diameter of the charge. For example, when a detonation wave propagates into a cylindrical PBX charge of diameter 45 mm, the detonation velocity is 7000 m s^{-1}. However, when the diameter is decreased to 20 mm, the detonation velocity decreases to 6700 m s^{-1}. No detonation occurs when the diameter is decreased below a certain limiting value. This minimum diameter is termed the critical diameter for detonation, d_c. The critical diameters and detonation velocities of some typical explosives are shown in Table 9.7. It is evident that d_c decreases as u_D increases. In other words, d_c decreases as the energy of the explosives increases. With decreasing strength of an explosive, i. e., low detonation velocity and low detonation pressure, the energy loss from the surface of a detonating charge increases due to the increased thickness of the detonation wave. In addition, d_c is dependent on the type of explosive case used. When a case made of hard material is used, d_c appears to be a small value compared to when a case made of soft material is used. The energy loss from the case made of hard material is smaller than that from the case made of soft material.

Table 9.7 Critical diameters of explosives.

Explosives	Critical diameter d_c (mm)	Detonation velocity u_D (m s^{-1})
TNT	8–10	6900
Picric acid	6	7350
RDX	1–1.5	8750
PETN	1–1.5	8400
AN	100	2700

9.4 Applications of Detonation Phenomena

9.4.1 Formation of a Flat Detonation Wave

In general, detonation of an explosive is initiated by a primer as a point-source. The detonation wave formed at the point-source propagates into the explosive in a spherical manner as shown in Fig. 9.1.

An explosive lens is used to generate a flat detonation wave. The explosive lens consists of two cone-shaped segments of explosives, an inner cone and an outer cone, which are fitted together as shown in Fig. 9.2. When detonation is initiated by an electric detonator, a booster charge positioned at the top-center of the inner-cone explosive detonates. Then, the inner cone detonates and the detonation wave propa-

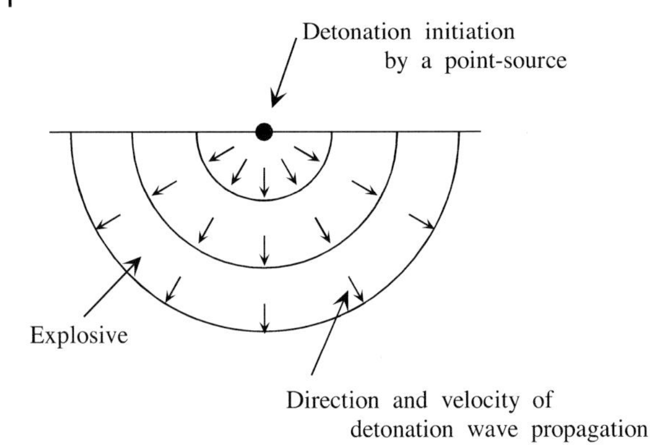

Fig. 9.1 Detonation wave propagation initiated by a point-source.

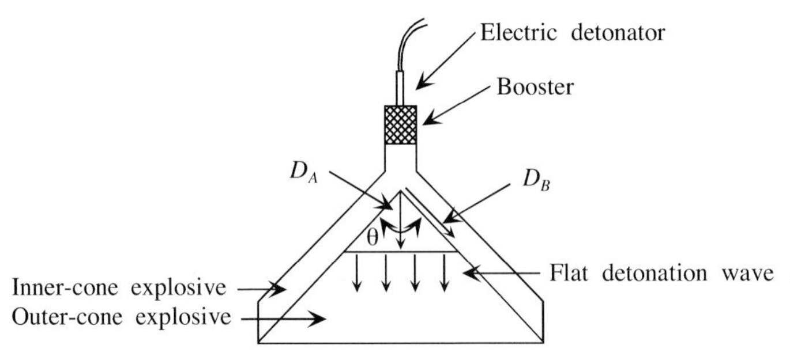

Fig. 9.2 Formation of a flat detonation wave by an explosive lens.

gates along the inner cone and the outer cone simultaneously. The detonation velocity of the outer cone, D_A, is given by

$$D_A = D_B \cos\theta/2 \tag{9.15}$$

where D_B is the detonation velocity of the inner cone and θ is the cone angle of both the inner-cone and outer-cone explosives. The detonation wave formed in the outer-cone explosive propagates one-dimensionally towards the bottom of the cone. The relative detonation velocities of the explosives must be $D_B > D_A$.

If the velocity of the flat detonation wave formed at the bottom of the outer-cone explosive is not sufficient for some objectives, a cylindrical-shaped high-explosive is attached to the bottom of the outer-cone explosive, as shown in Fig. 9.3. The detonation velocity of the attached high-explosive is higher than that of the outer-cone explosive. The flat detonation wave of the outer-cone explosive then initiates the detonation of the high-explosive and forms a reinforced flat detonation wave therein.

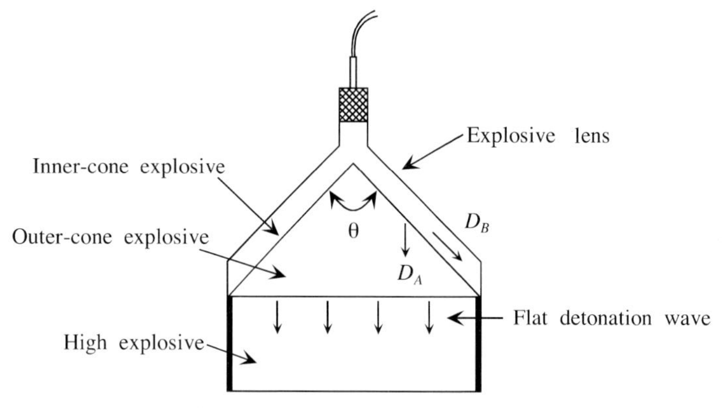

Fig. 9.3 Formation of a strong one-dimensional detonation wave with an explosive lens.

9.4.2
Munroe Effect

The detonative explosives used for different objectives, such as warheads, bombs, industrial mining, and civil engineering applications, are housed in various types of vessels. As a result, the performance of a given explosive depends not only on its constituent chemicals and its mass, but also on its physical shape. When a detonation is initiated at a point in an explosive charge, the detonation wave propagates spherically in all directions. When a detonation is initiated at a point at one end of an explosive charge, the detonation wave propagates semi-spherically within the charge. Thus, the strength of the detonation wave in the charge or of the shock wave in the atmosphere created by the detonation decreases rapidly with increasing distance from the initiation point of the detonation.

When a hollow space exists at the rear-end of an explosive charge, the detonation wave formed at the head-end of the charge propagates into the charge and is then deformed when it meets the hollow space. Due to this deformation, a concentrated shock wave accompanied by a high temperature and pressure is created in the hollow space, which propagates into the space when the shape thereof is appropriately designed. This detonation phenomenon is known as the Munroe effect, and is exploited to pierce thick metal plates. When the surface of the hollow space is lined with a thin metal plate, the deformed detonation wave creates a molten metal jet stream along the center line of the hollow. Such an explosive charge incorporating a lined metal hollow space is termed a shaped charge.

Fig. 9.4 shows the internal structure of a shaped charge used to obtain a high-energy jet stream.[11,12] The principal components of the shaped charge are an explosive, a detonator, a cone-shaped liner made of copper, and a surrounding casing for the explosive. When the detonator is initiated electrically, a detonation wave is formed, which propagates towards the liner. When the detonation wave reaches the top of the liner cone, the liner forms a molten jet stream that acquires high kinetic energy along the cone axis.

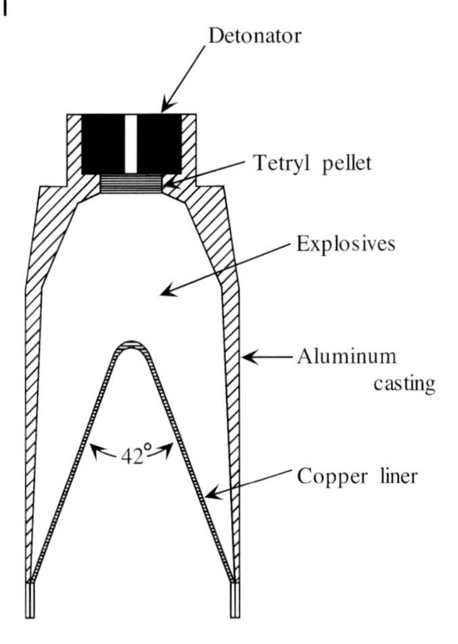

Fig. 9.4 Structure of a shaped charge.

The important parameters with regard to jet formation are the shape of the liner, the liner material, and the shaped cone angle. A Tetryl pellet is used to initiate detonation of the main explosives. A drawn copper liner is used as a cone, the angle of which is 42°, and its outer surface is machined precisely.[9,10] The shaped jet formed by the molten copper liner is ejected along the center line of the original cone. The minimum length of explosive charge needed to obtain effective strength of the jet stream is four times its diameter. To acquire the most effective kinetic energy, the distance from the end of the explosive charge to the target should be approximately one to three times the exit diameter of the cone. The effective cone angle is between 30 and 40°. High-density metal materials are chosen as cone-liner plates. The process of molten jet stream formation is illustrated in Fig. 9.5.

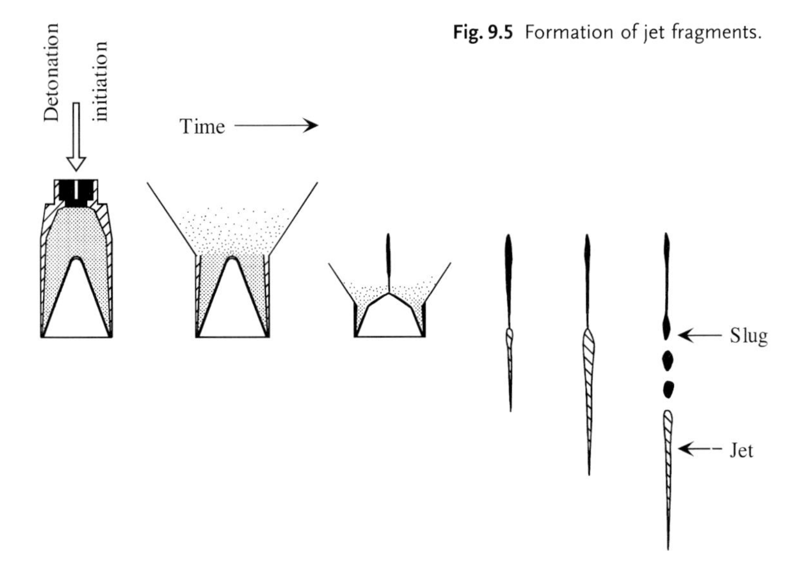

Fig. 9.5 Formation of jet fragments.

9.4.3
Hopkinnson Effect

When a shock wave propagates from one end (A) to the other end (B) of a solid body, a compression force is exerted in the wake of the shock wave that acts within the solid body. The shock wave is reflected at B and becomes an expansion wavethat propagates towards A. An expansion force is then exerted in the wake of the expansion wave that travels back from B. This process is shown schematically in Fig. 9.6.

When a compression wave travels into materials such as rock or concrete, no damage is inflicted on the materials because of their high compressive strength. However, when an expansion wave travels within the same materials, mechanical damage results near B. This is because rock and concrete are materials of low tensile strength. Fig. 9.7 shows a pair of photographs of the surface (A) and the reverse

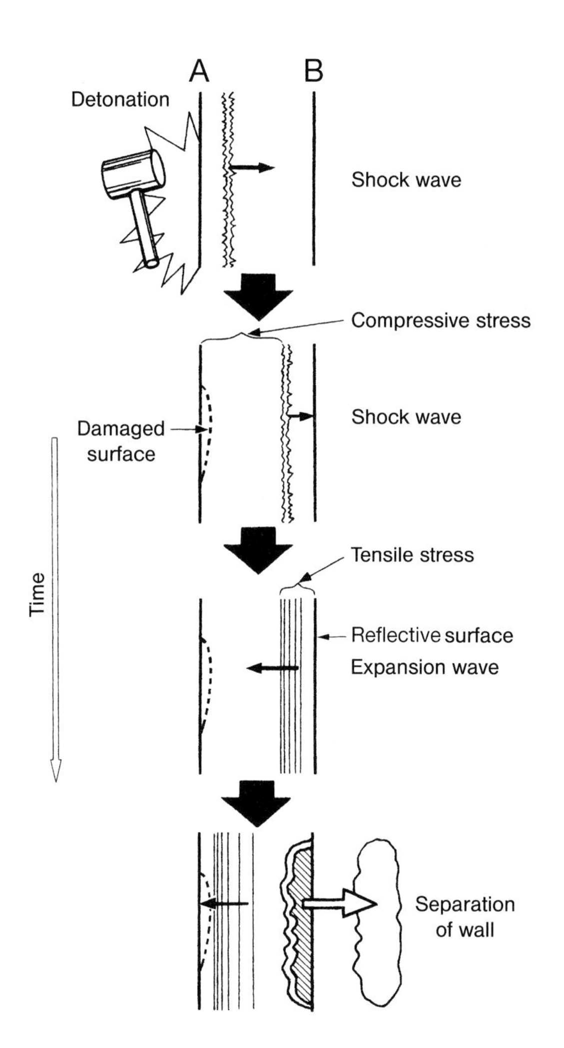

Fig. 9.6 Schematic representation of shock wave propagation and the generation of a reflected expansion wave within a solid wall.

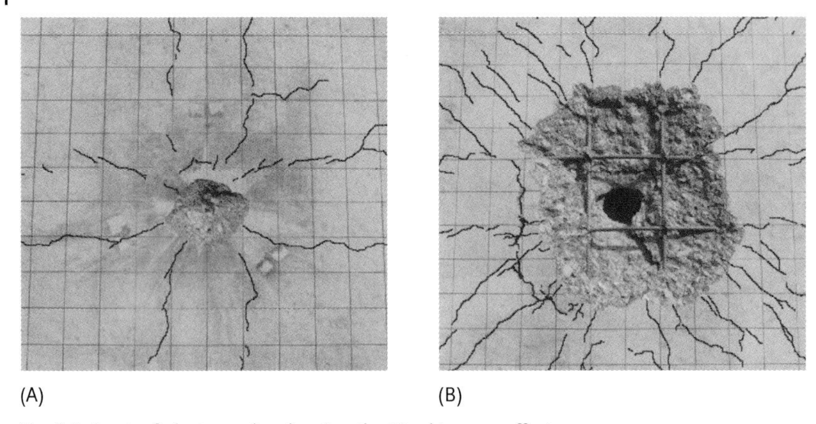

(A) (B)

Fig. 9.7 A set of photographs showing the Hopkinnson effect.

side (B) of a reinforced concrete wall.[13] An explosive is detonated at A and the shock wave propagates into the wall. The shock wave is reflected at B and becomes an expansion wave that travels back towards A. Though the surface of the wall at A remains largely undamaged despite the detonative explosion, a large portion of the reverse side B is damaged and separated by the expansion wave. This phenomenon is termed the Hopkinnson effect.

9.4.4
Underwater Explosion

When an explosive is detonated in water, a shock wave propagates through the water accompanied by a bubble. The chemical energy of the explosive is converted into shock wave energy and bubble energy. The volume of the bubble is increased by the expansion wave and decreased by the compression wave in an oscillatory fashion. The maximum size of the bubble is determined according to:[14–16]

$$r_{max} = \frac{3}{2}\frac{t_b}{2.24}\sqrt{\frac{2}{3}\frac{p_0}{\rho_w}} \tag{9.16}$$

where r_{max} is the maximum radius of the bubble (m), p_0 is the static pressure of the water (N m^{-2}), t_b is the periodic time of the bubble (s), and ρ_w is the density of water (kg m^{-3}).

The bubble energy E_b (J) is given by

$$E_b = (4/3)\ \pi r_{max}{}^3 p_0$$

$$= 68.4 p_0{}^{5/2} t_b{}^3 / m_e \tag{9.17}$$

and the shock wave energy E_s (J) is given by

$$E_s = 4\pi r^2 \int_0^{ts} p u_p \, dt \qquad (9.18)$$

$$= \left(4\pi r^2 / \rho_w a_w m_e\right) \int_0^{ts} p^2 dt$$

where r is the distance from the center of the explosion, p is the pressure at r, u_p is the particle velocity, a_w is the speed of sound in water, m_e is the explosive charge mass, and t_s is the time duration of the shock wave.

Experimental results of underwater explosion tests using an emulsion explosivecomposed of ammonium nitrate and hydrazine nitrate showed 0.85 MJ kg^{-1} for the shock wave energy and 2.0 MJ kg^{-1} for the bubble energy. The shock wave energy of underwater explosions is increased by the addition of aluminum powder to the explosives. The aluminum powder reacts with H_2O molecules in the bubble. Large amounts of H_2 molecules and heat are produced by the oxidation of the Al with H_2O according to:

$$2Al + 3H_2O \longrightarrow Al_2O_3 + 3H_2 + 949 \text{ kJ mol}^{-1}$$

The Al-H_2O reaction increases the temperature and the number of moles of gas in the bubble by the production of H_2 molecules. The pressure in the bubble is thereby increased. As a result, the bubble energy and shock wave energy are increased. It must be understood that the oxidation of aluminum powder is not like that of gaseous reactants. Reaction occurs at the surface of each aluminum particle and leads to the formation of an aluminum oxide layer that coats the particle. The oxidized layer prevents the oxidation of the interior particle. The combustion efficiency of aluminum particles increases with decreasing particle size.[16] The shock wave energy and bubble energy are increased by the use of nano-sized aluminum powders.

References

1 Zeldovich, Ia. B., and Kompaneets, A. S., Theory of Detonation, Academic Press, New York, 1960.

2 Strehlow, R. A., Fundamentals of Combustion, International Textbook, New York, 1968.

3 Fickett, W., and Davis, W. C., Detonation, University of California Press, CA, 1979.

4 Engineering Design Handbook, Principles of Explosive Behavior, AMPC 706–180, US Army Material Command, Washington, DC, 1972.

5 JANAF Thermochemical Tables, Dow Chemical Co., Midland, Michigan, 1960–1970.

6 Keshavard, M. H., and Pouretedal, H. R., Predicting the Detonation Velocity of CHNO Explosives by a Simple Method, *Propellants, Explosives, Pyrotechnics*, Vol. 30, 2005, pp. 105–108.

7 Mader, C. L., Detonation Properties of Condensed Explosives Computed Using the Becker–Kistiakowsky–Wilson Equation of State, Report LA-2900, Los Alamos Scientific Laboratory, Los Alamos, New Mexico, 1963.

8 Mader, C. L., Numerical Modeling of Explosives and Propellants, CRC Press, Boca Raton, FL, 1998.

9 Meyer, R., Explosives, Verlag Chemie, Weinheim, 1977.

10 Energetic Materials Handbook, Japan Explosives Society, Kyoritsu Shuppan, Tokyo, 1999.

11 Explosives Journal, No. 27, NOF Corporation, Tokyo, 1993.

12 Allison, R. E., and Vitali, R., An Application of the Jet Formation Theory to a 105 mm Shaped Charge, BRL Report No. 1165, 1962.

13 Hagiya, H., Morishita, M., Ando, T., Tanaka, H., and Matsuo, H., Evaluation of Explosive Damage of Concrete Wall by Numerical Simulation, Science and Technology of Energetic Materials, Japan Explosives Society, Vol. 64, No. 5, 2003, pp. 192–200.

14 Cole, R. H., Underwater Explosion, Dover Publications, Inc., New York, 1948.

15 Roth, J., Underwater Explosives, Encyclopedia of Explosives and Related Items, Vol. 10, US Army Research and Development Command, Dover, NJ, pp. U38–U81, 1983.

16 Kato, Y., Takahashi, K., Torii, A., Kurokawa, K., and Hattori, K., Underwater Explosion of Aluminized Emulsion Explosives, Energetic Materials, 30th International Annual Conference of ICT, 1999.

10
Formation of Energetic Pyrolants

10.1
Differentiation of Propellants, Explosives, and Pyrolants

When energetic materials react through combustion, chemical energy is released and high-temperature products are formed either by deflagration or by detonation. The difference between deflagration and detonation is significant, i. e., the combustion wave of a deflagration propagates at subsonic speed, whereas that of a detonation propagates at supersonic speed. Energetic materials used as propellants, explosives, and pyrolants are made of oxidizers and fuel components, which react to produce high-temperature combustion products. Propellants used for rocket propulsion are required to burn under deflagration to generate a high exhaust velocity from nozzles. Propellants used for gun propulsion are required to burn under deflagration to form high pressures. Explosives are required to burn under detonation to form a shock wave.

On the other hand, pyrolants are required to have various characteristics; some are slow burning with deflagration, while others are very fast burning with detonation. As shown in Table 10.1, energetic materials are divided into three major categories, i. e., propellants, explosives, and pyrolants, and are used for the indicated applications.

Though the physicochemical process of detonation is significantly different from that of deflagration, the two phenomena cannot be distinguished on the basis of the physicochemical properties of the energetic materials.[1–10] When energetic materials are slowly heated and ignited, in general, most are capable of deflagra-

Table 10.1 Differentiation of propellants, explosives, and pyrolants.

Energetic materials	Phenomenon	Application
Propellants	deflagration	rockets, guns, fireworks
Explosives	detonation	warheads, bombs, mines, blasting
Pyrolants	deflagration	gas generators, igniters, fireworks, squibs, safety fuses
	detonation	detonators, primers, initiators, detonating fuses

Propellants and Explosives. Naminosuke Kubota

ISBN: 978-3-527-31424-9

tion without detonation. When high heat flux and/or mechanical shock are supplied to energetic materials, most are capable of detonation. The criterion boundary is dependent on the physicochemical properties of the energetic materials. Detonation propagation fails when the energetic materials are small in size, and transient combustion from deflagration to detonation or from detonation to deflagration occurs.

10.1.1 Thermodynamic Energy of Pyrolants

Though pyrolants are used to produce high-temperature combustion products, similar to propellants and explosives, pyrolants generate neither propulsive forcesnor destructive forces. The high-temperature combustion products of pyrolants ignite propellants in rocket motors, initiate the detonation of explosives, or generate smoke and flare through deflagration or detonation phenomena. Metallic materials are commonly used as major ingredients of pyrolants in order to obtain high-temperature products of condensed metal oxides. When metal particles are incorporated into pyrolants, the combustion temperature is increased significantly due to the high heat of combustion of metal particles. However, since the molecular mass of the combustion products is increased by the increased molecular mass of the metal oxide particles, the mechanical force created by the gas pressure of the combustion products is not always increased by the addition of metal particles.

When a pyrolant is burned in a closed chamber, both gaseous and condensed-phase products are formed, and the pressure in the chamber is increased due to the increased number of molecules and the raised temperature. The pressure in the chamber is represented by the equation of state:

$$pV_0 = m(\mathrm{R}/M_g)T_g \tag{10.1}$$

where p is pressure, V_0 is the volume of the chamber, m is the mass of the pyrolant, R is the universal gas constant, M_g is the molecular mass, and T_g is the combustion temperature. However, Eq. (10.1) is only valid when the combustion products are entirely gaseous. The density of the burned gas, ρ_g, is defined by

$$\rho_g = m/V_0 \tag{10.2}$$

and the equation of state is also represented by the number of moles of the combustion products, n, given by $n = m/M_g$, according to

$$pV_0 = n\mathrm{R}T_g \tag{10.3}$$

When a pyrolant is composed of metallic particles and an oxidizer component, both gaseous molecules and metal oxides are formed as combustion products. Since the metal oxides are produced in the form of condensed-phase particles, the equation of state shown in Eq. (10.1) is no longer valid to evaluate the pressure in the chamber. Based on Eq. (10.1), the pressure that results from combustion of the metallized energetic material is represented by

$$p(V_0 - V_s) = (m - m_s)\frac{R}{M_g}T_g \tag{10.4}$$

where V_s is the total volume of the condensed particles, and m_s is their total mass. The density of the condensed particles, ρ_s, is given by

$$\rho_s = m_s/V_s \tag{10.5}$$

The pressure generated by the combustion of metallized pyrolants is sometimes lower than that generated by the combustion of non-metallized energetic materials.

10.1.2
Thermodynamic Properties

There are various types of pyrolants that generate gasless combustion products. The pyrolant composed of aluminum powder and iron oxide powder generates aluminum oxide and metallic iron as combustion products. This reaction represented by

$$2Al + Fe_2O_3 \rightarrow Al_2O_3 + 2Fe$$

is known to be highly exothermic and a gasless reaction. The aluminum oxide is produced as an agglomerated solid and the iron is obtained in the form of a molten liquid. Though the temperature is increased, there is no associated pressure increase, i. e., $m = m_s$ in Eq. (10.4). The equation of state given by Eq. (10.1) is not applicable for the combustion of this class of pyrolants. It must be noted, however, that the pressure in a closed chamber is increased by the increased temperature as long as the chamber contains air or an inert gas.

If one assumes a parameter, Θ, defined by

$$\Theta = T_g/M_g \tag{10.6}$$

Θ represents the thermodynamic energy state of an energetic material. Substituting Eq. (10.6) into the equation of state given by Eq. (10.1), one gets

$$pv_0 = mR\Theta \tag{10.7}$$

It can be seen that the pressure in a closed chamber is raised by an increase in Θ, even when the combustion temperature is low as long as the molecular mass of the reaction products is reduced. In other words, Θ indicates the work done by the combustion of pyrolants and is simply evaluated by the thermodynamic energy Θ defined in Eq. (10.6).

The combustion temperature of metallized pyrolants is high due to the high heat of combustion of metal particles, and the molecular mass of the combustion products is also high due to the metal oxides. Thus, Θ defined in Eq. (10.7) is not always high for metallized pyrolants compared with propellants used in rockets and guns. In order to increase the specific impulse of rocket propellants, a high value of Θ is required, rather than just a high combustion temperature, T_g. Both a high T_g and a low M_g of the combustion products are required for propellants in order to increase the specific impulse.

On the other hand, for the pyrolants used as igniters of AP composite propellants, for example, only high-temperature condensed products are required. In other words, neither gaseous products nor pressure is required. Metal particles are used as fuel components to obtain high combustion temperature, T_g, without consideration of the value of the molecular mass, M_g. When particulate metals such as Al, Mg, Ti, and Zr are oxidized, high-temperature metal oxides are formed. However, the molecular mass, M_g, of the reaction products is high due to the formation of metal oxides. Accordingly, the Θ value of pyrolants is not always high even if T_g is high.

10.2 Energetics of Pyrolants

10.2.1 Reactants and Products

When a material is heated to a certain temperature level, the chemical bonds between the atoms of the material are broken and then thermal decomposition occurs. The molecules and atoms produced by the decomposition react to form different molecules through many steps. For example, the reaction process may be represented by:

$$\Sigma R \rightarrow \Sigma P_1 \rightarrow \Sigma P_2 \rightarrow \rightarrow \Sigma P_n \rightarrow \rightarrow \Sigma P$$

where R is the original material defined as a reactant, P_1, P_2, and P_n are intermediate products, and P is the final reaction material defined as a product. When the reactant is composed of a mixture of two or more materials, the reaction process involves many more steps to reach a steady-state condition of the product. In other words, each component material of the reactant decomposes or combines to form intermediate molecular products, which also involves diffusional mixing processes in a physically separated space.

Though the molecular structures of the intermediate products are different from those of the reactants, and heat is released or absorbed at each reaction step, the energy of the initial reactant remains unchanged at each reaction step, and hence the energy of the initial reactant is equivalent to that of the final reaction product. Thus, the energy conservation is represented by Eq. (2.14), as described in Section 2.2, and the final reaction products and the adiabatic flame temperature are determined.

Since pyrolants are mixtures of various chemicals, such as crystalline particles, metal particles, metal oxide particles, and/or polymeric materials, the physicochemical properties of each constituent ingredient, such as melting point, decomposition temperature, phase change, or crystal transition, play important roles in determining the characteristics of the pyrolants. Since the structures of pyrolants are heterogeneous, their combustion waves are also heterogeneous and so the temperature is not uniform throughout the combustion wave. In other words, no thermodynamic equilibrium condition is attained in the combustion wave and the combustion remains incomplete.

10.2.2
Generation of Heat and Products

If one is able to collect the combustion products after a combustion experiment, the combustion temperature can be determined from the energy conservation relationship for the reactants and products. For example, when iron and potassium perchlorate react to produce heat, the reaction products and heat of reaction, $Q(r)$, can be determined by reference to thermochemical tables (NASA SP-273). In this case, the reaction of iron (0.84 mass fraction = 0.929 moles) and potassium perchlorate (0.16 mass fraction = 0.071 moles) is represented by

$$0.929\ Fe + 0.071\ KClO_4 \rightarrow n_{KCl}\ KCl + n_{FeO}\ FeO + n_{Fe}\ Fe$$

where n_{KCl}, n_{FeO}, and n_{Fe} show the stoichiometric amounts of KCl, FeO, and Fe, respectively. Each stoichiometry is determined by the mass conservation of elements between the reactants and products. The mass fractions of 0.84 Fe and 0.16 $KClO_4$ are equivalent to 0.929 and 0.071 moles, respectively. Table 10.2 shows computed results for the mole fractions of the products and the heats of formation of the reactants and the products. Table 10.3 also shows the enthalpy of the products, $Q(h)$, as a function of temperature and the heats of formation of the reactants and products, $Q(r)$. Based on the energy conservation law for a chemical reaction represented by Eqs. (2.15)–(2.17), the adiabatic flame temperature is determined for $Q(h) - Q(r) = 0$, and is obtained as 1845 K for the 0.929Fe + 0.071$KClO_4$ reaction. Though the identities of the combustion products are assumed in this computational process, the results are useful for estimating the overall thermochemical properties of complex pyrolants.

Table 10.2 Fractions of reactants and products and heats of formation.

		mass fraction	mole fraction	ΔH_f (kJ mol^{-1})
Reactants	Fe	0.84	0.929	0
	$KClO_4$	0.16	0.071	–430.3
Products	KCl	0.086	0.071	–436.7
	FeO	0.332	0.285	–272.0
	Fe	0.582	0.643	0

Table 10.3 Enthalpy and heat of reaction as a function of temperature.

T (K)	$H_T - H_0$ (kJ mol^{-1})			*Q*(*h*)	*Q*(*h*) – *Q*(*r*)
	KCl	FeO	Fe	(kJ mol^{-1})	(kJ mol^{-1})
298.15	0.000	0.000	0.000	0.000	–78
300	0.068	0.092	0.046	0.061	–78
400	3.749	5.187	2.674	3.467	–75
500	7.471	10.449	5.527	7.069	–71
600	11.278	15.865	8.613	10.867	–67
700	14.983	21.418	11.939	14.860	–63
800	18.761	27.093	15.553	19.073	–59
900	22.552	32.872	19.582	23.584	–54
1000	26.353	38.756	24.329	28.588	–49
1100	30.164	44.955	29.972	33.516	–44
1200	33.984	50.802	35.062	39.474	–39
1300	37.813	56.956	38.505	43.718	–34
1400	41.651	63.194	42.033	48.041	–30
1500	45.496	69.502	45.643	52.437	–26
1600	49.351	75.837	49.338	56.907	–21
1700	53.213	82.305	54.066	62.056	–16
1800	57.083	88.798	58.263	66.888	–11
1844	58.789	91.680	74.063	77.999	–0.04
1845	58.828	91.746	74.109	78.050	0.0
1846	58.867	91.811	74.155	78.101	0.06
1900	60.961	95.349	76.641	80.859	3
2000	64.847	101.955	81.243	85.981	8

10.3 Energetics of Elements

10.3.1 Physicochemical Properties of Elements

Fig. 10.1 shows part of the Periodic Table of the elements, highlighting the fact that some pure solid elements are used as fuel components of pyrolants. For example, magnesium (Mg) is oxidized by oxidizer fragments to produce magnesium oxides

Fig. 10.1 Elements in the Periodic Table used in pyrolants.

1	2	3	4	5	6	7	8	9	10	11	12	13	14	15	16	17	18
H																	He
Li	Be											B	C	N	O	F	Ne
Na	Mg											Al	Si	P	S	Cl	Ar
K	Ca	Sc	Ti	V	Cr	Mn	Fe	Co	Ni	Cu	Zn	Ga	Ge	As	Se	Br	Kr
Rb	Sr	Y	Zr	Nb	Mo	Tc	Ru	Rh	Pd	Ag	Cd	In	Sn	Sb	Te	I	Xe

through combustion, which is accompanied by high heat liberation and radiative emission. The reaction rate, heat release, and light emission processes are dependent on the various physicochemical characteristics of the magnesium, such as particle size, the type of oxidizer used, and the mixture ratio. The energetics of a mixture of elements used as fuel components and oxidizers is dependent on the combination of each fuel element and oxidizer.

The energetics of the reaction between the fuel element and the oxidizer is determined by the state of the outer electron orbits of the element and the oxidizer. The fuel elements are divided into two categories: metals and non-metals. Typical metals used as fuel components are Li, Mg, Al, Ti, and Zr, and typical non-metals used as fuel components are B, C, and Si. Some other metallic elements used in pyrolants, such as Ba, W, and Pt, are not shown in Fig. 10.1. The physicochemical properties of solid elements and their oxidized products are shown in Table 10.4.

Table 10.4 Physicochemical properties of the elements used in pyrolants and their oxidized products.

	n_a	m_a	o_{eo}	T_{mp}	T_{bp}	ϱ_p	Oxidized Products
(metals)							
Li	3	6.941	$2s^1$	453.7	1620	534	Li_2O, LiF, LiOH, Li_2CO_3, LiH
Be	4	9.012	$2s^2$	1551	3243	1848	BeO, BeF_2, BeH_2, $BeCO_3$
Na	11	22.990	$3s^1$	371.0	1156	971	NaO, Na_2O, NaF, Na_2CO_3
Mg	12	24.305	$3s^2$	922.0	1363	1738	MgO, MgF_2, $MgCO_3$, $Mg(OH)_2$
Al	13	26.982	$3s^23p^1$	933.5	2740	2698	Al_2O_3, AlF_3, $Al(OH)_3$
K	19	39.098	$4s^1$	336.8	1047	862	K_2O, K_2O_2, KO_2, KF, KOH
Ca	20	40.078	$4s^2$	1112	1757	1550	CaO, CaO_2, CaF_2, $Ca(OH)_2$
Ti	22	47.880	$3d^24s^2$	1998	3560	4540	TiO, Ti_2O_3, TiO_2, TiF_2, TiF_3
Cr	24	51.996	$3d^54s^1$	2130	2945	7190	CrO, Cr_2O_3, CrO_2, CrF_2, CrF_4
Mn	25	54.938	$3d^54s^2$	1517	2235	6430	MnO, Mn_2O_3, MnO_2, MnF_3
Fe	26	55.847	$3d^64s^2$	1808	3023	7874	FeO, Fe_2O_3, Fe_3O_4, FeF_2, FeF_3
Co	27	58.933	$3d^74s^2$	1768	3134	8900	CoO, Co_3O_4, $Co(OH)_2$, $Co_2(CO)_8$
Ni	28	58.690	$3d^84s^2$	1726	3005	8902	NiO, Ni_2O_3, NiO_2, NiF_2
Cu	29	63.546	$3d^{10}4s^1$	1357	2840	8960	Cu_2O, CuO, $CuCO_3$, $Cu(OH)_2$
Zn	30	65.390	$3d^{10}4s^2$	692.7	1180	7133	ZnO, ZnF_2, $Zn(OH)_2$
Sr	38	87.620	$5s^2$	1042	1657	2540	SrO, SrO_2, SrF_2, $Sr(OH)_2$
Zr	40	91.224	$4d^25s^2$	2125	4650	6506	ZrO_2, ZrF_4, $ZrCl_2$, $ZrCl_4$
(nonmetals)							
B	5	10.810	$2s^22p^1$	2573	3931	2340	B_2O_3, BF_3, B_2H_6, H_3BO_3
C	6	12.011	$2s^22p^2$	3820	5100	2260	CO, CO_2, CF_4, H_2CO_3
Si	14	28.086	$3s^23p^2$	1683	2628	2329	SiO_2, SiF_2, SiF_4, SiC
P	15	30.974	$3s^23p^3$	317.3	553	1820	P_4O_6, P_4O_{10}, PF_3, H_3PO_3
S	16	32.066	$3s^23p^4$	386.0	445	2070	S_2F_2, SF_2, SO_2, SF_4, SO_3

n_a = atomic number; m_a = atomic mass, kg kmol^{-1}; o_{eo} = outer electron orbit; T_{mp} = melting-point temperature, K; T_{bp} = boiling-point temperature, K; ϱ_p = density, kg m^{-3}

10.3.2
Heats of Combustion of Elements

The heat of combustion of an element (including the mass of reacted oxidizer) is equivalent to the value of the heat of formation of the products because the heats of formation of both the element and oxidizer are zero. Table 10.5 shows the heats of

Table 10.5 Heats of combustion (heats of formation) of elements used in pyrolants and of their combustion products [NASA SP-273].

Element	Product	Phase @298.15 K	M_m kg kmol^{-1}	Heat of formation kJ mol^{-1}	MJ kg^{-1}
Li	LiF	s	25.939	– 616.9	–23.78
	LiF	g	25.939	– 340.9	–13.14
	Li_2O	s	29.881	– 598.7	–20.04
	Li_2O	g	29.881	– 166.9	– 5.59
Be	BeF_2	s	47.009	–1027	–21.84
	BeF_2	g	47.009	– 796.0	–16.93
	BeO	s	25.012	– 608.4	–24.32
B	BF_3	g	67.806	–1137	–16.76
	B_2O_3	s	69.620	–1272	–18.27
	B_2O_3	g	69.620	– 836.0	–12.00
C	CF_2	g	50.008	– 182.0	– 3.64
	CF_4	g	88.005	– 933.2	–10.60
	CO	g	28.010	– 110.5	– 3.95
	CO_2	g	44.010	– 393.5	– 8.94
Na	NaF	s	41.988	– 575.4	–13.70
	NaF	g	41.988	– 290.5	– 6.92
	Na_2O	s	61.979	– 418.0	– 6.74
Mg	MgF_2	s	62.302	–1124	–18.05
	MgF_2	g	62.302	– 726.8	–11.67
	MgO	s	40.304	– 601.2	–14.92
Al	AlF_3	s	83.977	–1510	–17.99
	AlF_3	g	83.977	–1209	–14.40
	Al_2O_3	s	101.961	–1676	–16.44
Si	SiF_4	g	104.079	–1615	–15.52
	SiO_2	s	60.084	– 910.9	–15.16
	SiO_2	s(crystal)	60.084	– 908.3	–15.12
P	PF_3	g	87.969	– 958.4	–10.90
	P_4O_6	g	219.891	–2214	–10.07
S	SF_4	g	108.060	– 763.2	– 7.06

Phase: s = solid and g = gas at 298.15 K
M_m: molecular mass, kg kmol^{-1}

Table 10.5 Continued.

Element	Product	Phase @298.15 K	M_m kg kmol^{-1}	Heat of formation kJ mol^{-1}	MJ kg^{-1}
	SO_2	g	64.065	− 296.8	− 4.63
K	KF	s	58.097	− 568.6	− 9.79
	KF	g	58.097	− 326.8	− 5.63
	K_2O	s	94.196	− 361.5	− 3.84
Ca	CaF_2	s	78.075	−1226	−15.70
	CaF_2	g	78.075	− 784.5	−10.05
	CaO	s	56.077	− 635.1	−11.33
	CaO	g	56.077	43.9	0.78
Ti	TiF_2	g	85.877	− 688.3	− 8.02
	TiF_3	s	104.875	−1436	−13.69
	TiF_3	g	104.875	−1189	−11.33
	TiO	s	63.879	− 542.7	− 8.50
	TiO	g	63.879	54.4	0.85
	Ti_2O_3	s	143.758	−1521	−10.58
Cr	CrF_2	s	89.993	− 778.2	− 8.65
	CrF_3	s	108.991	−1173	−10.76
	CrO	g	67.995	188.3	2.77
	Cr_2O_3	s	151.990	−1140	− 7.50
Mn	MnF_2	s	92.935	− 849.4	− 9.14
	MnF_2	g	92.935	− 531.4	− 5.72
	MnF_3	s	111.933	−1071	− 9.57
	MnO	s	70.937	− 385.2	− 5.43
	Mn_2O_3	s	157.874	− 959.0	− 6.07
Fe	FeF_2	s	93.844	− 705.8	− 7.52
	FeF_2	g	93.844	− 389.5	− 4.15
	FeF_3	s	112.842	−1042	− 9.23
	FeF_3	g	112.842	− 820.9	− 7.28
	FeO	s	71.846	− 272.0	− 3.79
	FeO	g	71.846	251.0	3.49
	Fe_2O_3	s	159.692	− 824.2	− 5.16
Co	CoF_2	s	96.930	− 671.5	− 6.93
	CoF_2	g	96.930	− 356.5	− 3.68
	CoF_3	s	115.928	− 790.4	− 6.82
	CoO	s	74.933	− 237.9	− 3.18
	Co_3O_4	s	240.797	− 910.0	− 3.78
Ni	NiF_2	s	96.687	− 657.7	− 6.80
	NiF_2	g	96.687	− 335.6	− 3.47
	NiO	s	74.689	− 239.7	− 3.21
	NiO	g	74.689	309.6	4.15

Phase: s = solid and g = gas at 298.15 K
M_m: molecular mass, kg kmol^{-1}

Table 10.5 Continued.

Element	Product	Phase @298.15 K	M_m kg kmol^{-1}	Heat of formation kJ mol^{-1}	MJ kg^{-1}
Cu	CuF	s	82.544	– 280.3	– 3.40
	CuF	g	82.544	– 12.6	– 0.15
	CuF_2	s	101.543	– 538.9	– 5.31
	CuF_2	g	101.543	– 266.9	– 2.63
	CuO	s	79.545	– 156.0	– 1.96
	CuO	g	79.545	306.2	3.85
	Cu_2O	s	143.091	– 170.7	– 1.19
Zn	ZnF_2	s	103.387	– 764.4	– 7.39
	ZnF_2	g	103.387	– 494.5	– 4.78
	ZnO	s	81.389	– 350.5	– 4.31
Sr	SrF_2	s	125.617	–1217	– 9.69
	SrO	s	103.619	– 592.0	– 5.71
Zr	ZrF_2	s	129.221	– 962.3	– 7.45
	ZrF_2	g	129.221	– 558.1	– 4.32
	ZrF_4	s	167.218	–1911	–11.43
	ZrF_4	g	167.218	–1674	–10.01
	ZrO	s	107.223	58.6	0.55
	ZrO_2	s	123.223	–1098	– 8.91
	ZrO_2	g	123.223	– 286.2	– 2.32

Phase: s = solid and g = gas at 298.15 K
M_m: molecular mass, kg kmol^{-1}

combustion of typical elements used as fuel components of pyrolants when oxidized with fluorine gas or oxygen gas under standard conditions (298.15 K).

Based on the data shown in Table 10.5, Figs. 10.2 and 10.3 show the heats of combustion of the elements with oxygen gas and with fluorine gas, respectively, as a function of atomic number. Several peaks of heat of combustion appear periodically with increasing atomic number for oxidation with both oxygen and fluorine. The highest heat of combustion (MJ kg^{-1}) with oxygen is obtained with Be and the succeeding elements are Li, B, Al, Mg, Si, and Ti. When fluorine gas is used as an oxidizer, the highest heat of combustion (MJ kg^{-1}) is obtained with Li and the succeeding elements are Be, Al, Mg, B, Ca, and Si.

The density of the elements is also an important parameter with regard to obtaining a high heat of combustion in a reactor of limited volume. The volumetric heat of combustion (kJ m^{-3}) shows differences compared to the mass-based heat of combustion (kJ kg^{-1}). For example, the volumetric heat of combustion of Zr is the highest of the fuel elements and that of Ti is the next highest when oxidized by oxygen gas.

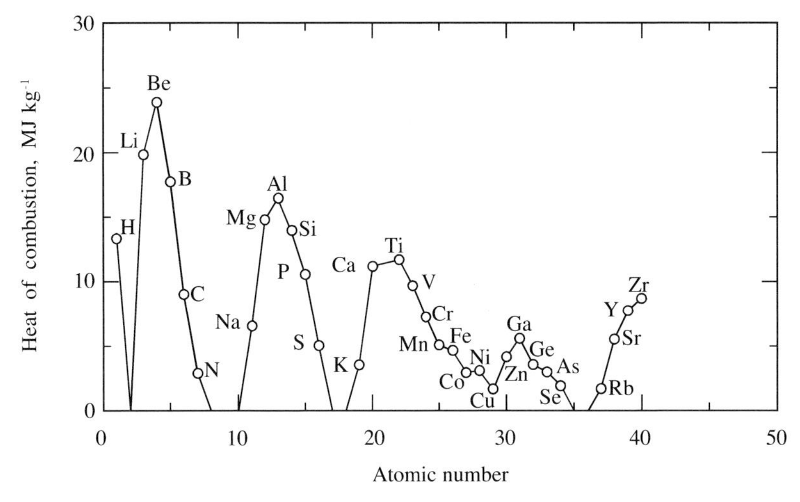

Fig. 10.2 Heats of combustion of the elements oxidized by oxygen gas as a function of atomic number.

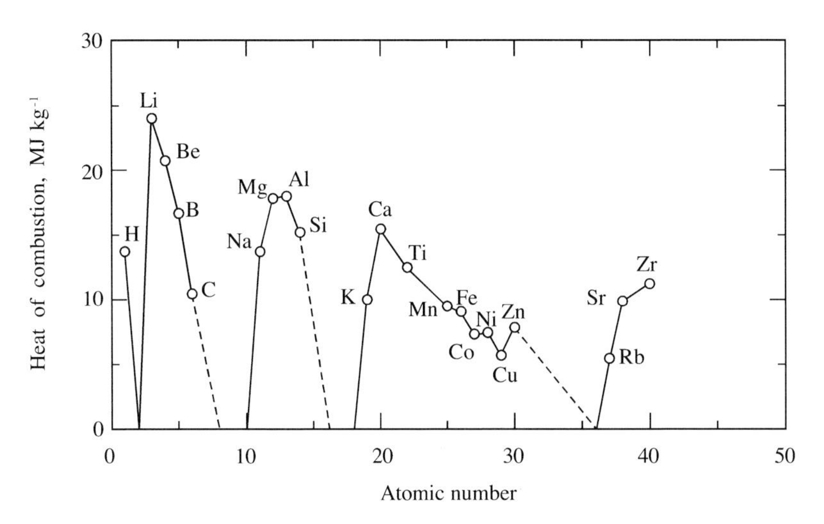

Fig. 10.3 Heats of combustion of the elements oxidized by fluorine gas as a function of atomic number.

10.4 Selection Criteria of Chemicals

10.4.1 Characteristics of Pyrolants

The energetic materials used as propellants, explosives, and pyrolants have distinct physicochemical properties. The burning pressure in the relevant combustion chamber ranges from 0.1 MPa to 20 MPa for rockets and from 10 MPa to 5000 MPa for guns. The burning time ranges from 1 s to 10^3 s for rockets and from 0.01 s to 0.1 s for guns. Nitropolymers such as nitrocellulose and nitroglycerin are used to formulate propellants because they contain both oxidizer and fuel components in their molecular structures. When nitropolymers burn, oxidizer and fuel fragments are formed by thermal decomposition and these react to produce high-temperature combustion products. Composite propellants composed of crystalline oxidizers-such as ammonium perchlorate or ammonium nitrate and polymeric hydrocarbon-fuels also burn to generate oxidizer and fuel fragments by thermal decomposition. These fragments diffuse into each other to form a premixed gas at the burning surface and react to form high-temperature combustion products. The polymeric hydrocarbon fuels also serve as binders for the crystalline oxidizer particles.

The pressure created by the combustion of explosives may be up to the order of 10^3 MPa and the burning time may be as short as 0.01 s. The combustion phenomena thus involve detonation accompanied by a shock wave. The chemicals used as explosives are therefore high energy density materials capable of detonation propagation. The typical materials used are TNT, HMX, RDX, and HNIW, which have adiabatic flame temperatures higher than 3000 K and densities in excess of 1600 kg m^{-3}. These crystalline materials are mixed with polymeric hydrocarbon binders to form plastic-bonded explosives. In these, unlike in propellants, the polymeric hydrocarbon binders are used to form rubber-like, rigid-shaped explosive grains, not to act as fuel components. Energetic materials composed of TNT, HMX, or RDX, ANFO explosives, or slurry explosives are detonable explosives used in warheads, bombs, industrial blasting, and mining.

Pyrolants are used in various types of pyrotechnic devices and equipment. Wide ranges of combustion temperatures and burning rates are required. Gasless combustion by metal oxidation is utilized in many cases. High-temperature particles need to be generated when igniters are used for rocket propellants. Detonating cords composed of PETN are used to transfer detonation shock waves at supersonic speeds. In general, non-toxic chemicals are used to formulate energetic materials. However, toxic combustion products are formed in some cases. Though some heavy metals are non-toxic, their oxides are sometimes toxic. Thus, the relevant reaction products always have to be taken into account in the formulation of pyrolants.

10.4.2
Physicochemical Properties of Pyrolants

The physicochemical characteristics that constitute selection criteria for the chemicals used in pyrolants can be summarized as follows:

Combustion products
- flame temperature
- gas volume (gas, gasless, or gas with solid or liquid particles)
- nature of flame (color and brightness)
- nature of smoke (smoky or non-smoky)
- environmentally friendly
- non-toxic

Ignitability
  by flame or by hot particles
Density
Melting-point temperature
Decomposition temperature
Volume change by phase transition
Physical stability
  mechanical strength and elongation
  shock sensitivity
  friction sensitivity
  electric spark sensitivity
Chemical stability
  anti-aging
  reactivity with other materials
Burning rate characteristics
  burning rate
  pressure sensitivity
  temperature sensitivity

Since pyrolants are composed of oxidizer and fuel components, their combustion characteristics are highly dependent on the combination of these components. When a mass fraction of HMX particles of 0.8 is mixed with a mass fraction of a hydrocarbon polymer of 0.2, the adiabatic flame temperature is reduced from 3200 K to 2600 K. HMX acts as a high-energy material, but not as a supplier of oxidizing species to oxidize the hydrocarbon polymer. The adiabatic temperature is reduced due to the formation of fuel-rich products. When a mass fraction of ammonium perchlorate (AP) of 0.8 is mixed with a mass fraction of a hydrocarbon polymer of 0.2, the adiabatic flame temperature is increased from 1420 K to 3000 K, even though AP is not a high-energy material. This is because AP serves as a supplier of oxygen to the hydrocarbon polymer. The mixture of AP and hydrocarbon polymer forms a stoichiometrically balanced pyrolant. Similar to AP, potassium perchlorate(KP), nitronium perchlorate (NP), potassium nitrate (KN), ammonium nitrate(AN), and hydrazinium nitroformate (HNF) act as suppliers of oxidizers and produce low-temperature oxidizing gaseous products. When these suppliers of oxidizers are mixed with fuel-rich components such as metal particles or hydrocarbon polymers, the combustion temperature is increased drastically.

When a fuel-rich pyrolant burns in the atmosphere, oxygen molecules from the atmosphere diffuse into the initial combustion products of the pyrolant. The combustion products burn further and generate heat, light, and/or smoke in the atmosphere. A typical example is the combustion process in ducted rockets: fuel-rich products generated in a gas generator are burnt completely in a combustion chamber after mixing with air pressurized by a shock wave that is taken in from the atmosphere.

In general, any energetic chemical may burn by way of deflagration or detonationphenomena. The burning mode is determined either by the chemical properties or by the ignition initiation process. When high-energy nitramines such as

HMX and RDX are heated, deflagration combustion occurs with a burning rate of about 1 mm s^{-1} at 1 MPa. However, when these nitramines are ignited by primers giving rise to shock waves, detonation combustion occurs with a burning rate of more than 7000 m s^{-1}. The characteristics of combustion wave propagation are determined by the Chapman–Jouguet relationship described in Refs. [1–5].

When pyrolants composed of metal particles and crystalline oxidizers, such as B-KNO_3 and Zr-NH_4ClO_4, burn in the atmosphere, additional oxygen is supplied to the combustion process from the ambient air. The oxygen diffuses into the combustion zone of the pyrolants and the combustion mode is changed. This diffusional process is an important process with regard to the design of pyrolants used in fireworks and flares. The temperature and emission of the combustion products are dependent on the combustion processes in the atmosphere. The ignitability and burning time of these pyrolants are dependent on the specific combination of metal particles and crystalline oxidizer particles. The mixture ratio, particle size, and volumetric densityof these mixtures are also important factors in the design of pyrolants.

10.4.3
Formulations of Pyrolants

Various types of pyrolants are formulated by combining various types of oxidizers and fuels. Some combinations are sensitive to mechanical shock, while other combinations may be detonated or ignited by an electric charge. Typical deflagrative and detonative energetic materials are shown in Table 2.3. Nitropolymer pyrolants, composite pyrolants, and black powder burn without detonative combustion. Pyrolants composed of metal particles and crystalline oxidizers, such as Mg-Tf, B-KNO_3, and Zr-NH_4ClO_4, are also deflagrative and are used as igniters. On the other hand, TNT, HMX, RDX, ANFO explosives, and slurry explosives are detonative, and are used in warheads and bombs. A pyrolant composed of PETN is also detonative and is used in detonating cords. Typical characteristics of deflagrative and detonative combustion are shown in Table 3.2.

Typical crystalline oxidizers used in energetic materials are KNO_3, $KClO_4$, and NH_4ClO_4, which produce gaseous oxidizing fragments. The fuel components are metal particles such as Li, B, Na, Mg, Al, Ti, or Zr, or polymeric hydrocarbon materials. Mixtures of these oxidizer and fuel components formulate energetic pyrolants. The heat produced by the oxidation of metals is high, and the combustion temperature is high when compared with the oxidation of hydrocarbons.

Two types of pyrolants are used in gas generators: (1) energetic polymers, and (2) crystalline oxidizer particles and inert polymers. Energetic polymers are nitropolymers, consisting of nitro groups and hydrocarbon structures, or azide polymers, consisting of azide groups and hydrocarbon structures. A colloidal mixture of nitrocellulose and nitroglycerin is a typical nitropolymer, and glycidyl azide polymer is a typical azide polymer, both of which generate gaseous products such as CO_2, H_2O, N_2, and CO when thermally decomposed. The burning of mixtures of crystalline oxidizer particles and inert polymers involves interdiffusion of the gaseous products of thermal decomposition of oxidizer fragments produced by the oxidizer particles and fuel fragments produced by the inert polymers. When no metal atoms are present within the crystalline oxidizers, as in the case of NH_4ClO_4 and NH_4NO_3, no solid particles are formed among the combustion products. However, when the crystalline oxidizers contain metal atoms, as in the case of KNO_3, $KClO_3$, $KClO_4$, $NaClO_3$, and $NaClO_4$, metal oxide particles are formed among the combustion products. Polybutadiene, polyester, and polyether are used as inert polymers.

When pyrolants are composed of fuel-rich mixtures, fuel-rich products are formed and the adiabatic flame temperature is relatively low. However, when these fuel-rich mixtures burn in air, oxygen molecules in the air oxidize the remaining fuel components of the products. The combustion of pyrolants occurs under heterogeneous reaction conditions, including gas, solid, and/or liquid phases. The diffusional process between the fuel and oxidizer components is dependent on the thermochemical properties and the particle sizes of the components. Thus, the reaction to produce heat and products is also heterogeneous in terms of time and space in the combustion wave of the pyrolants. Heat transfer from the high-temperature reacted region to the low-temperature unreacted region occurs in the combustion wave of the pyrolants. The rate-determining step for the burning of pyrolants appears to be different from that for the burning of propellants and explosives.

Though pyrolants produce heat through oxidation reactions, a luminous flame or bright particles are not always produced by these reactions. Reactions between iron particles and air generate heat either very slowly or very rapidly, depending on the particle size and the diffusion process of the air among the particles. No flame reaction occurs when large-sized iron particles are oxidized with air. Pyrolants that generate a luminous flame, smoke, and hot bright particles are used as igniters and gas generators. A very bright flame is generated by the combustion of magnesium in air or with crystalline oxidizers such as NH_4ClO_4, NH_4NO_3, or KNO_3. Smoke is generated by the combustion of phosphorus in air or with crystalline oxidizers. White smoke is also generated by the combustion of aluminum or magnesium in air. Colored flames are generated by the combustion of various types of metals.

Pyrolants are divided into seven categories:

(1) Energetic polymers
(2) Inert polymers + crystalline oxidizers
(3) Energetic polymers + energetic crystalline particles
(4) Metal particles + oxidizing polymers
(5) Metal particles + crystalline oxidizers
(6) Metal particles + metal oxide particles
(7) Energetic particles

The chosen combinations of these chemicals and metals depend on the requirements of the specific application. Gasless combustion prevents pressure increase in a closed combustion chamber. Some combinations of metal particles and metal oxide particles or of metal particles and crystalline oxidizers are chosen as chemical ingredients of gasless pyrolants. On the other hand, hydrocarbon polymers are used to obtain combustion products of low molecular mass, such as H_2O, CO, CO_2, and H_2. High pressure is thus obtained by the combustion of hydrocarbon polymers. Table 10.6 shows the chemical ingredients used to formulate various types of pyrolants.

Table 10.6 Chemical ingredients used to formulate the seven types of pyrolants.

(1) Energetic polymers
- Nitrocellulose (NC)
- Glycidyl azide polymer (GAP)
- Bis-azide methyloxetane (BAMO)
- 3-Azidomethyl-3-methyloxetane (AMMO)
- BAMO-NIMO
- NC-NG

(2) Inert polymers + crystalline oxidizers
- Inert polymers:
  - Hydroxy-terminated polybutadiene (HTPB)
  - Carboxy-terminated polybutadiene (CTPB)
  - Hydroxy-terminated polyurethane (HTPU)
  - Hydroxy-terminated polyether (HTPE)
  - Hydroxy-terminated polyester (HTPS)
  - Hydroxy-terminated polyacetylene (HTPA)
  - Polysulfide (PS)
- Crystalline oxidizers:
  - NH_4ClO_4, NH_4NO_3, KNO_3, $KClO_4$
  - Hydrazinium nitroformate (HNF)

(3) Energetic polymers + energetic crystalline particles
- Energetic polymers:
  - NC, GAP, BAMO, BAMO-NIMO, NC-NG
- Energetic crystalline oxidizers:
  - Cyclotetramethylene tetranitramine (HMX)
  - Cyclotrimethylene trinitramine (RDX)
  - Hexanitrohexaazatetracyclododecane (HNIW)
  - Ammonium dinitramide (ADN)

(4) Metal particles + oxidizing polymers
- Metal particles:
  - Mg, B, Zr, Al
- Oxidizing polymers:
  - Polytetrafluoroethylene (Tf)
  - Vinylidene fluoride hexafluoropropene polymer (Vt)

(5) Metal particles + crystalline oxidizers
- Metal particles:
  - B, Mg, Al, Ti, Zr
- Crystalline oxidizers:
  - $KClO_4$, NH_4ClO_4, $KClO_3$, KNO_3, $NaNO_3$, NH_4NO_3, HNF

(6) Metal particles + metal oxide particles
- Metal particles:
  - B, Al, Mg, Ti, Zr, Ni, W
- Metal oxide particles:
  - Fe_2O_3, CuO, BaO_2, PbO_2, Pb_3O_4, CoO, $BaCrO_4$, $PbCrO_4$, $CaCrO_4$, MnO_2

(7) Energetic particles
- Explosive particles:
  - HMX, RDX, HNS, DATB, TATB, TNT, HNIW, DDNP
- Metal azide particles:
  - PbN_3, NaN_3, AgN_3

10.5
Oxidizer Components

The oxidizers used in pyrolants are generally materials containing a relatively high mass fraction of oxygen atoms. The chemical potential of an oxidizer is determined by its combination with the fuel component. Metal oxides act as oxidizer components and react with fuel components. These reactions generate significant heat without the evolution of gaseous products. Pyrolants composed of metal oxides as oxidizer components are widely used in pyrotechnics. Materials containing fluorine atoms can also serve as oxidizers. SF_6 is a typical oxidizer containing fluorine atoms. The physicochemical characteristics that constitute selection criteria for the oxidizers used in pyrolants can be summarized as follows:

[OB] Oxygen balance
$\xi(O)$ Mass fraction of oxygen atoms
ΔH_f Heat of formation
ρ Density
T_{mp} Melting-point temperature
T_d Decomposition temperature
T_{bp} Boiling-point temperature
Crystalline particles or polymers
Phase transition
Shape of crystal
Shock sensitivity
Friction sensitivity

Table 10.7 shows the physicochemical properties of the crystalline materials used as oxidizers. Potassium and sodium are combined with nitrate or perchlorate to form stabilized crystalline oxidizers. Metal oxides are formed as their combustion products. On the other hand, ammonium ions are combined with nitrate or perchlorate to form stabilized crystalline oxidizers such as NH_4NO_3 and NH_4ClO_4 without metal atoms. When these oxidizers are decomposed, no solid products are formed. As discussed in Section 10.1.1, for the oxidizers used for propulsion, such as in propellants for rockets and guns, the molecular mass of the combustion products needs to be as low as possible.

Though metallic oxidizers such as potassium nitrate and potassium perchlorateare effective for producing heat and high-temperature combustion products, the molecular mass of the combustion products is high, which reduces the specific impulse when these oxidizers are used in propellants. In general, the generation of a high temperature is the major requirement for pyrolants used as igniters and smoke generators. Thus, metallic oxidizers are commonly used to formulate various types of pyrolants.

Though the chemical potentials of ammonium perchlorate and potassium perchlorate are high compared with those of other oxidizers, hydrogen chloride is formed as a combustion product. Hydrogen chloride is known to generate hydrochloric acid when combined with water vapor in the atmosphere. The chemical potentials of crystalline oxidizers are dependent on the fuel components with which they are combined to form pyrolants. The following physicochemical properties of the oxidizers should be considered in formulating pyrolants to meet the requirements of a given application:

Table 10.7 Physicochemical properties of crystalline oxidizers.

	$\xi(O)$	ΔH_f (kJ mol^{-1})	ϱ (kg m^{-3})	T_{mp} (K)	T_d (K)
Non-metallic oxidizers					
NH_4ClO_4	0.545	−296.0	1950	–	543
NH_4NO_3	0.5996	−366.5	1725	343	480
$N_2H_5C(NO_2)_3$	0.525	−53.8	1860	–	–
$NH_4N(NO_2)_2$	0.516	−151.0	1720	–	–
Metallic oxidizers					
KNO_3	0.4747	−494.6	2109	606	673
$KClO_4$	0.461	−431.8	2530	798	803
$KClO_3$	0.462	–	2340	640	–
$LiClO_4$	0.601	−382.2	2420	520	755
$NaClO_4$	0.522	−384.2	2530	750	–
$Mg(ClO_4)_2$	0.754	−590.5	2210	520	520
$Ca(ClO_4)_2$	0.535	−747.8	2650	540	540
$Ba(NO_3)_2$	0.3673	−995.6	3240	869	–
$Ba(ClO_3)_2$	0.3156	–	3180	687	–
$Ba(ClO_4)_2$	0.380	−809.8	3200	557	700
$Sr(NO_3)_2$	0.4536	−233.25	2986	918	–
$Sr(ClO_4)_2$	0.446	−772.8	–	–	–

10.5.1 Metallic Crystalline Oxidizers

10.5.1.1 Potassium Nitrate

Potassium nitrate, KNO_3, is well-known as the oxidizer component of black powder. The mass fraction of oxygen is 0.4747. Potassium nitrate is not hygroscopic and is relatively safe with regard to mechanical shock and friction. Experimental DTA and TG results show two endothermic peaks at 403 K and 612 K without a gasification reaction. The first endothermic peak at 403 K corresponds to a crystal structure transformationfrom trigonal to orthorhombic. The endothermic peak at 612 K corresponds to the melting point of KNO_3. The heat of fusion is 11.8 kJ mol^{-1}. Gasification starts at about 720 K and ends at about 970 K, the endothermic reaction being represented by:

$$2KNO_3 \rightarrow 2KNO_2 + O_2$$

$$2KNO_2 \rightarrow K_2O + 2NO + O_2$$

No solid residue remains in the high temperature region above 970 K.

10.5.1.2 **Potassium Perchlorate**

Potassium perchlorate, $KClO_4$, is a white, crystalline material used as an oxidizer component. Since potassium perchlorate is not hygroscopic and the mass fraction of oxygen is 0.46, it is mixed with various types of fuel components to formulate energetic pyrolants. Its melting point is 798 K, which is relatively high compared with those of other crystalline oxidizers. Potassium perchlorate decomposes thermally and produces oxygen at 803 K according to

$$KClO_4 \rightarrow KCl + 2O_2$$

Potassium perchlorate thus acts as a strong oxidizer and burns with fuel components accompanied by the emission of a white smoke of KCl particles.

10.5.1.3 **Potassium Chlorate**

Potassium chlorate, $KClO_3$, is a colorless, crystalline material with a melting point of 640 K. Just above the melting-point temperature, it decomposes to produce oxygen molecules according to:

$$KClO_3 \rightarrow KCl + 3/2O_2$$

Potassium chlorate is used as an oxidizer of chlorate explosives, primers, and matchheads. Mixtures of potassium chlorate and fuel components detonate relatively easily.

10.5.1.4 **Barium Nitrate**

Barium nitrate, $Ba(NO_3)_2$, is a colorless, crystalline material with a melting point of 865 K. $Ba(NO_2)_2$ and O_2 are produced in the low-temperature region, and $Ba(NO_2)_2$ decomposes to produce BaO, NO, and NO_2 in the high-temperature region according to:

$$Ba(NO_3)_2 \rightarrow Ba(NO_2)_2 + O_2$$

$$Ba(NO_2)_2 \rightarrow BaO + NO + NO_2$$

The gaseous products, O_2, NO, and NO_2, all act as oxidizer components when barium nitrate is mixed with fuel components. The decomposition process is endothermic, hence barium nitrate cannot be burned without fuel components.

10.5.1.5 **Barium Chlorate**

Barium chlorate, $Ba(ClO_3)_2$, is a colorless solid with a monoclinic crystal structure. It decomposes thermally above its melting-point temperature of 687 K according to:

$$Ba(ClO_3)_2 \rightarrow BaCl_2 + 3O_2$$

When barium chlorate burns with fuel components, the oxygen acts as an oxidizer, and the burning is accompanied by the light emission of a green-colored flame.

10.5.1.6 **Strontium Nitrate**

Strontium nitrate, $Sr(NO_3)_2$, is a high-density material with a melting point of 843 K. The initial thermal decomposition in the low-temperature region produces $Sr(NO_2)_2$, which decomposes to SrO in the high-temperature region according to:

$$Sr(NO_3)_2 \rightarrow Sr(NO_2)_2 + O_2$$

$$Sr(NO_2)_2 \rightarrow SrO + NO_2 + NO$$

The gaseous products, NO_2 and NO, also act as oxidizers when fuel components are present. When strontium nitrate burns with fuel components, a red-colored flame is formed.

10.5.1.7 **Sodium Nitrate**

Sodium nitrate, $NaNO_3$, is a colorless solid with a trigonal crystal structure and has a melting point of 581 K. The initial thermal decomposition starts at 653 K and produces $NaNO_2$, which then decomposes to NaO in the high-temperature region. Oxygen and nitrogen gases are formed by the thermal decomposition according to:

$$NaNO_3 \rightarrow NaNO_2 + 1/2O_2$$

$$NaNO_2 \rightarrow NaO + 1/2O_2 + 1/2N_2$$

The oxygen evolved acts as an oxidizer when fuel components are present, producing high-temperature combustion products.

10.5.2 **Metallic Oxides**

Several types of metallic oxides are used as oxidizers for pyrolants. A metallic oxide can oxidize a metal if the chemical potential of the resulting metal oxide is lower than that of the starting oxide. No gaseous reaction products are formed when metal particles are oxidized by metallic oxides in a closed system and very little pressure increase occurs since the reactants and products are all solids. Typical metallic oxides are MnO_2, Fe_2O_3, CoO, CuO, ZnO, and Pb_3O_4.

When a mixture of Fe_2O_3 powder and Al powder is ignited, a self-sustaining exothermic reaction occurs according to:

$$Fe_2O_3 + 2Al \rightarrow 2Fe + Al_2O_3$$

This reaction is used to obtain molten iron from iron oxide. Fe_2O_3 acts as an oxidizer of Al. The heat obtained from this reaction is gasless and the pressure remains unchanged.

When a mixture of Fe_2O_3 powder and Zr powder is ignited, Fe_2O_3 acts as an oxidizer of Zr, similar to the oxidation of Al. This reaction is exothermic and gasless, and is expressed by:

$$Fe_2O_3 + 3/2Zr \rightarrow 2Fe + 3/2ZrO_2$$

Since these oxidations of Zr and Al by Fe_2O_3 only occur when the mixtures are heated above 1500 K, the mixtures are safe materials. Pb_3O_4 is used as an oxidizer of Zr and B according to:

$$Pb_3O_4 + 2Zr \rightarrow 3Pb + 2ZrO_2$$

$$Pb_3O_4 + 8/3B \rightarrow 3Pb + 4/3B_2O_3$$

10.5.3 Metallic Sulfides

When a mixture of FeS_2 powder and Li powder is ignited, a self-sustaining exothermic reaction occurs according to:

$$FeS_2 + 2Li \rightarrow FeS + Li_2S$$

This reaction is similar to that between Fe_2O_3 powder and Li powder. Since it is a gasless exothermic reaction, a significant amount of heat is generated in a closed system under constant pressure conditions.

10.5.4 Fluorine Compounds

Though materials containing fluorine atoms can act as oxidizers, most fluorine compounds are gases or liquids at ambient temperatures and hence cannot be used as oxidizers in pyrolants. However, some solid fluorine compounds are utilized as oxidizer components of energetic pyrolants.

SF_6 is a crystalline material that acts as an oxidizer of metallic fuels. Since both sulfur and fluorine atoms act as oxidizer components, the oxidizing potential of SF_6 is high. Mixtures of SF_6 and Li generate significant heat as a result of the reaction:

$$SF_6 + 8Li \rightarrow 6LiF + Li_2S$$

Polytetrafluoroethylene (Tf) is a polymeric fluorine compound that consists of a $-C_2F_4-$ molecular structure,[11] which contains a mass fraction of fluorine of 0.75. Tf is insoluble in water and its specific mass is in the range 3550–4200 kg m^{-3} in pelletized form. Tf melts at approximately 600 K and decomposes thermally above 623 K to produce hydrogen fluoride (HF). The auto-ignition temperature in air is in the range 800–940 K. Tf acts as an oxidizer when mixed with metals and generates solid carbon as a reaction product. Tf is known by the commercial name "Teflon".[11]

Similar to Tf, vinylidene fluoride hexafluoropropane polymer (Vt) is a polymeric fluorine compound; its molecular structure consists of $C_5H_{3.5}F_{6.5}$ units.[12] Vt acts as an oxidizer when mixed with metals. The chemical potential of Vt is lower than that of Tf due to its lower mass fraction of fluorine atoms. Vt is also known by the commercial name "Viton" of the Du Pont company.[12]

Vt is insoluble in water and its specific mass is in the range 1770–1860 kg m^{-3} in pelletized form. Its flash point is higher than 477 K. When Vt is heated above 590 K, hydrogen fluoride is formed and the polymer burns abruptly. When Vt is ignited in air, hydrogen fluoride, carbonyl fluoride, carbon monoxide, and low molecular mass fluorocarbons are formed.

10.6 Fuel Components

The selection of fuel components to be mixed with oxidizer components is also an important issue in the development of pyrolants for various applications. Metal particles are used as fuel components to develop small-scale pyrolant charges as deployed in igniters, flares, and fireworks. Non-metal particles such as boron and carbon are used to formulate energetic pyrolants. Polymeric materials are commonly used as fuel components to develop relatively large-scale pyrolant charges, such as gas generators and fuel-rich propellants.

10.6.1 Metallic Fuels

Metals are used as energetic fuel components of pyrolants in the form of fine particles. Most metal particles are mixed with oxidizer components such as crystalline particles or fluorine-containing polymers. Some metals react with gaseous oxidizer fragments while others react with molten oxidizer fragments. As shown in Table 10.5, Li, Be, Na, Mg, Al, Ti, Fe, Cu, and Zr are typical metals that generate high heat when oxidized. Since oxidation occurs at the surface of each metal particle, the total surface area of the particles is the dominant factor with regard to rates of combustion.

Lithium (Li) is a silver-colored soft metal, and the lightest of all the metallic elements. Li is oxidized by atmospheric nitrogen to form Li_3N. Though Li melts at 453.7 K, its boiling point temperature is very much higher at 1620 K. A deep-violet flame is formed when Li is burned in air. Its standard potential is about 3.5 V and a relatively high electric current is formed when it is used in batteries.

Beryllium (Be) is a gray-colored brittle metal. Be burns in air to form BeO or Be_3N_2, releasing a high heat of combustion. However, Be and its compounds are known as highly toxic materials.

Sodium (Na) is a silver-colored soft metal and the lightest after Li and K. Na reacts violently and exothermically with water according to:

$$Na + H_2O \rightarrow NaOH + 1/2H_2$$

Na burns in air when heated above its melting point temperature of 371 K according to:

$$2Na + O_2 \rightarrow Na_2O_2$$

When Na reacts with O_2 at high pressure and temperature, Na_2O_2 is decomposed to form Na_2O. A luminous yellow flame is formed when Na is heated in a flame as a result of emission of the characteristic D-line at 588.997 and 589.593 nm.

Magnesium (Mg) is a silver-colored light and soft metal used as a major component of magnesium alloys, which are used to construct aircraft and vehicles. Mg burns in air to produce bright white light, and so Mg particles are used as an ingredient of pyrolants. Mg reacts with nitrogen gas to produce Mg_3N_2 according to:

$$3Mg + N_2 \rightarrow Mg_3N_2$$

It also reacts with F_2, Cl_2, and Br_2 to produce MgF_2, $MgCl_2$, and $MgBr_2$, respectively. These reactions are all exothermic. Mg is also exothermically oxidized by CO_2 to produce MgO and carbon according to:

$$2Mg + CO_2 \rightarrow 2MgO + C$$

Aluminum (Al) is a silver-colored light and soft metal used as a major component of aluminum alloys, which are used to construct aircraft and vehicles, similar to Mg alloys. However, Al is known as a readily combustible metal. Thus, Al particles are used as major fuel components of pyrolants. Al particles are mixed with ammonium perchlorate particles and polymeric materials to form solid propellants and underwater explosives. The reaction between aluminum powder and iron oxide is known as a high-temperature gasless reaction and is represented by:

$$2Al + Fe_2O_3 \rightarrow Al_2O_3 + 2Fe$$

Iron oxide acts as an oxidizer and aluminum acts as a fuel. The reaction is highly exothermic, such that it is used for the welding of iron structures in civil engineering.

Titanium (Ti) is used as a major component of Ti alloys, which are used as heat-resistant light metals for the construction of aircraft and rockets. Though Ti is also used as an anti-acid metal, it burns in air when heated above 1500 K. $TiCl_4$ is a liquid used to form colored smoke in air.

Iron (Fe) is the most popularly used metal for various objectives. Though Fe is oxidized only very slowly in air to form iron oxides, the reaction rate increases drastically with increasing temperature. Fe powder is used to generate heat in various applications, including fireworks.

Copper (Cu) is a bright brown-colored metal. When Cu powder or wire is heated in a high-temperature flame, its characteristic blue-colored emission is observed. Thus, Cu particles are commonly used as a component of aerial shells of fireworks.

Zirconium (Zr) is a silver-colored metal with high density (6505 kg m^{-3}). The amorphous Zr powder obtained by reduction of zirconium oxide is black in color. Zr particles are easily ignited by static electricity compared with other metallic particles. Zr powder burns very rapidly in air and generates significant heat and bright light emission when the temperature is raised above 1500 K. Igniters used for rocket propellants are made from mixtures of Zr powder and crystalline oxidizers. High-temperature ZrO_2 particles are dispersed on the surface of the propellant for hot-spot ignition.

10.6.2 Non-metallic Solid Fuels

As shown in Table 10.5, non-metallic fuels used as ingredients of pyrolants are boron, carbon, silicon, phosphorus, and sulfur. Similarly to metal particles, non-metal particles are oxidized at their surfaces. The processes of diffusion of oxidizer fragments to the surface of a particle and the removal of oxidized fragments therefrom are the rate-controlling steps for combustion.

10.6.2.1 Boron

Boron (B) is a dark-gray non-metal, which has the hardest structure after diamond and is chemically stable and difficult to oxidize. Boron exists in two structural types, namely crystalline and amorphous, the physicochemical properties of which are different. Crystalline boron has a density of 2340 kg m^{-3}, a melting point temperature of 2573 K, and a boiling point temperature of 3931 K. Boron particles burn in oxygen gas when heated at about 1500 K to form B_2O_3 and also burn in nitrogen gas at above 2000 K to form BN. The heat of combustion of boron to form B_2O_3 is 18.27 kJ kg^{-1} and the volumetric heat of reaction is 42.7 MJ m^{-3}, which is the highest of all the elements. Thus, boron particles are used as a valuable fuel component of solid-fueled ramjets and ducted rockets. Nevertheless, the combustion of boron particles is difficult because the oxide layer formed on their burning surface prevents sustained oxidation towards the interior of the particles. The boiling point temperature of boron oxide is 2133 K.

Boron is an important ingredient of pyrolants. A mixture of boron and potassium nitrate particles forms a pyrolant used as an igniter in rocket motors. The stoichiometric mixture of B and KNO_3 reacts according to:

$$10B + 6KNO_3 \rightarrow 5B_2O_3 + 3K_2O + 3N_2$$

However, unlike gas-phase reactions, the boron particles are oxidized at their surfaces, so that some unreacted boron remains after combustion. A significant difference between B and Al is that the amount of oxygen gas (O_2) needed to produce boron oxide is 2.22 kg [O_2]/kg [B] whereas the amount required to produce aluminum oxide is just 0.89 kg [O_2]/kg [Al]. Since the amount of oxygen needed to burn boron particles is approximately 2.5 times larger than that needed to burn aluminum particles, boron particles are favorably used as fuel components in ducted rockets that take in air from the atmosphere rather than as constituents of rocket propellants.

10.6.2.2 **Carbon**

Carbon (C) is a unique material used for various objectives in pyrolants. There are several structural forms of carbon, namely diamond, graphite, fullerene (C_{60}), and amorphous carbon. The unit cell of the diamond structure contains three-dimensional tetrahedral arrangements of covalent bonds involving all four outer orbit electrons. Thus, diamond is a single-crystal material, the density of which of 3513 kg m^{-3} is the highest among the carbon allotropes, and its the melting point temperature is 3820 K. The hardness of diamond is the highest of all materials. The unit cell of the graphite structure contains a two-dimensional hexagonal arrangement composed of the covalent bonds of three outer orbit electrons. The gasification temperature of graphite is 5100 K and its density is 2260 kg m^{-3}. Since the density of diamond is higher than that of amorphous carbon or graphite, very fine diamond particles made for industrial purposes are also used as a fuel component of pyrolants.

Fullerene is composed of spherical C_{60} structural units, with pentagonal and hexagonal arrangements of carbon atoms on the surface of the sphere. The density of C_{60} is higher than that of graphite. Molecular-sized carbon particles such as carbon nanotubes and nanohorn are made artificially for various objectives.

Amorphous carbon comprises various combinations of carbon atoms. Charcoalis a typical amorphous form of carbon and is used as a major component of black powder and ballistic modifiers of rocket propellants. Charcoal contains a large number of tiny pores and the total surface area within the structure is approximately 1–3 m^2 mg^{-1}. This surface area plays a significant role as a catalytic surface in various chemical reactions. It is well known that the burning rate of black powder is very fast because of the large surface area of the carbon structure.

10.6.2.3 **Silicon**

Silicon (Si) forms silicone polymers similar to hydrocarbon polymers. Though both the heat of reaction and the density of Si are higher than those of C, the ignition temperature of Si is much higher than that of C.

10.6.2.4 **Sulfur**

There are several types of crystalline sulfur (S). Sulfur is a major oxidizer component of black powder. Metals react with sulfur to form various types of sulfides such as FeS, FeS_2, ZnS, CdS, and Li_2S.

10.6.3 Polymeric Fuels

Polymeric materials used as fuel components of pyrolants are classified into two types: active polymers and inert polymers. Typical active polymers are nitropolymers, composed of nitrate esters containing hydrocarbon and oxidizer structures, and azide polymers, containing azide chemical bonds. Hydrocarbon polymers such as polybutadiene and polyurethane are inert polymers. When both active and inert polymers are mixed with crystalline oxidizers, polymeric pyrolants are formed.

10.6.3.1 Nitropolymers

Nitrocellulose (NC) is a nitropolymer composed of a hydrocarbon structure with -O–NO_2 bonds. The hydrocarbon structure acts as a fuel component and the -O–NO_2 bonds act as an oxidizer. Since the physical structure of NC has a fine, fiber-like nature, it absorbs liquid nitroglycerin (NG) to form a gelatinized energetic material. Though NC is a fuel-rich material, the incorporation of NG, which contains high mole fractions of -O–NO_2 bonds, forms a stoichiometrically balanced energetic material. The resulting gelatinized NC-NG is known as a double-base propellant, as used in rockets and guns. The physicochemical properties, such as combustion temperature, burning rate, pressure sensitivity, and temperature sensitivity, as well as the mechanical properties of these materials, can be varied by changing the mixture ratio of NC and NG or by the addition of chemical stabilizers or burning rate catalysts, in order to formulate pyrolants used as gas generators or smokeless igniters. TMETN and TEGDN are liquid nitrate esters containing -O–NO_2 bonds similar to NG. Mixtures of NC-TMETN or NC-TEGDN form polymeric pyrolants.

10.6.3.2 Polymeric Azides

Azide polymers contain –N_3 bonds within their molecular structures and burn by themselves to produce heat and nitrogen gas. Energetic azide polymers burn very rapidly without any oxidation reaction by oxygen atoms. GAP, BAMO, and AMMOare typical energetic azide polymers. The appropriate monomers are cross-linked and co-polymerized with other polymeric materials in order to obtain optimized properties, such as viscosity, mechanical strength and elongation, and temperature sensitivities. The physicochemical properties GAP and GAP copolymers are described in Section 4.2.4.

10.6.3.3 Hydrocarbon Polymers

Hydrocarbon polymers (HCP) are used not only as fuel components but also as binders of crystalline oxidizers and metal powders in the formulation of pyrolants, similar to composite propellants and plastic-bonded explosives. There are many types of HCP, the physicochemical properties of which are dependent on their molecular structures. The viscosity, molecular mass, and functionality of the polymers are the major parameters that determine the mechanical properties of the pyrolants. In general, high mechanical strength at high temperatures and high mechanical elongation at low temperatures are needed in the design of pyrolants. The chemical properties of HCP determine the mechanical characteristics of the pyrolants.

Typical examples of HCP are hydroxy-terminated polybutadiene (HTPB), carboxy-terminated polybutadiene (CTPB), hydroxy-terminated polyether (HTPE), hydroxy-terminated polyester (HTPS), and hydroxy-terminated polyacetylene (HTPA). The physicochemical properties of various types of HCP are described in Section 4.2.3.

10.7
Metal Azides

Azides, which are characterized by an $-N_3$ bond, are energetic materials that produce heat and nitrogen gas when they are decomposed. Typical metal azides are lead azide ($Pb(N_3)_2$), sodium azide (NaN_3), and silver azide (AgN_3). Lead azide is susceptible to detonation by friction, mechanical shock, or sparks. It is used to initiate ignition of various types of pyrolants and also as a pyrolant in blasting caps for detonating explosives. Its ignition temperature is about 600 K and its detonation velocity is approximately 5000 m s^{-1}. Its heat of formation is 1.63 MJ kg^{-1} and its density is 4800 kg m^{-3}. The physicochemical properties of silver azide are similar to those of lead azide. It is additionally sensitive to light. Its ignition temperature is about 540 K and its density is 5100 kg m^{-3}.

Sodium azide is not as sensitive as lead azide or silver azide to friction or mechanical shock. Since sodium azide reacts with metal oxides to generate nitrogen gas, mixtures of sodium azide and metal oxides are used as pyrolants in gas generators. However, sodium azide reacts with copper and silver to form the corresponding azides, both of which are detonable pyrolants.

References

1 Mayer, R., Explosives, Verlag Chemie, Weinheim, 1977.
2 Sarner, S. F., Propellant Chemistry, Reinhold Publishing Co., New York, 1966.
3 Japan Explosives Society, Energetic Materials Handbook, Kyoritsu Shuppan, Tokyo, 1999.
4 Kosanke, K. L., and Kosanke, B. J., Pyrotechnic Ignition and Propagation: A Review, Pyrotechnic Chemistry, Journal of Pyrotechnics, Inc., Whitewater, CO, 2004, Chapter 4.
5 Kosanke, K. L., and Kosanke, B. J., Control of Pyrotechnic Burn Rate, Journal of Pyrotechnics, Inc., Whitewater, CO, 2004, Chapter 5.
6 Urbanski, T., Black Powder Chemistry and Technology of Explosives, Vol. 3, Pergamon Press, New York, 1967.
7 Merzhonov, A. G., and Abramov, V. G., Thermal Explosion of Explosives and Propellants, A Review, *Propellants and Explosives*, Vol. 6, 1981, pp. 130–148.
8 Comkling, J. A., Chemistry of Pyrotechnics, Marcel Dekker, 1985.
9 Kubota, N., Propellant Chemistry, Pyrotechnic Chemistry, Journal of Py-

rotechnics, Inc., Whitewater, CO, 2004, Chapter 11.

10 Propellant Committee, Propellant Handbook, Japan Explosives Society, Tokyo, 2005.

11 Algoflon F5, Material Safety Data Sheet, Ausimont USA, Inc.

12 Viton Free Flow, Material Safety Data Sheet, VIT027, Du Pont Company.

11
Combustion Propagation of Pyrolants

11.1
Physicochemical Structures of Combustion Waves

11.1.1
Thermal Decomposition and Heat Release Process

Numerous experimental studies have been carried out on the burning rate characteristics of propellants and pyrolants.[1–14] The heats of reaction of energetic materials are evaluated by measurements obtained using combustion bombs, where the measured values are dependent not only on the chemical components but also on the method of burning in the bombs. Furthermore, the burning rate is dependent on both the mixture ratio of oxidizer and fuel components and the size of the crystalline oxidizer particles. In general, the burning rate of an energetic material is increased or decreased by the addition of a small amount of a catalyst.[1]

The heat produced by the reaction of a pyrolant is dependent on various physicochemical properties, such as the chemical nature of the fuel and oxidizer, the fractions in which they are mixed, and their physical shapes and sizes. Metal particles are commonly used as fuel components of pyrolants. When a metal particle is oxidized by gaseous oxidizer fragments, an oxide layer is formed that coats the particle. If the melting point of the oxide layer is higher than that of the metal particle, the metal oxide layer prevents further supply of the oxidizer fragments to the metal, and so the oxidation remains incomplete. If, however, the melting point of the oxide layer is lower than that of the metal particle, the oxide layer is easily removed and the oxidation reaction can continue.

Pyrolants composed of polymeric materials and crystalline oxidizers generate gaseous products of relatively low molecular mass, as in the case of propellants. The polymeric materials decompose to generate gaseous fuel fragments, and these react with gaseous oxidizer fragments generated by the decomposition of the crystalline oxidizer particles. Diffusional flamelets are formed above each oxidizer particle and the final combustion products are formed far downstream of the pyrolant surface.

When metal oxide particles are used as oxidizers, metal particles are used as fuel components. Since the melting point and gasification temperatures of metal oxide particles are very high, the oxidation reaction only occurs when the metal particles

Propellants and Explosives. Naminosuke Kubota

ISBN: 978-3-527-31424-9

melt and there is sufficient contact surface surrounding each metal oxide particle. Since the reaction between a metal and a metal oxide is a gasless reaction, there is no associated pressure increase in a closed chamber.

Various types of ignition system are used to ignite pyrolants: electrical heating, laser heating, convective heating by a hot gas flow, and conductive heating from a hot material. Once ignition is established, the heat produced at the ignition surface of the pyrolant is conducted to the bulk pyrolant. This heat serves to increase the temperature of the interior portion of the pyrolant. The heated surface zone is melted and/or gasified and reaction takes place here to produce the final combustion products. This successive physicochemical process continues until the pyrolant is completely consumed.

11.1.2 Homogeneous Pyrolants

Homogeneous pyrolants are composed of homogeneous materials that contain oxidizer and fuel fragments in their molecular structures. When a homogeneous pyrolant is heated above its decomposition temperature, reactive oxidizer and fuel fragments are formed, which react together to produce high-temperature combustion products. Nitrocellulose (NC) is a homogeneous pyrolant consisting of hydrocarbon fragments and -O–NO_2 oxidizer fragments. NC produces NO_2 as an oxidizer component and aldehydes as a fuel component by the breakage of -O–NO_2 bonds in the molecular structure. In general, NC is mixed with liquid nitrate esters-such as NG and TMETN to form gelatinized homogeneous pyrolants, similar to double-base propellants. The NO_2 molecules react with hydrocarbon fragments to produce heat and combustion products. The burning rate characteristics and the combustion wave structure for this class of pyrolants are described in Chapter 6.

Azide polymers such as GAP and BAMO are also homogeneous pyrolants consisting of hydrocarbon fragments and azide bonds, -N_3. When azide polymers are heated, heat is released by -N_3 bond breakage without oxidation.[15,16] The remaining hydrocarbon fragments are thermally decomposed and produce gaseous products. The burning rate characteristics and the combustion wave structure for azide polymers are described in Chapter 5.

11.1.3 Heterogeneous Pyrolants

Energetic crystals such as ammonium perchlorate (AP), potassium perchlorate(KP), potassium nitrate (KN), and ammonium nitrate (AN) are typical oxidizers used to formulate heterogeneous pyrolants. When these crystalline oxidizers are mixed with hydrocarbon polymers such as hydroxy-terminated polybutadiene(HTPB) or hydroxy-terminated polyurethane (HTPU), heterogeneous pyrolants are formulated. When these pyrolants are heated, thermochemical changes occur in the crystalline oxidizers and polymers, leading to the generation of oxidizer and fuel fragments separately in space on the heated surface of the pyrolant. These fragments are produced through physical and chemical changes such as melting, sublimation, decomposition, or gasification, and diffuse into each other to ultimately generate high-temperature combustion products.

Heterogeneous pyrolants composed of AP particles and HTPB are equivalent to AP composite propellants used for rocket motors, as described in Chapter 7. The physicochemical properties of AP composite pyrolants, such as combustion temperature, burning rate, pressure sensitivity, and temperature sensitivity, including with regard to mechanical properties, are varied by changing the size of the AP particles and the mixture ratio of AP particles and HTPB. Aluminized AP composite pyrolants are commonly used as igniters of AP composite propellants. The combustion temperature is increased by approximately 15% by the addition of 10% aluminum. However, when the residence time in the combustion chamber is very short, less than about 50 ms, incomplete combustion of the aluminum particles occurs, especially when they are greater than 10 μm in diameter. Agglomeration of molten aluminum particles occurs on and above the burning surface of the aluminized AP pyrolants.

Mixtures of crystalline and other solid particles, including metal particles, are widely used as energetic ingredients of igniters. Boron is known as a non-metal fuel ingredient used in igniters. When boron particles are oxidized with KN particles, high-temperature solid particles are formed as combustion products. The physical structure of the pyrolant is heterogeneous and the combustion wave structure is also heterogeneous due to the formation of oxidized solid and/or liquid particles with or without gaseous products.

11.1.4
Pyrolants as Igniters

Igniters are used to supply sufficient heat energy for the ignition of combustible energetic materials. When heat is supplied to the surface of a material, the surface temperature is increased from the initial temperature to the ignition temperature. Igniters used for the ignition of pyrolants are typically initiated by means of an electrically heated fine platinum wire. When a small region of the energetic material is heated by the wire, it is ignited and produces heat and/or a shock wave. This heat and/or shock wave propagate into the main charge of the igniter. For detonation, the igniter is required to supply not only heat but also mechanical shock, which serves to initiate shock propagation into the detonable material. The energy of the shock wave is converted into thermal energy to initiate gasification and chemical reaction of the detonable material. This successive process constitutes ignition for detonation. If the heat and/or the mechanical shock are insufficient, the detonable material burns without detonation and the ignition for detonation fails. The combustion propagates into the detonable material as a deflagration wave. In some cases, the deflagration wave tends to become a detonation wave, the so-called deflagration to detonation transition, DDT. In other cases, once the detonation wave propagates into the detonable material, it tends to become a deflagration wave, the so-called detonation transition to deflagration, DTD.

Two types of igniters are used for ignition in solid rocket motors, those giving: (i) high-volumetric and (ii) high-temperature combustion products. The combustion of nitropolymer propellants composed of nitrate esters such as nitrocellulose and nitroglycerin becomes unstable if the pressure is reduced to below about 5 MPa. In order to maintain stable combustion in the combustion chamber of a rocket motor, the pressure established by the igniter is thus required to be higher than 5 MPa. The igniters used for this class of propellants are black powders, which produce high-volumetric combustion products. On the other hand, composite propellantscomposed of ammonium nitrate or ammonium perchlorate with polymeric materials are ignited with metallized energetic materials used as igniter pyrolants. The metal oxides produced by the combustion of the metallized energetic materials act on the propellant surface as dispersed hot particles for ignition.

When a metallized energetic material is burned as a propellant igniter in a rocket chamber, a consequence of the aforementioned production of metal oxides as hot condensed particles is that there is very little associated pressure increase. However, the surface of the propellant grain in the chamber is ignited by the hot particles and a stable burning pressure is established. Typical metallized pyrolants used as igniters are shown in Table 11.1.

Table 11.1 Chemical compositions of metallized pyrolants used as igniters.

Composition	Mixture ratio by mass	Metal oxides	T_g, K
B-KNO_3	40:60	B_2O_3	3000
Mg-Tf	30:70	MgF_2	3700
Al-AP-PB	15:70:15	Al_2O_3	3100
B-AP-PB	10:75:15	B_2O_3	2200

Tf: polytetrafluoroethylene; PB: hydroxy-terminated polybutadiene; T_g: combustion temperature

11.2 Combustion of Metal Particles

Metal particles are major fuel ingredients of pyrolants. Their oxidation and combustion processes are largely dependent on their physicochemical properties, such as the type of metal, the particle size, the surface structure, and so on. In addition, a further important parameter is the type of oxidizer used to react with the metal particles. The reaction process of a metal particle with a gaseous oxidizer can be described as follows: gaseous oxidizer molecules heated by the ignition energy attack the surface of the particle. The temperature of the particle surface is increased and the heat penetrates into the particle. When the temperature is high enough to first melt and then gasify the surface, oxygen molecules react with the metal atoms on and above the molten surface. This reaction is exothermic and the heat liberated is also supplied to the particle. The combustion products, i. e., metal oxide species, diffuse out from the particle. Fresh oxidizer molecules are continually supplied from the surroundings.

The most important process with regard to metal oxidation and combustion is the formation of metal oxides on the surface of the metal particles. Some metal particles become coated with an oxide layer that surrounds the unreacted metal, whereas in other cases finely divided metal oxides are formed that are expelled from the surface of the metal particle. When solid shells of metal oxides are formed, no additional supply of the oxidizer fragments (or molecules) to the metal particles is possible. The oxidation process is thus interrupted and incomplete combustion occurs. On the other hand, when the metal oxide is expelled from the surface, oxidizer fragments continue to be supplied to the underlying unreacted surface of the metal.

11.2.1 Oxidation and Combustion Processes

11.2.1.1 Aluminum Particles

When an aluminum particle is oxidized with a gaseous oxidizer, it becomes coated with a solid layer of Al_2O_3. Since this oxide layer covers the entire surface of the particle, no further oxidizer molecules can be supplied to the underlying surface of unreacted aluminum. No successive oxidation reaction and combustion can occur and hence there is incomplete combustion of the aluminum particles. However, if the unreacted aluminum beneath the Al_2O_3 layer is melted and vaporized by heat supplied through the Al_2O_3 layer, the layer is broken due to the vapor pressure of the molten aluminum. The molten and/or vaporized aluminum is then forced out through the broken Al_2O_3 layer. Many empty shells of Al_2O_3 remain after such burnout of aluminum particles.

11.2.1.2 Magnesium Particles

Magnesium is easily vaporized and ignited, even below its melting point, to generate a brilliant flame in air. Magnesium reacts exothermically with hot water to produce $Mg(OH)_2$ and H_2 gas. Magnesium burns well under a wide range of oxidizer concentrations, i. e., from extremely limited to rich mixture ratios. When Mg particles are mixed with MnO_2 particles, Mg-MnO_2 pyrolants are formed. These pyrolants react exothermically when heated according to

$$MnO_2 + 2Mg \rightarrow 2MgO + Mn$$

Molten metallic manganese is formed if the reactor is thermally insulated.

The incorporation of magnesium particles into fluorine-containing polymeric materials, such as polyfluoroethylene (Tf) or vinylidene fluoride hexafluoropropene polymer (Vt), generates energetic pyrolants. The magnesium particles are oxidized by fluorine molecules eliminated from these polymers to produce high-temperature magnesium fluoride.

11.2.1.3 **Boron Particles**

The combustion efficiency of boron particles is extremely low when they are burned in air. A boron oxide layer coats the particles, preventing successive supply of oxygen molecules to the underlying unoxidized boron. Thus, the combustion efficiency increases with decreasing particle size. Though crystalline boron is extremely inert, amorphous boron burns in oxygen and reacts with sulfur at 900 K. The oxidation of boron with KNO_3 is extremely effective and high-temperature boron oxide is formed. A mixture of amorphous boron particles and KNO_3 particles in a ratio of 75:25 burns rapidly with a brilliant green flame being emitted by the reaction product of high-temperature BO_2. Such a mixture is easily ignited by a flame from black powder.

11.2.1.4 **Zirconium Particles**

Zirconium particles in air are sensitive to ignition by static electricity. This sensitivity increases with decreasing particle size. When Zr particles are heated in air, reaction with oxygen occurs at their surface. This reaction proceeds very violently to produce high-temperature zirconium oxide. A large number of bright light streams are emitted from the particles when they come asunder. The reaction process is represented by

$$Zr + O_2 \rightarrow ZrO_2$$

The ignition temperature in air is low at between 450 K and 470 K.

11.3 Black Powder

11.3.1 Physicochemical Properties

Black powder is the oldest explosive in history, dating back to the eighth century. Its chemical composition is well-known as a mixture of potassium nitrate, sulfur, and charcoal. The mixture ratio is varied according to the purpose for which it is to be used, with the ranges $\xi_{KNO3}(0.58–0.79)$, $\xi_S(0.08–0.20)$, and $\xi_C(0.10–0.20)$. Black powder composed of particles less than 0.1 mm in diameter is used for shell burst of fireworks and fuses. The grade with diameter 0.4–1.2 mm is used for the launch of spherical shells of fireworks, while that with diameter 3–7 mm is used in stone mines. Since black powder is sensitive to sparks caused by mechanical impact, friction, and static electricity, black powder containers should be made of brass or aluminum alloys rather than iron or steel. When Cl and Ca or Mg are present as impurities, $CaCl_2$ or $MgCl_2$ is formed and the thermal performance of KN is reduced. Contamination with NaCl also needs to be avoided for the same reason.

11.3.2
Reaction Process and Burning Rate

Black powder burns very rapidly, with a deflagration wave rather than a detonation-wave, because the burning surface is large due to its porosity. The deflagration wavepenetrates into the pores and ignites the interior surface. Thus, the mass-burning rate of black powder is highly dependent on the type of charcoal used. Charcoal is made from wood, the porosity of which is dependent on the species of tree from which it is derived.

The linear burning rate of black powder is not well characterized because its measurement is difficult and it is also dependent on the ignition method used. In contrast to propellants, the porous black powder has to be packed into a bed, the burning surface of which is not flat and so the burning flame spreads three-dimensionally. Furthermore, the mass-burning rate is dependent on the porosity and the particle size of the KNO_3, S, and C used. Moreover, the water vapor contained within black powder as an impurity significantly affects its burning rate.

11.4
Li-SF_6 Pyrolants

11.4.1
Reactivity of Lithium

Lithium is a light metal known to generate heat when oxidized by oxygen, fluorine, chlorine, and sulfur. This is clear from the Periodic Table of the elements, the metal being akin to sodium, magnesium, and aluminum, and from the heats of combustion shown in Figs. 10.2 and 10.3. When lithium is oxidized by fluorine, the heat generated is the highest of all the metals. The reaction scheme is represented by:

$$2Li + F_2 \rightarrow 2LiF$$

The initially liquid reaction product, LiF, becomes an agglomerated solid below about 500 K.

When lithium is oxidized by sulfur, the exothermic reaction is represented by:

$$2Li + S \rightarrow Li_2S$$

As in the case of LiF, the initially liquid Li_2S becomes an agglomerated solid below about 500 K.

11.4.2
Chemical Characteristics of SF_6

Since fluorine is a gas at room temperature, it does not constitute a practical oxidizer of pyrolants, not least from an environmental point of view. SF_6, on the other hand, is a solid oxidizer containing both fluorine and sulfur. When lithium is mixed with SF_6, an exothermic oxidation occurs according to:

$$8Li + SF_6 \rightarrow 6LiF + Li_2S$$

Since both Li and SF_6 are solid at room temperature, external heat is needed to melt these reactants in order to initiate their exothermic reaction. When the reaction between Li and SF_6 occurs in a closed chamber, the reaction products of LiF and Li_2S are obtained in liquid form due to the heat generated.

An Li-SF_6 pyrolant is made by mixing Li particles and SF_6 particles and is pressed into a rod-shaped grain. When one end of the grain is ignited in a closed chamber, the reaction wave propagates one-dimensionally along the rod. The reaction products of LiF and Li_2S remain in the same location in the chamber as a liquid-phase agglomerate, retaining the rod shape. The propagation velocity is of the order of several mm s^{-1}. The combustion is a gasless reaction and so the pressure remains unchanged in the closed chamber. This reactor is therefore used as a heat source.

11.5 Zr Pyrolants

Zirconium is used as an energetic fuel component of pyrolants because its heat of combustion is high and its density is also high compared with those of other metal fuels such as magnesium, aluminum, and titanium.[3–5] Zirconium reacts not only with gaseous oxidizers but also with metal oxides such as $BaCrO_4$ and Fe_2O_3. The reactions between zirconium and metal oxides generate heat without producing gaseous products and so there is no pressure increase in a closed chamber. Thus, pyrolants made of zirconium and metal oxide particles have been used as heat sources of thermal batteries.

Though zirconium particles are easily ignited by gaseous oxidizers and crystalline oxidizers such as KNO_3 and $KClO_4$, ignition of zirconium-metal oxide mixtures is only initiated in the high-temperature region above 2000 K. Thus, a two-stage ignition system is required for zirconium-metal oxide pyrolants.

11.5.1 Reactivity with $BaCrO_4$

Zr particles react violently with $BaCrO_4$ particles, similarly to the burning of Zr particles in air. However, a mixture of Zr particles and $BaCrO_4$ particles is relatively insensitive to friction or mechanical shock. DTA and TG data indicate that an exothermic reaction accompanied by a small mass loss occurs at 542 K. The maximum exothermic peak is at 623 K and the rate of the exothermic reaction-gradually decreases as the temperature is increased further. The observed characteristics are highly dependent on the size of the Zr particles because the reaction occurs at their surface. The exothermic peak observed at 623 K becomes higher with decreasing size of the Zr particles. The combustion reaction of the mixture is represented by:

$$Zr + BaCrO_4 \rightarrow ZrO_2 + BaO_2 + Cr$$

When the reactor is thermally insulated, molten chromium metal and BaO are formed.

11.5.2 Reactivity with Fe_2O_3

Zr powder reacts with Fe_2O_3 powder at high temperatures above 2000 K according to:

$$3Zr + 2Fe_2O_3 \rightarrow 3ZrO_2 + 4Fe$$

This reaction is highly exothermic and produces molten iron similarly to the well-known reaction between Al and Fe_2O_3:

$$2Al + Fe_2O_3 \rightarrow Al_2O_3 + 2Fe$$

A high heat input is needed to initiate the reaction. Since Zr particles react with $KClO_4$ particles at a relatively low temperature of around 650 K, immediately after melting of the $KClO_4$, small amounts of $KClO_4$ particles are incorporated into mixtures of Zr and Fe_2O_3 particles. The Zr and $KClO_4$ particles react exothermically according to:

$$2Zr + KClO_4 \rightarrow 2ZrO_2 + KCl$$

Since the reaction products of ZrO_2 and KCl are condensed-phase materials, the reaction is a gasless one and the pressure in the reaction chamber remains unchanged. This exothermic reaction increases the temperature of both the Zr and Fe_2O_3 particles and initiates the reaction between them. In this way, $KClO_4$ particles serve as an ignition initiator for pyrolants composed of Zr and Fe_2O_3 particles. Pyrolant systems composed of Zr + Fe_2O_3 as major reactants and Zr + $KClO_4$ as an ignition initiator are used as gasless heat sources in closed chambers.

11.6 Mg-Tf Pyrolants

11.6.1 Thermochemical Properties and Energetics

The heat produced by the oxidation of magnesium (Mg) with fluorine (F_2) is 16.8 MJ kg^{-1} (Mg) and is higher than that produced by the oxidation of Mg with oxy-

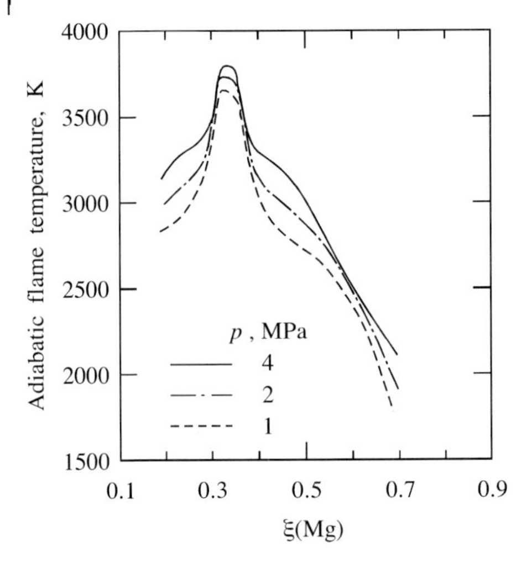

Fig. 11.1 Adiabatic flame temperature of Mg-Tf pyrolants as a function of ξ_{Mg}.

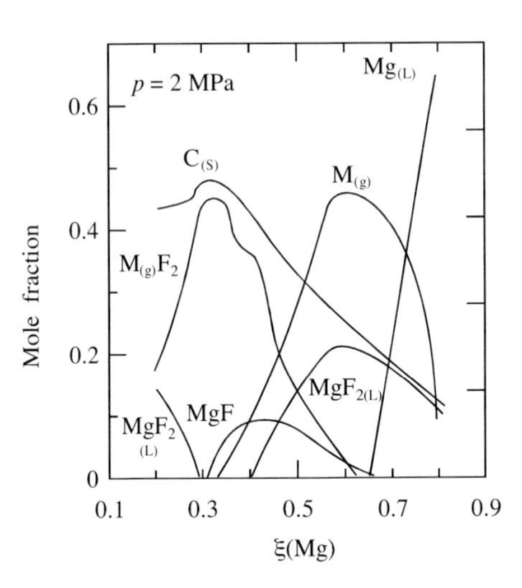

Fig. 11.2 Combustion products of Mg-Tf pyrolants as a function of ξ_{Mg}.

gen (O_2). The combustion of Mg with polytetrafluoroethylene (Tf) is a typical example of rapid oxidation of magnesium with fluorine.[6–12] Tf is composed of a -C_2F_4- molecular structure, which contains a mass fraction of fluorine of $\xi_F(0.75)$. Since Mg and Tf are both solid materials, the reaction process is complex, being accompanied by phase changes from solid to liquid and to gas.

Figs. 11.1 and 11.2 show the adiabatic flame temperature, T_f, and the mole fractions of combustion products, respectively, as a function of the mass fraction of Mg designated by ξ_{Mg}. The maximum T_f is obtained at $\xi_{Mg}(0.33)$. The major combustion products at the maximum T_f are $C_{(s)}$ and $MgF_{2(g)}$. The mass fractions $\xi_{C(s)}$ and $\xi_{MgF2(g)}$ decrease while $\xi_{Mg(g)}$ and $\xi_{MgF2(L)}$ increase with increasing ξ_{Mg}. When ξ_{Mg} is increased further, above $\xi_{Mg}(0.66)$, $\xi_{Mg(g)}$ and $\xi_{MgF2(L)}$ start to decrease.

11.6.2 Reactivity of Mg and Tf

The thermal decomposition of Tf begins at about 750 K and is complete at about 900 K, as shown by TG and DTA measurements on Tf particles at a heating rate of 0.167 K s^{-1} in an argon atmosphere at 0.1 MPa.[6,7] No undecomposed residue remains above 900 K. When Mg particles are mixed with Tf particles, gasification accompanied by an exothermic reaction occurs between about 750 K and 800 K. This exothermic reaction is between the Mg particles and the gaseous decomposition products of the Tf particles. An endothermic peak observed at 923 K can be ascribed to the heat of fusion of the Mg particles.

A residue remains undecomposed above 893 K. The mass fraction of this residue is 0.656, whereas the mass fraction of Mg contained within the pyrolant is $\xi_{Mg}(0.600)$. This indicates that the residue is produced by oxidation of the Mg by the fluorine produced upon thermal decomposition of the Tf. An X-ray analysis revealed that the remaining residue above 893 K in the TG experiments consisted of Mg and MgF_2. The oxidation reaction during the decomposition of Mg-Tf pyrolants is thus:

$$Mg + F_2 \rightarrow MgF_2$$

One mole of Mg (24.31 g) reacts with one mole of fluorine gas (38.00 g) to form one mole of magnesium fluoride (62.31 g). Since the oxidation reaction occurs at the surface of each Mg particle, the particles become coated with an oxidized surface layer of MgF_2. The thickness of such an MgF_2 surface layer has been calculated as 0.19 μm.[6]

11.6.3 Burning Rate Characteristics

The burning rates of Mg-Tf pyrolants with particle diameters of $d_{Mg} = 22$ μm and $d_{Tf} = 25$ μm as a function of ξ_{Mg} are shown in Fig. 11.3. Table 11.2 shows the chemical compositions tested to examine the effects of ξ_{Mg}/ξ_{Tf} and d_{Mg}/d_{Tf}, where d_{Mg} and d_{Tf} are the average diameters of the particles of Mg and Tf, respectively. The burning rate increases on going from $\xi_{Mg}(0.3)$ to $\xi_{Mg}(0.7)$ at constant pressure. The pressure exponent, n, defined by $n = d \ln r / d \ln p$ at constant initial temperature, is relatively

Table 11.2 Chemical compositions of Mg-Tf pyrolants.

Pyrolant	ξ_{Mg}/ξ_{Tf}	d_{Mg}/d_{Tf}	ξ_{Vt}
A	0.30/0.70	22/25	0.03
B	0.40/0.60	22/25	0.03
C	0.60/0.40	22/25	0.03
D	0.70/0.30	22/25	0.03
E	0.60/0.40	200/25	0.03
F	0.60/0.40	200/450	0.03
G	0.60/0.40	22/450	0.03

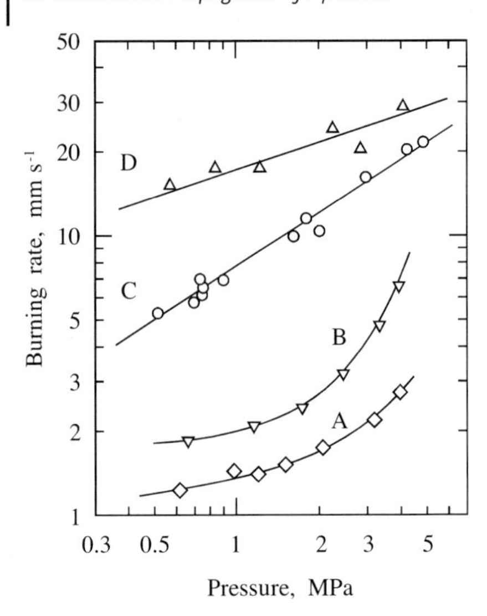

Fig. 11.3 Burning rate characteristics of Mg-Tf pyrolants.

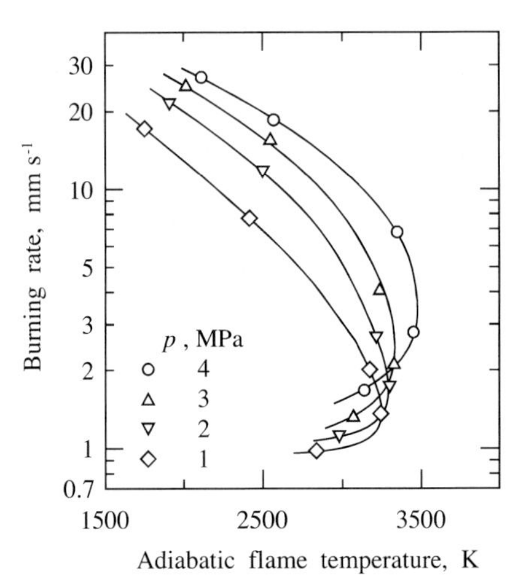

Fig. 11.4 Burning rate characteristics of Mg-Tf pyrolants as a function of adiabatic flame temperature.

constant for high ξ_{Mg} in the pressure range 0.5–5 MPa; n = 0.30 and 0.60 for $\xi_{Mg}(0.7)$ and $\xi_{Mg}(0.6)$, respectively. Though the burning rate becomes relatively pressure-independent as ξ_{Mg} is decreased from $\xi_{Mg}(0.4)$ to $\xi_{Mg}(0.3)$ in the pressure range below 1 MPa, the pressure exponent increases rapidly when the pressure is increased beyond 1 MPa.[6,7]

In Fig. 11.4, burning rates are plotted as a function of T_f at different pressures. The burning rate increases with increasing T_f in the range $\xi_{Mg} < 0.3$ at constant pressure. However, the burning rate increases with decreasing T_f in the range $\xi_{Mg} > 0.3$ and decreases with increasing energy density when ξ_{Mg} is higher than the stoichiometric ratio. In general, the burning rates of solid propellants, such as double-base propellants and composite propellants, increase with increasing

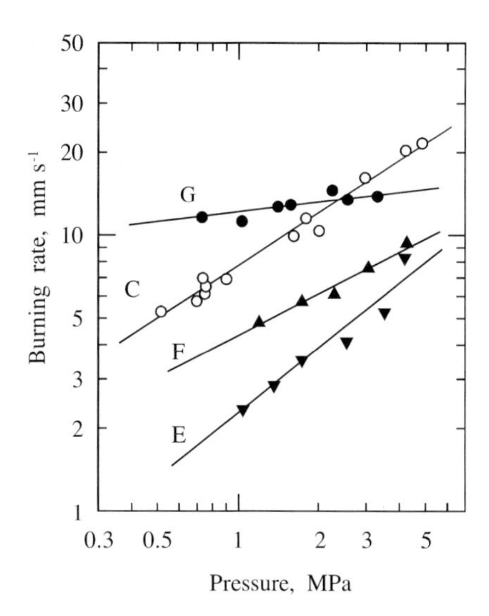

Fig. 11.5 Burning rate characteristics of Mg-Tf pyrolants composed of $\xi_{Mg}/\xi_{Tf} = 0.60/0.40$.

energy density. This is a significant difference between the burning rate characteristics of Mg-Tf pyrolants and conventional propellants.

The effects of the particle size of Mg and Tf on burning rate are shown in Fig. 11.5. The Mg-Tf pyrolants used to obtain the data shown in Fig. 11.5 all have a mixture ratio of $\xi_{Mg}/\xi_{Tf} = 0.60/0.40$. The burning rate increases with decreasing Mg particle size. Furthermore, the burning rate is dependent not only on d_{Mg} but also on d_{Tf}; the burning rate is higher for the pyrolant containing larger Tf particles ($d_{Tf} = 450$ μm) than for that containing smaller Tf particles ($d_{Tf} = 25$ μm).

The burning rates of Mg-Tf pyrolants are highly dependent on the particle sizes of both the Mg and Tf. The burning rate at $\xi_{Mg}(0.60)$ is high when small-sized Mg and large-sized Tf particles are used in the pressure range below 2 MPa. Fig. 11.6 shows the relationship between the burning rate and the total surface area of the

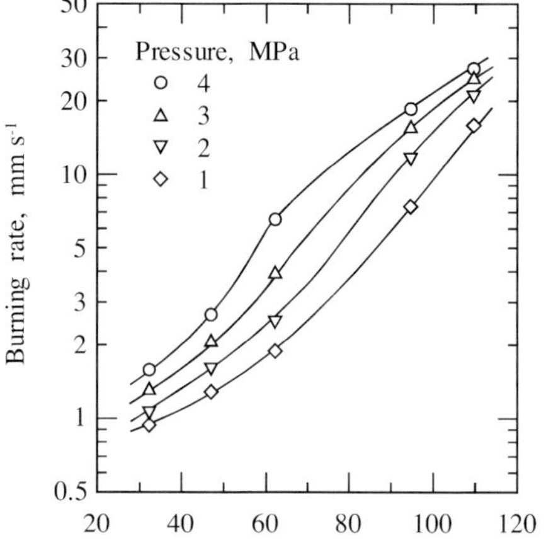

Fig. 11.6 Burning rates of Mg-Tf pyrolants as a function of the total surface area of the Mg particles.

Mg particles. The burning rate increases with increasing surface area. In other words, the burning rate increases with either decreasing particle size of Mg at constant ξ_{Mg} or increasing ξ_{Mg} at constant Mg particle size.[6,7]

11.6.4 Combustion Wave Structure

The temperature of an Mg-Tf pyrolant composed of $\xi_{Mg}(0.30)$ rises from its initial value to the gas-phase temperature as the burning surface of the sample regresses. The temperature increases exponentially from the initial value to the burning surface temperature as a result of thermal conduction. The temperature in the gas phase just above the burning surface increases rapidly. Since the burning surface is highly heterogeneous because the pyrolant consists of discrete particles of Mg and Tf, there is scatter in the data concerning the burning surface temperature. However, it is evident that the burning surface temperature exceeds both the melting point of Mg (923 K) and the decomposition temperature of Tf (800–900 K). Thus, the Mg particles incorporated into the pyrolants are considered to be molten on and above the burning surface. The burning surface temperature tends to decrease with increasing ξ_{Mg} at constant pressure.

The Mg particles melt at the burning surface and are partially oxidized by the fluorine produced by thermal decomposition of the Tf particles. Meanwhile, the Tf particles decompose completely to produce fluorine and other gaseous fragments. During decomposition of the Tf particles, some of the Mg particles melt and form agglomerates on and above the burning surface, while others are ejected into the gas phase whereupon they are rapidly oxidized by the fluorine. The oxidation of each Mg particle occurs at the molten layer on its surface.

Referring to Section 3.5, the burning mode of an Mg-Tf pyrolant can be understood by the use of Eqs. (3.73)–(3.76). If one assumes a homogeneous condensed phase, the heat balance at the burning surface is represented by

$$r = \alpha_s \phi / \psi \tag{3.73}$$

The temperature gradient, ϕ, in the gas phase at the burning surface is dependent on physicochemical parameters such as the diffusional mixing of the oxidizer and fuel components and the reaction rate of these components. Since the temperature in the gas phase just above the burning surface of Mg-Tf pyrolants increases rapidly, the determination of ϕ is not possible because the size of the thermocouple junction is too large to measure ϕ accurately. If one assumes that the Mg particles melt and that the Tf particles decompose by gasification at the burning surface of Mg-Tf pyrolants, the heat of reaction at the burning surface is given by

$$Q_s = Q_{s,Mg} + (1 - \xi_{Mg}) Q_{s,Tf} \tag{11.1}$$

where $Q_{s,Mg}$ is the heat of fusion of Mg and $Q_{s,Tf}$ is the heat of decomposition of Tf. Substituting $Q_{s,Mg} = -379$ kJ kg^{-1} and $Q_{s,Tf} = -6578$ kJ kg^{-1} into Eq. (11.1), Q_s is determined for Mg-Tf pyrolants as a function of ξ_{Mg}.

The burning-surface temperature data and Q_s data[6] are substituted into Eq. (3.73) to determine the condensed-phase parameter, ψ_i. The heat flux feedback from the gas phase, $q_{g,s}$, is given by

$$q_{g,s} = \lambda_g \phi = \rho_p c_p \psi r \quad (11.2)$$

where $\rho_p = 1.80 \times 10^3$ kg m^{-3} and $c_p = 10.5$ kJ kg^{-1} K^{-1}. It is shown that $q_{g,s}$ increases with increasing ξ_{Mg} and that the burning rate is strongly dependent on the heat flux feedback from the gas phase to the burning surface. In other words, the rate of heat production just above the burning surface increases with increasing total burning surface area of the Mg particles incorporated into unit mass of the pyrolant.[6,7]

Thus, the rate of heat generation on and above the burning surface of Mg-Tf pyrolants is seen to depend on the surface area of the Mg particles and ξ_{Mg}. It is evident that the surface area of the Mg particles available for oxidation by fluorine is insufficient when the Mg particle size is large because only a thin surface layer of each Mg particle is involved in the reaction on and just above the burning surface. Accordingly, the rate of heat transfer from the gas phase to the burning surface of the Mg-Tf pyrolant increases with decreasing Mg particle size and/or with increasing ξ_{Mg}, and hence the burning rate of the Mg-Tf pyrolant is increased.

11.7
B-KNO_3 Pyrolants

11.7.1
Thermochemical Properties and Energetics

Boron is oxidized by crystalline oxidizers such as KNO_3, $KClO_4$, and NH_4ClO_4. KNO_3 is most commonly used to obtain high heat energy and high burning rate. As shown in Fig. 11.7, the adiabatic flame temperature of B-KNO_3 pyrolants varies

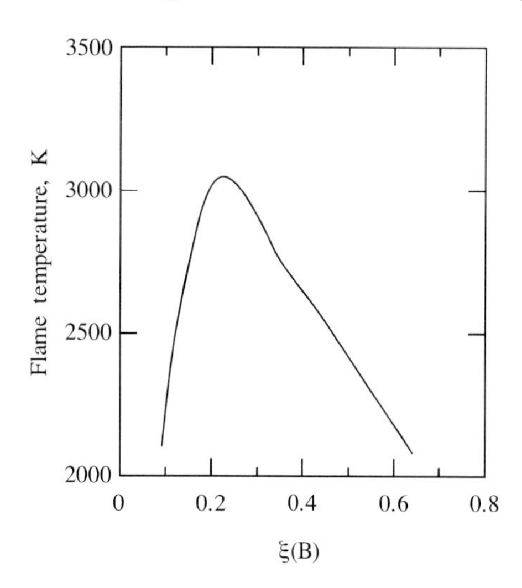

Fig. 11.7 Adiabatic flame temperature of B-KNO_3 pyrolants as a function of ξ_B.

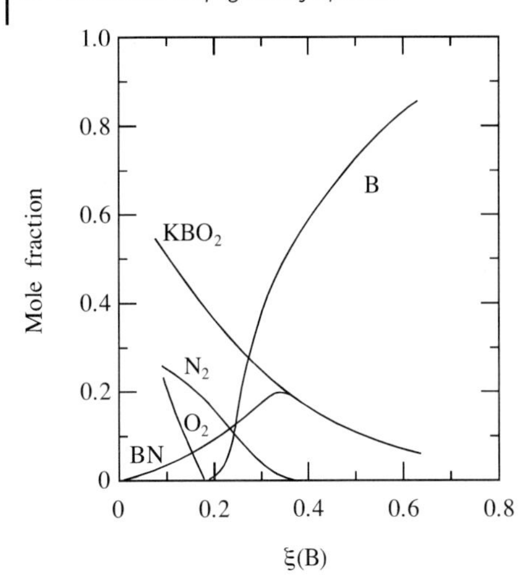

Fig. 11.8 Combustion products of B-KNO_3 pyrolants as a function of ξ_B.

with ξ_B and the maximum flame temperature of 3070 K is obtained at $\xi_B(0.22)$. The major combustion products at 0.1 MPa are KBO_2, N_2O, and BN, as shown in Fig. 11.8. The boron particle size is the dominant factor in determining the reaction rate. The oxidation of a boron particle occurs at its surface and so the particle becomes coated with a molten layer of B_2O_3. The oxidizer fragments, NO_2, produced by decomposition of the KNO_3 particles, are prevented from reaching the underlying unreacted boron by this molten layer and so the interior part of the boron particle remains unoxidized. It is therefore important to use fine boron particles in order to obtain high combustion efficiency.

11.7.2
Burning Rate Characteristics

The burning rates of B-KNO_3 pyrolants are dependent not only on ξ_B but also on θ_B. The maximum burning rate is seen for relatively boron-rich mixtures, even though the theoretical maximum flame temperature is obtained at $\xi_B(0.22)$. The total surface area of the boron particles is the important factor for obtaining a high combustion temperature and hence for obtaining a high burning rate. The burning rate behavior in the low-pressure region below 0.1 MPa is somewhat different from that in the high-pressure region, as shown in Fig. 11.9. The burning rates of B-KNO_3 pyrolants composed of d_B (5 μm) show plateau burning when $\xi_B(0.6)$ or $\xi_B(0.8)$ is used. The observed plateau burning is caused by incomplete combustion of the boron particles due to the low concentration KNO_3 in the pyrolants. A high proportion of the boron particles remains unreacted.

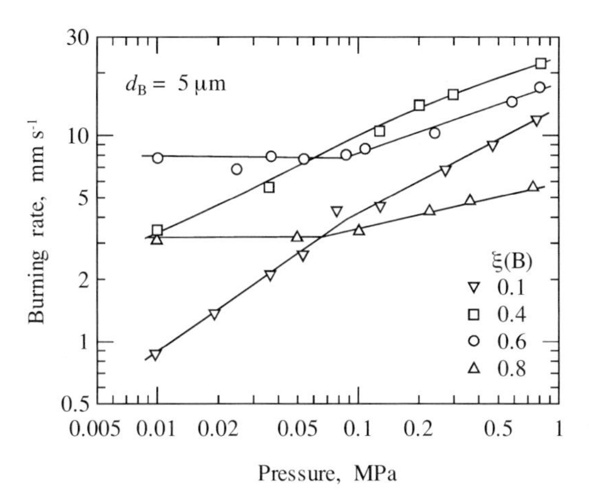

Fig. 11.9 Burning rates of B-KNO_3 pyrolants in the low-pressure region as a function of ξ_B.

11.8
Ti-KNO_3 and Zr-KNO_3 Pyrolants

11.8.1
Oxidation Process

When titanium particles are mixed with KNO_3 particles, a Ti-KNO_3 pyrolant is formed. Thermal decomposition of a Ti-KNO_3 pyrolant composed of $\xi_{Ti}(0.50)$ is accompanied by two endothermic peaks at 403 K and 612 K. The first peak corresponds to a crystal transformation, while the second corresponds to the melting of KNO_3. The processes of gasification and thermal decomposition of the KNO_3 particles are unaffected by the presence of Ti particles. No chemical reaction occurs between the Ti particles and the molten KNO_3. When the temperature is increased above 970 K, an exothermic gasification reaction occurs, which is complete at about 1200 K. This reaction process is the oxidation of the Ti particles by the gaseous decomposition products of KNO_3. Through this oxidation process, Ti-KNO_3 pyrolants produce heat.

When a mixture of Zr particles and KNO_3 particles is heated, the same two endothermic processes of KNO_3 as mentioned above occur: crystal transformation at 403 K and melting at 612 K. An exothermic reaction between 630 K and 750 K is the oxidation of the Zr particles by the molten KNO_3. No gasification occurs during this exothermic reaction. The oxidation process of Zr particles with KNO_3 is fundamentally different from that of Ti particles with KNO_3. The oxidation of Ti particles occurs in the gas phase, whereas that of Zr particles occurs in the liquid phase of molten KNO_3. The activation energy of the exothermic reaction is 200 kJ mol^{-1} for the Ti-KNO_3 pyrolant and 105 kJ mol^{-1} for the Zr-KNO_3 pyrolant. The results indicate that Zr particles are more easily oxidized by KNO_3 than Ti particles. Both activation energies are less dependent on the mass fractions of these particles.

11.8.2
Burning Rate Characteristics

The burning rate of a Ti-KNO_3 pyrolant composed of $\xi_{Ti}(0.33)$ is pressure-dependent and the pressure exponent is approximately 0.8 in the pressure region between 0.2 MPa and 1.0 MPa. The burning rate is 1.0 mm s^{-1} at 0.2 MPa and 4.1 mm s^{-1} at 1.0 MPa. On the other hand, the burning rate of a Zr-KNO_3 pyrolant composed of $\xi_{Zr}(0.33)$ is less dependent on pressure and the pressure exponent is 0.0 throughout the same pressure region. The burning rate is 50 mm s^{-1}. It should be noted that the burning rate of the Zr-KNO_3 pyrolant is more than ten times higher than that of the Ti-KNO_3 pyrolant at 1.0 MPa. This difference can be ascribed to the different oxidation processes as mentioned above: Zr particles are oxidized in the condensed phase of molten KNO_3 whereas Ti particles are oxidized in the gas phase of the gaseous decomposition products of KNO_3. In general, liquid-phase reactions are less pressure-sensitive than gas-phase reactions, as one might expect.

11.9
Metal-GAP Pyrolants

11.9.1
Flame Temperature and Combustion Products

Though GAP generates no oxidizer fragments among its combustion products, a mixture of metal particles and GAP forms a metal-GAP pyrolant. The thermal decomposition and burning rate are dependent on the type of metal used. Typical metals used with GAP are Al, Mg, B, Ti, and Zr. GAP produces relatively high mole fractions of N_2, H_2, and CO, which react with metal particles to produce the metal nitrides, hydrides, and oxides, respectively. These reactions are highly exothermic and generate high-temperature liquid and/or solid particles. Though the flame temperature of GAP without metals is about 1400 K, this increases with increasing ξ_M. The maximum flame temperatures obtained with metal-GAP pyrolants are shown in Table 11.3.

The highest flame temperature for a B-GAP pyrolant is obtained at $\xi_B(\sim 0.2)$, whereupon BN is produced, while that for an Al-GAP pyrolant is obtained at $\xi_{Al}(\sim 0.4)$, whereupon AlN is produced. Al reacts with N_2 generated by the decom-

Table 11.3 Maximum flame temperatures of metal-GAP pyrolants.

Metal	Al	Mg	Zr	B	Ti
ξ_M	0.4	0.2	0.5	0.2	0.3
T_f (K)	2752	2102	2566	2725	2004

position of GAP to form AlN when the mass fraction exceeds $\xi_{Al}(0.4)$. The mole fraction of H_2 decreases with increasing ξ_{Al}.

The flame temperature of Mg-GAP pyrolant is 2100 K in the range $\xi_{Mg}(0.1)$–$\xi_{Mg}(0.3)$. Further addition of Mg decreases the flame temperature. When the mass fraction of Mg is higher than $\xi_{Mg}(0.4)$, no CO is produced. The reaction of Mg with CO is represented by:

$$Mg + CO \rightarrow MgO + C$$

The reaction between Ti and N_2 occurs in the low-temperature region at below $\xi_{Ti}(0.2)$ for Ti-GAP pyrolants. On the other hand, the reaction between Ti and C occurs in the high-temperature region at above $\xi_{Ti}(0.2)$. The reactions of Ti with N_2 and C are represented by:

$$2Ti + N_2 \rightarrow 2TiN$$

$$Ti + C \rightarrow TiC$$

The maximum mole fraction of TiN is obtained at about $\xi_{Ti}(0.2)$, which also gives rise to the maximum flame temperature. The reaction of Ti and C occurs in the region above $\xi_{Ti}(0.2)$. The mole fraction of TiC increases with increasing ξ_{Ti}. It is noted that the reaction between Zr and N_2 produces ZrN in the region above $\xi_{Zr}(0.5)$.

11.9.2
Thermal Decomposition Process

The results of thermochemical experiments reveal that an exothermic reaction of GAP occurs at about 526 K and that no other thermal changes occur. When Mg or Ti particles are incorporated into GAP to formulate Mg-GAP or Ti-GAP pyrolants, two exothermic reactions are seen: the first is the aforementioned exothermic decomposition of GAP, and then a second reaction occurs at 916 K for the Mg-GAP pyrolant and at 945 K for the Ti-GAP pyrolant. There is no reaction between either Mg or Ti and GAP at the temperature of the first exothermic reaction. Both Mg and Ti particles within GAP are ignited by the heat generated by the respective second exothermic reactions.

11.9.3
Burning Rate Characteristics

Ti-GAP pyrolant composed of $\xi_{Ti}(0.2)$ burns in an inert atmosphere (argon gas) without visible light emission from the gas phase below 5 MPa, whereas Mg-GAP pyrolant composed of $\xi_{Mg}(0.2)$ burns and generates a flame accompanied by a bright emission. This emission is generated by oxidation of the Mg particles that are ejected from the burning surface of the Mg-GAP pyrolant. Since the combustion temperature of GAP is high enough to ignite the Mg particles, the oxidation process of each Mg particle with the N_2 gas generated by GAP occurs on and above the burning surface of the pyrolant.

On the other hand, the ignition temperature of the Ti particles is higher than that of the Mg particles, and no ignition of the Ti particles occurs in the gas phase below 5 MPa. However, when the pressure is increased above 5 MPa, Ti-GAP pyrolant composed of $\xi_{Ti}(0.2)$ is ignited and produces a luminous flame and a solid residue that retains the shape of the original pyrolant sample. Though the burning rate of GAP is 9 mm s^{-1} at 3 MPa, this is decreased to 7.5 mm s^{-1} by the addition of $\xi_{Ti}(0.2)$ and to 5.0 mm s^{-1} by the addition of $\xi_{Mg}(0.2)$. The burning rate of GAP is higher than that of Mg-GAP or Ti-GAP, even though its adiabatic flame temperature is lower than those of the metal-GAP pyrolants. The pressure exponent of burning rate, n, defined by $r = ap^n$, is 0.36 for GAP, 0.32 for Mg-GAP pyrolant, and 0.58 for Ti-GAP pyrolant.

X-ray analysis results show the formation of MgN as a combustion product of Mg-GAP pyrolants. The reaction occurs with nitrogen gas formed by the decomposition of GAP according to:

$$2Mg + N_2 \rightarrow 2MgN$$

However, no MgO is seen in the residue, indicating that this compound is not produced by reaction between Mg and CO. On the other hand, no TiN is seen in the residue of the Ti-GAP pyrolant, implying that there is no reaction between Ti and N_2 to produce this compound. Ti particles react exothermically with carbon formed by the decomposition of GAP according to:

$$Ti + C \rightarrow TiC$$

Thus, the combustion temperature of Ti-GAP pyrolants reaches the maximum value of 2000 K.

11.10
Ti-C Pyrolants

11.10.1
Thermochemical Properties of Titanium and Carbon

Pyrolants composed of titanium (Ti) and carbon (C) react to form titanium carbide (TiC) according to:

$$Ti + C \rightarrow TiC + 184.1 \text{ kJ mol}^{-1}$$

This reaction is highly exothermic without yielding gaseous products, i. e., it is a gasless reaction, and the adiabatic flame temperature is 3460 K at the stoichiometric mixture ratio. However, this reaction only occurs at high temperatures, above 2000 K. In order to initiate the combustion, a high heat input is needed for ignition.

It is shown that T_f increases with increasing ξ_{Ti} and that the maximum T_f is obtained at $\xi_{Ti}(0.80)$. Since a Ti-C pyrolant is a heterogeneous mixture, the reaction rate is highly dependent on ξ_{Ti}, as well as on the particle sizes of both Ti and C.

11.10.2
Reactivity of Tf with Ti-C Pyrolants

The ignition temperature of a mixture of Ti and C is relatively high compared with those of other pyrolants. When a small amount of polytetrafluoroethylene (Tf) is added to a Ti-C pyrolant, the ignition temperature is significantly lowered due to the exothermic reaction between Ti and Tf. Since Tf consists of a $-C_2F_4-$ chemical structure, the oxidizer gas, F_2, is formed by thermal decomposition of Tf according to:

$$-C_2F_4- \rightarrow F_2 + C_{(s)}$$

Ti particles react with F_2 gas according to:

$$Ti + 3/2F_2 \rightarrow TiF_3 \quad \text{and} \quad Ti + 2F_2 \rightarrow TiF_4$$

These processes are accompanied by heat release, similar to the reaction of Tf and Mg, i. e., Mg particles are oxidized to form MgF_2. It has been reported that the exothermic reaction between Ti and C occurs only when the mixture is heated above 2000 K. Since the reaction between Ti and Tf occurs at a lower temperature of about 830 K, the remaining Ti within the Ti-C pyrolant readily reacts with C with the aid of the heat generated by this reaction.

11.10.3
Burning Rate Characteristics

The burning rates of Ti-C pyrolants composed of $\xi_{Ti}(0.6)$–$\xi_{Ti}(0.8)$ increase linearly in log (burning rate) versus log (pressure) plots. These pyrolants are made of Ti particles (20 μm), C particles (0.5 μm), and Tf (5 μm), plus a small amount of Vt (Viton: $C_5H_{3.5}F_{6.5}$), which serves as a binder for the Ti and C particles.[5]

The pressure exponent of burning rate, n, defined in $r = ap^n$, is determined to be relatively independent of the mass fraction of ξ_{Ti}: n = 0.45 for $\xi_{Ti}(0.6)$ and n = 0.40 for $\xi_{Ti}(0.8)$. Though the burning rate increases with increasing ξ_{Ti} at a pressure of 0.1 MPa, the burning rate of a pyrolant composed of $\xi_{Ti}(0.4)$ is very low, and no self-sustaining combustion occurs when a pyrolant composed of $\xi_{Ti}(0.2)$ is ignited. The self-sustaining combustion limit is about $\xi_{Ti}(\sim 0.3)$ at pressures below 1.0 MPa. The burning rate is also dependent on the particle size of the Ti used in the pyrolants. The burning rate of a pyrolant composed of 20 μm Ti particles is higher than that of one composed of 50 μm Ti particles at 0.1 MPa.

Since the reaction of Ti occurs at the surface of each particle, the reaction of the Ti and C particles is dependent on the total surface area of the Ti particles, β_{Ti}, incorporated into the pyrolant. The relationship between the burning rate, r, and β_{Ti}

shows that r increases with increasing β_{Ti} at constant pressure. The burning rate at 0.1 MPa increases with increasing β_{Ti} when the same-sized Ti particles are used. In addition, the burning rate is high when large-sized Ti particles are used at the same β_{Ti}.

11.11 NaN_3 Pyrolants

11.11.1 Thermochemical Properties of NaN_3 Pyrolants

Sodium azide, NaN_3, is a crystalline material that produces high-temperature nitrogen gas as a decomposition product. When metal oxide particles are mixed with NaN_3 particles, sodium oxide is formed along with the nitrogen gas and heat is produced. Pyrolants composed of NaN_3 and metal oxides are called NaN_3 pyrolants and are used as N_2 gas generators.[15] The gas generation process of NaN_3 pyrolants is highly dependent on the oxides mixed with the NaN_3. In order to obtain superior NaN_3 pyrolants, it is necessary to understand the role of oxides in the gasification process of NaN_3. Typical metal oxides mixed with NaN_3 are shown in Table 11.4.

Table 11.4 Metal oxides mixed with NaN_3 to formulate NaN_3 pyrolants.

Oxides	Chemical formula
Iron oxides	α-Fe_2O_3, γ-Fe_2O_3, Co-γ-Fe_2O_3, FeOOH
Cobalt oxides	CoO, Co_3O_4
Copper oxide	CuO
Manganese oxide	MnO_2
Aluminum oxide	Al_2O_3
Boron oxide	B_2O_3
Potassium oxide	K_2O
Titanium oxide	TiO_2

11.11.2 NaN_3 Pyrolant Formulations

In order to enhance their energetics and to obtain high burning rates, various materials are added to NaN_3 pyrolants, as shown in Table 11.5. Nitrates such as sodium nitrate, $NaNO_3$, and strontium nitrate, $Sr(NO_3)_2$, are useful oxidizers because of their high oxygen balances ($NaNO_3$: +47.1%; $Sr(NO_3)_2$: +37.8%). These nitrates serve to oxidize sodium atoms liberated from NaN_3 and to produce heat. Graphite, silicon whiskers, or metal fibers are also added to enhance the mechanical proper-

Table 11.5 Typical chemicals used to formulate NaN_3 pyrolants.

Sodium nitrate	$NaNO_2$
Strontium nitrate	$Sr(NO_3)_2$
Graphite	C
Whisker	SiC, Si_3N_4
Fiber	aluminia, zirconium

ties of the NaN_3 pyrolants. These materials are pulverized by the use of a jet-mill, ball-mill, or locking mixer to make particles as fine as possible that can be homogeneously dispersed within the pyrolants. The mixtures are pressed into strand samples that are used to measure the burning rate. The mass fraction of NaN_3 in these NaN_3 pyrolants ranges from $\xi_{NaN3}(0.75)$ to $\xi_{NaN3}(0.58)$, with the remaining fractions being made up by the oxides and other additives.

11.11.3 Burning Rate Characteristics

The burning rate, *r*, of a pyrolant composed of NaN_3 and CoO is 19 mm s^{-1} at $\xi_{CoO}(0.30)$. The burning rate decreases rapidly with increasing CoO in the range above $\xi_{CoO}(0.30)$, and becomes 12 mm s^{-1} at $\xi_{CoO}(0.40)$. When $\xi_{NaNO2}(0.02)$ is added, the burning rate is increased in the range below $\xi_{CoO}(0.30)$ and reaches a maximum of 22 mm s^{-1} at $\xi_{CoO}(0.27)$. The burning rate of a pyrolant composed of NaN_3, Co_3O_4, and $NaNO_2$ is 20 mm s^{-1} at $\xi_{Co3O4}(0.35)$. This decreases rapidly with increasing Co_3O_4 in the range above $\xi_{Co3O4}(0.35)$, and is 15 mm s^{-1} at $\xi_{Co3O4}(0.40)$. When $\xi_{NaNO2}(0.02)$ is added, the burning rate is increased in the range below $\xi_{Co3O4}(0.34)$ and reaches a maximum of 25 mm s^{-1} at $\xi_{Co3O4}(0.30)$. Table 11.6 shows the effect of metal oxides on the burning rates of NaN_3 pyrolants.

Table 11.6 Burning rates of various types of NaN_3 pyrolants at 7 MPa.

r (mm s^{-1})	ξ_{NaN_3}	$\xi_{Fe_2O_3}$	ξ_{MnO_2}	ξ_{CuO}	ξ_{NaNO_2}	$\xi_{9Al_2O_32B_2O_3}$	ξ_C
43	0.60	0.26			0.14		
50	0.61			0.39			
52	0.62			0.30		0.08	
42	0.62		0.30			0.08	
39	0.62	0.10	0.20			0.08	
26	0.58	0.22		0.15			0.05

11.11.4
Combustion Residue Analysis

The mass fraction of the residue of an NaN_3 pyrolant containing Fe_2O_3 and $NaNO_2$ with $\xi_{NaNO2}(0.10)$ is 0.65. When $\xi_{NaNO2}(0.12)$ is added, the mass fraction of the residue is reduced significantly to 0.15. Furthermore, only a very small amount of residue, mass fraction 0.02, remains when $\xi_{NaNO2}(0.14)$ is added.

It is evident that a large amount of residue, mass fraction 0.95–1.0, remains in the absence of $NaNO_2$. The addition of more than $\xi_{NaNO2}(0.1)$ leads to the formation of very tiny pieces at the burning surface of the NaN_3 pyrolants during burning. This fragmentation process reduces the amount of residue that remains after combustion.

11.12
GAP-AN Pyrolants

11.12.1
Thermochemical Characteristics

In general, pyrolants composed of a polymeric material and AN particles are smokeless in character, their burning rates are very low, and their pressure exponents of burning rate are high. However, black smoke is formed as ξ_{AN} is decreased and carbonaceous layers are formed on the burning surface. These carbonaceous layers are formed from the undecomposed polymeric materials used as the matrix of the pyrolant. When crystalline AN particles are mixed with GAP, GAP-AN pyrolants are formed. Since GAP burns by itself, the GAP used as a matrix for AN particles decomposes completely and burns with the oxidizer gases generated by the AN particles.

The combustion products of aluminized GAP-AN pyrolants at 10 MPa are H_2O, H_2, N_2, CO, CO_2, and Al_2O_3. The mass fraction of Al_2O_3 increases linearly while that of H_2O decreases linearly with increasing ξ_{Al}. This is caused by the overall reaction of $2Al + 3H_2O \rightarrow Al_2O_3 + 3H_2$.

11.12.2
Burning Rate Characteristics

The burning rate decreases with increasing ξ_{AN} at constant pressure, especially in the range $\xi_{AN}(0.3–0.5)$. The effect of the addition of AN on the pressure exponent of burning rate, n, is significant: $n = 0.70$ at $\xi_{AN}(0.0)$, 1.05 at $\xi_{AN}(0.3)$, and 0.78 at $\xi_{AN}(0.5)$. The burning rate decreases drastically over a relatively small T_f range of 1410–1570 K. However, it remains relatively unchanged when T_f is increased from 1570 K to 2060 K. In general, the burning rate increases as the energy content per unit mass of pyrolant increases for conventional pyrolants and propellants.[16]

11.12.3 Combustion Wave Structure and Heat Transfer

The AN particles incorporated into GAP-AN pyrolants form a molten layer on the burning surface and decompose to form oxidizer fragments. The fuel-rich gas produced by the decomposition of GAP interdiffuses with these oxidizer fragments on and above the burning surface and produces a premixed flame. A luminous flameis formed above the burning surface.

When AP particles are added to GAP-AN pyrolants, a number of luminous flamelets are formed above the burning surface. These flamelets are produced as a result of diffusional mixing between the oxidizer-rich gaseous decomposition products of the AP particles and the fuel-rich gaseous decomposition products of the GAP-AN pyrolants. Thus, the temperature profile in the gas phase increases irregularly due to the formation of non-homogeneous diffusional flamelets.

When Al particles are added to GAP-AN pyrolants, agglomerated Al fragments are formed on the burning surface. However, when Al particles are mixed with pyrolants composed of GAP, AN, and AP, numerous flame streams are formed in the gas phase. The Al particles are oxidized by the gaseous decomposition products evolved by the AP particles. The combustion efficiency of the Al particles is improved significantly by the addition of the AP particles.[16]

11.13 Nitramine Pyrolants

11.13.1 Physicochemical Properties

HMX and RDX are energetic materials that produce high-temperature combustion products at about 3000 K. If one assumes that the combustion products at high temperature are H_2O, N_2, and CO, rather than CO_2, both nitramines are considered to be stoichiometrically balanced materials and no excess oxidizer or fuel fragments are formed. When HMX or RDX particles are mixed with a polymeric hydrocarbon, a nitramine pyrolant is formed. Each nitramine particle is surrounded by the polymer and hence the physical structure is heterogeneous, similar to that of an AP composite pyrolant.

11.13.2 Combustion Wave Structures

When a nitramine pyrolant is formulated using HTPE, HTPS, or HTPA, each of which has a relatively high oxygen content, the nitramine particles and the polymer melt and mix together to form an energetic liquid material that decomposes to produce a premixed flame above the burning surface of the pyrolant. Thus, the flame structure appears to be homogeneous, similar to that of a nitropolymer pyrolant. On the other hand, when HTPB is used to formulate nitramine pyrolants, the flame structure becomes heterogeneous due to the formation of carbonaceous materials on the burning surface. Since HTPB is a non-melting polymer, there is no mixing of the decomposed fragments of the nitramine particles and HTPB on the burning surface. The gaseous decomposition products of the nitramine particles and HTPB only interdiffuse above the burning surface.

Since nitramine pyrolants are fuel-rich materials, the flame temperature decreases with increasing hydrocarbon polymer content. The polymers act as coolants and generate thermally decomposed fragments as a result of the exothermic heat of the nitramine particles. The major decomposition products of the polymers are H_2, HCHO, CH_4, and $C_{(s)}$. When AP particles are incorporated into nitramine pyrolants, AP-nitramine composite pyrolants are formed. AP particles produce excess oxidizer fragments that oxidize the fuel fragments of the polymers that surround them. Thus, the addition of AP particles to nitramine pyrolants forms stoichiometrically balanced products and the combustion temperature increases.

Since the adiabatic flame temperature of a pyrolant composed of $\xi_{HMX}(0.70)$, $\xi_{AP}(0.10)$, and $\xi_{HTPB}(0.20)$ is 3200 K at 10 MPa, this composition is used as a high specific impulse and reduced-smoke propellant for rockets. The adiabatic flame temperature of HMX is 3400 K at 10 MPa, and the burning rate of a pyrolant composed of $\xi_{HMX}(0.80)$ and $\xi_{HTPB}(0.20)$ is of the order of 3 mm s^{-1} at this pressure. The burning rates of nitramine pyrolants are less dependent on the particle size of the nitramines used. This is because the nitramine particles melt and mix with the molten polymers that surround them. On the other hand, each AP particle forms a diffusional flamelet with the gaseous decomposition products of the polymers on and above the burning surface. The heat flux feedback from the gas phase to the burning surface is then dependent on the size of these flamelets. The burning rates of AP-nitramine pyrolants increase with decreasing size of the AP particles, as in the case of AP pyrolants.

11.14 B-AP Pyrolants

11.14.1 Thermochemical Characteristics

Boron (B) is a unique non-metallic substance that produces significant heat when it burns in air or with crystalline oxidizers. Though the density of boron is 2.33×10^3 kg m^{-3} and that of aluminum (Al) is 2.69×10^3 kg m^{-3}, the heat of combustion of boron is 18.27 MJ kg^{-1} and that of aluminum is 16.44 MJ kg^{-1}. A mixture of boron particles and crystalline oxidizers such as KNO_3 and NH_4ClO_4 forms a boron-based pyrolant used in gas generators and igniters.[17] Boron particles burn with glycidyl azide polymer (GAP) as a typical energetic polymer. Indeed, a mixture of boron particles and GAP constitutes a polymeric gas-generating pyrolant used in ducted rockets and solid ramjets.[17]

A mixture of B and AP particles formulates an energetic B-AP pyrolant. A small amount of polymeric material is added to serve as a binder of the B and AP parti-

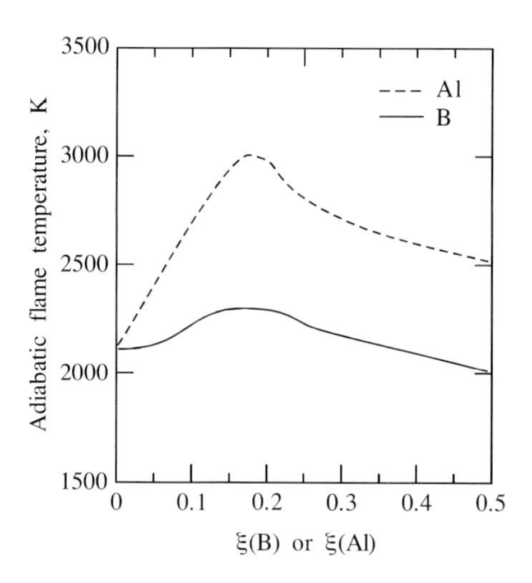

Fig. 11.10 Adiabatic flame temperature as a function of ξ(B) and ξ(Al).

cles. Carboxy-terminated polybutadiene (CTPB) is chosen as a binder and is cured with an epoxy resin. Fig. 11.10 shows the adiabatic flame temperature, T_f, as a function of the mass fraction of boron particles, ξ_B. The base matrix of the B-AP pyrolants is composed of a mixture of $\xi_{AP}(0.79)$ and $\xi_{CTPB}(0.21)$.

The adiabatic flame temperature, T_f, is 2220 K at ξ(0.0), i. e., without B or Al particles. T_f remains relatively unchanged in the region of ξ_B less than 0.05, but then rises to a maximum of 2260 K at $\xi_B(0.15)$. On the other hand, when Al is mixed with AP, T_f reaches a maximum of 3000 K at $\xi_{Al}(0.18)$. The maximum flame temperature of a B-AP pyrolant is thus approximately 740 K lower than that of an Al-AP pyrolant.[17]

11.14.2 Burning Rate Characteristics

B-AP pyrolants made with CTPB are cured with epoxy resin as in the case of conventional AP-CTPB composite propellants. The mixture ratio of large-sized AP particles (200 μm in diameter) and small-sized particles (20 μm in diameter) is 0.30/0.70. The mass fraction of boron is variously 0.010, 0.050, 0.075, or 0.150, and the diameter of the boron particles, d_B, is either 0.5 μm, 2.7 μm, or 9 μm.

Fig. 11.11 shows the burning rates of the aforementioned B-AP pyrolants as a function of pressure. The burning rates are seen to increase linearly with increasing pressure in plots of ln (burning rate) versus ln (pressure). The burning rate increases with increasing ξ_B and with decreasing size of the boron particles at constant pressure. However, the pressure exponent of burning rate, n, defined in Eq. (3.71), is determined as 0.5 for each of the tested pyrolants. It should be noted that n remains unchanged even when the size of the added boron particles is changed.

Fig. 11.12 shows a comparison of the burning rates of B-AP and Al-AP pyrolantsas a function of pressure. In analogy to the B-AP pyrolants, the burning rates of the Al-AP pyrolants increase with increasing ξ_{Al} and with decreasing size of the alum-

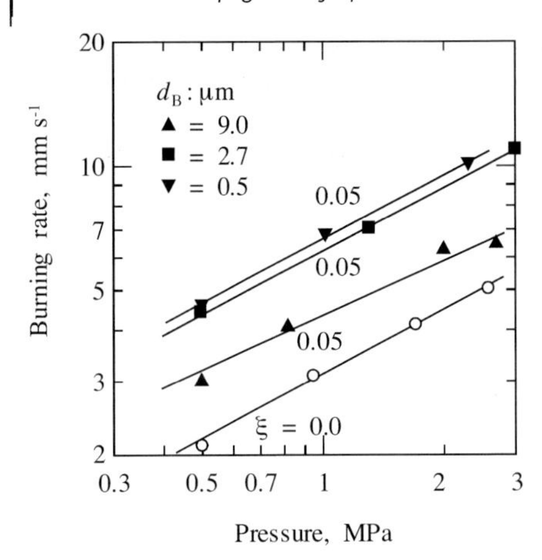

Fig. 11.11 Effect of ξ_B on burning rate as a function of pressure.

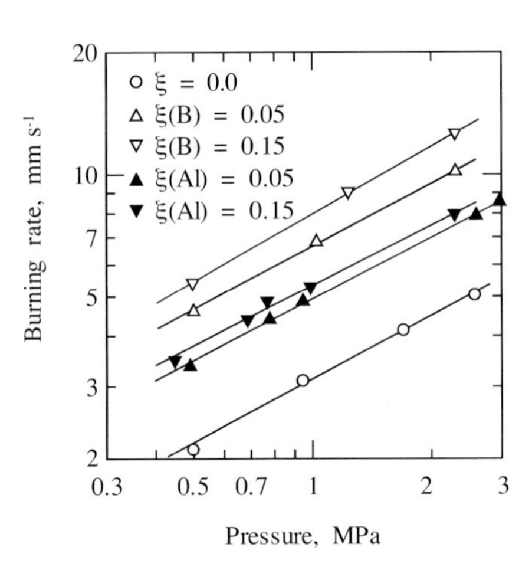

Fig. 11.12 Effect of ξ_B and ξ_{Al} on the burning rates of B-AP and Al-AP pyrolants as a function of pressure.

inum particles. Though the burning rate increases linearly in an ln (burning rate) versus ln (pressure) plot, n remains unchanged at 0.5 even when aluminum particles are added. However, it is evident that the burning rate of the boron pyrolant is higher than that of the aluminum pyrolant at equal ξ_B or ξ_{Al}.[17]

Fig. 11.13 shows the temperature profiles in the combustion waves of AP pyrolants with and without B particles at 1 MPa. It is evident that the temperature gradient above the burning surface is increased by the addition of B particles. Thus, the heat flux transferred back from the gas phase to the propellant is increased, and hence the burning rate is increased. The heat of reaction is also increased as the mass fraction of B particles is increased, as shown in Fig. 11.14. The results indicate that the B particles act as a fuel component in the gas phase, undergoing oxidation just above the burning surface.

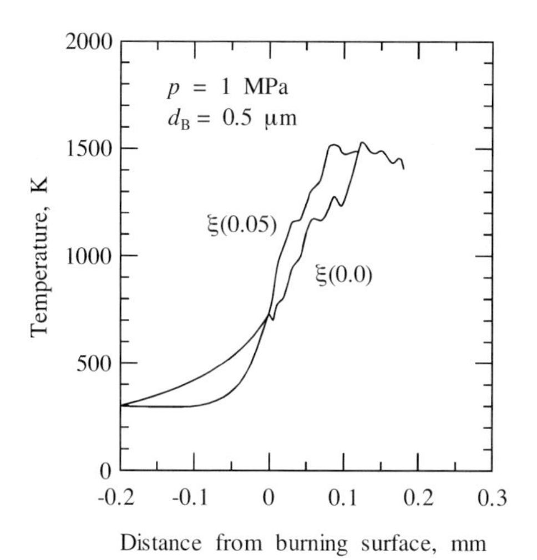

Fig. 11.13 Temperature profiles in the combustion waves of AP pyrolants with and without boron particles.

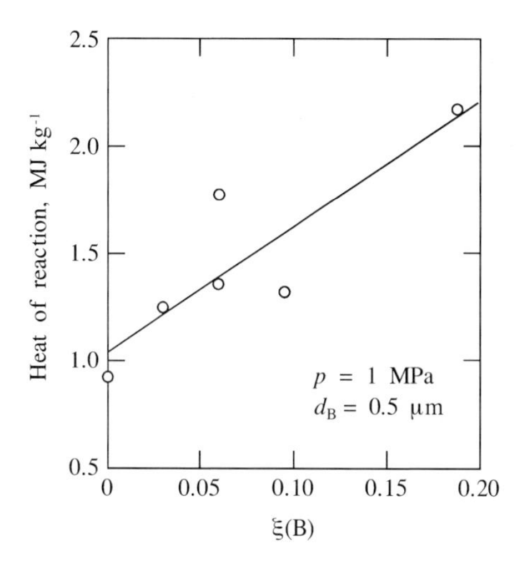

Fig. 11.14 Heat of reaction as a function of ξ_B.

11.14.3 Burning Rate Analysis

Fig. 11.15 shows the level of burning rate augmentation, defined according to $\varepsilon_B = (r_B - r_0)/r_0$, as a function of ξ_B, where r_0 is the burning rate without additives and r_B is the burning rate of a boron pyrolant composed of boron particles of $d_B = 0.5$ μm. Though ε_B increases with increasing ξ_B, it is less dependent on pressure. Since the oxidation of the boron particles in the combustion wave occurs at the burning surface or in the gas phase, the surface area of the particles is an important parameter with regard to the rate of oxidation. The oxidation of the boron particles proceeds from their surfaces and an oxidized layer beneath the surface penetrates into the

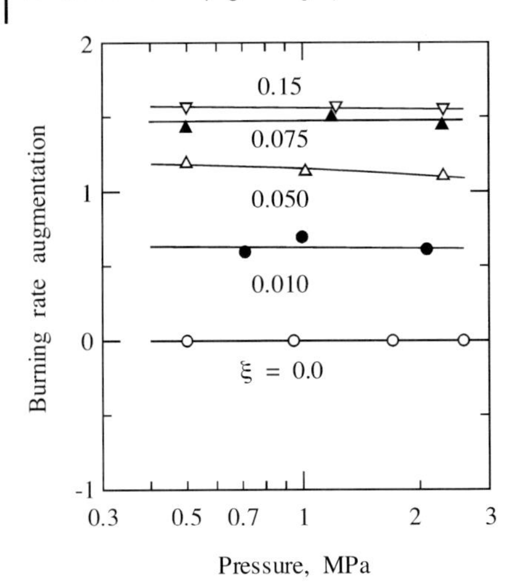

Fig. 11.15 Burning rate augmentation as a function of pressure for B-AP pyrolants containing different mass fractions of boron particles.

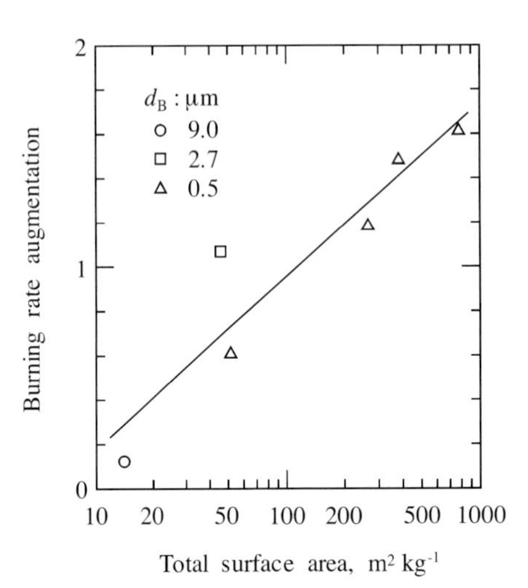

Fig. 11.16 Burning rate augmentation as a function of the total surface area of the boron particles in B-AP pyrolants composed of boron particles of different sizes.

particle. As shown in Fig. 11.16, ε_B increases with increasing total surface area of the boron particles in unit mass of the pyrolant. The results indicate that the production of heat by the oxidation of the boron particles occurs in the gas phase just above the burning surface, and hence the heat flux feedback from the gas phase to the burning surface is increased. Consequently, the burning rate is increased by the addition of boron particles.

Fig. 11.17 shows burning rate augmentation, ε_B, as a function of the adiabatic flame temperatures of B-AP and Al-AP pyrolants. The incorporation of aluminum particles into a base matrix composed of AP-CTPB pyrolant increases ε_B. However, the effect of the addition becomes saturated for adiabatic flame temperatures higher than about 2500 K. On the other hand, the incorporation of boron particles into the same base matrix increases ε_B more effectively, even though the adiabatic

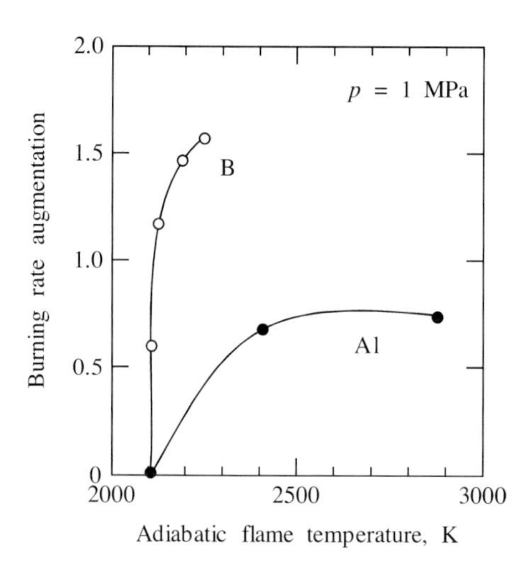

Fig. 11.17 Burning rate augmentation as a function of adiabatic flame temperature of B-AP and Al-AP pyrolants.

flame temperature remains unchanged. The results indicate that the heat flux transferred back from the final reaction zone, which is located at a greater distance from the burning surface, does not play a significant role in determining the burning rate of a boron pyrolant.[17]

11.14.4
Site and Mode of Boron Combustion in the Combustion Wave

As shown in Fig. 11.13, the temperature in the gas phase increases rapidly when boron particles are added. This is due to the higher burning rate of the boron pyrolant compared with that of the pyrolant without boron. Though the burning surface temperature is approximately 700 K for both pyrolants, the temperature in the gas phase just above the burning surface increases more rapidly when boron is added. The observed results indicate that the boron is oxidized just above the burning surface, within a distance of 0.1 mm. The heat flux transferred back from the gas phase to the burning surface is increased by the increased temperature in the gas phase.

Based on the heat balance at the burning surface, the burning rate is represented by

$$r = \alpha_s \phi / \psi \tag{3.73}$$

The boron particles are thermally inert in the solid phase beneath the burning surface of the pyrolant. The oxidation of the boron particles occurs just above the burning surface. This implies that the temperature gradient in the gas phase, ϕ, increases and hence the burning rate is increased accordingly.

Thus, the burning rate is increased by the addition of boron particles. Though the heat flux increases with increasing pressure for both pyrolants with and without boron particles, the heat flux at constant pressure is increased when boron particles are added. Furthermore, the heat flux also increases with decreasing d_B at constant

ξ_B and with increasing ξ_B at constant d_B. Since the oxidation of the boron particles proceeds from their surfaces, the surface area of the boron particles plays a dominant role in determining the reaction rate. In fact, the total surface area of the boron particles in unit volume of pyrolant is seen to correlate with the burning rate, as shown in Fig. 11.16.

11.15
Friction Sensitivity of Pyrolants

11.15.1
Definition of Friction Energy

The friction sensitivity of pyrolants is an important parameter with regard to avoiding unexpected spontaneous ignition. Heat is produced when there is mechanical friction between pyrolant grains or pyrolant grains and a metal part. This heat serves to raise the temperature of the friction surface. When the surface temperature is raised sufficiently such as to gasify the surface layer of the grains, gaseous species are formed between the two friction surfaces. Since the temperature of the gaseous species is also high, an exothermic reaction occurs and ignites them. The process from the applied mechanical force causing frictional heating to ignition determines the friction sensitivity of the pyrolants.[1]

Though a mechanical force applied to a pyrolant in the form of friction is converted into thermal energy, the evaluation of this thermal energy is difficult because the actual amount of energy produced is dependent on the surface roughness, its hardness, and the nature of the materials (homogeneous or composite materials). Thus, the friction sensitivity is evaluated from the mechanical energy supplied to the pyrolant. Similarly to friction sensitivity, shock sensitivity to a strike by a fall-hammer is used to evaluate the level of safety. The mechanical shock imparted to the surface of a pyrolant by a falling mass is also converted into thermal energy. Since the thermal energy resulting from the mechanical shock is also difficult to evaluate, the potential energy of the fall-hammer is assumed to be equal to the thermal energy.

11.15.2
Effect of Organic Iron and Boron Compounds

Fig. 11.18 shows experimental results concerning the friction sensitivity of mixtures of AP particles and catalysts.[1] The AP particles are trimodal mixtures with diameters of 200 μm (33 %), 35 μm (33 %), and 5 μm (34 %). Catocene ($C_{27}H_{32}Fe_2$) and ferrocene ($C_{10}H_{10}Fe$) are both liquid organic compounds with iron atoms in their molecular structures and carborane is a liquid compound that contains boron atoms.[1] The physicochemical properties of *n*-hexyl carborane are shown in Table 11.7. These catalysts are used to increase the burning rate of AP pyrolants.

Friction sensitivity is evaluated using a BAM friction apparatus according to the Japanese safety test standard JIS K 4810. The friction energy imparted to the mixtures is determined as the product of the applied mass and the moving distance of

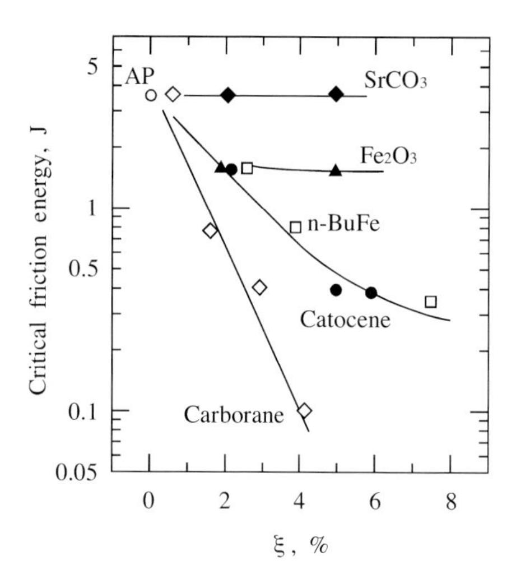

Fig. 11.18 Critical friction energy of mixtures of AP + liquid organoiron and AP + boron catalysts is decreased with increasing mass fraction of these catalysts.

Table 11.7 Physicochemical properties of *n*-hexyl carborane.

Carborane	*n*-hexyl carborane
Molecular structure	$B_{10}H_{10}C_2HC_6H_{13}$
Density	0.90×10^3 kg m^{-3} (at 293 K)
Boiling point	400 K (at 0.1–0.2 mmHg)
Color	colorless liquid
Stability	stable below 500 K

the friction board. The critical friction energy, $E_{f,c}$, is defined as the minimum energy that has to be supplied to the mixtures for ignition. As shown in Fig. 11.18, the decreasing $E_{f,c}$ of the AP + catalyst mixtures reflects increasing sensitivity with increasing mass fraction of the catalysts. The $E_{f,c}$ of the AP + carborane mixtures decreases linearly with increasing mass fraction of carborane.[1]

Fig. 11.19 shows the relationship between $E_{f,c}$ and ignition temperature of the mixtures. The ignition temperature is defined as the temperature at which an ignition time of 4 seconds is required in an isothermal closed chamber. Both ignition temperature and $E_{f,c}$ are lowered simultaneously as the mass fraction of the catalysts is increased. The $E_{f,c}$ is also lowered as the metal atom content within the catalysts is increased. The $E_{f,c}$ of an AP + catocene mixture is equivalent to that of an AP + ferrocene mixture as long as the same number of iron atoms is contained within both mixtures. The effect of boron atoms on $E_{f,c}$ is lower compared with that of iron atoms when equal numbers of the respective atoms are contained within the mixtures.

When the catalysts are in the form of crystalline solid particles, the characteristics of $E_{f,c}$ are different compared to when liquid organoiron catalysts are used. As shown in Fig. 11.18, the critical friction energy is independent of the mass fraction of the catalysts $SrCO_3$ and Fe_2O_3. Though Fe_2O_3 reduces the $E_{f,c}$ similarly to the

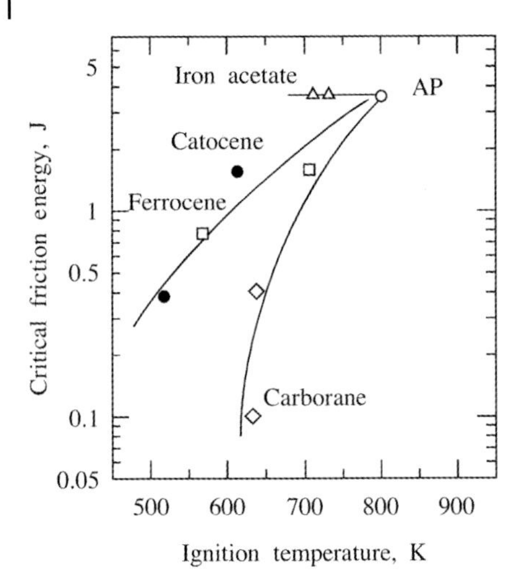

Fig. 11.19 Critical friction energy is decreased when the critical ignition temperature is lowered for both organoiron and boron catalysts.

liquid organoiron compounds catocene or ferrocene, the $E_{f,c}$ of the AP + Fe_2O_3 mixture remains unchanged even when the mass fraction of Fe_2O_3 is increased from $\xi_{Fe2O3}(0.03)$ to $\xi_{Fe2O3}(0.05)$. The burning rate of a reference AP pyrolant is increased significantly when 5 % catocene is added, as shown in Fig. 11.20. However, very little effect on the burning rate is observed when carborane (5 %) is added to the reference AP pyrolant, even though the critical friction energy of the AP + carborane mixture is decreased. The AP pyrolant used as a reference is composed of $\xi_{AP}(0.87)$ [trimodal, with particle diameters 200 μm (33 %), 35 μm (33 %), and 5 μm (34 %)] and $\xi_{HTPB}(0.13)$.

Fig. 11.20 shows the burning rates of an AP pyrolant without a catalyst and with 5 % of three different catalysts. Catocene is seen to be the most effective catalyst for increasing the burning rate as compared with iron acetate and carborane. The burning rate is increased threefold from 5 mm s^{-1} to 15 mm s^{-1} at 1 MPa and the

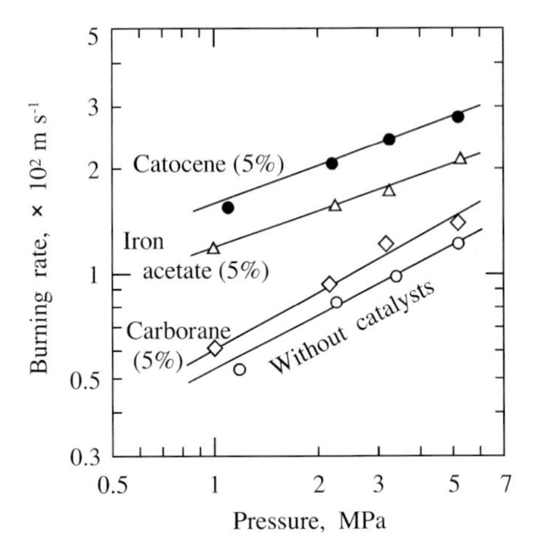

Fig. 11.20 Effect of catalysts on the burning rate of an AP pyrolant.

pressure exponent is decreased from 0.51 to 0.35 by the addition of catocene. Catocene acts most effectively in the low-pressure region and its effect diminishes with increasing pressure. Similarly to catocene, iron acetate is also effective in increasing the burning rate and its effect is most pronounced in the low-pressure region. However, the effect of the addition of 5 % carborane is very small and the pressure exponent remains unchanged. Nevertheless, as shown in Fig. 7–24, the burning rate of AP pyrolants is increased approximately ninefold when 10 % carborane is added. This increased burning rate is due to the oxidation reaction of boron atoms liberated from the carborane.

Fig. 11.21 shows the results of TG and DTA measurements on mixtures of AP particles and catalysts. The endothermic peak observed at 513 K is caused by the crystal structure transformation of AP from orthorhombic to cubic. A two-stage exothermic decomposition occurs in the range 573–720 K. The decomposition of the AP is seen to be drastically accelerated by the addition of catocene. The exothermic peak accompanied by mass loss occurs before the AP crystal transformation. Although the AP is sensitized by the addition of carborane, no effect is seen on the AP decomposition. The results indicate that carborane acts as a fuel component in the gas phase but does not catalyze the decomposition of AP. Thus, the critical friction energy is lowered due to the increased reaction rate in the gas phase. The results imply that the initiation of ignition by friction is caused by the ignition of the gaseous products of the AP pyrolants.[1]

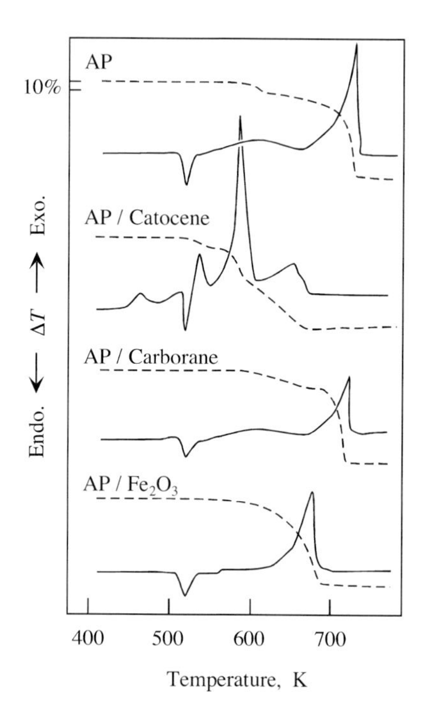

Fig. 11.21 Thermal decomposition processes of mixtures of AP and catalysts.

References

1 Bazaki, H., and Kubota, N., Friction Sensitivity of Ammonium Perchlorate Composite Propellants, *Propellants, Explosives, Pyrotechnics*, Vol. 16, 1991, pp. 43–47.
2 Chapman, D., Sensitiveness of Pyrotechnic Compositions, Pyrotechnic Chemistry, Journal of Pyrotechnics, Inc., Colorado, 2004, Chapter 17.
3 Ilyin, A., Gromov, A., An, V., Faubert, F., de Izarra, C., Espagnacq, A., and Brunet, L., Characterization of Aluminum Powders: I. Parameters of Reactivity of Aluminum Powders, *Propellants, Explosives, Pyrotechnics*, Vol. 27, 2002, pp. 361–364.
4 Fedotova, T. D., Glotov, O. G., and Zarko, V. E., Chemical Analysis of Aluminum as a Propellant Ingredient and Determination of Aluminum and Aluminum Nitride in Condensed Combustion Products, *Propellants, Explosives, Pyrotechnics*, Vol. 25, 2000, pp. 325–332.
5 Takizuka, M., Onda, T., Kuwahara, T., and Kubota, N., "Thermal Decomposition Characteristics of Ti/C/Tf Pyrolants", AIAA Paper 98–3826, 1998.
6 Kubota, N., and Serizawa, C., Combustion of Magnesium/Polytetrafluoroethylene, *J. of Propulsion and Power*, Vol. 3, No. 4, pp. 303–307, 1987.
7 Kubota, N., and Serizawa, C., Combustion Process of Mg/Tf Pyrotechnics, *Propellants, Explosives, Pyrotechnics*, 12, 145–148, 1987.
8 Peretz, A., "Investigation of Pyrotechnic MTV Compositions for Rocket Motor Igniters", AIAA Paper 82–1189, 1982.
9 Kuwahara, T., Matsuo, S., and Shinozaki, N., Combustion and Sensitivity Characteristics of Mg/Tf Pyrolants, *Propellants, Explosives, Pyrotechnics*, 22, 198–202, 1997.
10 Kuwahara, T., and Ochiai, T., "Burning Rate of Mg/Tf Pyrolants", 18th International Pyrotechnics Seminar, Colorado, 539–549, 1992.
11 Koch, E.-C., Metal-Fluorocarbon Pyrolants III: Development and Application of Magnesium/Teflon/Viton (MTV), *Propellants, Explosives, Pyrotechnics*, Vol. 27, 2002, pp. 262–266.
12 Koch, E.-C., Metal-Fluorocarbon Pyrolants IV: Thermochemical and Combustion Behavior of Magnesium/Teflon/Viton (MTV), *Propellants, Explosives, Pyrotechnics*, Vol. 27, 2002, pp. 340–351.
13 Engelen, K., Lefebvre, M. H., and Hubin, A., Properties of a Gas-Generating Composition Related to the Particle Size of the Oxidizer, *Propellants, Explosives, Pyrotechnics*, Vol. 27, 2002, pp. 290–299.
14 Kubota, N., and Aoki, I., "Characterization of Heat Release Process of Energetic Materials", 30th International Annual Conference of the ICT, Karlsruhe, 1999.
15 Kubota, N., and Sonobe, T., Combustion Mechanism of Azide Polymer, *Propellants, Explosives, Pyrotechnics*, Vol. 13, 1988, pp. 172–177.
16 Kubota, N., Sonobe, T., Yamamoto, A., and Shimizu, H., Burning Rate Characteristics of GAP Propellants, *J. of Propulsion and Power*, Vol. 6, No. 6, 1990, pp. 686–689.
17 Kuwahara, T., and Kubota, N., Role of Boron in Burning Rate Augmentation of AP Composite Propellants, *Propellants, Explosives, Pyrotechnics*, Vol. 14, 1989, pp. 43–46.

12
Emission from Combustion Products

12.1
Fundamentals of Light Emission

12.1.1
Nature of Light Emission

When a solid material is heated, the color changes from its original color to red, then to yellow, and finally to white, as seen for heated charcoal. The color of metals such as iron, copper, and aluminum changes in a similar way. However, when the metals are gasified and metal atoms are formed, the color emitted from the atoms is different from that of the original metals. The light emitted by hot materials is electromagnetic radiation. Since radiation is emitted in the form of waves, the radiation energy is dependent not only on the number of emitters but also on the wavelength of the radiation. Radiation from combustion products is divided into several zones according to wavelength, ranging from ultraviolet to infrared, as shown in Table 12.1.

Table 12.1 Zones of visible light and their wavelength ranges.

wavelength (nm)							
ultraviolet	visible						infrared
below 390	390–425 purple	425–445 blue	500–575 green	575–585 yellow	585–620 orange	620–750 red	above 750

Only the visible wavelength zone can be discerned by the human eye, the sensitivity of which is wavelength-dependent. The color of visible light is sensed according to its wavelength; the shortest visible wavelength is in the purple zone at 390 nm and the longest visible wavelength is in the red zone at 750 nm. The sensitivity of the human eye is not proportional to the emission intensity emitted by a body. The human eye is most sensitive in the yellow zone at about 580 nm and decreases at both higher and lower wavelengths.

When the temperature of atoms or molecules is raised, they are promoted to excited states.[1–3] Radiation energy is then emitted when they fall back to their normal

Propellants and Explosives. Naminosuke Kubota

ISBN: 978-3-527-31424-9

energy states. The wavelength of this radiation is characteristic of each individual excited atom or molecule. A line spectrum is emitted by an excited atom and a band spectrum is emitted by an excited molecule.

The radiation emitted by solid particles is a continuous emission consisting of all wavelengths. For example, bright light is emitted when aluminum, magnesium, or zirconium particles burn in air. These metal particles are melted and gasified to produce excited metal atoms, radicals, ions, and/or oxide particles. During these processes, various types of spectra, such as line spectra, band spectra, and a continuous spectrum, are emitted. The processes of recombination of radicals or ionsalso emit continuous-wavelength radiation.

12.1.2
Black-Body Radiation

Black-body radiation is the radiation emitted by a black-colored solid material, a so-called black body, that absorbs and also emits radiation of all wavelengths. A black body emits a continuous spectrum of radiation, the intensity of which is dependent on its wavelength and on the temperature of the black body. Though a black body is an idealized system, a real solid body that absorbs and emits radiation of all wavelengths is similar to a black body. The radiation intensity of a black body, $I_{\lambda B}$, at wavelength λ is given by Planck's radiation law:[1]

$$I_{\lambda B} = 2Ac_1\lambda^{-5}\exp(-c_2/\lambda T)d\lambda \tag{12.1}$$

where $c_1 = 5.88 \times 10^{-11}$ W mm^{-2}, $c_2 = 14.38$ mm K, λ is in mm, $I_{\lambda B}$ is in W per unit solid angle normal to the surface, T is temperature, and A is the surface area of the black body in mm^2. A real body absorbs and emits a lower intensity compared with a black body, and the radiation intensity of a real body at the wavelength λ is expressed by:

$$I_\lambda = \varepsilon_\lambda I_{\lambda B} \tag{12.2}$$

where I_λ is the radiation intensity of a real body at wavelength λ and ε_λ is the emissivity of the real body at wavelength λ. Since the radiation intensity of a real body is always less than that of a black body, the emissivity at any wavelength is always less than 1.

The radiation intensity of a black body represented by Eq. (12.1) is increased with increasing temperature throughout the wavelength range. The maximum intensity appears at a certain wavelength, which shifts as the temperature is changed. The wavelength of maximum intensity is expressed by Wien's displacement law according to:

$$\lambda_m T = 2.58 \text{ mm K} \tag{12.4}$$

where λ_m is the wavelength of maximum energy density and T is the temperature of the black-body emitter. The total radiation energy emitted from unit surface area of a black body, W, is expressed by the Stefan–Boltzmann law as:[1]

$$W = \sigma T^4 \qquad (12.5)$$

where σ is the Stefan–Boltzmann constant given by 5.67×10^{-8} W m^{-2} K^{-4}.

12.1.3 Emission and Absorption by Gases

Light emission from high-temperature gases ranges from the ultraviolet, through the visible, and into the infrared. The light is recognized either directly by the human eye or with the aid of optical or electromagnetic instruments. Though visible light has only a narrow wavelength range from 0.35 μm to 0.75 μm, ultraviolet light ranges from 1 nm to 0.35 μm and infrared light ranges from 0.75 μm to 1 mm. Gaseous atoms and molecules emit light at their own characteristic wavelengths as line spectra and band spectra. For example, sodium atoms emit yellow light of wavelengths 588.979 nm and 589.593 nm in a line spectrum, the so-called D-line doublet. These emissions occur as a result of a ^2S-^2P electron orbit transition.[1]

Infrared light is divided into three zones: the near-infrared, the mid-infrared, and the far-infrared, as shown in Table 12.2. The wavelength of the near-infrared is below 2.5 μm, that of the mid-infrared ranges from 2.5 μm to 25 μm, and that of the far-infrared is above 25 μm. Infrared emission between 3 μm and 30 μm is caused by vibrational modes of the molecules, while that above 30 μm is caused by rotational modes.

Table 12.2 Zones of infrared radiation and their associated wavelengths.

Wavelength	0.35 μm	0.75 μm	3 μm	5 μm	14 μm
		Visible		Infrared	
			near-infrared	mid-infrared	far-infrared

Gaseous molecules in the atmosphere absorb radiation of the infrared, visible, and ultraviolet ranges. Water vapor, H_2O, in the atmosphere absorbs through vibrational modes at wavelengths of 1.8, 2.7, and 6.3 μm, and through a rotational mode in the range from 10 μm to 100 μm. CO absorbs at wavelengths of 2.3 and 4.5 μm due to its vibrational modes, while CO_2 absorbs at wavelengths of 2.8, 4.3, and 15 μm, also through vibrational modes.[1]

The absorption of light by the atmosphere is wavelength-dependent. There are numerous condensed materials in the atmosphere, such as water mists, sands, fine organic materials, and so on. These fine particles absorb or reflect light, and the transmitted energy is affected by the physicochemical properties of the particles.

The major absorbers of infrared radiation in the atmosphere are H_2O, CO_2, and O_3. The infrared wavelength ranges of 2–2.5 μm, 3–5 μm, and 8–12 μm are known as atmospheric windows. The radiative energies transmitted from the Sun to the Earth's surface within these atmospheric windows are shown in Table 12.3.

Table 12.3 Radiative energy to the Earth's surface.

Wavelength range, μm	Radiative energy W m^{-2}
1.8–2.5	24
3.0–4.3	10
4.4–5.2	1.6
7.6–13	1.0

Infrared emissions are dependent on the nature of the emissive body and its temperature. For example, a jetfighter emits strongly in the infrared around the wavelength ranges of 2 μm and 3–5 μm from the exhaust nozzle, 4–5 μm from the exhaust gases, and 3–5 μm and 8–12 μm from the flight body. Infrared sensors used in the aforementioned atmospheric windows are semiconductors composed of the chemical compounds PbS for 2.5 μm, InSb for 3–5 μm, and HgCdTe for 3–5 μm and 10 μm.

12.2 Light Emission from Flames

12.2.1 Emission from Gaseous Flames

Gaseous high-temperature flames emit both visible and ultraviolet light due to energy level transitions of the electrons in the molecules.[1–3] Such transitions are observed as band spectra as a result of changes in the vibrational and rotational energies of the molecules. In general, hydrocarbon-air flames produce CO_2, H_2O, CO, N_2, O_2, radicals, and ions. These molecules, radicals, and ions each emit radiation according to their individual spectra. The hydroxyl radical, OH, in flames emits radiation as a band spectrum commencing at 306.4 nm in the ultraviolet region. CH, HCO, and NH radicals in flames emit radiation in the visible and near-ultraviolet regions. However, no gaseous flames give appreciable emission or absorption of radiation in both the visible and ultraviolet regions unless they contain OH radicals.[1]

12.2.2
Continuous Emission from Hot Particles

When metallized pyrolants are burned, metal particles and metal oxide particles are produced in the flames. The combustion products of these pyrolants emit continuous-wavelength radiation owing to the presence of hot particles and a multitude of atoms and free electrons. Not only the emission energy but also the wavelength of the light changes with the temperature of the particles.[3–10] The emissivity and absorbtivity of the particles are also dependent on temperature. The radiative energy from hot particles is given by Eq. (12.2).

12.2.3
Colored Light Emitters

When small amounts of alkali metals, alkaline earth metals, or their salts are introduced into a gaseous flame, flame reactions occur relatively easily at low temperatures.[3–5] The liberated metal atoms are promoted to excited states and then return to their normal states. Radiation corresponding to the characteristic line spectra of the individual metal atoms is emitted as a result of this energy transition. Colored radiation discernible by the human eye, ranging from red to blue, is dependent on the type of metal atoms, as shown in Table 12.4.[3–6]

Table 12.4 Colored emissions from metals.

Metal	Color
Li	red
Rb	dark-red
Sr	crimson
Ca	orange-red
Na	yellow
Tl	yellow-green
Ba	green-yellow
Mo	green-yellow
Cu	blue-green
Ga	blue
As	light-blue
Sb	light-blue
Sn	light-blue
Pb	light-blue
In	indigo-blue
Cs	green-purple
K	lilac
Mg	white
Al	white

In general, color emitters used as components of pyrolants are metallic compounds rather than metal particles. Metal particles agglomerate to form liquid metal droplets and liberation of metal atoms in flames occurs only at the surface of the droplets. On the other hand, metallic compounds decompose at relatively low temperatures compared with metal particles and liberate dispersed metal atoms. Table 12.5 shows typical salts used to obtain emissions of the requisite colors.

Table 12.5 Chemical compounds used as color emitters.

a. Red Emitters
- Strontium compounds
  - $Sr(NO_3)_2$, Sr_2O_4, $SrCO_3$
- Calcium compounds
  - $CaCO_3$, $CaSO_4$

b. Yellow Emitters
- Sodium compounds
  - $Na_2C_2O_4$, Na_2CO_3, $NaHCO_3$, $NaCl$

c. Green Emitters
- Barium compounds
  - $Ba(NO_3)_2$, $BaCO_3$, BaC_2O_4

d. Blue Emitters
- Copper compounds
  - $CuHAsO_3$, $CuSO_4$, $CuCO_3 \cdot Cu(OH)_2$

12.3 Smoke Emission

12.3.1 Physical Smoke and Chemical Smoke

One of the major applications of pyrolants is to produce smoke clouds or smoke curtains by chemical reactions.[3–6] Smoke is defined as condensed particles that can remain in the atmosphere for at least several seconds. The radiative emission from smoke itself is small because of its low-temperature nature. Thus, no visible emission is seen in the dark by the human eye. The applications of smoke are:
a. As a color display for various amusements, including fireworks in the daytime.
b. Concealment of arms such as tanks and vehicles at the front line.
c. To create decoys for arms.

In general, the chemicals used to create color displays in the daytime are various types of dyes and oils. Though dyes and oils do not fall into the category of pyrolants that generate colored smoke by combustion reactions, they are dispersed in the atmosphere by the combustion or decomposition gases of pyrolants. Typical examples of color dyes are indigo for blue, rhodamine for red, and auramine for yellow. Dyes are mixed with potassium chlorate ($KClO_3$) and sulfur (S) or sucrose to obtain low-temperature gases that are used to disperse the dyes and to generate small airborne particles ranging in size from several microns to several hundred microns. The combustion reaction with S occurs according to:

$$2KClO_3 + 3S \rightarrow 2KCl + 3SO_2$$

The temperature of the gases produced by the reaction between $KClO_3$ and S is low enough to prevent burning in air. A mass fraction of $NaHCO_3$ of approximately 0.2 is also incorporated as a coolant that prevents a flame reaction between the dyes and air.

12.3.2
White Smoke Emitters

A flow stream produced from boiling water appears white in color. Similar to cloud in the sky, condensed water vapor shows a white color in the atmosphere. Humid air leads to condensation when nucleating materials are present in the atmosphere, producing a white-colored fog. However, condensed water vapor and fog appear as black smoke when the background is brighter than the foreground.

When phosphorus is burned with oxygen, phosphorus pentoxide (P_4O_{10}) is formed. P_4O_{10} immediately absorbs humidity from the air to form phosphoric acid, $OP(OH)_3$, which gives rise to a white fog or smoke. A mixture of C_2Cl_6, Zn, and ZnO reacts to produce zinc chloride ($ZnCl_2$) according to:

$$2C_2Cl_6 + 3Zn + 3ZnO \rightarrow 6ZnCl_2 + 3CO + C$$

Finely dispersed $ZnCl_2$ absorbs moisture from humid air and forms a white fog.

A mixture of ammonium perchlorate (AP: NH_4ClO_4) and a hydrocarbon polymer (BDR) used as fuel binder forms an AP pyrolant that generates white smoke when it burns in a humid atmosphere. The polymer acts as a binder of the AP particles to form a rubber-like material. When the AP pyrolant burns, the AP particles oxidize the hydrocarbon polymer according to:

$$NH_4ClO_4 + \text{hydrocarbon} \rightarrow N_2 + H_2O + CO_2 + HCl$$

Combustion with complete gasification occurs when an AP pyrolant is composed of $\xi_{AP}(0.86)$ and $\xi_{BDR}(0.14)$. The mass fraction of hydrogen chloride (HCl) among the combustion products is about 0.3. It is well known that HCl molecules combine with water vapor in the atmosphere to generate a white smoke. It is for this reason that AP pyrolants act as white smoke generators in a humid atmosphere.

When fine aluminum particles are incorporated into AP pyrolants, aluminum oxide (Al_2O_3) particles are formed when they burn. Dispersal of these aluminum oxide particles in the atmosphere generates white smoke even when the atmosphere is dry. The mass fraction of aluminum particles added is approximately 0.2 for the complete combustion of AP pyrolants. Though an excess of aluminum particles can be burned with atmospheric oxygen, the addition of excess aluminum particles decreases the combustion temperature of the aluminized AP pyrolant. Incomplete combustion of the aluminum particles then results and molten aluminum particles agglomerate without oxidation in the atmosphere.

12.3.3 Black Smoke Emitters

Soot is formed from carbon-containing materials through incomplete combustion. When an oil in a container burns in the atmosphere, soot is formed, which consists of fine carbon particles. The size of the particles is dependent on the chemical structure and the thermal decomposition process of the oil. Solid hydrocarbon materials produce soot when thermally decomposed. When anthracene ($C_{14}H_{10}$) or naphthalene ($C_{10}H_8$) burns with $KClO_4$ under fuel-rich conditions, black sooty smoke is formed. The soot is easily burned when sufficient air and ignition energy are supplied to the mixture. A mixture of hexachloroethane (C_2Cl_6) and magnesium particles reacts to produce magnesium chloride ($MgCl_2$) and solid carbon ($C_{(s)}$) according to:

$$3Mg + C_2Cl_6 \rightarrow 3MgCl_2 + 2C_{(s)}$$

The solid carbon formed appears as soot, which is dispersed as a black smoke in the atmosphere without its combustion. Similar reactions of mixtures of Al-ZnO-C_2Cl_6, Al-TiO_2-C_2Cl_6, and Mn-ZnO-C_2Cl_6 generate black smoke as follows:

$$2Al + 3ZnO + C_2Cl_6 \rightarrow Al_2O_3 + 3ZnCl_2 + 2C_{(s)} \qquad +2.71\ kJ\ kg^{-1}$$

$$4Al + 3TiO_2 + 2C_2Cl_6 \rightarrow 3TiCl_4 + 2Al_2O_3 + 4C_{(s)} \qquad +2.51\ kJ\ kg^{-1}$$

$$3Mn + 3ZnO + C_2Cl_6 \rightarrow 3ZnCl_2 + 3MnO + 2C_{(s)} \qquad +1.53\ kJ\ kg^{-1}$$

12.4 Smokeless Pyrolants

12.4.1 Nitropolymer Pyrolants

Physical mixtures of nitropolymers are known as smokeless propellants, as used in rockets and guns. The -O–NO_2 chemical bonds contained within the nitropolymersact as oxidizer components and the remaining hydrocarbon structures act as fuel components. The major combustion products are CO_2, H_2O, N_2, and CO, and additional small amounts of radicals such as ·OH, ·H, and ·CH are also formed. These products are fundamentally smokeless without solid particles.

Several types of nitrate esters are used to formulate smokeless nitropolymers. Nitrocellulose (NC) is mixed with nitroglycerin (NG), trimethylolethane trinitrate(TMETN), triethyleneglycol dinitrate (TEGDN), or diethyleneglycol dinitrate(DEGDN) to formulate colloidal mixtures that are used as smokeless nitropolymer pyrolants. Their burning rate characteristics are changed by the addition of small amounts of lead compounds that act as burning rate catalysts to increase the burning rate and to reduce the pressure exponent, giving rise to the effects known as super-rate burning and plateau burning. Nitropolymer pyrolants are used as smokeless rocket propellants and gas generators.

Though nitropolymer pyrolants are known as smokeless pyrolants, a large amount of black smoke is generated when nitropolymer pyrolants burn at low pressures below about 3 MPa due to incomplete combustion. For example, double-base propellants composed of nitrocellulose and nitroglycerin can no longer be called "smokeless propellants" under low-pressure burning conditions. The combustion wave of a nitropolymer pyrolant consists of successive reaction zones: surface reaction zone, fizz zone, dark zone, and flame zone. A two-stage temperature profile is thus seen for the gas phase. In the fizz zone, heat is produced by the reduction of NO_2, which is generated by the decomposition reaction at the burning surface. The NO produced in the fizz zone by the reduction of NO_2 reacts relatively slowly in the dark zone. In the flame zone, NO is reduced to N_2 and the temperature of the pyrolant system reaches its maximum value.

Since the reaction involving NO is termolecular in nature, the oxidation of gaseous fuel species by NO is very slow at low pressures and the reaction only occurs when the temperature is well above 1000 K. It is for this reason that flameless burning of nitropolymer pyrolants occurs in the low-pressure region.

It has been reported that the flameless reaction involving NO at low pressures is accelerated significantly by the addition of a small amount of metallic nickel or an organonickel compound. The nickel acts as a catalyst to promote the dark-zone reaction of nitropolymer pyrolants. The flameless burning of nitropolymer pyrolants becomes luminous-flame burning on the addition of the catalyst. The gas-phase temperature of the flameless burning (about 1300 K) is increased drastically to the temperature of the flame burning (higher than 2500 K) at 0.1 MPa.

12.4.2
Ammonium Nitrate Pyrolants

Ammonium nitrate (AN) is a crystalline oxidizer that produces an oxidizer-rich gas when thermally decomposed according to:

$$NH_4NO_3 \rightarrow N_2 + 2H_2O + 1/2O_2$$

When AN powder is mixed with a polymeric material, the oxygen gas produced by the decomposition of the AN powder reacts with the hydrocarbon fragments of the thermally decomposed polymeric material. The major combustion products are CO_2 and H_2O. Nitropolymers are not used as fuel components of AN pyrolants because of the reaction between the NO_2 formed by their decomposition and the AN powder. This reaction occurs very slowly and damages the physical structure of the AN pyrolant. Instead, polymeric materials containing relatively high mass fractions of oxygen in their structures are used. Hydroxy-terminated polyether (HTPE) is commonly used to formulate AN pyrolants. Hydroxy-terminated polybutadiene(HTPB) is not favorable for use as a fuel component of AN pyrolants because of its low mass fraction of oxygen. When HTPB is used, the AN pyrolant burns incompletely and forms a large amount of carbonaceous fragments on and above the burning surface.

The burning rates of AN pyrolants are of the order of 1–3 mm s^{-1} at 10 MPa, which is very slow compared with AP pyrolants. Decomposition catalysts for AN powder, such as chromium trioxide (Cr_2O_3) and ammonium dichromate($(NH_4)_2Cr_2O_7$), need to be incorporated into AN pyrolants to aid their successive burning. Very unstable burning occurs without these catalysts.

Table 12.6 shows the physicochemical properties and the combustion characteristics of an AN pyrolant composed of $\xi_{AN}(0.83)$ and $\xi_{HTPB}(0.15)$ with ammonium dichromate as a burning rate catalyst. The burning rate of the AN pyrolant is 0.8 mm s^{-1} at 1 MPa and 2.0 mm s^{-1} at 10 MPa. The pressure exponent of burning rate, as defined in Eq. (3.68), is 0.4 in the pressure range 1–10 MPa.

Table 12.6 Physicochemical properties of an AN pyrolant (mass fraction).

Oxidizer	Fuel	Catalyst
AN	HTPE	$(NH_4)_2Cr_2O_7$
0.83	0.15	0.02
Density		1570 kg m^{-3}
Combustion temperature		1500 K
Pressure exponent		0.4

Since the combustion temperature of AN pyrolants is very low compared with other composite pyrolants, their specific impulse when used as rocket propellants is also low. However, they are used as gas generators for the control of various types of mechanics owing to the low temperature and low burning rate characteristics.

12.5 Smoke Characteristics of Pyrolants

The smoke characteristics of three types of pyrolants, namely nitropolymer pyrolants composed of NC-NG with and without a nickel catalyst, and a B-KNO_3 pyrolant, have been examined in relation to the use of these pyrolants as igniters of rocket motors. Though nitropolymer pyrolants are fundamentally smokeless in nature, a large amount of black smoke is formed when they burn at low pressures below about 4 MPa due to incomplete combustion. Metallic nickel or organonickel compounds are known to catalyze the gas-phase reaction of nitropolymer pyrolants, denoted here by NP. A comparative study on the smoke generation by NP with and without a nickel catalyst has been conducted to determine the effectiveness of smoke reduction. An igniter composed of B-KNO_3, denoted here by BK, was also tested as a reference smoky igniter. The chemical compositions of the pyrolants examined are shown in Table 12.7.

Table 12.7 Chemical formulations of the pyrolants used to evaluate smoke characteristics (mass fraction).

Pyrolant	NC	NG	DEP	Ni	B	KNO_3	HTPS
NP	0.518	0.385	0.097	–	–	–	–
NP-Ni	0.513	0.381	0.096	0.01	–	–	–
BK	–	–	–	–	0.236	0.708	0.056

NC: nitrocellulose
NG: nitroglycerin
DEP: diethyl phthalate
B: boron
KNO_3: potassium nitrate
HTPS: hydroxy-terminated polyester
Ni: nickel (particle size: 0.1 μm in diameter)

The mass fraction of nickel powder incorporated into the NP pyrolant was 0.01 and the diameter of the nickel particles was 0.1 μm. The NP pyrolants with and without nickel particles were pressed into pellet-shaped grains 1 mm in diameter and 1 mm in length. The BK pyrolant was pressed into pellet-shaped grains 3 mm in diameter and 3 mm in length.

Fig. 12.1 shows theoretical computational results for the adiabatic flame temperature, T_f, and the specific gas volume, v_c, of the combustion products of NP and BK pyrolants. Both T_f and v_c are important parameters for increasing the heat flux supplied to the propellant ignition surface and for building up a rocket motor-

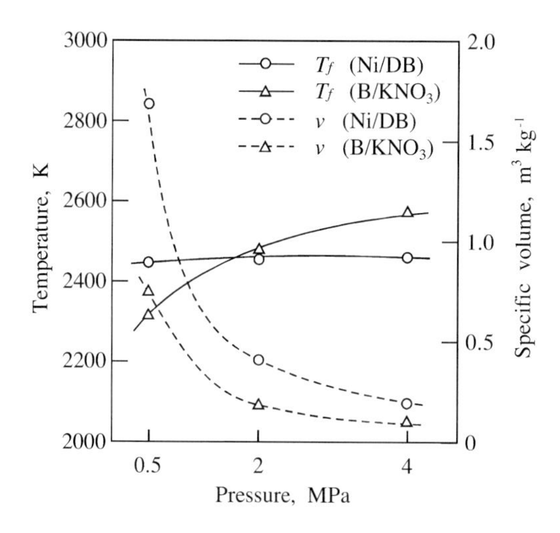

Fig. 12.1 Adiabatic flame temperature and specific gas volume of NP and BK pyrolants.

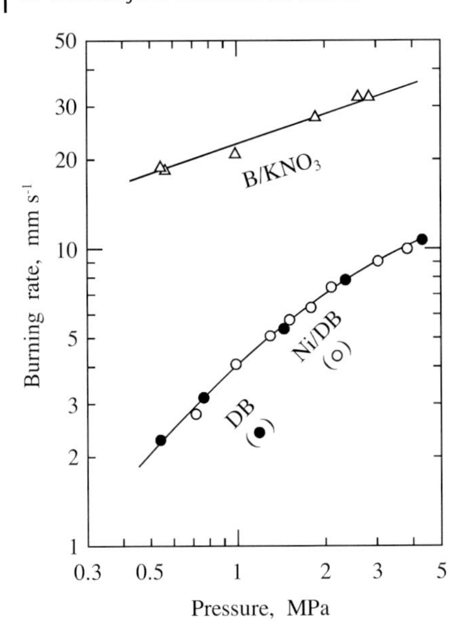

Fig. 12.2 Burning rate characteristics of BK and NP pyrolants, showing that the burning rate of the NP pyrolant remains unchanged by the addition of nickel particles.

pressure. It is shown that T_f of NP is higher than that of BK at low pressures below 2 MPa and that v_c of NP is greater than that of BK. The theoretical computational results for T_f and v_c of the NP-Ni pyrolant are approximately equivalent to those of the NP pyrolant without nickel particles.

The burning rate of pyrolants is also an important parameter for the design of igniter grains. The linear burning rates of the three types of pyrolant grains are shown in Fig. 12.2. The burning rate increases with increasing pressure for the three types of pyrolant. The burning rate of BK is approximately ten times higher at 0.4 MPa and four times higher at 4 MPa than that of the NP pyrolants with or without nickel particles. It is clearly evident that the burning rate of the NP pyrolant remains unchanged by the addition of nickel particles. The pressure exponent, n, defined according to n = ln (burning rate)/ln (pressure), is 0.33 for BK between 0.5 MPa and 3 MPa. The pressure exponent for NP and NP-Ni decreases with increasing pressure from 0.87 at 1 MPa to 0.45 at 4 MPa.

The heats of explosion, H_{exp}, of the pyrolants are shown as a function of pressure in Fig. 12.3. It is evident that H_{exp} of the NP pyrolant is increased by the addition of nickel particles in the low-pressure region below about 2 MPa. The measured H_{exp} of the NP-Ni pyrolant becomes less pressure-dependent and reaches approximately 97 % of the theoretical value. The results indicate that the lower value of H_{exp} of the NP pyrolant is caused by incomplete combustion at about 4 MPa and that the increased H_{exp} of the NP-Ni pyrolant is caused by a catalytic effect on the gas-phase reaction which increases the temperature. The gas-phase reduction of NO to N_2 in the dark zone of the NP pyrolant in the low-pressure region is promoted by the addition of the nickel particles.

The smoke characteristics and the ignition capabilities of igniters are evaluated by using a micro-rocket motor. Ignition delay time and rate of pressure rise are

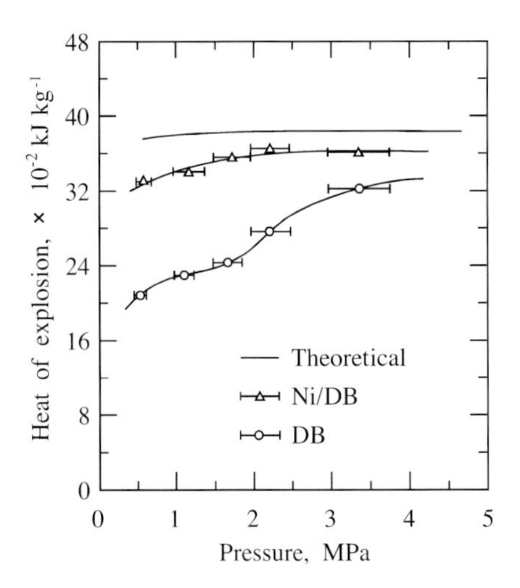

Fig. 12.3 Heat of explosion of the NP pyrolant is increased significantly in the low-pressure region by the addition of nickel particles.

measured by means of a pressure transducer attached to the combustion chamber. As shown in Fig. 12.4, a visible light source (halogen-tungsten lamp) and a carbon dioxide laser source are positioned so as to intercept the smoke expelled from the exhaust nozzle of the rocket motor. A light detector (phototransistor) and a laser light detector (pyrometer) are positioned opposite the visible light and laser sources such that the smoke expelled from the rocket nozzle passes between the light sources and the detectors.[7]

The propellant used for the micro-rocket motor is a conventional double-base propellant composed of NC, NG, and DEP. The propellant is formed into an end-burning-type grain with dimensions of 110 mm in diameter and 20 mm in length. The nozzle throat area is selected as to obtain a steady-state chamber pressure of about 5 MPa. The igniter consists of an initiator squib, a small amount of booster charge powder, and the main igniter charge. The smoke concentration is determined as a function of the mass of the main igniter charge.

When the aforementioned NP pyrolant is used as the main igniter charge, the ignition fails and no burning of the propellant occurs. However, when the NP-Ni pyrolant is used, the NC-NG propellant is ignited and the chamber pressure is raised. The pressure profiles and light attenuation of the BK and NP-Ni igniters without rocket propellant grains are shown in Figs. 12.5 and 12.6, respectively. It is evident that the light attenuation of the NP-Ni igniter is lower than that of the BK igniter. The results indicate that much less smoke is generated by the NP-Ni grain than by the BK grain. The test results of the pressure profiles and light attenuation with propellant grains are shown in Figs. 12.7 and 12.8. No clear difference in the ignition time delays and the pressure build-up processes in the rocket motor is seen when the BK and NP-Ni igniters are used. However, the light attenuation is drastically reduced when the BK igniter is replaced with the NP-Ni igniter. There are no major differences in the attenuation data between visible-light and laser measurements.

Fig. 12.9 shows the results of measurements of light attenuation versus igniter grain mass for the BK and NP-Ni igniters with propellant grains in a rocket motor. Though the attenuation increases with increasing igniter grain mass for both the

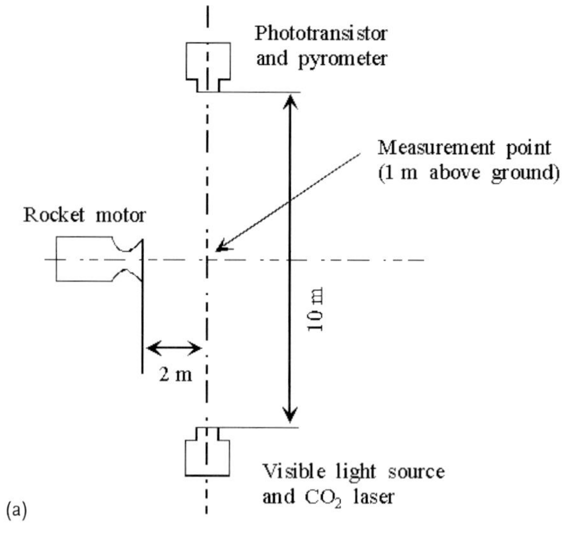

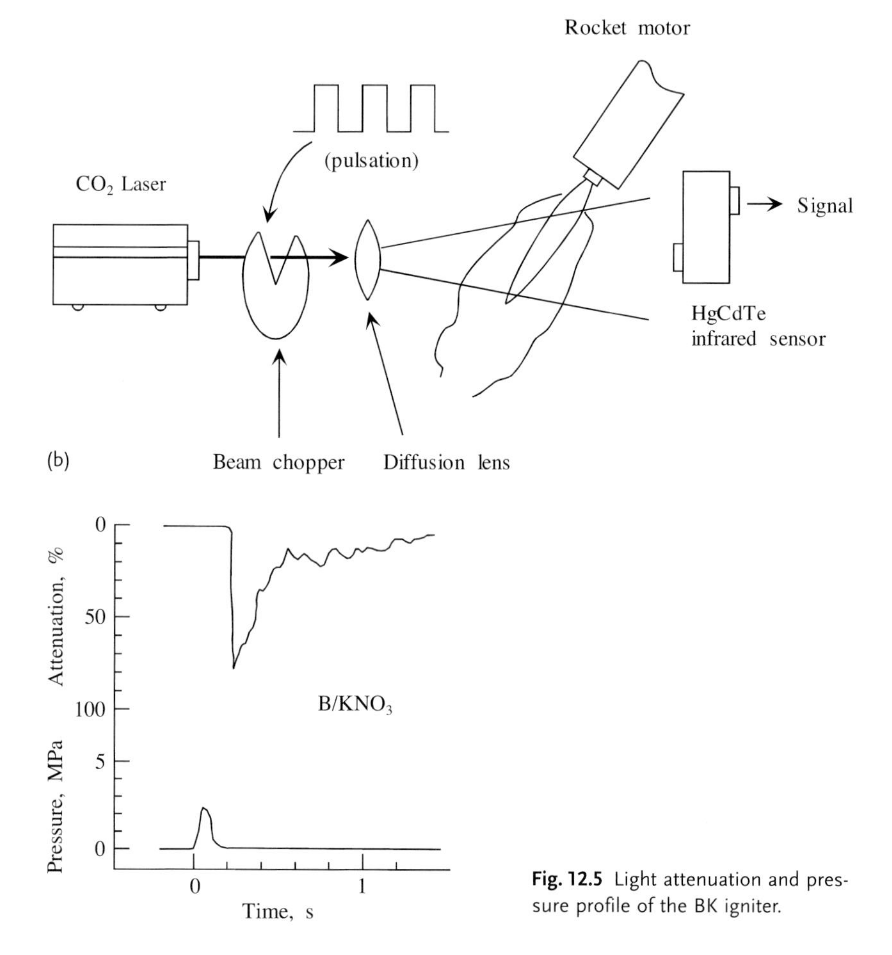

Fig. 12.4 Experimental set-up for smoke and flame measurements.

Fig. 12.5 Light attenuation and pressure profile of the BK igniter.

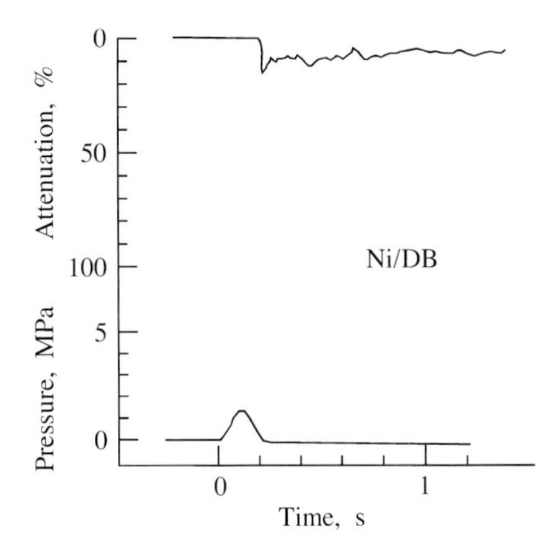

Fig. 12.6 Light attenuation and pressure profile of the NP-Ni igniter.

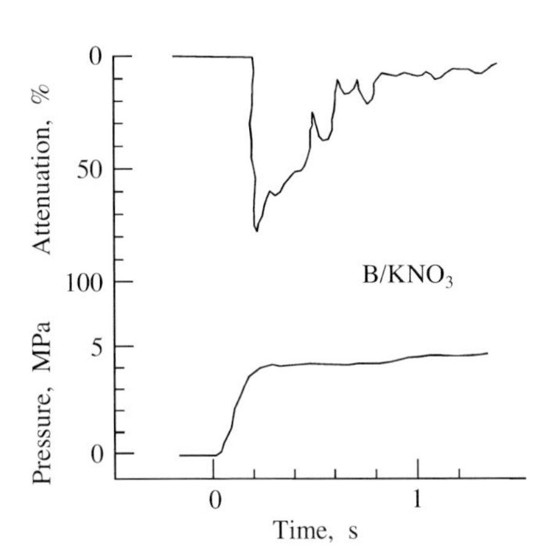

Fig. 12.7 Light attenuation and pressure profile of the ignition process of a micro-rocket motor with the BK igniter.

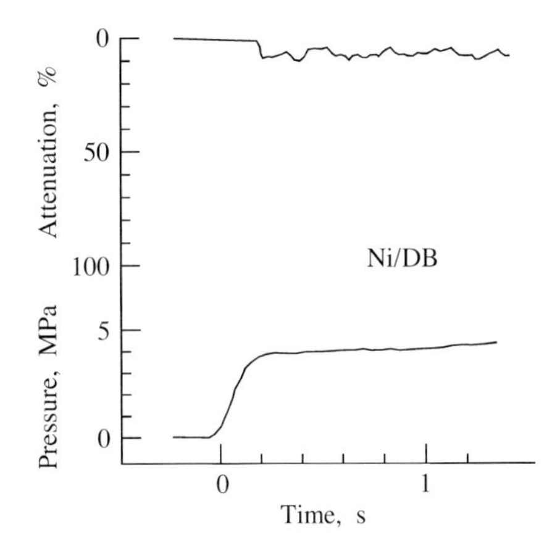

Fig. 12.8 Light attenuation and pressure profile of the ignition process of a micro-rocket motor with the NP-Ni igniter.

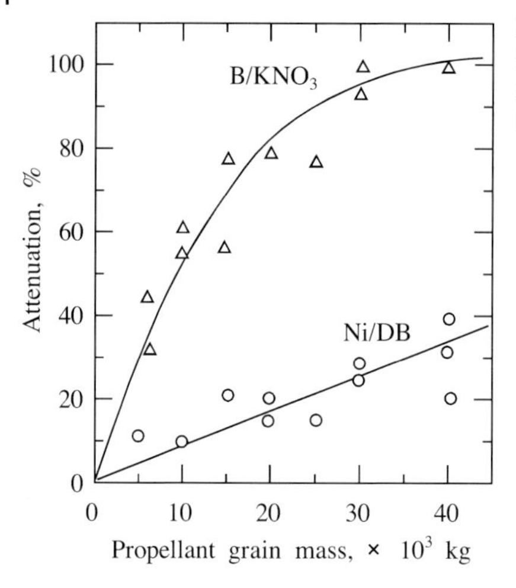

Fig. 12.9 Light attenuation by smoke from the BK and NP-Ni igniters, showing that the NP-Ni igniter produces less smoke than the BK igniter.

BK and NP-Ni igniters, the attenuation by the NP-Ni igniter is much lower than that by the BK igniter for the same mass of igniter grain.

12.6 Smoke and Flame Characteristics of Rocket Motors

12.6.1 Smokeless and Reduced Smoke

Since NC is a fuel-rich nitrate ester, a nitropolymer propellant with a high NC content generates black smoke as a combustion product. In addition, the combustion of nitropolymer propellants becomes incomplete at low pressures below about 3 MPa and black smoke composed of solid carbon particles is formed. This incomplete combustion is caused by the slow rates of the reactions of NO with aldehydes and CO in the combustion wave. Thus, the nitropolymer propellants are no longer smokeless propellants under low-pressure burning conditions.

When a nitropolymer propellant is composed of stoichiometrically balanced ingredients, the combustion products expelled from the exhaust nozzle of a rocket form a plume that is smokeless and shows very low visible emission. However, the plume becomes a yellowish flame when the nitropolymer propellant is a fuel-rich mixture. This yellowish flame is caused by afterburning of the carbon particles generated in the combustion chamber. The carbon particles burn with air that diffuses into the plume from the atmosphere. Diffusional mixing of the flowing combustion products expelled from the exhaust nozzle with air forms a diffusional flame. Through this afterburning, combustion is completed with the formation of CO_2, H_2O, and N_2 downstream of the plume. Thus, no smoke trail is seen for a rocket projectile trajectory. Fig. 12.10 (a) shows a typical flight trajectory of a rocket fuelled by a nitropolymer propellant.

(a) (b)

Fig. 12.10 Rocket flight trajectories assisted by (a) an NC-NG double-base propellant and (b) an aluminized AP composite propellant.

When a composite propellant composed of ammonium perchlorate (AP) and a hydrocarbon polymer burns in a rocket motor, HCl, CO_2, H_2O, and N_2 are the major combustion products and small amounts of radicals such as $\cdot$OH, $\cdot$H, and $\cdot$CH are also formed. These products are smokeless in nature and the formation of carbon particles is not seen. The exhaust plume emits weak visible light, but no afterburning occurs because AP composite propellants are stoichiometrically balanced mixtures and, in general, no diffusional flames are generated.

HCl molecules form visible white fog when water vapor is present in the atmosphere. An HCl molecule acts as a nucleus, becoming surrounded by H_2O molecules, which forms a fog droplet large enough to be visible. When the combustion products of an AP composite propellant are expelled from a rocket nozzle into the atmosphere, a white smoke trail is seen as a rocket projectile trajectory whenever the relative humidity of the air is above about 40%. Furthermore, if the temperature of the atmosphere is below 0 °C (below 273 K), the H_2O molecules generated among the combustion products form a white fog with the HCl molecules even if the relative humidity is less than 40%. Thus, the amount of white fog generated by the combustion of an AP composite propellant is dependent not only on the humidity but also the temperature and pressure of the atmosphere.

In general, AP composite propellants contain aluminum (Al) particles as a fuel component to increase their specific impulse. These Al particles react with oxidizer components according to

$$2Al + 3/2O_2 \rightarrow Al_2O_3 \quad \text{and} \quad 2Al + 3H_2O \rightarrow Al_2O_3 + 3H_2$$

and so form aluminum oxide (Al_2O_3). This oxidation process occurs at the surface of each aluminum particle in the liquid and/or solid phase. A large number of the aluminum particles agglomerate to form large Al_2O_3 particles, with diameters of the order of 0.1 mm to 1 mm. When these particles are dispersed from a nozzle into the atmosphere, a dense white smoke is formed as a trail of the rocket projectile trajectory. In general, the mass fraction of aluminum particles contained within high-energy AP composite propellants ranges from 0.10 to 0.18, and the majority of their white smoke is caused by the dispersal of aluminum oxide particles in the atmosphere. Fig. 12.10 (b) shows a typical smoke trail of a rocket projectile fuelled by an aluminized AP composite propellant.

AP composite propellants without aluminum particles are termed "reduced-smoke propellants" and are employed in tactical missiles to conceal their launch site and flight trajectory. No visible smoke is formed when the relative humidity of the atmosphere is less than about 40%. However, since high-frequency combustion oscillation tends to occur in the combustion chamber in the absence of solid particles that serve to absorb the oscillatory energy, a mass fraction of 0.01–0.05 of metallic particles is still required for the reduced-smoke propellants. These particles and/or their oxide particles generate thin smoke trails. The white smoke trail includes the white fog generated by the HCl molecules and the condensed water vapor of the humid atmosphere.

12.6.2
Suppression of Rocket Plume

Afterburning occurs when the combustion products of propellants are expelled from an exhaust nozzle into the atmosphere.[11–13] Since the combustion products emanating from the nozzle are fuel-rich, containing CO, H_2, $C_{(s)}$, and/or molten metal particles, these gases and particles are spontaneously ignited and combusted upon mixing with ambient air and form an afterburning flame, i. e., a plume. Various problems are encountered as a result of an afterburning flame, such as its infrared emission, luminosity, and ionization of molecules present at the periphery of the flame. The afterburning is dominated by the energetics of propellants and by the aerodynamic mixing process in the atmosphere.[11]

A rocket nozzle is composed of a convergent-divergent nozzle. The flow velocity in the combustion chamber increases from a subsonic flow in the convergent part to the sonic speed at the throat and to a supersonic flow in the divergent part. The flow velocity reaches its maximum at the nozzle exit. On the other hand, both the temperature and pressure in the combustion chamber decrease successively along the nozzle flow direction and reach their minimum values at the nozzle exit. The temperature of the combustion products in the nozzle varies as the expansion area ratio of the nozzle, ε, is varied, where $\varepsilon = A_e/A_t$; A_e is the nozzle exit area and A_t is the nozzle throat area. Two types of nozzle flows are assumed to determine the chemical compositions of the combustion products along the flow direction: equilibrium flow and frozen flow. In the case of equilibrium flow, the chemical composition of the combustion gas in the nozzle varies with changing pressure and temperature. In the case of frozen flow, no chemical reaction occurs in the nozzle

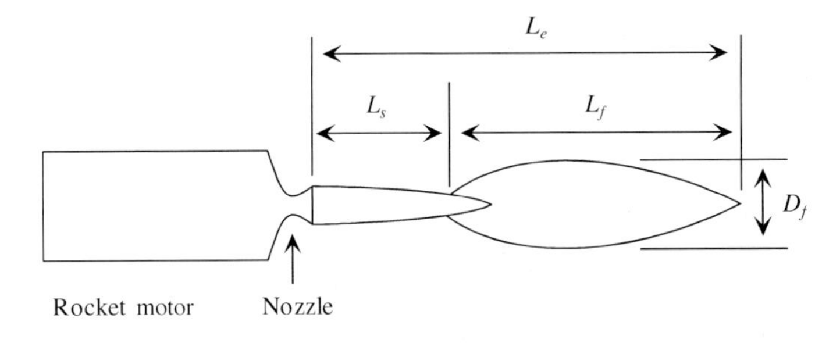

L_s : Primary flame L_f : Secondary flame

Fig. 12.11 Definition of the primary flame and the secondary flame of a rocket plume.

and so the chemical composition remains unchanged throughout the nozzle flow and the temperature of the combustion products decreases in an adiabatic expansion flow.

Fig. 12.11 shows the structure of a rocket plume generated downstream of a rocket nozzle. The plume consists of a primary flame and a secondary flame.[11] The primary flame is generated by the exhaust combustion gas from the rocket motor without any effect of the ambient atmosphere. The primary flame is composed of oblique shock waves and expansion waves as a result of interaction with the ambient pressure. The structure is dependent on the expansion ratio of the nozzle, as described in Appendix C. Therefore, no diffusional mixing with ambient air occurs in the primary flame. The secondary flame is generated by mixing of the exhaust gas from the nozzle with the ambient air. The dimensions of the secondary flame are dependent not only on the combustion gas expelled from the exhaust nozzle, but also on the expansion ratio of the nozzle. A nitropolymer propellant composed of $\xi_{NC}(0.466)$, $\xi_{NG}(0.369)$, $\xi_{DEP}(0.104)$, $\xi_{EC}(0.029)$, and $\xi_{PbSt}(0.032)$ is used as a reference propellant to determine the effect of plume suppression. The burning rate characteristics of the propellants are shown in Fig. 6–31. Since the nitropolymer propellant is fuel-rich, the exhaust gas forms a combustible gaseous mixture with the ambient air. This gaseous mixture is ignited and afterburning occurs somewhat downstream of the nozzle exit. The major combustion products in the combustion chamber are CO, H_2, CO_2, N_2, and H_2O. The fuel components are CO and H_2, the mole fractions of which at the nozzle throat are $\xi_{CO}(0.47)$ and $\xi_{H2}(0.24)$.

12.6.2.1 Effect of Chemical Reaction Suppression

Potassium salts are known to act as suppressants of spontaneous ignition of hydrocarbon flames arising from interdiffusion with ambient air. It has been reported that potassium salts act to retard the chemical reaction in the flames of nitropolymer propellants. Two types of potassium salts used as plume suppressants are potassium nitrate (KNO_3) and potassium sulfate (K_2SO_4). The concentration of the salts is varied to determine their region of effectiveness as plume suppressants.

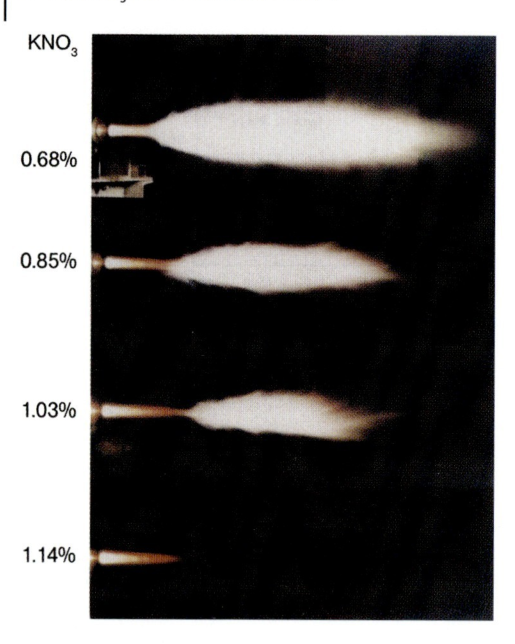

Fig. 12.12 Flame photographs of rocket plumes, showing that the dimensions of the secondary flame decrease with increasing concentrations of KNO_3.

Fig. 12.12 shows a typical set of flame photographs of a nitropolymer propellant treated with potassium nitrate. From top to bottom, the photographs represent KNO_3 contents of 0.68 %, 0.85 %, 1.03 %, and 1.14 %. Each of these experiments was performed under the test conditions of 8.0 MPa chamber pressure and an expansion ratio of 1. Though there is little effect on the primary flame, the secondary flame is clearly reduced by the addition of the suppressant. The secondary flame is completely suppressed by the addition of 1.14 % KNO_3. The nozzle used here is a convergent one, i. e., the nozzle exit is at the throat.

Fig. 12.13 shows the extent of the secondary flame zone as a function of the concentration of KNO_3 at a chamber pressure of 4 MPa and with Dt = 5.0 mm with nozzle area expansion ratios of ε = 6.3 and 11.7. No clear difference is seen for the different values of ε. It is evident that the zone shrinks with increasing concentration of KNO_3 and thus also with increasing mass fraction of potassium atoms contained within the propellant. Fig. 12.14 shows the extent of the secondary flame zone as a function of the concentration of K_2SO_4 at a chamber pressure of 4 MPa with D_t = 5.0 mm and ε = 1. Like KNO_3, K_2SO_4 is seen to be effective as a plume suppressant. The length of the secondary flame shortens as $\xi(K_2SO_4)$ or $\xi(K)$ is increased. Fig. 12.15 shows the extent of the secondary flame zone as a function of the concentration of K_2SO_4 at a chamber pressure of 12 MPa with D_t = 9.3 mm, and ε = 4. Fig. 12.16 shows the temperature along the center axis of the exhaust gas measured by means of suitably positioned thermocouples under the conditions relating to Fig. 12.15. The temperature decreases with increasing distance from the nozzle exit and also decreases with increasing $\xi(K_2SO_4)$ at the same distance from the nozzle. It is important to note that the exhaust gas temperature is decreased by the addition of K_2SO_4. It is decreased from 910 K to about 700 K by the addition of

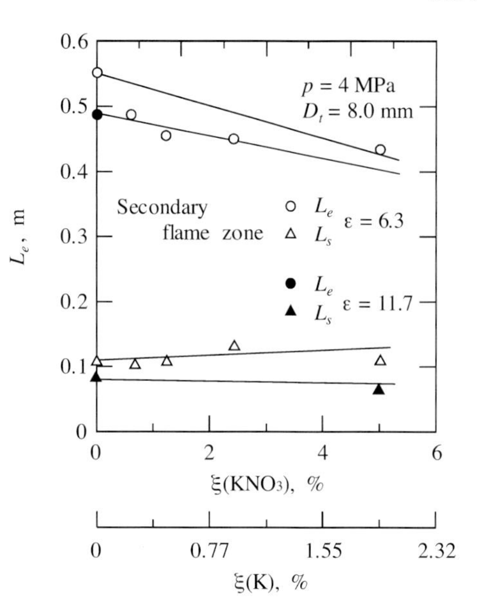

Fig. 12.13 Shrinkage of secondary flame zone by suppression with KNO_3 at ε = 6.3 and 11.7.

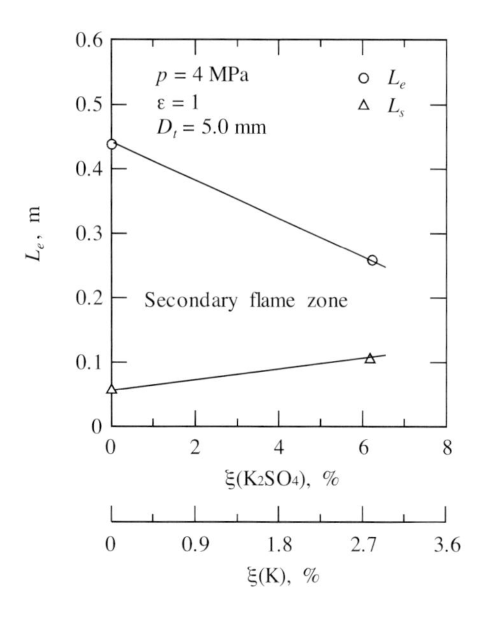

Fig. 12.14 Shrinkage of secondary flame zone by suppression with K_2SO_4 at ε = 1.

2% K_2SO_4. This temperature decrease is sufficient to inhibit ignition of the interdiffused mixture of exhaust gas and ambient air, thereby preventing the generation of a secondary flame.

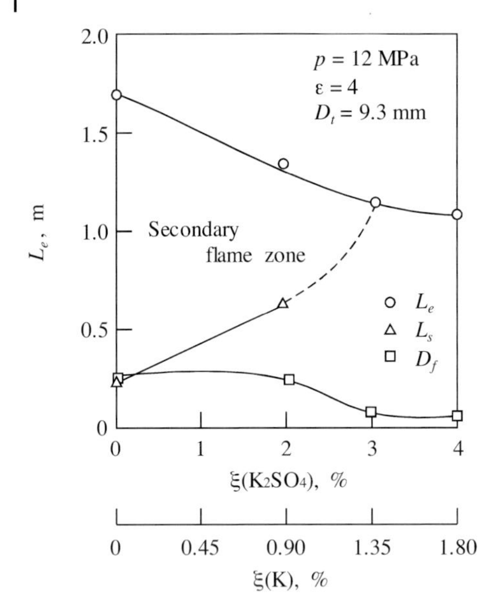

Fig. 12.15 Shrinkage of secondary flame zone by suppression with K_2SO_4 at ε = 4.

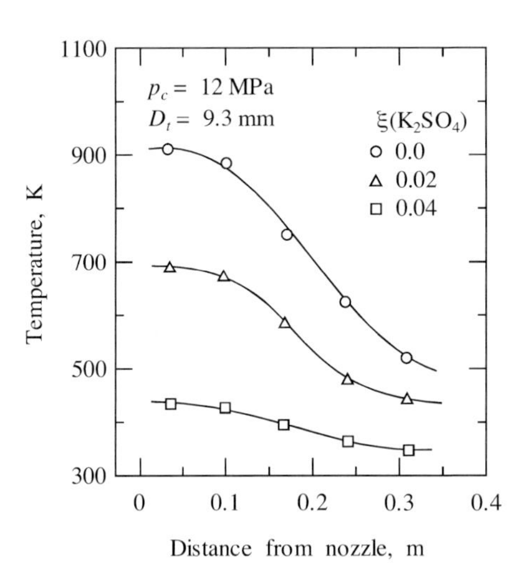

Fig. 12.16 Temperature along the exhaust nozzle flow and effect of the concentration of K_2SO_4.

12.6.2.2 **Effect of Nozzle Expansion**

Fig. 12.17 shows a typical set of afterburning flame photographs obtained when a nitropolymer propellant without a plume suppressant is burned in a combustion chamber and the combustion products are expelled through an exhaust nozzle into the ambient air. The physical shape of the luminous flame is altered significantly by variation of the expansion ratio of the nozzle. The temperature of the combustion products at the nozzle exit decreases and the flow velocity at the nozzle exit increases with increasing ε at constant chamber pressure.

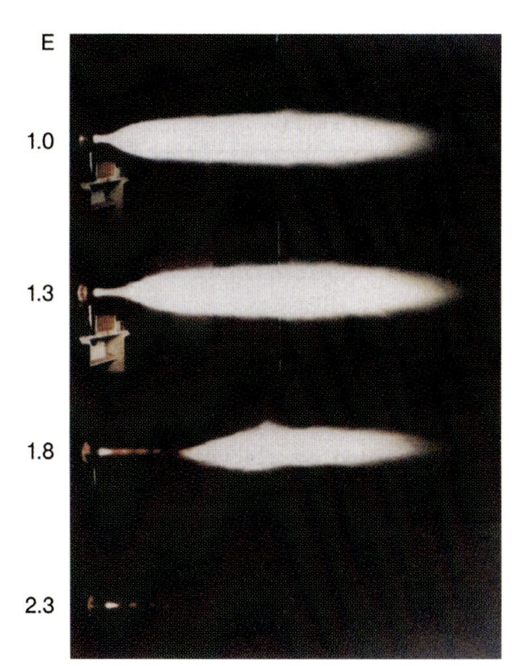

Fig. 12.17 Flame photographs of rocket plumes, showing that the dimensions of the secondary flame decrease as the nozzle expansion area ratio is increased.

When a convergent nozzle is used, i. e., the flow is at sonic velocity at the nozzle exit, afterburning occurs and a relatively large secondary flame is formed.[11] When a divergent nozzle is attached to the nozzle throat so that the expansion ratio, ε, is increased, the gas temperature at the nozzle exit is decreased and the flow velocity is increased due to an adiabatic expansion process. However, the gas temperature tends to increase or decrease through the formation of a diamond array-downstream of the nozzle exit, as shown in Fig. C-6, Appendix C. The luminous flame is distended from the nozzle exit and the size of the secondary flame is also decreased with increasing ε. However, the location of the end of the luminous flame remains unchanged. When the expansion ratio is $\varepsilon = 2.3$, no afterburning occurs and the secondary flame disappears. These results indicate that the creation of afterburning is dependent on the temperature of the expanded gas at the nozzle exit. When the temperature is high, the exhaust gas from the nozzle is ignited immediately upon mixing with the ambient air downstream of the nozzle exit, thereby generating a luminous flame.

12.7
HCl Reduction from AP Propellants

12.7.1
Background of HCl Reduction

Ammonium perchlorate (AP) is one of the most important oxygen carriers for the formulation of composite propellants. Furthermore, AP composite propellants demonstrate superior ballistic properties, such as burning rate, pressure exponent, and temperature sensitivity, when compared with all other propellants. In addition, AP is used as an oxidizer component of pyrolants used to formulate gas generators, igniters, and flares by virtue of their advantageous production of high heat flux and their high burning rates compared with other pyrolants. Unlike alkali metal perchlorates, such as potassium perchlorate ($KClO_4$) and sodium perchlorate ($NaClO_4$), AP has the advantage of being completely convertible to gaseous reaction products. Propellants composed of crystalline AP particles and polymeric hydrocarbons thus produce only gaseous combustion products and no condensed particles. The overall reaction scheme can be represented by:

$$NH_4ClO_4 + \text{hydrocarbon polymer} \rightarrow N_2, CO_2, CO, H_2O, HCl$$

Thus, AP is a valuable oxidizer for formulating smokeless propellants or smokeless gas generators. However, since the combustion products of AP composite propellants contain a relatively high concentration of hydrogen chloride (HCl), white smoke is generated when they are expelled from an exhaust nozzle into a humid atmosphere. When the HCl molecules diffuse into the air and collide with H_2O molecules therein, an acid mist is formed which gives rise to visible white smoke. Typical examples are AP composite propellants used in rocket motors. Based on experimental observations, white smoke is formed when the relative humidity exceeds about 40%. Thus, AP composite propellants without any metal particles are termed reduced-smoke propellants. On the other hand, a white smoke trail is always seen from the exhaust of a rocket projectile assisted by an aluminized AP composite propellant under any atmospheric conditions. Thus, aluminized AP composite propellants are termed smoke propellants.

The U.S. Space Shuttle, for example, has two large solid-rocket boosters for its launch stage. The booster propellant has the following composition:

Chemical components	Mass fraction	
Ammonium perchlorate (AP)	0.70	Oxidizer
Polybutadiene acrylonitrile (PBAN)	0.14	Fuel and binder
Aluminum	0.16	Fuel

Thus, the two boosters generate large amounts of gaseous and condensed products, including HCl, which are expelled into the atmosphere. When the exhaust HCl gas diffuses into humid air, the HCl condenses to form hydrochloric acid mist, which grows to produce acid rain if it is present at a sufficient level to produce large droplets in the air. Table 12.8 shows an example computation of the amount of HCl gas generated and the area subjected to acid rain when sufficient humidity is provided by the atmosphere. Though the amount of HCl produced is large, the Space Shuttle is launched from the east coast of the Florida peninsula and the exhaust gases are dispersed over the Atlantic Ocean.

Table 12.8 Acid mist formation by the Space Shuttle boosters.

Propellant mass of the two booster propellants:	500 kg /booster × 2 = 1.000×10^3 kg
Mass fraction of HCl contained within the two booster propellants:	0.217
Total mass of HCl within the combustion products:	$1.000 \times 10^3 \times 0.217 = 0.217 \times 10^3$ kg
Hydrochloric acid:	0.372% in water = 538×10^3 kg = 493 m^3
Area subject to 1 mm of acid rain composed of 0.1% hydrochloric acid:	22 km × 22 km

Since AP is a valuable oxidizer because of its high oxidizer performance, safe handling, and cost effectiveness compared with other crystalline oxidizers, it is difficult to replace it with other oxidizers that might circumvent its disadvantage of HCl gas generation. Various types of propellants have been proposed to eliminate the HCl content from the combustion products.

12.7.2
Reduction of HCl by the Formation of Metal Chlorides

High-temperature HCl molecules tend to react with metal particles. When particles of Na compounds or Mg particles are incorporated into AP composite propellants, sodium chloride or magnesium chloride are formed. In general, aluminum particles are incorporated into AP composite propellants. However, Cl atoms or Cl_2 molecules generated by the decomposition of AP react with H_2O molecules to produce HCl molecules. Chemicals containing Na or Mg atoms react with HCl after their thermal decomposition.

When sodium nitrate ($NaNO_3$) is incorporated into an AP propellant, the concentration of HCl molecules is reduced by reaction according to:

$$NH_4ClO_4 + NaNO_3 \rightarrow NH_4NO_3 + NaClO_4 \rightarrow \text{production of NaCl}$$

$$\rightarrow Na_2O + HCl \rightarrow \text{production of NaCl}$$

Thus, the number of HCl molecules is reduced when sodium nitrate particles are incorporated into AP propellants. This class of propellants is termed "scavenged propellants". Fig. 12.18 shows the results of thermal equilibrium computations of scavenged AP propellants as a function of the mass fraction of sodium nitrate,

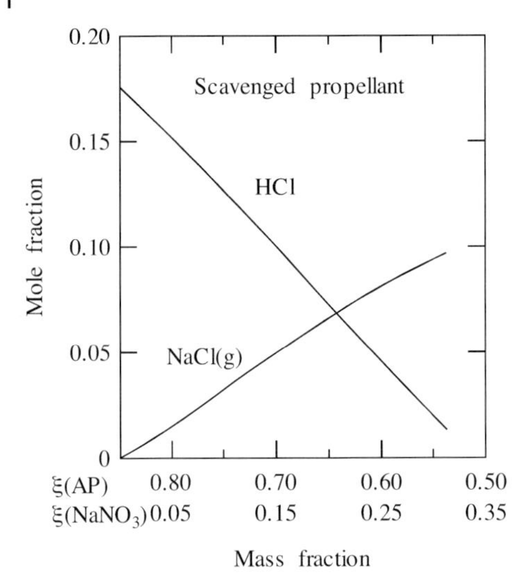

Fig. 12.18 Mole fractions of HCl and NaCl produced by scavenged AP propellants as a function of ξ($NaNO_3$).

ξ($NaNO_3$). The AP propellant without $NaNO_3$, ξ(0.0), is composed of mass fractions of ξ(AP) = 0.85 and ξ(HTPB) = 0.15. The HCl gas among the combustion products of the AP propellant is converted into NaCl, which is a stable and environmentally benign material.

When magnesium particles are incorporated into AP propellants, these react with HCl molecules generated in the combustion chamber according to:

$$
\begin{aligned}
NH_4ClO_4 + Mg &\rightarrow MgO + MgCl_2 + HCl \\
&\rightarrow MgO + H_2O \rightarrow Mg(OH)_2 \\
&\rightarrow Mg(OH)_2 + 2HCl \rightarrow \text{production of } MgCl_2
\end{aligned}
$$

and magnesium chloride is formed. Thus, the number of HCl molecules is reduced when magnesium particles are incorporated into AP pyrolants. This class of propellants is termed "neutralized propellants". Fig. 12.19 shows the combustion products of neutralized AP propellants as a function of ξ(Mg). The AP propellant without magnesium, ξ(0.0), is composed of mass fractions of ξ(AP) = 0.85 and ξ(HTPB) = 0.15. The HCl gas among the combustion products of the AP propellant is converted into magnesium chloride, which is a stable and environmentally benign material. The specific impulses of the scavenged and neutralized propellants are shown as a function of ξ($NaNO_3$) or ξ(Mg) in Fig. 12.20. Since Mg particles act as a fuel component, similarly to Al particles, the specific impulse of neutralized propellants increases with increasing ξ(Mg). Though $NaNO_3$ acts as an oxidizer component, the specific impulse cannot be increased by the replacement of AP with $NaNO_3$. Thus, the specific impulse of scavenged propellants decreases with increasing ξ($NaNO_3$).

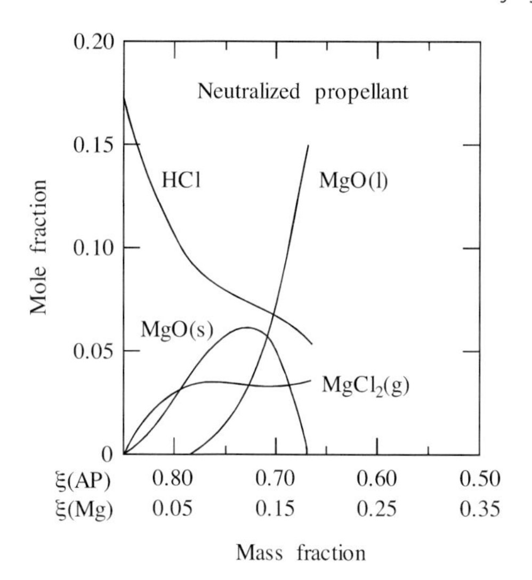

Fig. 12.19 Mole fractions of the combustion products of neutralized AP propellants as a function of ξ(Mg).

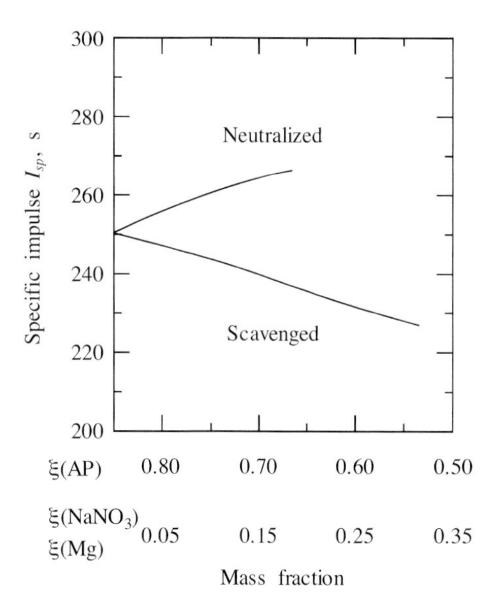

Fig. 12.20 Specific impulses of scavenged and neutralized AP composite propellants as a function of ξ($NaNO_3$) or ξ(Mg)

12.8 Reduction of Infrared Emission from Combustion Products

The principal infrared emissions from gaseous combustion products of propellants are caused by the high-temperature CO_2 and H_2O molecules. When nitropolymer propellants or AP composite propellants burn, large amounts of high-temperature CO_2 and H_2O molecules are formed. If these propellants burn incompletely due to their fuel-rich compositions, large amounts of hydrocarbon fragments and solid

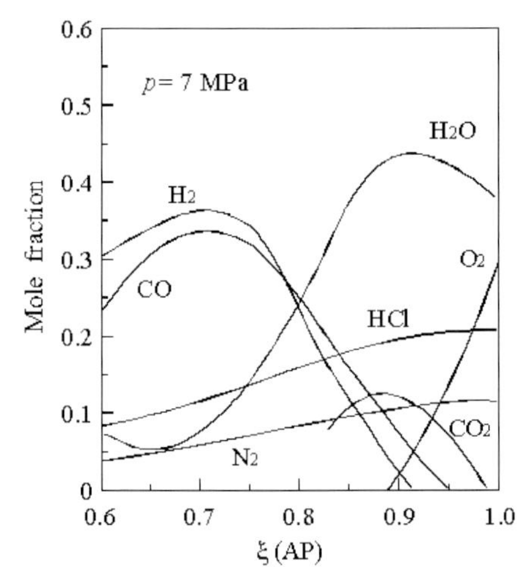

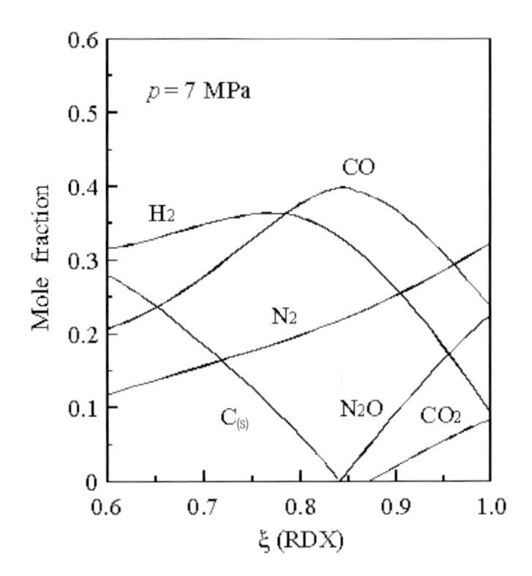

Fig. 12.21 Mole fractions of the combustion products formed by AP-HTPB and RDX-HTPB composite propellants.

carbon particles are formed instead of CO_2 and H_2O molecules. These fragments and particles emit continuous-spectrum radiation, including infrared radiation. It is not possible to reduce infrared emission from these combustion products.

Fig. 12.21 shows the combustion products of AP-HTPB and RDX-HTPB composite propellants. Large amounts of H_2O, HCl, and CO_2 are formed when an AP-HTPB propellant composed of $\xi_{AP}(0.85)$ is burnt. The molecules of H_2O, HCl, and CO_2 each emit infrared radiation. On the other hand, no CO_2 or $C_{(s)}$ is formed when an RDX-HTPB propellant composed of $\xi_{RDX}(0.85)$ is burnt. Instead, large amounts of CO, H_2, and N_2 molecules are formed as its major combustion products. However, no infrared radiation is emitted from H_2 or N_2 molecules. Though CO molecules are formed at $\xi_{RDX}(0.85)$, the infrared radiation emitted from these is less than that from H_2O or CO_2 molecules.

References

1 Gaydon, A. G., and Wolfhard, H. G., Flames: Their Structure, Radiation and Temperature, Third Edition (revised), Chapman and Hall, London (1970).

2 Sturman, B., An Introduction to Chemical Thermodynamics, Pyrotechnic Chemistry, Journal of Pyrotechnics, Inc., Whitewater, CO (2004), Chapter 3.

3 Kosanke, K. L., and Kosanke, B. J., The Chemistry of Colored Flame, Pyrotechnic Chemistry, Journal of Pyrotechnics, Inc., Whitewater, CO (2004), Chapter 9.

4 Hosoya, M., and Hosoya, F., Science of Fireworks, Thokai University Press, Tokyo (1999).

5 Shimizu, T., Chemical Components of Fireworks Compositions, Pyrotechnic Chemistry, Journal of Pyrotechnics, Inc., Whitewater, CO (2004), Chapter 2.

6 Energetic Materials Handbook, Japan Explosives Society, Kyoritsu Shuppan, Tokyo (1999).

7 Koch, E.-C., Dochnahl, A., IR Emission Behavior of Magnesium/Teflon/Viton

(MTV) Compositions, *Propellants, Explosives, Pyrotechnics*, Vol. 25, 2000, pp. 37–40.

8 Koch, E.-C., Review on Pyrotechnic Aerial Infrared Decoys, *Propellants, Explosives, Pyrotechnics*, Vol. 26, 2001, pp. 3–11.

9 Gillard, P., and Roux, M., Study of the Radiation Emitted During the Combustion of Pyrotechnic Charges. Part I: Non-Stationary Measurement of the Temperature by Means of a Two-Color Pyrometer, *Propellants, Explosives, Pyrotechnics*, Vol. 27, 2002, pp. 72–79.

10 Taylor, M. J., Spectral Acquisition and Calibration Techniques for the Measurement of Radiative Flux Incident upon Propellant, *Propellants, Explosives, Pyrotechnics*, Vol. 28, 2003, pp. 18.

11 Iwao, I., Kubota, N., Aoki, I., Furutani, T., and Muramatsu, M., Inhibition of Afterburning of Solid Propellant Rocket, Explosion and Explosives, Industrial Explosives Society Japan, Vol. 42, No. 6, pp. 366–372 (1981) or Translation by Heimerl, J. M., Klingenberg, G., and Seiler, F., T 3/86, Ernst-Mach-Institut Abteilung für Ballistik, Fraunhofer-Gesellschaft, Weil-am-Rhein (1986).

12 Gillard, P., de Izarra, C., and Roux, M., Study of the Radiation Emitted During the Combustion of Pyrotechnic Charges. Part II: Characterization by Fast Visualization and Spectroscopic Measurements, *Propellants, Explosives, Pyrotechnics*, Vol. 27, 2002, pp. 80–87.

13 Blanc, A., Deimling, L., and Eisenreich, N., UV- and IR-Signatures of Rocket Plumes, *Propellants, Explosives, Pyrotechnics*, Vol. 27, 2002, pp. 185–189.

13
Transient Combustion of Propellants and Pyrolants

13.1
Ignition Transient

Ignition of an energetic material such as a propellant or pyrolant is initiated by heat supplied to its surface through conductive, convective, and/or radiative heat transfer.[1,2] Conductive heat is supplied to an energetic material by means of hot condensed particles dispersed on the ignition surface. For example, an igniter made of boron particles and potassium nitrate generates a large number of boron oxide particles, the temperature of which exceeds 3000 K. Conductive heat transfer from the hot oxide particles to the surface of the energetic material increases the temperature of the latter. An exothermic gasification reaction starts and reactive gases are given off from the surface of the energetic material. These reactive gases then form high-temperature combustion products and ignition is established.

When high-temperature gases produced by an igniter flow over the surface of an energetic material, convective heat transfer occurs and the surface temperature is raised. As in the case of conductive heating, an exothermic reaction occurs and reactive gases are formed. When a radiative light emitter irradiates the surface of an energetic material, the surface temperature is also raised and an exothermic gasification occurs at the surface. Ignition of the energetic material is then established. Ignition delay arises not only as a result of the heat-transfer process, but also as a result of various physicochemical processes.

13.1.1
Convective and Conductive Ignition

When a high-temperature gas flow passes over the surface of an energetic material, convective heat transfer occurs and the surface temperature increases. When the temperature reaches the decomposition value, a gasification reaction occurs either endothermically or exothermically. The heat transferred to the surface is also transferred to the interior of the energetic material and increases the temperature therein simultaneously. The gaseous decomposition products on and above the surface react to produce heat, and then part of this heat is transferred back to the decomposing surface. When the heat is sufficient to cause self-decomposition at the surface, ignition of the energetic material is established. The process re-

Propellants and Explosives. Naminosuke Kubota

ISBN: 978-3-527-31424-9

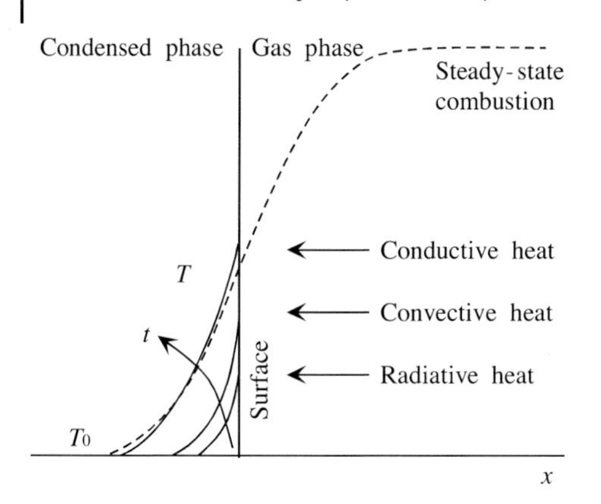

Fig. 13.1 Ignition transient process by conductive, convective, and radiative heat.

sponsible for the temperature increase in the condensed phase is illustrated in Fig. 13.1.

Since initiation of the decomposition is dependent on the heat flux supplied by the high-temperature gas flow, the ignition process is dependent on the various gas-flow parameters, such as temperature, flow velocity, pressure, and the physico-chemical properties of the gas.

The heat flux transferred from the gas flow to the surface of the energetic material, q, is represented by

$$q = h_g(T_g - T_s) \tag{13.1}$$

where h_g is the heat-transfer coefficient, T_g is the temperature of the gas flow that flows parallel to the surface, and T_s is the surface temperature of the energetic material. The heat-transfer coefficient is dependent on various parameters, such as the gas-flow velocity, u_g, the physical dimensions of the flow channel, the properties of the gas, and the surface roughness of the energetic material. The heat-transfer coefficient is given in terms of dimensionless parameters as

$$\mathrm{Nu} = c\mathrm{Re}^m\mathrm{Pr}^n \tag{13.2 a}$$

where Nu is the Nusselt number, Re is the Reynolds number, Pr is the Prandtl number, and c, m, and n are constants. In the case of a gas flow above the flat surface of an energetic material, the Nu, Re, and Pr numbers are represented by

$$\mathrm{Nu} = h_{gx}x/k_g \tag{13.2 b}$$

$$\mathrm{Re} = \rho_g u_g x/\mu_g \tag{13.3}$$

$$\mathrm{Pr} = c_g\mu_g/\lambda_g \tag{13.4}$$

where c_g is the specific heat, μ_g is the viscosity, and λ_g is the thermal conductivity of the gas flow. In the case of a gas flow above a flat surface, the Nu number is represented by

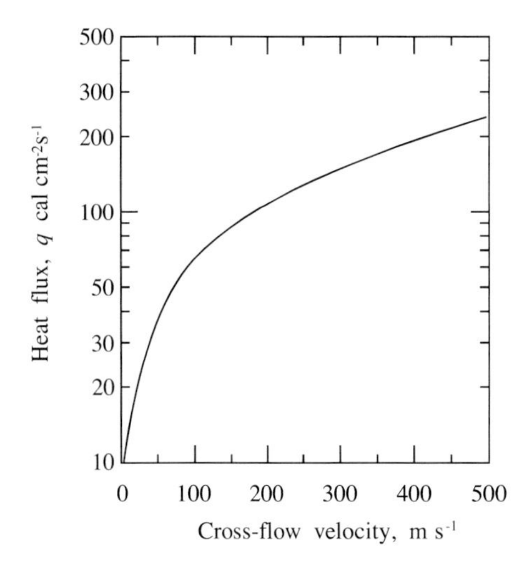

Fig. 13.2 Heat flux transferred to a flat plate in a boundary layer, plotted as ln q, as a function of u_g.

$$\mathrm{Nu} = 0.322\mathrm{Pr}^{1/3}\mathrm{Re}^{1/2} \tag{13.2 c}$$

where x is the distance from the leading edge of the flat surface, and h_{gx} is the heat-transfer coefficient from the gas flow to the flat surface at the distance x from the leading edge of the flat surface. In the case of a circular port,

$$\mathrm{Nu} = 0.0395\mathrm{Pr}^{1/3}\mathrm{Re}^{3/4} \tag{13.2 d}$$

where $\mathrm{Nu} = h_g d/\lambda_g$ and $\mathrm{Re} = u_g \rho_g d/\mu_g$, with d being the inner diameter of the perforated circular port. Combining the heat-transfer equation, Eq. (13.2), with appropriate initial and boundary conditions, and also with the heat-conduction equationin the solid phase of the energetic material, the heat flux transferred from the gas phase to the solid phase is obtained.

Fig. 13.2 shows the result of a computation of the heat flux transferred from the main cross-flow stream to the flat plate as a function of the cross-flow velocity based on Eqs. (13.1) and (13.2 c). The parameter values used for the computation are x = 0.4 m, c_g = 1.55 kJ kg^{-1} K^{-1}, $\lambda_g = 8.4 \times 10^{-4}$ kJ s^{-1} m^{-1} K^{-1}, $\mu_g = 4.5 \times 10^{-5}$ kg s^{-1} m^{-1}, ρ_g = 2.25 kg m^{-3}, and M_g = 24.6 kg kmol^{-1} at p = 2 MPa. The gas temperature of the main flow stream is T_g = 2630 K and the surface temperature on the flat plate is T_s = 600 K. The heat flux increases with increasing cross-flow velocity.

13.1.2
Radiative Ignition

When high-intensity radiative energy is supplied to the surface of an energetic material, the surface absorbs the heat and its temperature increases. If the energetic material is optically translucent, part of the radiation energy penetrates into the interior of the propellant, where it is absorbed. When the surface reaches its decomposition temperature, reactive gaseous materials are formed on and above the surface through an endothermic or exothermic decomposition reaction.[1,2] When the reaction in the gas phase is established and the temperature is increased

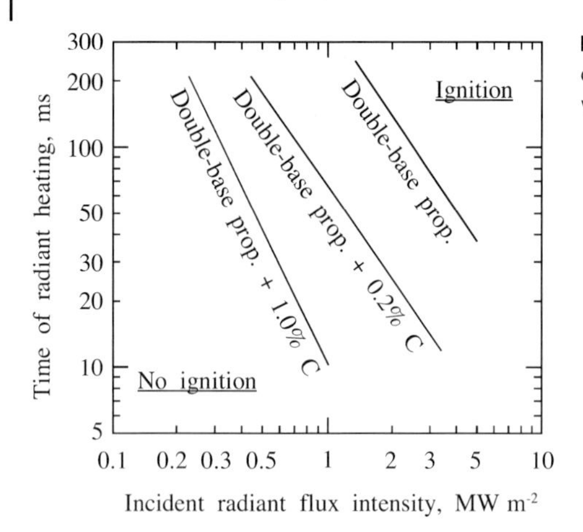

Fig. 13.3 Radiative ignition criteria of double-base propellants with and without carbon black.

by the reaction, the temperature beneath the surface is increased through heat conduction. When the surface reaction continues after the radiative heat energy from the external heat source is cut off, ignition of the energetic material is established.

Fig. 13.3 shows the ignition boundaries of double-base propellants under radiative heating. As the incident radiant flux intensity, I_f, increases, the time of radiant heating needed to achieve ignition, τ_{ig}, becomes shorter. As the propellant is rendered opaque by the addition of carbon black, τ_{ig} becomes shorter at constant I_f or I_f becomes smaller at constant τ_{ig}. The in-depth absorption is reduced and the majority of the radiative energy is absorbed at the ignition surface of the propellant.

In a simplified analysis based on the conductive heating beneath the burning surface, the time of ignition is represented by[2]

$$\tau_{ig} = \frac{\alpha_p \pi}{4}\left[\frac{\rho_p c_p (T_g - T_0)}{I_f}\right]^2 \tag{13.5}$$

The total incident radiant intensity, E_f, is given by

$$E_f = I_f \tau_{ig} = (\pi/4)\lambda_p \rho_p c_p (T_g - T_0)^2 / I_f \tag{13.6}$$

The linear relationship of $\ln \tau_{ig} \sim -2 \ln I_f$, and hence $\tau_{ig} \sim I_f^{-2}$, is seen for double-base propellants, as shown in Fig. 13.3.

13.2
Ignition for Combustion

13.2.1
Description of the Ignition Process

Ignition involves a reaction process of the oxidizer and fuel fragments that are produced at the surface of an energetic material that produces the heat needed to achieve steady-state burning. The surface temperature is first increased by additional heat provided externally by means of an igniter. When the temperature on or just beneath the surface of the energetic material reaches the decomposition temperature, reactive oxidizer and fuel fragments are produced on and above the heated surface. These fragments then react to produce heat and high-temperature reaction products.

One assumes that the oxidizer component, O, and the fuel component, F, react to produce reaction product, P, along the burning direction, x. The reaction is expressed by

$$aO + F \rightarrow bP$$

where a and b show the stoichiometry of the reaction. The mass species equation in the gas phase is given by

$$(\partial Y_O/\partial t) + u_g(\partial Y_O/\partial x) = D(\partial^2 Y_O/\partial x^2) - aY_O Y_F\, Z_g \exp(-E_g/RT) \quad (13.7)$$

$$(\partial Y_F/\partial t) + u_g(\partial Y_F/\partial x) = D(\partial^2 Y_F/\partial x^2) - Y_O Y_F\, Z_g \exp(-E_g/RT) \quad (13.8)$$

where x is distance along the reaction direction, t is time, Y_O and Y_F are the concentrations of the oxidizer and fuel fragments, respectively, u_g is the flow velocity of the reactive gas, E_g is the activation energy in gas phase, R is the universal gas constant, Z_g is a pre-exponential factor, and D is the diffusion coefficient between O and F.

The energy equation of the reactive gas is given by

$$\frac{\partial T}{\partial t} + u_g \frac{\partial T}{\partial x} = \lambda_g \frac{\partial^2 T}{\partial x^2} + (Q_g/\rho_g c_g) Y_O Y_F Z_g \exp(-E_g/RT) \quad (13.9)$$

where ρ is the density, c is the specific heat, Q is the heat of reaction, and the subscript g denotes the gas phase.

The energy equation in the solid phase is expressed by

$$\frac{\partial T}{\partial t} + u_p \frac{\partial T}{\partial x} = \frac{\lambda_g}{\rho_p c_p} \frac{\partial^2 T}{\partial x^2} + \frac{Q_p}{c_p} Z_p \exp(-E_p/RT) + \frac{1}{\rho_p c_p} \frac{\partial I}{\partial x} \quad (13.10)$$

where u_p is the regression velocity of the condensed phase, I is the radiation absorption in the solid phase, and the subscript p denotes the condensed phase. The rate of thermal decomposition at the surface is expressed by an Arrhenius-type law; the regressing velocity is given by

$$u_p = Z_s \exp(-E_s/RT_s) \quad (13.11)$$

where the subscript s denotes the regressing surface. Equation (13.11) indicates that the regressing velocity increases with increasing regressing surface temperature, T_s.

When the ignition process occurs under conditions of constant pressure, the momentum equation is expressed by

$$p = \text{constant} \quad (13.12)$$

and the mass continuity equation is expressed by

$$\rho_p u_p = \rho_g u_g \quad (13.13)$$

Equations (13.7)–(13.13) are used to evaluate the ignition processes of energetic materials with appropriate initial and boundary conditions. In general, the conditions in the thermal field for ignition are given by

$$T(x, 0) = T_0$$

$$T(-\infty, t) = T_0$$

$$T(0^-, t) = T(0^+, t)$$

$$\lambda_g(\partial T/\partial x)_{0+,t} = \lambda_p(\partial T/\partial x)_{0-,t}$$

and the conditions in the concentration field for ignition are given by

$$Y_O(x, 0) = Y_O(+\infty, t) = Y_o^\infty$$

$$Y_F(x, 0) = Y_F(+\infty, t) = 0$$

$$(\rho_p u_p Y_O)_{0-} = \{\rho_g u_g Y_O - \rho_g D(\partial Y_O/\partial x)\}_{0+}$$

$$(\rho_p u_p Y_F)_{0-} = \{\rho_g u_g Y_F - \rho_g D(\partial Y_F/\partial x)\}_{0+}$$

where 0^+ denotes the gas phase at the regressing surface and 0^- denotes the condensed phase at the regressing surface.

In the solid phase, well below the regressing surface, the second and third terms of Eq. (13.10) may be neglected because no chemical reaction occurs and no irradiation energy penetrates through the regressing surface. Thus, Eq. (13.10) is expressed by

$$(\partial T/\partial t) + u_p(\partial T/\partial x) = (\lambda_p/\rho_p c_p)(\partial^2 T/\partial x^2) \quad (13.14)$$

The boundary conditions are given by

$$T(x, 0) = T_0 \quad (13.15)$$

$$\partial T/\partial x \rightarrow 0: \quad x \rightarrow -\infty \quad (13.16)$$

13.2.2 Ignition Process

The initial conditions vary with time because the physicochemical process of ignition varies according to the ignition energy supplied to the ignition surface of the energetic material. A typical example of a radiative ignition process is shown below:

$$0 \leq t < t_v: \quad \lambda_p(\partial T/\partial x)_{0-} = I(t) \quad (13.17)$$

$$t_v \leq t < t_g: \quad \lambda_p(\partial T/\partial x)_{0-} = I(t) + \rho_p u_p Q_p \quad (13.18)$$

$$t_g \leq t < t_f: \quad \lambda_p(\partial T/\partial x)_{0-} = I(t) + \rho_p u_p Q_p + \lambda_g(\partial T/\partial x)_{0+} \quad (13.19)$$

$$t_f \leq t < t_{ss}: \quad \lambda_p(\partial T/\partial x)_{0-} = \rho_p u_p Q_p + \lambda_g(\partial T/\partial x)_{0+} \quad (13.20)$$

$$t_{ss} \leq t: \quad \lambda_p(\partial T/\partial x)_{0-} = I(t) + \rho_p u_p Q_p + \lambda_g(\partial T/\partial x)_{0+} \quad (13.21)$$

where t_v is the onset time of the thermal decomposition at the surface, t_g is the onset time of the gasification at the surface, t_f is the cut-off time of the irradiation, and T_v, T_g, and T_f are the surface temperatures at t_v, t_g, and t_f, respectively. During the time $t_f \leq t < t_{ss}$, ignition is completed, combustion occurs without external heating, the surface temperature reaches T_s, and the burning rate attains a steady-state value with regression velocity u_p at t_{ss}. Fig. 13.4 shows a schematic diagram of the ignition process expressed by Eqs. (13.17)–(13.21).

Real ignition processes are rather complicated and heterogeneous because igniters contain various types of metal particles and crystalline oxidizer particles. The metal particles are oxidized by the gaseous oxidizer fragments and produce high-

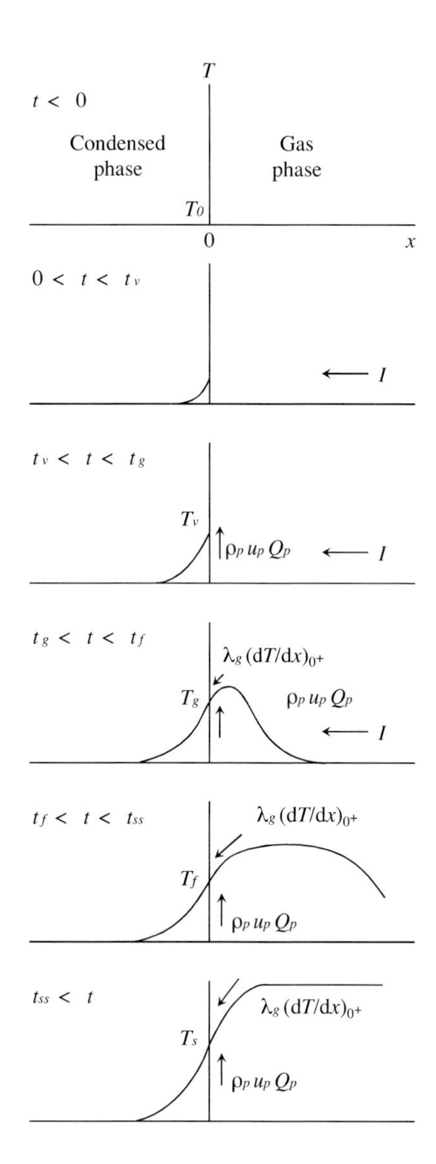

Fig. 13.4 An ignition scheme of an energetic material by irradiation.

temperature oxide particles. These hot particles are dispersed on the propellant surface and provide heat for ignition through solid-to-solid heat conduction. This type of ignition is called "hot-spot ignition". This ignition process is not described by the equations shown above. Furthermore, the burning surface of an energetic material such as a composite propellant or pyrolant is not homogeneous. The decomposition and gasification processes at the surface of a non-homogeneous energetic material are subject to spatial and temporal variations.

13.3 Erosive Burning Phenomena

13.3.1 Threshold Velocity

When an energetic material burns under high-temperature cross-flow conditions, erosive burning occurs.[3,4] Fig. 13.5 shows the burning rates without cross-flow, r_0, for high-, reference-, and low-energy double-base propellants.[4] The chemical compositions and the adiabatic flame temperatures, T_f, of the propellants are shown in Table 13.1. The erosive ratio, defined according to $\varepsilon = r/r_0$, increases with increasing cross-flow velocity for the three types of propellants. As shown in Fig. 13.6, there exists a threshold velocity for each propellant, approximately 70 m s^{-1} for the low-, 100 m s^{-1} for the reference-, and 200 m s^{-1} for the high-energy propellant. The erosive ratio is about 2.4 for the low-energy propellant at 300 m s^{-1}, and the Mach number is approximately 0.3.

The flow field of a double-base propellant during erosive burning is shown schematically in Fig. 13.7. The flow in an internal burning is turbulent and a turbulent boundary layer is established. The luminous flame of the high-temperature zone is distended from the burning surface and the fizz zone, which is the important zone with regard to determining the burning rate, is just above the burning surface. When the velocity is low, the fizz zone lies within the viscous sublayer, wherein the velocity is very low. Thus, the fizz zone is not affected by the cross-flow, and hence the burning rate remains unchanged. However, when the cross-flow velocity is increased, the luminous flame zone approaches the burning surface and the dark zone diminishes due to the increased turbulent intensity. The heat flux then increases and so too does the burning rate.

Table 13.1 Chemical compositions and adiabatic flame temperatures of high-, reference-, and low-energy propellants.

Propellant	ξ_{NC}	ξ_{NG}	ξ_{DEP}	T_f (K)
High energy	0.556	0.404	0.040	2720
Reference energy	0.504	0.366	0.130	2110
Low energy	0.475	0.345	0.180	1780

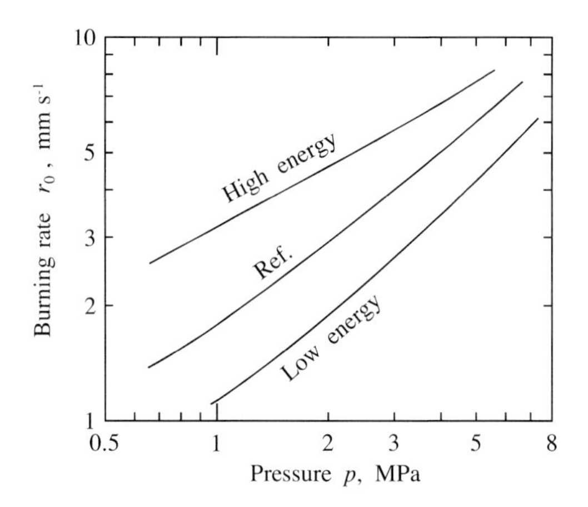

Fig. 13.5 Burning rates of high-energy, reference, and low-energy double-base propellants.

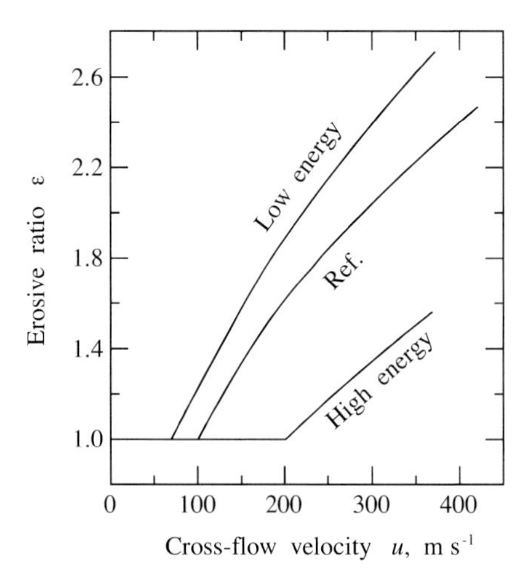

Fig. 13.6 Erosive ratio and threshold velocity of erosive burning for high-energy, reference, and low-energy double-base propellants, showing that the low-energy propellant is most sensitive to the convective heat flux.

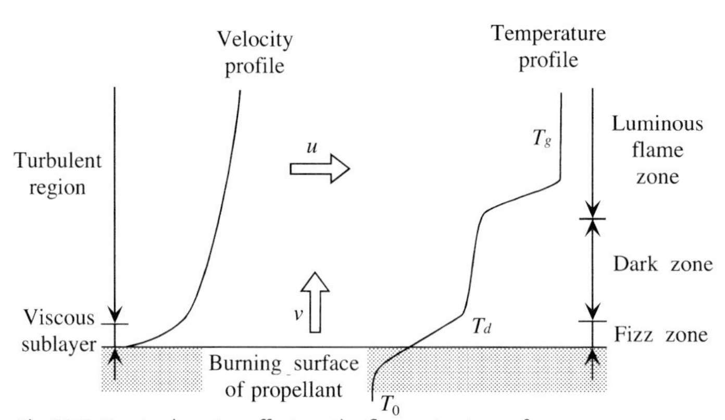

Fig. 13.7 Erosive burning effect on the flame structure of a double-base propellant.

The velocity of the burned gas emanating from the burning surface is high for the high-energy propellant since the burning rate is high compared with that of the low-energy propellant at constant pressure. In addition, the temperature gradient in the fizz zone is steep and the thickness of the fizz zone is small. Thus, the burning rate is less sensitive to the cross-flow in the case of the high-energy propellant.[4] This aerochemical process determines the observed threshold velocity shown in Fig. 13.6.

13.3.2 Effect of Cross-Flow

On igniting the surface of a propellant slab set in a rectangular tube, of which one end is closed and the other end is open, the combustion gas flows out from the open end. Though there is no flow velocity at the closed end, the flow velocity increases in the tube along the flow direction. Furthermore, the mass flow rate also increases along the flow direction and reaches its maximum at the open end. The combustion gas generated from the burning surface of the propellant slab flows toward the open end, perpendicular to the burning surface, x, and enters the cross-flow of the main stream. The temperature (T_g) and the flow velocity (u) of the gas formed by the burning of the propellant slab increase in the boundary layer formed above the burning surface from the closed end toward the open end when a double-base propellant is used.

Heat transfer from the cross-flow to the burning surface occurs through the boundary layer. The increased heat flux due to the cross-flow increases the burning rate expressed by Vieille's law, $r = ap^n$. This combustion phenomenon is called erosive burning. The burning rate is also given by the flow parameters in the boundary layer. The erosive burning rate is expressed by the cross-flow parameters in the boundary layer. However, this erosive burning only occurs when the cross-flow velocity exceeds a certain critical velocity, termed the threshold velocity. Below this threshold velocity, the cross-flow has no effect on burning rate. Furthermore, erosive burning is dependent on various parameters, such as the type of pyrolant, the pressure, and the temperature of the cross-flow gases. This is because erosive burning phenomena are closely related to the combustion flame structures of the relevant propellants adjacent to their burning surfaces.

An understanding of the erosive burning mechanism is important to optimize the thrust versus pressure design used for rocket motors, guns, and various types of pyrotechnics. Since erosive burning occurs under high-velocity and high-temperature conditions, experimental determination of its rate-controlling factors is very difficult.

13.3.3 Heat Transfer through a Boundary Layer

Though the burning rate of a propellant is represented by Eq. (3.68) at constant pressure, Eq. (3.68) is no longer valid when a cross-flow is applied to the burning propellant. The heat flux transferred from the cross-flow to the burning surface of the propellant increases the burning rate. In general, the heat flux from the gas flow to the wall surface of a flow channel, q, is represented by Eq. (13.1). Though the heat flux without blow-off gas perpendicular to the main flow direction is determined by Eqs. (13.2)–(13.4), these equations are no longer valid due to the additional flow from the burning surface. The heat-transfer coefficient along the flow channel in a circular port is given by the semi-empirical equation:[3]

$$h_0 = 0.0288 c_g \mu_g^{0.2} \mathrm{Pr}^{-0.667}\ kG^{0.8}/L^{0.2} \tag{13.22}$$

where G is the mass flow flux in the circular port, k is an experimentally determined constant, and L is the distance from the leading edge of the fluid flow. The heat-transfer coefficient is correlated with the physical and flow properties of the fluid by:

$$\mathrm{St} = 0.0288\ \mathrm{Re}^{-0.2}\ \mathrm{Pr}^{0.667} \tag{13.23 a}$$

where St is the Stanton number, and the Reynolds number is based on the distance x from the leading edge of the flat plate. These numbers are defined by the physical properties of the fluid in the boundary layer according to

$$\mathrm{St} = h_0/\rho_g u_g c_g \tag{13.23 b}$$

In a number of experimental studies, the heat-transfer coefficient with cross-flow, h, is determined according to

$$h = h_0 \exp(-\beta\rho_p r/G) \tag{13.24}$$

where β is a blow-off parameter from the burning surface that is determined experimentally. It is evident that Pr is a function of the physical properties of the cross-flow gas and that Re is also a function of the kinematic viscosityof the gas and the flow velocity; the heat flux is properly given as a function of u_g, T_w, T_g, and the physical properties of the combustion gas.

If one assumes that the increased burning rate, r_e, caused by the cross-flow is attributable to the heat flux through the boundary layer, the overall burning rate of the energetic material, r, is given by

$$r = r_0 + r_e \tag{13.25}$$

where r_0 is the burning rate without cross-flow. Thus, one obtains the burning rate equation with cross-flow, the so-called erosive burning rate equation, as

$$r = ap^n + kh_0 \exp\{-\beta\rho_p r/G\} \tag{13.26 a}$$

$$= ap^n + (0.0288\ c_g \mu_g^{0.2}\ \mathrm{Pr}^{-0.667})\ (kG^{0.8}/L^{0.2})\ \exp(-\beta\rho_p r/G) \tag{13.26 b}$$

$$= ap_n + \alpha(G^{0.8}/L^{0.2})\ \exp(-\beta\rho_p r/G) \tag{13.26 c}$$

where k is a constant dependent on the flow interaction between the flow parallel to the burning surface and the blow-off gas flow from the burning surface. The parameter α is given by

$$\alpha = 0.0288\ c_g \mu_g^{0.2} \mathrm{Pr}^{-0.667}\ k \tag{13.27}$$

This burning rate equation with cross-flow is derived on the basis of experimental data and is known as the Lenoir–Robilard equation.[3]

13.3.4
Determination of Lenoir–Robilard Parameters

Fig. 13.8 shows typical erosive burning-rate data for a nitropolymer propellant. The propellant is composed of $\xi_{NC}(0.504)$, $\xi_{NG}(0.366)$, and $\xi_{DEP}(0.130)$. The burning rate increases with increasing mainstream velocity, u, or with increasing mass velocity, G, when u_g is increased beyond about 100 m s^{-1}. The burning rate remains unchanged when the flow velocity is below 100 m s^{-1}. The experimental data are correlated with $\alpha = 0.8 \times 10^{-4}$ m$^{2.8}$ kg$^{-0.8}$s$^{-0.2}$ and $\beta = 270$ for a distance from the fore-end of $L = 80$ mm. Fig. 13.9 shows the relationship between the blow-off parameter and the cross-flow velocity. The parameter values used for the computation are $\beta = 270$, $\rho_p = 1.57 \times 10^3$ kg m^{-3}. The blow-off parameter increases with increasing cross-flow velocity at constant burning rate, and decreases with increasing burning rate at constant cross-flow velocity.

When the momentum of the cross-flow above the burning surface is much smaller than the momentum of the combustion gas flow perpendicular to the burning surface, the heat flux from the cross-flow to the burning surface remains unchanged. No erosive burning occurs below a certain cross-flow velocity. This cross-flow velocity is termed the threshold velocity of erosive burning. In general, the threshold velocity increases with increasing burning rate without cross-flow at constant pressure.

Though erosive burning is highly dependent on the cross-flow velocity, the physical structure of the propellant also plays a dominant role in determining the erosive

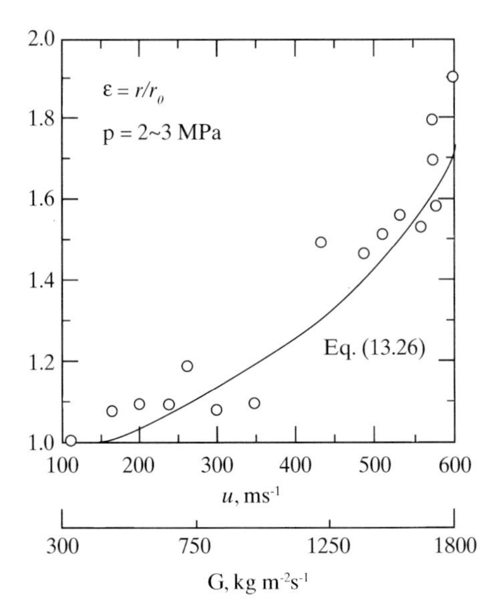

Fig. 13.8 Erosive burning model calculation and experimental data for erosive ratio as a function of gas flow velocity or mass flow velocity.

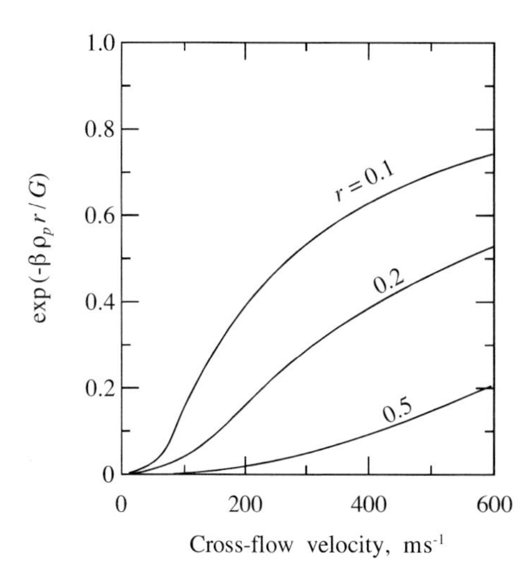

Fig. 13.9 Computed blow-off parameter as a function of cross-flow velocity.

burning characteristics. When a propellant is composed of a binder that readily melts prior to its thermal decomposition, the pressure exponent of burning ratetends to decrease, i. e., plateau burning characteristics are seen, without a cross-flow velocity. A typical melting binder is polyurethane (PU). Fig. 13.10 shows the burning rate of a propellant composed of $\xi_{AP}(0.8)$ and $\xi_{PU}(0.2)$ without cross-flow velocity. Plateau burning occurs between 3 MPa and 7 MPa. When the AP-PU propellant is burned under a cross-flow velocity at 3.3 MPa, the burning rate decreases with increasing cross-flow velocity and reaches a minimum at about 370 m s^{-1}, as shown in Fig. 13.11. The burning rate is decreased from 7.4 mm s^{-1} without cross-flow to 5.7 mm s^{-1} at a cross-flow velocity of 370 m s^{-1}, i. e., a decrease of 23 % in the burning rate. This erosive burning is termed "negative erosive burning" and is never observed for the plateau burning of double-base propellants or AP-HTPB composite propellants. Upon further increase of the cross-flow velocity beyond 370 m s^{-1}, the burning rate starts to increase once more, in a similar manner as the erosive burning characteristics of conventional propellants. The heat flux trans-

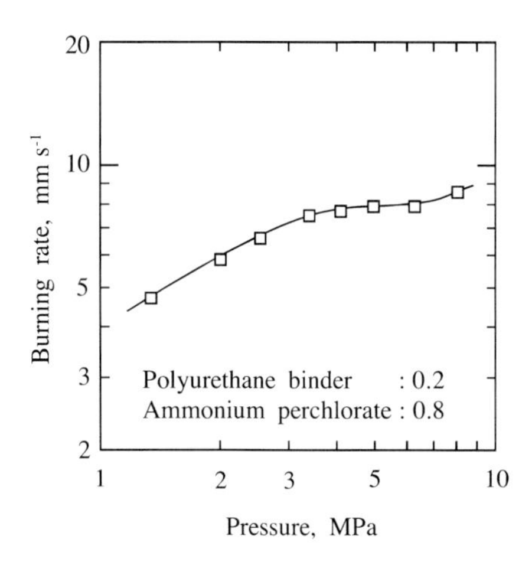

Fig. 13.10 Burning rate characteristics of an AP-PU composite propellant, showing plateau burning.

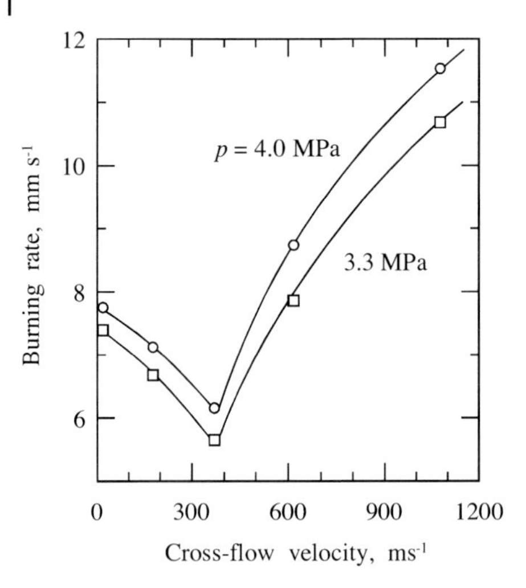

Fig. 13.11 Negative erosive burning of an AP-PU composite propellant.

ferred from the main cross-flow stream to the burning surface increases as long as a high-temperature flow-velocity is applied to the burning surface, as shown in Fig. 13.2. Close observation of the burning surface during burning reveals that it is partly covered with a molten layer of the binder used as a fuel component of the propellant under cross-flow conditions. This molten layer prevents decomposition of the AP particles incorporated into the propellant and the heat flux transferred back from the gas phase to the condensed phase is decreased. When the cross-flow velocity is increased beyond 170 m s^{-1}, the shear stress exerted on the burning surface by the laminar sublayer in the boundary layer removes the molten layer. The heat flux to the burning surface is then increased and normal erosive burning is established as for other propellants.

13.4 Combustion Instability

13.4.1 T^* Combustion Instability

If the combustion products of a propellant attain a state of thermal equilibrium, the combustion temperature may be determined theoretically, as described in Chapter 2. However, the combustion in a rocket motor is incomplete and so the flame temperature remains below the adiabatic flame temperature.[5] If one assumes that the flame temperature, T^*, varies with pressure, p_c, in a rocket motor, T^* is expressed by[5]

$$T^* = bp_c^{2m} \qquad (13.28)$$

where m is the pressure exponent of the flame temperature and b is a constant within a certain pressure range. The mass balance for rocket motor operation in a steady state is represented by

$$\rho_p r A_b = \zeta A_t p_c / T^{*1/2} \tag{13.29}$$

where ζ is a parameter of the combustion gas given by

$$\zeta = \sqrt{\frac{\gamma}{R_g}\left(\frac{2}{\gamma+1}\right)^{\frac{\gamma+1}{2(\gamma-1)}}} \tag{13.30}$$

Equation (13.29) is derived in Section 14.1.3. Using the burning rate of a propellant given by Eqs. (3.68) and (13.29), the criterion for stable burning is:[5]

$$n + m < 1 \tag{13.31}$$

This criterion is the so-called T^* combustion instability. The stability criterion expressed by $n < 1$ is not sufficient to obtain stable combustion when the flame temperature is dependent on pressure.[1]

In general, m is approximately zero in the high-pressure region for most propellants. However, T_f of nitropolymer propellants such as single-base and double-base propellants decreases with decreasing pressure below about 5 MPa. Since direct determination of m is difficult, the heat of explosion, H_{exp}, is evaluated as a function of pressure. The flame temperature, denoted by T^*, is determined by assuming an average specific heat of the combustion products, c_p, according to:

$$T^* = H_{exp}/c_p + T_0 \tag{13.32}$$

where T_0 is the initial propellant temperature.

Combustion tests carried out for a rocket motor demonstrate a typical T^* combustion instability. Double-base propellants composed of NC-NG propellants with and without a catalyst (1% nickel powder) were burned. Detailed chemical compositions of both propellants are given in Section 6.4.6 and the burning rate characteristics are shown in Fig. 6.29. The addition of nickel is seen to have no effect on burning rate and the pressure exponent is $n = 0.70$ for both propellants.

The heat of explosion of the uncatalyzed NC-NG propellant decreases rapidly when the pressure is decreased below about 4 MPa, as shown in Fig. 13.12. However, the heat of explosion of the catalyzed NC-NG propellant remains relatively unchanged, even below 2 MPa.

Substituting the measured H_{exp} data into Eqs. (13.32) and (13.28), one obtains an experimentally determined m value as a function of pressure. The results indicate that m of the uncatalyzed NC-NG propellant is approximately 0.35 at 0.6 MPa, gradually decreases with increasing pressure, and becomes zero at 5 MPa and above; m of the catalyzed NC-NG propellant is approximately zero over the same tested pressure range. Substituting the results for m and n into Eq. (13.35), one gets

$$m + n > 1 \quad \text{at } p < 1.2\ \text{MPa}$$

$$m + n < 1 \quad \text{at } p > 1.2\ \text{MPa}$$

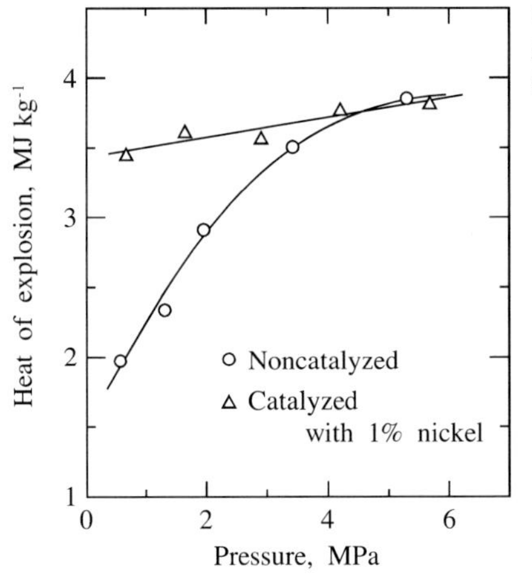

Fig. 13.12 Effect of the addition of nickel powder on the heat of explosion of a double-base propellant.

for the uncatalyzed NC-NG propellant, and

$m + n < 1$ throughout the pressure range tested

for the catalyzed NC-NG propellant.

The combustion tests conducted for a rocket motor show that the combustion becomes unstable below 1.7 MPa and that the burning acquires a chuffing mode in the case of the uncatalyzed propellant. However, as expected, the combustion is stable even below 0.5 MPa for the nickel-catalyzed NC-NG propellant, as shown in Fig. 13.13. Propellants for which the flame temperature decreases with decreasing pressure tend to exhibit T^* combustion instability.

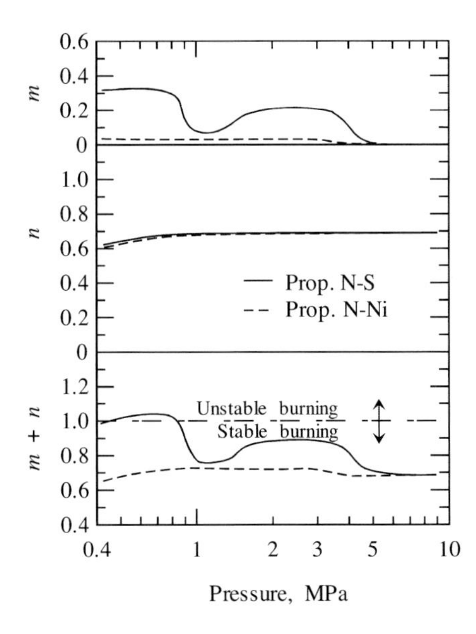

Fig. 13.13 T^* combustion instability evaluated on the basis of the $m + n$ stability criterion.

13.4.2
L^* Combustion Instability

Low-frequency oscillation in a rocket motor depends on the pressure exponent of propellant burning rate and the free volume of the chamber.[6,7] When a double-base propellant composed of $\xi_{NC}(0.510)$, $\xi_{NG}(0.355)$, and $\xi_{DEP}(0.120)$, with $\xi_{PbSa}(0.015)$ as a platonizing catalyst, burns in a strand burner, the burning rate characteristicsmay be divided into four zones, as shown in Fig. 13.14. The pressure exponent varies between the pressure zones: $n = 0.44$ in zone I above 3.7 MPa, $n = 1.1$ in zone II between 3.7 MPa and 2.1 MPa, $n = 0.77$ in zone III between 2.1 MPa and 1.1 MPa, and $n = 1.4$ in zone IV below 1.1 MPa.

When the propellant grain burns in a rocket motor in end-burning mode, L^* changes from 4 m to 20 m during the burning. Typical pressure–time traces are shown in Fig. 13.15. In zone I, at pressures above 3.7 MPa, the burning is very stable as expected. In zone II, in which the pressure exponent is approximately unity, a sinusoidal oscillatory burning with frequency in the range 6–8 Hz occurs when L^* is short.[7] The highest amplitude of the oscillation is about 20% of the time-averaged pressure and the oscillatory burning diminishes as L^* increases. In zone III, between 2.1 MPa and 1.1 MPa, in which the pressure exponent is 0.77, the burning is stable without pressure oscillation. In zone IV, in which the pressure exponent is 1.4, stable burning is not possible because the mass discharge rate from the nozzle is always higher than the mass generation rate in the chamber. The domains of stable burning, oscillatory burning, and unstable burning are shown in Fig. 13.16.

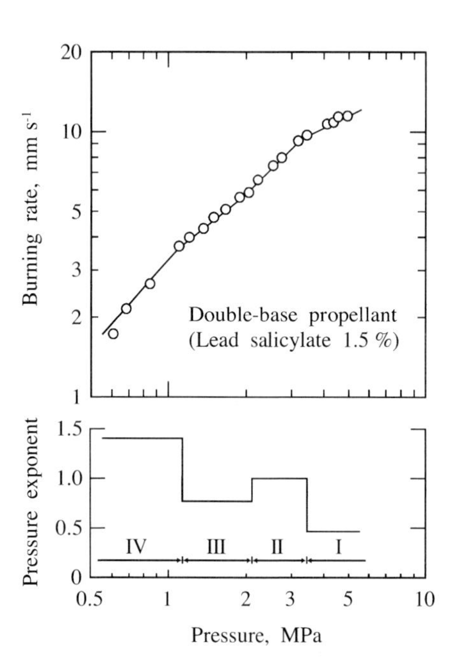

Fig. 13.14 Burning rate and pressure exponent of a lead-catalyzed double-base propellant.

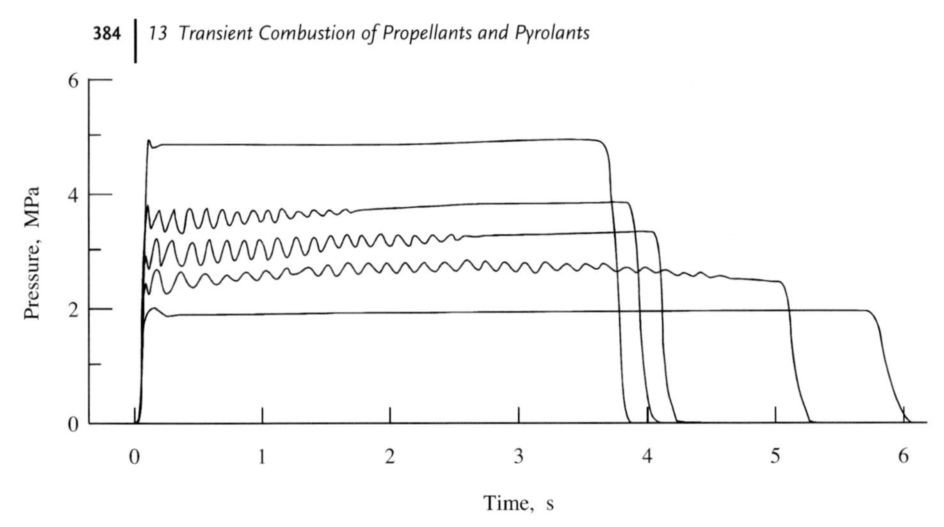

Fig. 13.15 Pressure versus time curves for a lead-catalyzed propellant in a rocket motor.

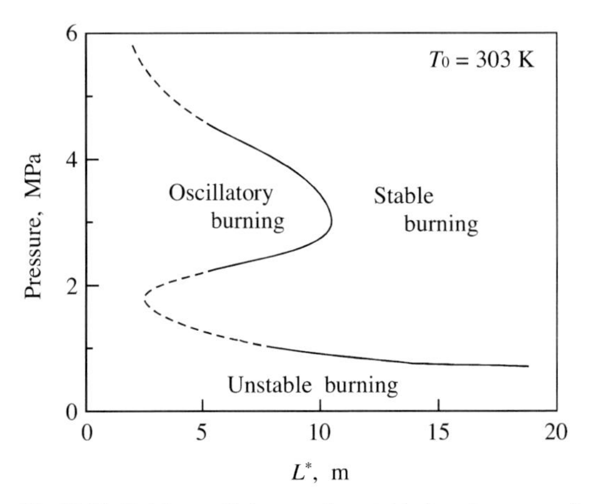

Fig. 13.16 Stable, oscillatory, and unstable burning zones for a lead-catalyzed double-base propellant.

Assuming that the gas in the chamber behaves as an ideal gas and neglecting the density of the gas compared to that of the propellant, the mass balance for a rocket motor is given by

$$\tau_{ch} dp/dt + p - (\rho_p K_n / c_D) r = 0 \tag{13.33}$$

where τ_{ch} is the chamber time constant given by $\tau_{ch} = L^* / c_D R_g T_f$. It is assumed that the amplitude of the pressure oscillation is sinusoidal according to:

$$p(t) = p_c + \phi e^{\alpha t} \cos \omega t \tag{13.34}$$

where α is the exponential growth constant, ϕ is the amplitude of pressure oscillation, and ω is the frequency. In the case of a very slow oscillation, the period of the oscillation, $2\pi/\omega$, is much larger than the characteristic time of the thermal wave of the propellant, τ_{th}, thus a linear burning rate law, $r = ap^n$, may apply during oscillatory burning. Assuming that the propellant burns with a lead time, τ, relative to the pressure oscillation, the instantaneous burning rate is

$$r(t) = a\{p(t + \tau)\}^n = a\{p_c + \phi e^{\alpha(t+\tau)}\cos\omega(t + \tau)\}^n \tag{13.35}$$

In general, the amplitude of the pressure oscillation is much smaller than the mean pressure as $\phi/p_c << 1$, and thus the burning rate is given by

$$r(t) = ap_c^{\,n} + anp_c^{\,n-1}\,\phi e^{\alpha(t+\tau)}\cos\omega(t + \tau) \tag{13.36}$$

Substituting Eqs. (13.34) and (13.36) into Eq. (13.33), the following set of equations is obtained:

$$\cos\omega\tau = (1 + \alpha\tau_{ch})/n \tag{13.37 a}$$

$$\sin\omega\tau = \omega\tau_{ch}/n \tag{13.37 b}$$

Combining Eqs. (13.37 a) and (13.37 b), one obtains the following relationship between α and ω:

$$(1 + \alpha\tau_{ch})^2 + (\omega\tau_{ch})^2 = n^2 \tag{13.38}$$

Since the pressure exponent of burning rate is less than unity for conventional propellants, α becomes negative, and the burning becomes stable. In the case of n being greater than unity, α becomes positive, and increasingly oscillatory burning may occur. When n is very close to unity, α becomes approximately zero, and ω can be determined from the approximation

$$\omega \approx (n^2 - 1)^{1/2}/\,\tau_{ch} \tag{13.39}$$

Though α and ω are not uniquely determined by Eq. (13.39) at given τ_{ch} and p_c, Eq. (13.37 a) gives the relationship:

$$\alpha \leq (n - 1)/\,\tau_{ch} \tag{13.40}$$

and indicates that α decreases with increasing τ_{ch} at a given pressure, p_c. In other words, oscillatory burning diminishes with increasing L^*, as given by the relationship $L^* = (c_D R_g T_f)\,\tau_{ch} = V_c/A_t$.

This predicted trend is consistent with the observed oscillatory burning behavior in zone II. Since α in zone II is determined to be of the order of 1 s^{-1} or less, ω may be calculated from Eq. (13.38). For example, the calculated ω is 46 rad s^{-1} when τ_{ch} is 0.01 s ($L^* \approx 4$ m) at $p_c = 3.5$ MPa. Furthermore, the term $\omega\,\tau_{ch}/n$ is of the order of

0.1 rad in the oscillatory zone; it is obtained from the approximation of Eq. (13.37 b) to be equal to τ_{ch}.

13.4.3
Acoustic Combustion Instability

13.4.3.1 Nature of Oscillatory Combustion

When an energetic material burns in a combustion chamber fitted with an exhaust nozzle for the combustion gas, oscillatory combustion occurs. The observed frequency of this oscillation varies widely from low frequencies below 10 Hz to high frequencies above 10 kHz. The frequency is dependent not only on the physical and chemical properties of the energetic material, but also on its size and shape. There have been numerous theoretical and experimental studies on the combustion instability of rocket motors. Experimental methods for measuring the nature of combustion instability have been developed and verified.[6] However, the nature of combustion instability has not yet been fully understood because of the complex interactions between the combustion wave of propellant burning and the mode of acoustic waves.

When combustion instability occurs for an internal burning grain of a rocket motor, the burning rate of the grain varies with time and so does the pressure in the rocket motor. The pressure versus time curve shows oscillations of a certain frequency. When the propellant burning mode is not in harmony with the pressure oscillation mode, the combustion instability tends to decay. However, when the burning mode is in harmony with the oscillation mode, the pressure oscillation is amplified.

There are many oscillatory modes for an internal-burning propellant grain in a rocket motor.[6] The acoustic wave travelling from the fore-end of the motor to the rear-end is called the longitudinal mode. The wave travelling in a radial direction in the port of the internal propellant grain is called the radial mode, and that travelling in a tangential direction in the port is called the tangential mode. Oscillatory combustion appears to become coupled with an acoustic wave travelling over the burning surface of the propellant in the motor. In pressure-coupled oscillatory combustion, the combustion chamber cavity acts as an acoustic oscillator in establishing acoustic pressure waves. In some cases, velocity-coupled oscillatory combustion occurs due to the oscillating gas flow parallel to the burning surface. This velocity-coupled oscillation increases the local burning rate of the propellant grain. The oscillatory gas flow increases the rate of heat transfer from the gas flow to the burning surface, similarly to the combustion phenomenon of local erosive burning. As a result of pressure- or velocity-coupled oscillation, the mean chamber pressure often rises to more than twice the designed chamber pressure. However, when the propellant burning is not coupled with the acoustic mode, no oscillatory combustion occurs and the burning rate is given by Vieille's law, $r = ap^n$, i. e., as a function of the static pressure in the rocket motor.

If one assumes a simplified sinusoidal oscillation, the frequency of the oscillation, ν, is represented by

$$\nu = a/\lambda = na/2x \qquad (13.41)$$

where λ is wavelength, a is acoustic velocity, n is the number of standing waves, and x is the distance from one end of the wave to the other. The pressure in the chamber, represented by $p(r, t)$, is dependent on the position, r, and time, t. One assumes that the pressure in the motor is given by

$$p(r, t) = p_c + \delta_p \tag{13.42}$$

where p_c is the time-averaged pressure and δ_p is the pressure variation with position and time due to the acoustic pressure wave travelling in the chamber.

13.4.3.2 Combustion Instability Test

Fig. 13.17 shows the structure and principle of a T-burner, as used to measure the response function of propellants. Two propellant samples are placed at the respective ends of the T-burner. The burner is pressurized with nitrogen gas to the test pressure level. The acoustic mode of the burning established in the burner is uniquely determined by the speed of sound therein and the distance between the burning surfaces of the two samples. When the propellant samples are ignited, pressure waves travel from one end to the other between the burning surfaces of the samples. When a resonance pressure exists for a certain length of the T-burner, the propellant is sensitive to the frequency.[6] The response function is determined by the degree of amplification of the pressure level.

Combustion of a propellant in a rocket motor accompanied by high-frequency pressure oscillation is one of the most harmful phenomena in rocket motor operation. There have been numerous theoretical and experimental studies on the acoustic mode of oscillation, concerning both the medium-frequency range of 100 Hz–1 kHz and the high-frequency range of 1 kHz–30 kHz. The nature of oscillatory combustion instability is dependent on various physicochemical parameters, such

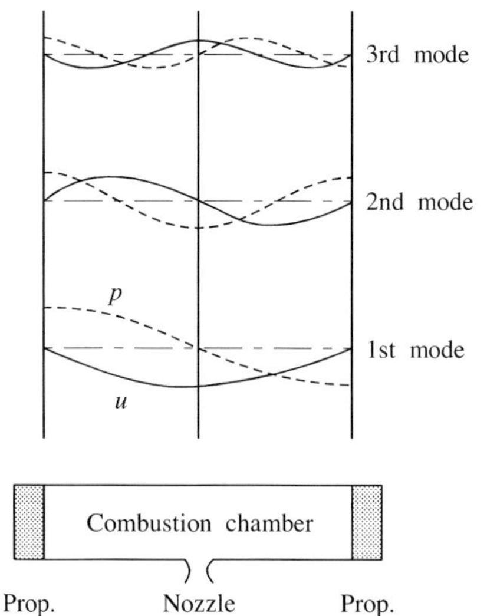

Fig. 13.17 Combustion modes in a T-burner.

as the burning rate characteristics, mechanical properties, and energy density of the propellants, and the physical shape and dimensions of the propellant grain.[8,9]

In general, when an internal burning grain is relatively long, more than 2 m, a longitudinal mode is set up from the head-end to the nozzle. When the port of an internal burning grain is small, less than 0.2 m, a tangential mode or a radial mode across the interior surface of the port occurs. In order to evaluate the sensitivity of burning rate to pressure oscillation for various types of propellants, the response function, i. e., burning rate sensitivity to burning pressure, is determined for each propellant grain experimentally.

The shape of the propellant grain used for combustion instability tests is a six-pointed-star geometry with internal burning or a circular port with internal and twofold end burning, in order to obtain a flat pressure burning with time. An igniter is placed at the head-end of the motor and a convergent-divergent nozzle is placed at the rear-end. The pressure traces in the motor are measured by means of two types of transducers, both of which are mounted on the head-end closure. One type, to measure the DC components of the pressure, consists of a strain gauge pick-up, while the other type, to measure the AC components of the pressure, consists of a piezoelectric pick-up. The recorded data are displayed on a triggered sweep dual-beam oscilloscope that records the DC and AC components of the pressure traces simultaneously. The recorded AC pressure data are analyzed by using a bandpass filter to determine the power spectral density during combustion instability.[8,9]

Fig. 13.18 shows DC and AC pressure curves for an RDX-AP composite propellant having a six-pointed-star geometry burning at 9 MPa. The combustion sud-

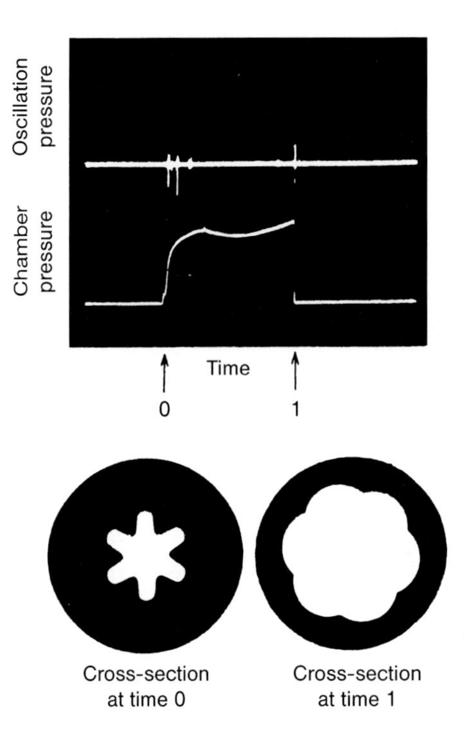

Fig. 13.18 Combustion instability of a six-pointed-star grain, showing DC and AC components of pressure and the cross-section of the grain (1) before ignition and (2) after burning interruption.

denly becomes unstable after ignition and is interrupted by the nozzle ejection caused by the pressure peak. This rapid pressure decay acts to extinguish the propellant grain in the motor. Cross-sections of the grain before ignition and after burning interruption are also shown in Fig. 13.18. Fig. 13.19 shows a typical set of pressure–time curves for stable and unstable oscillatory combustion in a rocket motor. It has been well established that solid particles in a combustion chamber absorb the kinetic energy of the oscillatory motion of the combustion gas and hence suppress this motion. The tested RDX-AP composite propellants were composed of $\xi_{RDX}(0.43)$ of 120 μm in diameter, $\xi_{AP}(0.43)$ of 20 μm in diameter, and $\xi_{HTPB}(0.14)$, with and without 0.5–2.0 % Al particles (5 μm in diameter) or 0.5–2.0 % Zr particles (5 μm in diameter). The burning rate remains unchanged by the addition of Al or Zr particles, as shown in Fig. 13.20. The pressure exponents of the propellants are all $n = 0.46$, with or without Al or Zr particles.[9] The interior grain geometry is a cylindrical, and the grain burns at its internal cylindrical surface and also at both ends, in order to maintain a flat pressure versus time curve. As shown in Fig. 13.19 (a), burning of the propellant grain without Al particles is accompanied by high-frequency pressure oscillation. The oscillatory burning pressure returns to the nor-

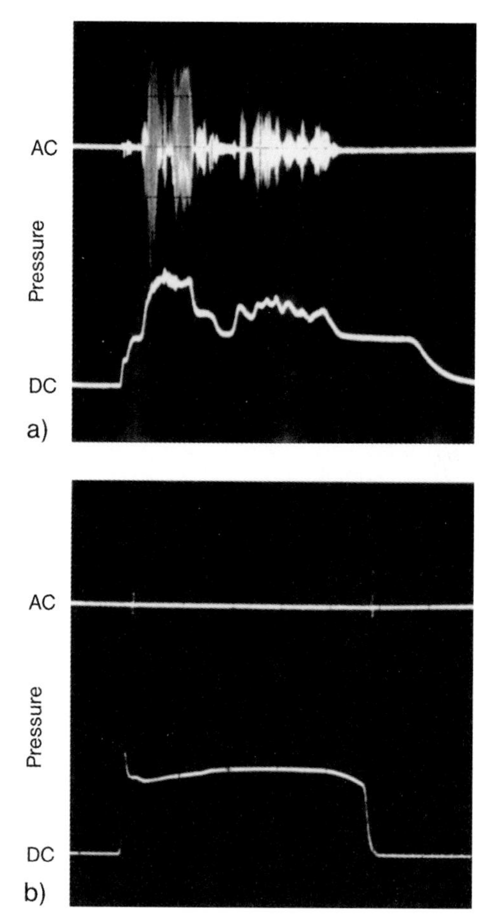

Fig. 13.19 Pressure–time curves of stable and unstable combustion in a rocket motor.

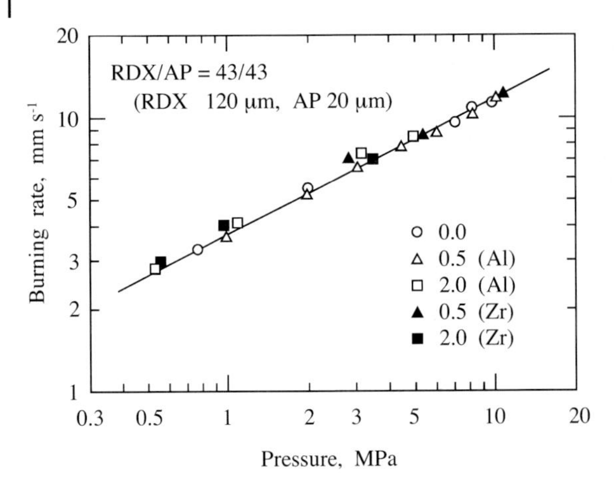

Fig. 13.20 Burning rate characteristics of RDX-AP composite propellants with or without metal particles.

mal burning pressure after a certain time and the propellant continues to burn steadily until it is completely consumed. On the other hand, the propellant grain containing Al particles burns normally after ignition and the pressure versus burning time curve appears as expected, as shown in Fig. 13.19 (b).

Fig. 13.21 shows another example of oscillatory burning of an RDX-AP composite propellant containing 0.40% Al particles. The combustion pressure chosen for the burning was 4.5 MPa. The DC component trace indicates that the onset of the instability is 0.31 s after ignition, and that the instability lasts for 0.67 s. The pressure instability then suddenly ceases and the pressure returns to the designed pressure of 4.5 MPa. Close examination of the anomalous bandpass-filtered pressure traces reveals that the excited frequencies in the circular port are between 10 kHz and 30 kHz. The AC components below 10 kHz and above 30 kHz are not excited, as shown in Fig. 13.21. The frequency spectrum of the observed combustion instability is shown in Fig. 13.22. Here, the calculated frequency of the standing waves in the rocket motor is shown as a function of the inner diameter of the port and frequency. The sonic speed is assumed to be 1000 m s^{-1} and L = 0.25 m. The most excited frequency is 25 kHz, followed by 18 kHz and 32 kHz. When the observed frequencies are compared with the calculated acoustic frequencies shown in Fig. 13.23, the dominant frequency is seen to be that of the first radial mode, with possible inclusion of the second and third tangential modes. The increased DC pressure between 0.31 s and 0.67 s is considered to be caused by a velocity-coupled oscillatory combustion. Such a velocity-coupled oscillation tends to induce erosive burning along the port surface. The maximum amplitude of the AC component pressure is 3.67 MPa between 20 kHz and 30 kHz.[8,9]

When two propellant grains with and without Al particles are arranged along the axis of a motor, i. e., one grain without Al particles is placed at the fore-end and the other grain is placed at the rear-end, the motor combustion instability occurs at about 0.3 s after ignition, as shown in Fig. 13.24 (a). The motor is broken by the rapid pressure increase caused by the instability and the propellant combustion is

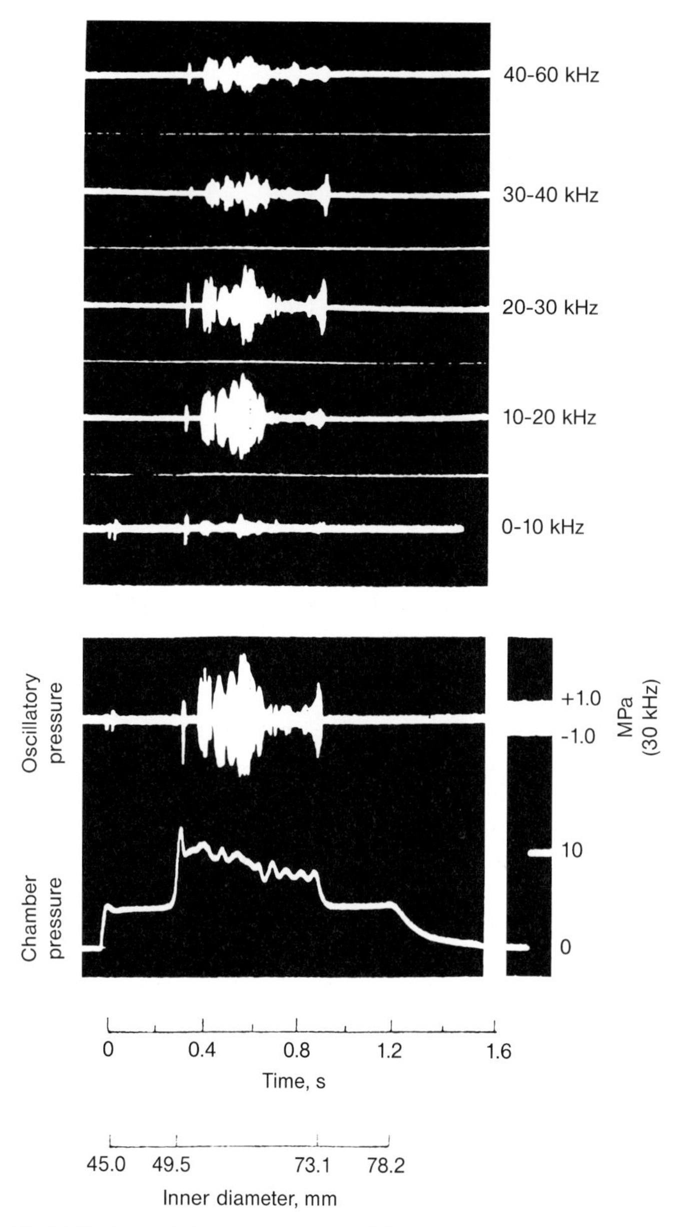

Fig. 13.21 DC- and AC-pressure curves of the combustion instability of an RDX-AP composite propellant containing 0.4% Al particles.

terminated. This is similar to the combustion instability seen when one propellant grain without Al particles is tested. When the arrangement of the two propellant grains is reversed, i. e., the grain with Al particles is placed at the fore-end and the grain without Al particles is placed at the rear-end, no combustion instability occurs

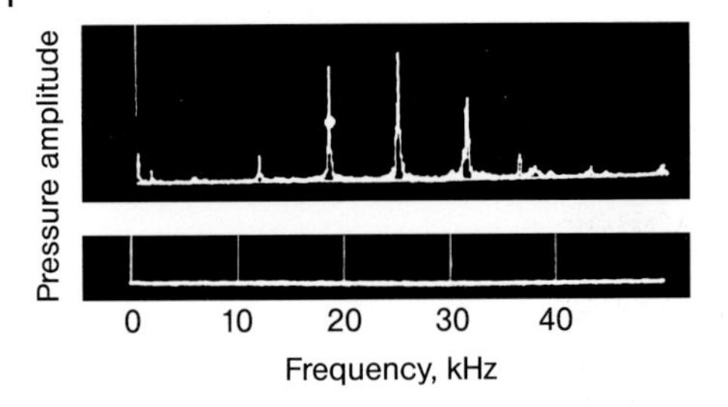

Fig. 13.22 Frequency spectrum of the oscillatory burning shown in Fig. 13–21.

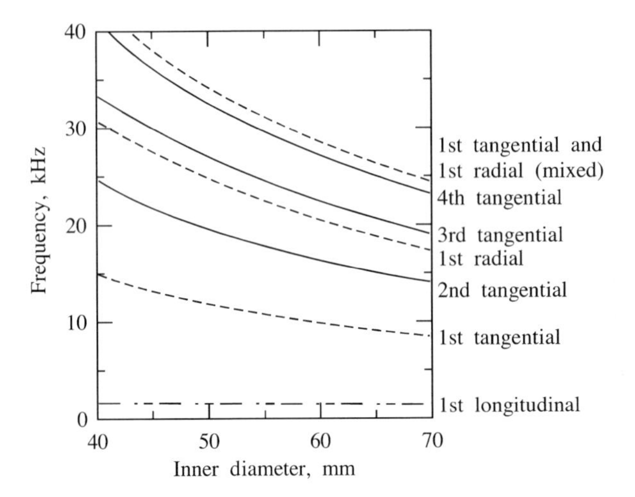

Fig. 13.23 Acoustic modes of the combustion test shown in Fig. 13–21, shown as frequency versus port diameter.

and the normal steady-state combustion pressure is attained until the two propellant grains are consumed, as shown in Fig. 13.24 (b). These results indicate that this type of combustion instability is suppressed by the dispersal of Al particles (possibly Al_2O_3 particles) throughout the internal free volume in the combustion chamber. When the propellant grain without Al particles is placed at the fore-end of the motor, no particles are emitted into the free volume of the chamber during burning. The observed combustion instability is caused in the free volume around the grain without Al particles, even though Al particles are present in the free volume around the grain positioned at the rear-end.

It is evident that the standing pressure wave in a rocket motor is suppressed by solid particles in the free volume of the combustion chamber. The effect of the pressure wave damping is dependent on the concentration of the solid particles, and the size of the particles is determined by the nature of the pressure wave, such as the frequency of the oscillation and the pressure level, as well as the properties of the combustion gases. Fig. 13.25 shows the results of combustion tests to determine the effective mass fraction of Al particles. When the propellant grain without Al particles is burned, there is breakdown due to the combustion instability. When

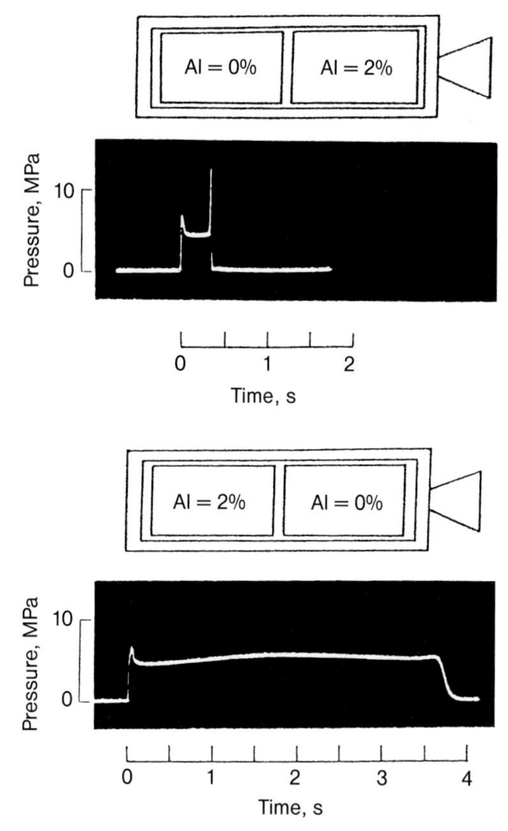

Fig. 13.24 The combustion of a non-aluminized RDX-AP composite propellant becomes stable when an aluminized RDX-AP propellant is positioned at the fore-end of the combustion chamber.

0.47 % Al particles is incorporated into the propellant grain, a strong pressure peak is observed at about 0.3 s after ignition. However, the burning then returns to its normal pressure. When 0.50 % Al particles is incorporated, no combustion instability is seen, and the propellant is consumed completely and smoothly. It is evident that the minimum content of Al particles lies between 0.47 % and 0.50 %. Fig. 13.26 shows the regions of stable and unstable combustion in a plot of metal particle concentration versus pressure. The combustion instability of the tested RDX-AP composite propellants is seen to be suppressed by the addition of 0.5 % of Al or Zr particles. However, these results cannot be generalized or extrapolated to the suppression of combustion instability encountered with other types of propellants. Suppression of combustion instability is dependent on various physicochemical characteristics, such as the chemical composition of the propellant, the size and geometry of the propellant grains, the size and concentration of the solid particles incorporated into the propellant, the propellant burning rate, and the mechanical properties of the propellant.

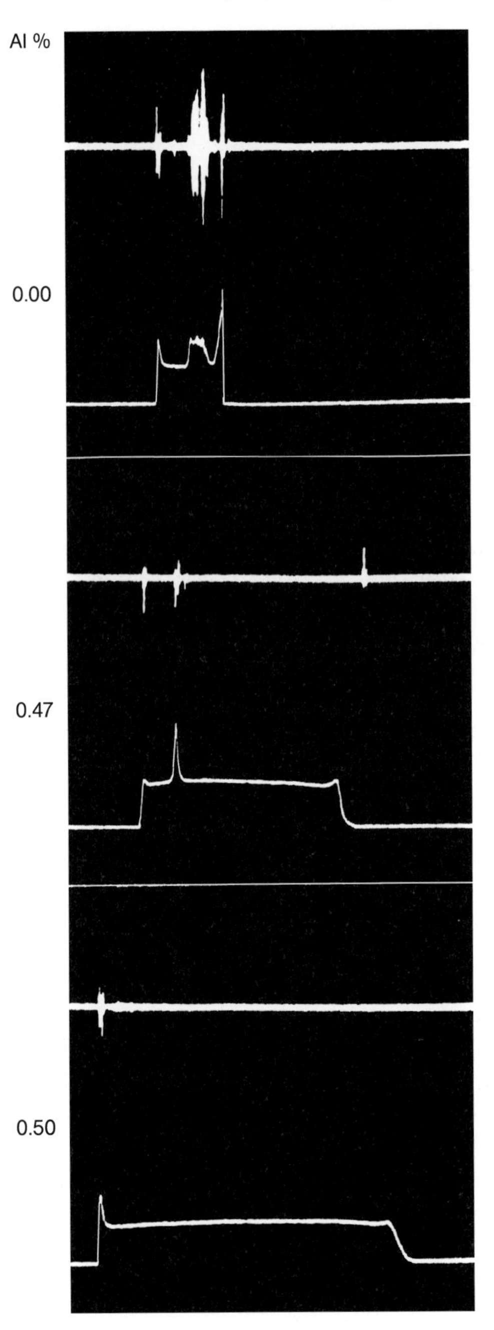

Fig. 13.25 Combustion instability is suppressed as the concentration of aluminum particles is increased (the average designed chamber pressure is 4.5 MPa).

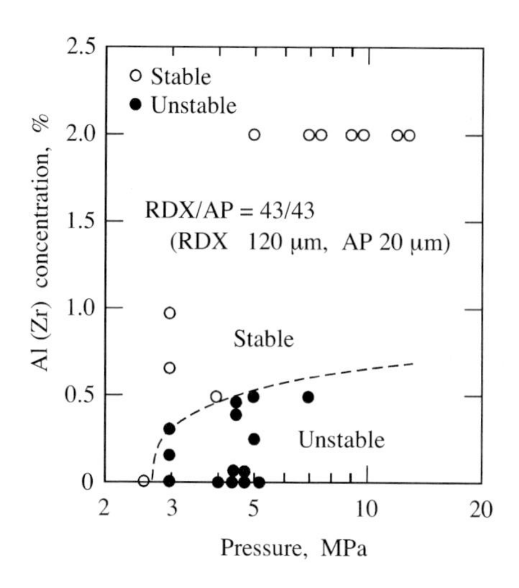

Fig. 13.26 Stable and unstable combustion regions for an RDX-AP composite propellant.

13.4.3.3 **Model for Suppression of Combustion Instability**

If the pressure oscillation represented by Eq. (13.42) is sinusoidal in shape, the pressure variation at a fixed position in the combustion chamber is expressed by

$$\delta_p = p_0 \exp(\alpha + i\omega)t \tag{13.43}$$

where α is the growth constant and ω is the angular frequency of the oscillation. The real number of α can be represented by

$$\alpha = \alpha_g + \alpha_d \tag{13.44}$$

where α_g is an instability driving constant and α_d is a damping constant. Note that α_d depends on various physicochemical parameters, such as the geometry of the combustion chamber and nozzle, the propellant elasticity, and the viscosity of the combustion products. However, since the observed combustion instability consists of transverse modes generated in the circular port, the damping effects are considered to be small compared with the particle damping effect. If one assumes that the major damping effect is due to acoustic attenuation of the particles in the gas phase,[1] then combustion instability occurs when $\alpha_g + \alpha_p = 0$, where α_p is the particle damping constant. The criterion for the occurrence of combustion instability is then defined as $\alpha_g = -\alpha_p$.

According to the theory proposed by Horton and McGie,[10] the particle damping constant is determined according to:

$$-\alpha_p = \sum_D 3\pi n_D R\nu(1+Z)\left\{\frac{16Z^4}{16Z^4 + 72\delta Z^3 + 81\delta^2(1+2Z+2Z^2)}\right\} \tag{13.45}$$

where n_D = number of particles per unit volume with diameter D, ν = kinematic viscosity of the combustion gas, δ = ratio of the densities of the combustion gas and the particles, $Z = (\omega R^2/2\nu)^{1/2}$, and R = radius of the particles.[11] The damping factor increases with increasing frequency at constant particle size, and the maximum damping factor is achieved by optimized selection of the diameter and the number of particles at constant particle density contained within the propellant.

13.5
Combustion under Acceleration

13.5.1
Burning Rate Augmentation

When a propellant or pyrolant containing metal particles burns under the influence of an acceleration field along the burning direction, the combustion modeis altered compared with that under no acceleration and hence the burning rate is altered.[12–14] Since the acceleration force acts on the mass of the metal particles, the size of the particles is an important parameter in determining the effect on the burning rate. In general, the metal particles, such as those of aluminum and magnesium, incorporated into a propellant melt and agglomerate at the burning surface. On the other hand, the oxidizer particles, such as ammonium perchlorate(AP) or potassium perchlorate (KP), decompose thermally to generate oxidizer gas. The molten metal particles react with the oxidizer gas that surrounds them to form metal oxide particles. When the propellant is subject to an acceleration force along the direction of the burning surface, the formation process of the molten particles is modified compared to that without acceleration. Thus, the heat-transfer process is altered by the acceleration and hence the burning rate of the propellant is also altered.

When an aluminized propellant burns under acceleration in the direction of the regressive burning, the acceleration force acts on the molten aluminum or on its oxide formed on the burning surface. Since the force acting on the solid or condensed materials is much larger than that acting on the gaseous combustion products, some of the aluminum particles incorporated into the propellant remain on the burning surface and accumulate to form molten aluminum agglomerates.

When a propellant grain burns with internal burning in a spinning chamber, a centrifugal acceleration acts on both the propellant grain and its combustion products. When solid particles are among the combustion products, the centrifugal acceleration also acts on these. The combustion gas and the solid particles generated in the spinning chamber are expelled from the nozzle attached to the end of the chamber. The combustion process of the metal particles is altered by the centrifugal acceleration and hence the burning rate of the propellant grain is also altered.[12–14]

The combustion gas in the port of an internal-burning grain flows along the burning surface of the port towards the nozzle, where it is no longer subject to centrifugal acceleration. When the combustion gas is subject to centrifugal acceleration within the port, a radial pressure gradient is created in the port and hence the burning rate is increased due to the increased pressure. However, the effect of this centrifugal acceleration is trivial when the combustion products are composed only of gaseous species.

The acceleration force in the radial direction acting on the solid particles, F_r, is represented by

$$F_r = mr\omega^2 \tag{13.46}$$

where m is the mass of each solid particle, r is the distance between the center of the spinning chamber and the burning surface of the propellant, and ω is the number of spins per unit time. Since the density of the combustion gas in the chamber is much less than that of the solid particles, the combustion gas flows along the internal burning surface and flows out from the nozzle attached to the end of the spinning chamber. However, if F_r is larger than the pressure force acting in the radial direction towards the spinning center of the chamber, the solid particles remain on the internal burning surface of the propellant grain. This pressure force is created by the combustion between the propellant burning surface and the surface of each solid particle.

13.5.2
Effect of Aluminum Particles

When aluminum particles are incorporated into a pyrolant grain, a number of molten aluminum agglomerates are formed on the burning surface under a centrifugal force.[12–14] This agglomeration process occurs due to the higher density of the molten metal particles compared with that of the gaseous products. The size of the molten agglomerates is increased as the burning of the propellant grain proceeds. Since the temperature of the molten agglomerates is high, the heat flux transferred from the agglomerates to the propellant burning surface is increased. Accordingly, the local burning rate of the propellant is increased, as a result of which a number of pits are formed on the burning surface, as shown in Fig. 13.27.[13]

When a grain of a pyrolant consisting of $\xi_{AP}(0.768)$, $\xi_{CTPB}(0.150)$, and $\xi_{Al}(0.082)$ in the shape of a six-pointed star is burned in a spinning combustion chamber, many pits due to agglomerated aluminum or aluminum oxide particles are formed along the outline of the star. The diameter of the aluminum particles in this example was 48 μm. The combustion test was conducted at 4 MPa under a centrifugal acceleration of 60 g.[14] When a tubular-shaped grain is burned under a spinning

Fig. 13.27 An extinguished burning surface of a six-pointed-star aluminized AP-CTPB propellant grain. Many pits are formed along a tip of the star.

motion, many pits are formed non-uniformly over the whole surface. It is important to note that the burning surface area is increased due to the formation of these pits, and consequently the pressure in the chamber is increased further.

13.6
Wired Propellant Burning

13.6.1
Heat-Transfer Process

The heat flux transferred back from the gas phase to the condensed phase plays a dominant role in determining the burning rates of propellants. The burning rate increases with increasing pressure because of the increased reaction rate in the gas phase and the associated increased heat flux transferred back to the propellant. When a fine metal wire is embedded within a propellant along the burning direction, heat is transferred back to the condensed phase through the wire by conduction.[15,16] Since the heat conductivities and thermal diffusivities of metals are much greater than those of gases, the rate of heat transfer through metal wires is much greater than that through gases. Therefore, the heat flux transferred back from the gas phase to the burning surface via a protruding metal wire is greater than the heat flux transferred back by the combustion gas. The temperature of the propellant surrounding the wire is increased due to this conducted heat and a local portion of the propellant is ignited. Thus, the apparent burning rate of the propellant surrounding the wire is faster than the actual overall burning rate of the propellant.[15,16] Fig. 13.28 shows the burning surface of a propellant strand with an embedded silver wire.[16] A cone-shaped burning surface is seen in the vicinity of the silver wire, rather than the usual flat burning surface, and the combustion progresses rapidly along the wire. For observation of the burning surface structure, an NC-NG double-base propellant strand is used, which is relatively transparent and thus allows observation of the regression of the cone-shaped burning surface. Since the burning surface area is increased due to the presence of the embedded silver wire, the propellant grain is subject to an intermediate type of burning between end burning and internal burning.

The heat produced in the gas phase is transferred back to the burning surface of the propellant and is also transferred back to the condensed phase beneath the burning surface by conduction. In other words, the temperature of the propellant beneath the burning surface is increased by heat conducted through the propellant material prior to burning. When a metal wire is embedded within the propellant along the burning direction, the heat produced in the gas phase is transferred through the wire that protrudes into the gas phase. Heat is conducted from the high-temperature zone of the wire protruding into the gas phase to the low-temperature zone of the wire beneath the burning surface. Since the heat transfer within the wire occurs along the wire direction, x, and in the radial direction of the wire, r, the temperature distribution in the wire is represented by

$$(r_w/\alpha_w)\partial T/\partial x = \partial^2 T/\partial x^2 + \partial^2 T/\partial r^2 + (1/r)\partial T/\partial r \tag{13.47}$$

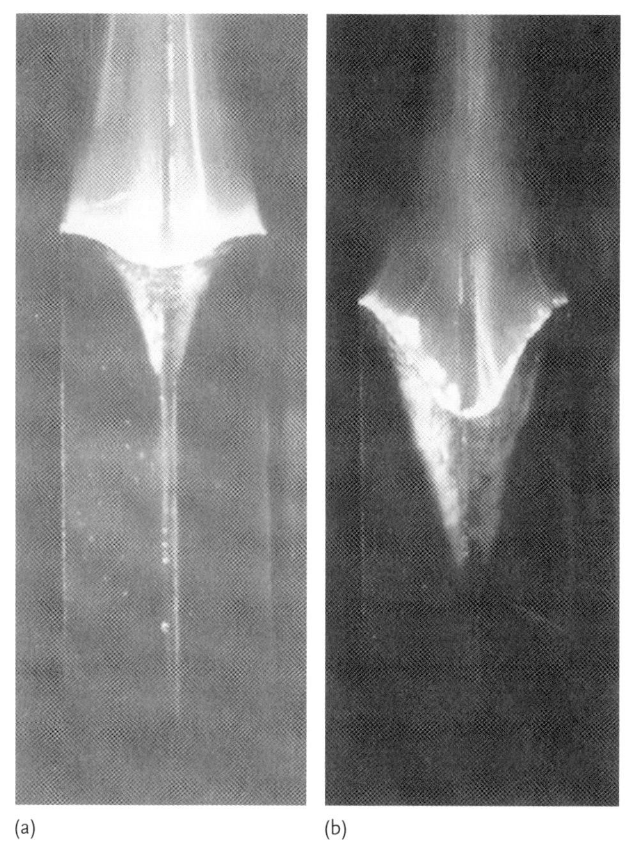

(a) (b)

Fig. 13.28 Combustion photographs of a wired propellant: a cone-shaped burning surface is formed around the silver wire (0.8 mm in diameter) (a) 0.8 s after ignition and (b) 1.2 s after ignition.

where r_w is the burning rate of the propellant along the wire, T is temperature, and α_w is the thermal diffusivity of the wire. For steady-state burning of the propellant, r_w is constant and the temperature distribution along the wire becomes constant. Equation (13–47) can be separated at the interface of the gas phase and the condensed phase, $x = 0$. For the portion of the wire in the condensed phase ($x \leq 0$), heat is transferred from the wire to the propellant by conduction. For the portion of the wire in the gas phase ($x > 0$), heat is transferred from the gas phase to the wire. Appropriate heat conduction between the wire and propellant at the interface must be specified in Eq. (13–47) as boundary conditions. In any case, the thermal diffusivity α_w, given by $\lambda_w/c_w\rho_w$, plays a dominant role in determining the rate of heat transfer from the wire to the propellant, where c_w is the specific heat and ρ_w is the density of the wire.

13.6.2
Burning Rate Augmentation

When a propellant strand with an embedded metal wire is ignited at the top, the burning surface regresses in a cone shape with the metal wire as its axis. The burning rate perpendicular to the cone surface remains unchanged, while the burning rate along the metal wire is increased. The cone angle of the burning surface around the metal wire is large at first, but gradually becomes smaller as the burning progresses. A larger burning surface transition rate is obtained with a finer silver wire. Furthermore, the burning surface transition rates caused by silver, copper, and iron wires reflect the relative thermal diffusivities of the metals. The thermal diffusivities of silver, copper, and iron are 1.65×10^{-4} m^2 s^{-1}, 1.30×10^{-4} m^2 s^{-1}, and 0.18×10^{-4} m^2 s^{-1}, respectively. These values are consistent with the results obtained by Caveny et al.[15]

Fig. 13.29 shows the burning rates of a propellant with three types of wires embedded, namely silver (Ag), tungsten (W), and nickel (Ni) wires.[7] The propellant is a double-base propellant composed of $\xi_{AP}(0.44)$, $\xi_{NG}(0.43)$, $\xi_{DEP}(0.11)$, and $\xi_{EC}(0.2)$. The burning rate of the non-wired propellant, r_0, is increased approximately 4 times by the Ag wire, 2.5 times by the W wire, and 1.5 times by the Ni wire at 1 MPa. The pressure exponent is 0.60 for the non-wired, 0.51 for the Ag-wired, 0.55 for the W-wired, and 0.51 for the Ni-wired propellant. No significant effect of the various wires on the pressure exponent is thus seen. The higher the thermal diffusivity of the metal wire, the higher the burning rate along the wire, r_w.[16]

The heat flux transferred by a metal wire is also dependent on its diameter. The rate of heat transfer from a high-temperature gas to a wire protruding from a burning surface is determined by the thickness of the wire. When the wire is thin, its rate of heat gain is fast because its heat capacity is low. However, the rate of heat loss

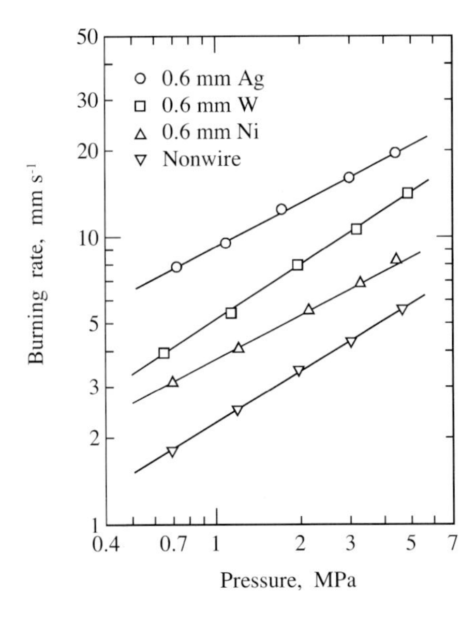

Fig. 13.29 Effect of various metal wires on the burning rate of a propellant along the wires.

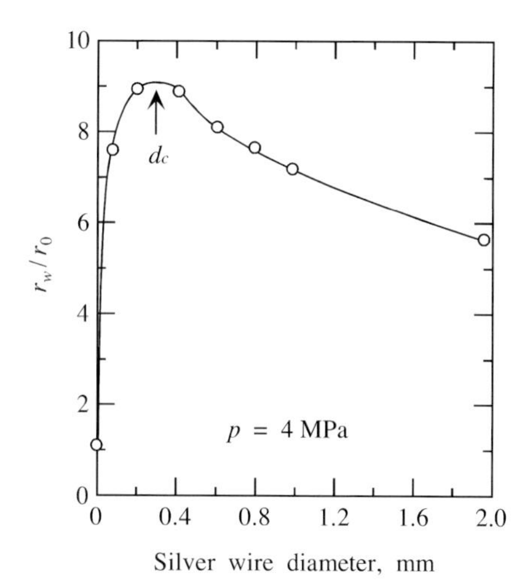

Fig. 13.30 Effect of silver wire diameter on burning rate along the wire, showing that the maximum burning rate is obtained at a diameter of about 0.3 mm.

from the heated wire to the propellant beneath the burning surface is also low. Fig. 13.30 shows the variation of r_w/r_0 with the diameter of a silver wire at 4 MPa.[16] As the diameter of the wire is decreased, r_w/r_0 increases in the region above the optimal diameter, d_c. The ratio r_w/r_0 reaches a maximum when the diameter of the wire is d_c, approximately 0.3 mm in the present case. When the silver wire is very thin, 5 µm or 50 µm in diameter, no cone-shaped burning surface is seen along its length.

The temperature of the propellant around the silver wire is increased smoothly from its initial value, T_0, to the temperature in the gas phase. The temperature at the burning surface of a nitropolymer propellant is approximately 570 K.[4] The temperature starts to rise at about 10 mm below the burning surface of the propellant with the embedded silver wire, whereas the temperature of the propellant without the wire only starts to rise within 1 mm below the burning surface. The thermal influence of an embedded silver wire in the radial direction extends at most to 70–100 µm around the wire.[2] The temperature gradient in the silver wire inside the propellant at the burning surface increases when the diameter of the wire is increased from 0.6 mm to 1.0 mm. This indicates that the heat flux penetrates deeper inside the propellant through the wires as the wire thickness is made smaller. Thus, the temperature of the propellant around the wire increases. It must be noted that the temperature gradient in the wire inside of the propellant at the burning surface is significantly lower than that in the propellant at the burning surface.

When a mass fraction of 0.10 of ammonium perchlorate (AP: 200 µm in diameter) is incorporated into a nitropolymer propellant, i. e., to form an AP-CMDB propellant, numerous diffusional flamelets are formed above the burning surface due to the mixing of the oxidizer-rich gaseous decomposition products of the AP and the fuel-rich gaseous decomposition products of the nitropolymer propellant.[16] The luminous flame zone approaches the burning surface and the average flame temperature increases from 2270 K to 2550 K following the addition of the AP particles. However, the fizz zone structure remains relatively unchanged and the heat flux transferred back from the fizz zone to the burning surface remains unchanged.

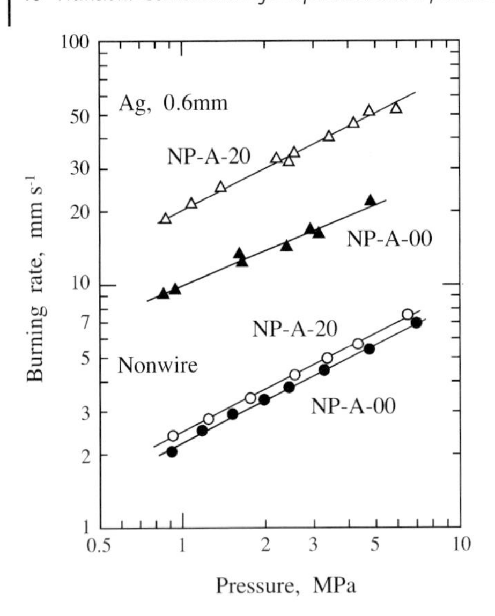

Fig. 13.31 Burning rate augmentation of an NC-NG double-base propellant and an AP-CMDB propellant along silver wires.

Thus, the burning rate remains almost unchanged by the addition of the AP particles, as shown in Fig. 13.31. The chemical compositions of the propellants with and without AP particles are shown in Table 13.2. The absence of an effect is due to the decomposition of AP particles on the burning surface, as a result of which the heat flux transferred back to the burning surface remains almost unchanged. However, the decomposition products of the AP react above the burning surface and generate diffusional flamelets. Consequently, the gas-phase temperature in the dark zone is increased. When a silver wire (0.8 mm in diameter) is embedded within both propellants, a significant difference in the degree of burning rate augmentation between the double-base propellant (NP-A-00) and the AP-CMDB propellant (NP-A-20) is observed, as also shown in Fig. 13.31. The silver wire that protrudes into the dark zone is heated by the high-temperature diffusional flamelets. Thus, the burning rate along the silver wire is increased, even though the average burning rate of the AP-CMDB propellant remains unchanged. When silver wires of different diameters are embedded within the AP-CMDB propellant (NP-A-20), there are no significant differences in the degree of the burning rate augmentation, as shown in Fig. 13.32.

Higher burning rates are also obtained by incorporating chopped fine wires or chopped fine metal sheets into propellant grains during casting processes. Though

Table 13.2 Chemical compositions of the propellants with or without AP particles.

Propellant	ξ_{NC}	ξ_{NG}	ξ_{DEP}	ξ_{EC}	ξ_{AP}
NP-A-00	44.0	43.0	11.0	2.0	–
NP-A-20	35.2	34.4	8.8	1.6	20.0

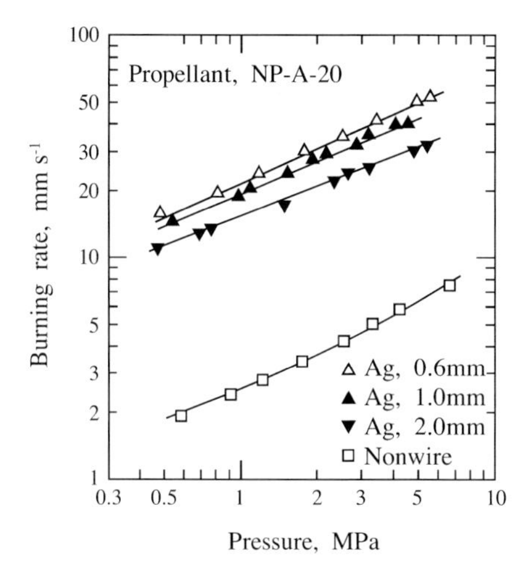

Fig. 13.32 Effect of silver wire diameter on burning rate augmentation of an AP-CMDB propellant.

the wires and sheets are mixed randomly within the grains, the average thermal diffusivity of the grains is increased. Additional heat is conducted from the gas phase to the propellant through the wires and sheets that protrude from the burning surface, and hence the burning rate is increased.

References

1 Kubota, N., Rocket Combustion, Nikkan Kogyo Press, Tokyo (1995).

2 Hermance, C. E., Solid Propellant Ignition Theories and Experiments, Fundamentals of Solid-Propellant Combustion, Chapter 5, Vol. 90, Progress in Astronautics and Aeronautics, AIAA (Eds.: Kuo, K. K., and Summerfield, M.), New York, 1984, pp. 239–304.

3 Razdan, M. K., and Kuo, K. K., Erosive Burning of Solid Propellants, Fundamentals of Solid-Propellant Combustion, Chapter 10, Vol. 90, Progress in Astronautics and Aeronautics, AIAA (Eds.: Kuo, K. K., and Summerfield, M.), New York, 1984, pp. 515–598.

4 Ishihara, A., and Kubota, N., Erosive Burning Mechanism of Double-Base Propellants, 21 st Symposium (International) on Combustion, The Combustion Institute, Pittsburgh, PA (1986), pp. 1975–1981.

5 Kubota, N., Role of Additives in Combustion Waves and Effect on Stable Combustion Limit of Double-Base Propellants, *Propellants and Explosives*, Vol. 3, 1978, pp. 163–168.

6 Price, E. W., Experimental Observations of Combustion Instability, Fundamentals of Solid-Propellant Combustion, Chapter 13, Vol. 90, Progress in Astronautics and Aeronautics, AIAA (Eds.: Kuo, K. K., and Summerfield, M.), New York, 1984, pp. 733–790.

7 Kubota, N., and Kimura, J., Oscillatory Burning of High Pressure Exponent Double-Base Propellants, *AIAA Journal*, Vol. 15, No. 1, 1977, pp. 126–127.

8 Kubota, N., Kuwahara, T., Yano, Y., Takizuka, M., and Fukuda, T., Unstable Combustion of Nitramine/Ammonium Perchlorate Composite Propellants, AIAA-81–1523, AIAA, New York (1981).

9 Kubota, N., Yano, Y., and Kuwahara, T., Particulate Damping of Acoustic Instability in RDX/AP Composite Propellant Combustion, AIAA-82–1223, AIAA, New York (1982).

10 Horton, M. D., and McGie, M. R., Particulate Damping of Oscillatory Combus-

tion, *AIAA Journal*, Vol. 1, No. 6, 1963, pp. 1319–1326.

11 Epstein, P. A., and Carhart, R. R., The Absorption of Sound in Suspensions and Emulsions, I: Water Fog in Air, *J. of Acoustic Soc. Am.*, Vol. 25, No. 3, 1953, pp. 553–565.

12 Niioka, T., and Mitani, T., Independent Region of Acceleration in Solid Propellant Combustion, *AIAA Journal*, Vol. 12, No. 12, 1974, pp. 1759–1761.

13 Niioka, T., Mitani, T., and Ishii, S., Observation of the Combustion Surface by Extinction Tests of Spinning Solid Propellant Rocket Motors, Proceedings of the 11 th International Symposium on Space Technology and Science, Tokyo, 1975, pp. 77–82.

14 Niioka, T., and Mitani, T., An Analytical Model of Solid Propellant Combustion in an Acceleration Field, *Combustion and Science and Technology*, Vol. 15, 1977, pp. 107–114.

15 Caveny, L. H., and Glick, R. L., The Influence of Embedded Metal Fibers on Solid Propellant Burning Rate, *Journal of Spacecraft and Rockets*, Vol. 4, No. 1, 1967, pp. 79–85.

16 Kubota, N., Ichida, M., and Fujisawa, T., Combustion Processes of Propellants with Embedded Metal Wires, *AIAA Journal*, Vol. 20, No. 1, 1982, pp. 116–121.

14
Rocket Thrust Modulation

14.1
Combustion Phenomena in a Rocket Motor

14.1.1
Thrust and Burning Time

Rocket propellants are used to generate high-temperature and high-pressure combustion products in a rocket motor. The principal design criterion of a rocket motor is to obtain a desired pressure versus burning time curve, so as to obtain a required thrust versus time curve. Two types of propellant burning are used, as shown in Fig. 14.1: internal burning type and end burning type. Internal burning is used to obtain high thrust and a short burning time, whereas end burning is used to obtain low thrust and a long burning time. Some combinations of internal and end burnings are used to obtain medium thrust and a medium burning time. However, there exists a "not possible zone" in the thrust versus burning time relationship, as shown in Fig. 14.2, due to the limited burning rate characteristics of propellants.

The burning rate of the propellant is one of the important parameters in designing an optimized rocket motor. The burning surface area of a propellant grain in a combustion chamber is equally as important as its burning rate. When an internal-burning or internal-external-burning grain is used, a high mass burning rate is ob-

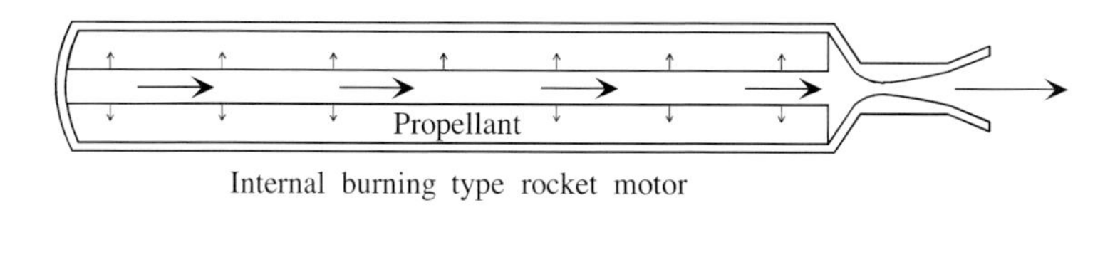

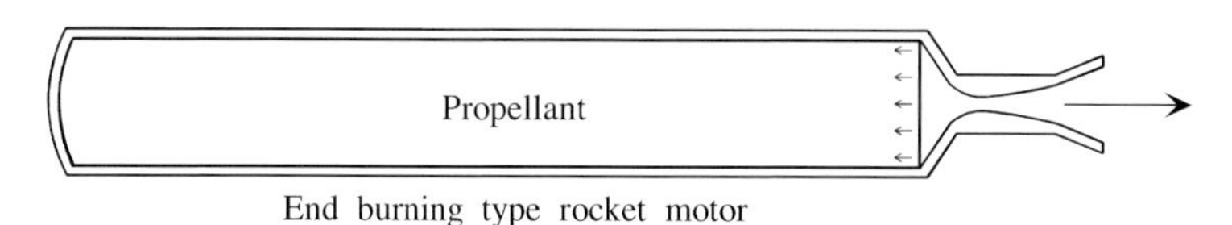

Fig. 14.1 Propellant burning in rocket motors.

Propellants and Explosives. Naminosuke Kubota

ISBN: 978-3-527-31424-9

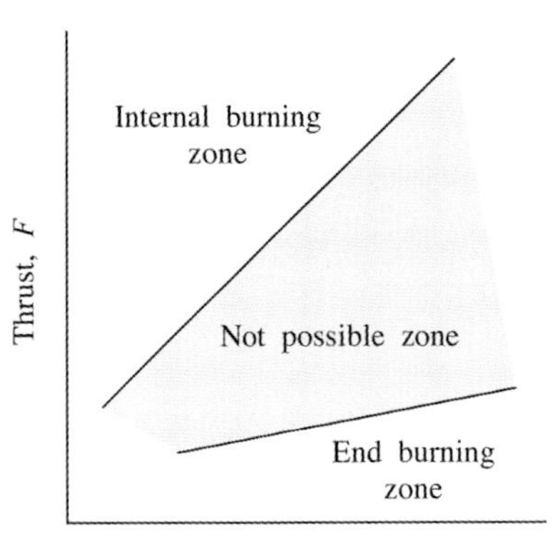

Fig. 14.2 Zones of burning in rocket motors.

tained. On the other hand, when an end-burning grain is used, a low mass burning rate is obtained. As described in Section 13.6, a burning surface area intermediate between that of internal-burning and end-burning may be obtained by using a wired propellant grain. Fig. 14.3 shows three pressure versus burning time curves for a rocket motor, representing external-internal-burning, end-burning, and wired-burning propellants. The propellant grains are made of an AP-HTPB composite propellant. Three silver wires ($d = 1.2$ mm) are embedded in the wired-burning propellant grain. The pressure versus burning time curve for the wired-burning grain lies between the curves for the external-internal-burning and end-burning grains. The "not possible zone" shown in Fig. 14.2 becomes a "possible zone" by the use of wired propellant grains.

Since the burning rate of a propellant is dependent on the burning pressure, the mass balance between the mass generation rate in the chamber and the mass discharge rate from the nozzle is determined by the pressure.[1–4] In addition, the propellant burning rate in a rocket motor is affected by various phenomena that influence the mass balance relationship. Fig. 14.4 shows typical combustion phenomena encountered in a rocket motor, from pressure build-up by ignition to pressure decay upon completion of the combustion.

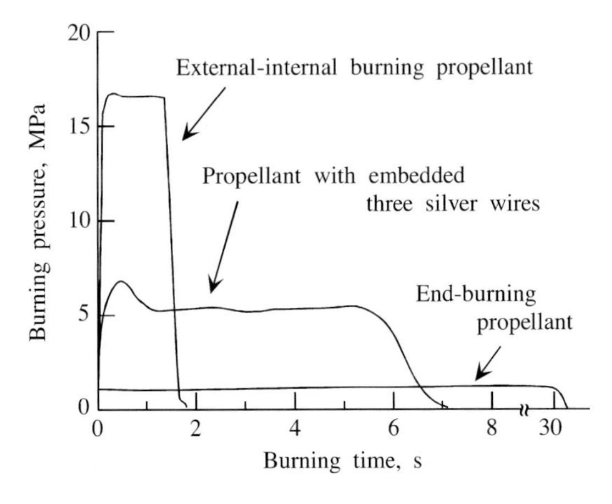

Fig. 14.3 Pressure versus burning time curves for a wired burning, an external-internal burning, and an end burning.

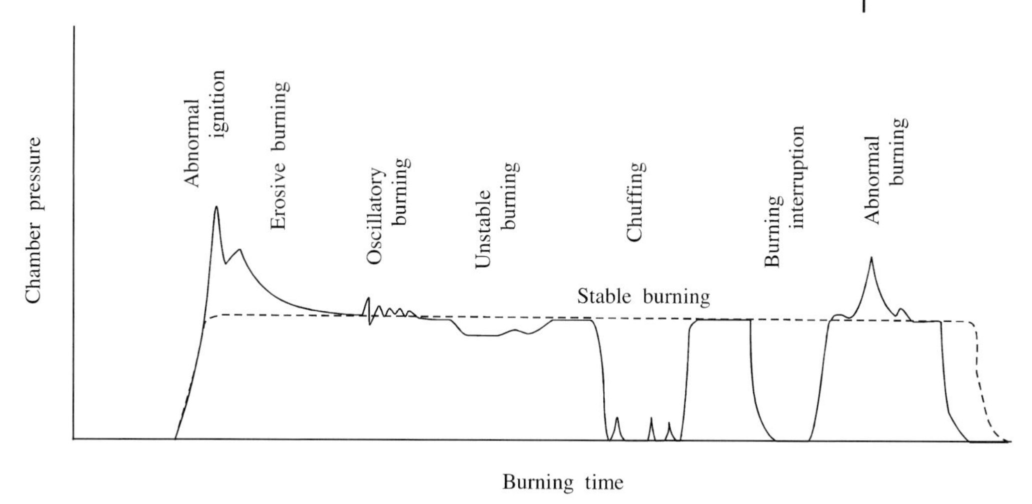

Fig. 14.4 Combustion phenomena in a rocket motor, from ignition to completion of the combustion.

Once a propellant grain is ignited, there is convective, conductive, and/or radiative heat transfer to the unburned portion of the propellant grain. If excess heat flux is supplied to the unburned material, the propellant grain generates an excess of combustion gas and an over-pressure is created, i. e., abnormal ignition occurs. Immediately after ignition of an internal burning grain as shown in Fig. 14.1, the gas flow velocity along the flow direction is high and the heat flux transferred from the combustion gas flow to the grain surface is also high. As a result, the burning rate of the grain perpendicular to the flow direction increases drastically, i. e., erosive burning occurs. The pressure in the rocket motor then increases abnormally. The erosive burning soon diminishes as the port area of the internal burning increases.

14.1.2 Combustion Efficiency in a Rocket Motor

The combustion performance of a rocket motor is dependent on various physicochemical processes that occur during propellant burning. Since the free volume of a rocket motor is limited for practical reasons, the residence time of the reactive materials that produce the high temperature and high pressure for propulsion is too short to allow completion of the reaction within the limited volume of the motor as a reactor. Though rocket motor performance is increased by the addition of energetic materials such as nitramine particles or azide polymers, sufficient reaction time for the main oxidizer and fuel components is required.

Metal particles, most commonly aluminum particles, are also known as additives for propellants and pyrolants that increase the combustion temperature and hence also the specific impulse. However, a heat-transfer process from the high-temperature gas to the aluminum particles is required to melt the particles and then a subsequent diffusional process of oxidizer fragments toward each aluminum particle is required to complete the combustion. When the residence time of the aluminum particles is insufficient, the oxidation cannot be completed and the overall combustion efficiency in the rocket motor becomes much lower than the theoretically expected value.

It is well understood that AP-HTPB composite propellants burn very effectively to produce high-temperature combustion products. When energetic materials are incorporated into AP-HTPB propellants, the combustion performance is increased by the increased temperature and/or increased gas volume. However, in many cases, the experimental performance of rocket propellants containing nitramine or aluminum particles falls short of the theoretical performance owing to incomplete reaction of these additives. The reaction processes of these additives include various physicochemical changes, such as thermal decomposition, melting, sublimation, diffusional processes, and the oxidation of gaseous decomposition products by the oxidizer gases generated by decomposition of the AP particles. When aluminum particles are incorporated into AP-HTPB propellants, their combustion remains incomplete in a rocket motor.

The specific impulse of a rocket motor, I_{sp}, as defined in Eq. (1.75), is dependent on both propellant combustion efficiency and nozzle performance. Since I_{sp} is also defined by Eq. (1.79), rocket motor performance can also be evaluated in terms of the characteristic velocity, c^*, defined in Eq. (1.74) and the thrust coefficient, c_F, defined in Eq. (1.70). Since c^* is dependent on the physicochemical parameters in the combustion chamber, the combustion performance can be evaluated in terms of c^*. On the other hand, c_F is dependent mainly on the nozzle expansion process, and so the nozzle performance can be evaluated in terms of c_F. Experimental values of c^*_{exp} and c_{Fexp} are obtained from measurements of chamber pressure, p_c, and thrust, F:

$$c^*_{exp} = p_c A_t / \dot{m}_g \tag{14.1}$$

$$c_{Fexp} = F / A_t p_c \tag{14.2}$$

where A_t is the nozzle throat area, and $\dot{m}_g$ is the mass generation rate in the combustion chamber. Since the total propellant mass in the combustion chamber, M_p, and the burning time, t_b, are known, $\dot{m}_g$ is determined as the time-averaged mass generation rate, $\dot{m}_g = M_p/t_b$. These values can be measured experimentally with relatively high accuracy.

Figs. 14.5 and 14.6 show theoretical and experimental values of c^* and c_F for aluminized AP-RDX-HTPB propellants as a function of the mass fraction of aluminum particles ξ_{Al}. The propellant consists of $\xi_{AP+Al}(0.73)$, $\xi_{RDX}(0.15)$, and $\xi_{HTPB}(0.12)$. The AP particles are a bimodal mixture of 400 μm (40 %) and 20 μm (60 %) in diameter. The average RDX particle size is 120 μm in diameter. The average size of the aluminum particles is 20 μm in diameter. The chamber pressure is 7 MPa and the nozzle expansion ratio is $\varepsilon = 6$.

The theoretical c^*_{th} increases with increasing ξ_{Al} up to $\xi_{Al}(0.20)$ and then decreases with increasing ξ_{Al} in the region $\xi_{Al} > 0.20$; the theoretical c_{Fth} increases linearly with increasing ξ_{Al}. The burning test results indicate that both c^*_{exp} and c_{Fexp} are lower than the theoretical c^*_{th} and c_{Fth} in the examined ξ_{Al} range. As ξ_{Al} is in-

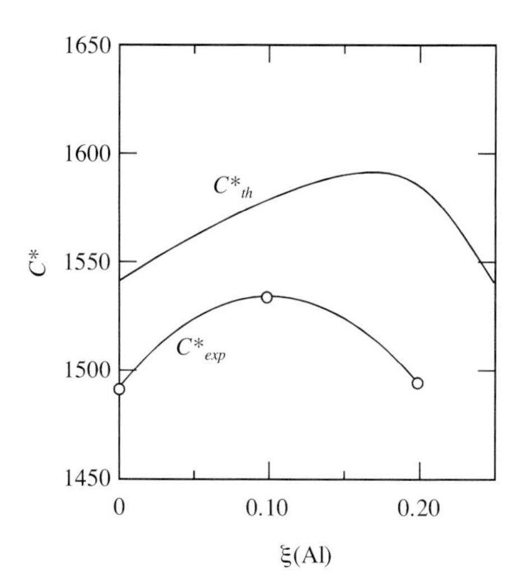

Fig. 14.5 Theoretical and experimental characteristic velocity of AP-RDX-HTPB propellants as a function of ξ(Al).

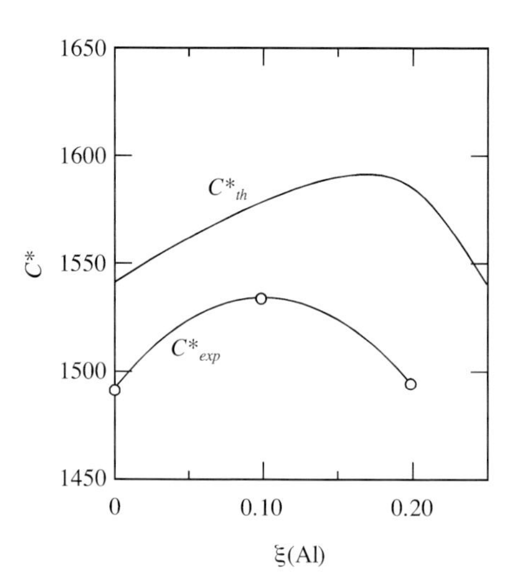

Fig. 14.6 Theoretical and experimental thrust coefficient of AP-RDX-HTPB propellants as a function of ξ(Al) .

creased, c^*_{exp} first increases, reaching a maximum at $\xi_{Al}(0.10)$, and thereafter decreases once more. The fact that c^*_{exp} decreases in the region $\xi_{Al} > 0.10$ indicates that the combustion efficiency of the aluminum particles becomes low when ξ_{Al} becomes high. This is due to agglomeration of the aluminum particles to form relatively large particles that cannot be completely oxidized. The specific impulse is determined according to Eq. (1.79) as $I_{sp} = c_F c^*/g$. Fig. 14.7 shows the variation of the theoretical specific impulse $I_{sp,th}$, the experimental specific impulse, $I_{sp,exp}$, and the I_{sp} efficiency, $\eta = I_{sp,exp}/I_{sp,th}$, with propellant composition. It is evident that η; decreases with increasing ξ(Al) due to incomplete combustion of the Al particles and a two-phase flow of gas and solid and/or liquid particles in the nozzle.

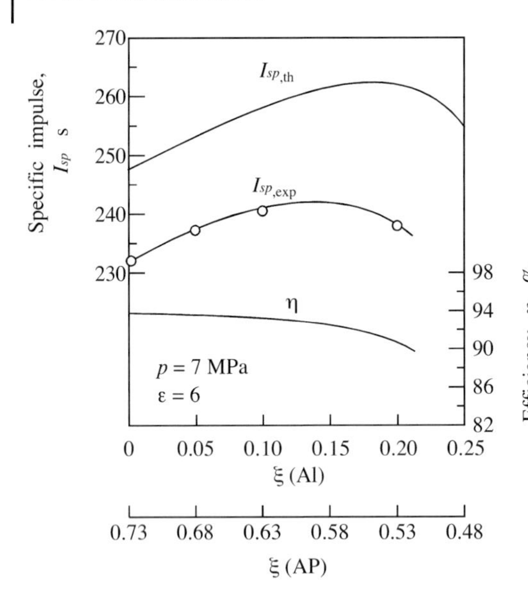

Fig. 14.7 Theoretical and experimental specific impulse and Isp efficiency of aluminized AP-RDX-HTPB propellants as a function of ξ(Al).

14.1.3
Stability Criteria for a Rocket Motor

Internal burning of a propellant grain in a rocket motor leads to pressure oscillationin the motor. If the strength of this pressure oscillation is high, the rocket motor may burst or be damaged. When the pressure exponent of burning rate, n, is high, unstable burning such as low-frequency pressure oscillation, chuffing, or burning interruption occurs.[4]

Let us consider a propellant burning in a rocket motor as shown in Fig. 14.8. The mass generation rate in the chamber, $\dot{m}_g$, is given by

$$\dot{m}_g = \rho_p A_b r \tag{14.3}$$

where A_b is the burning surface area and ρ_p is the propellant density. The mass discharge rate from the nozzle, $\dot{m}_d$, is given by Eq. (1.60) and is represented by

$$\dot{m}_d = c_D A_t p_c \tag{14.4}$$

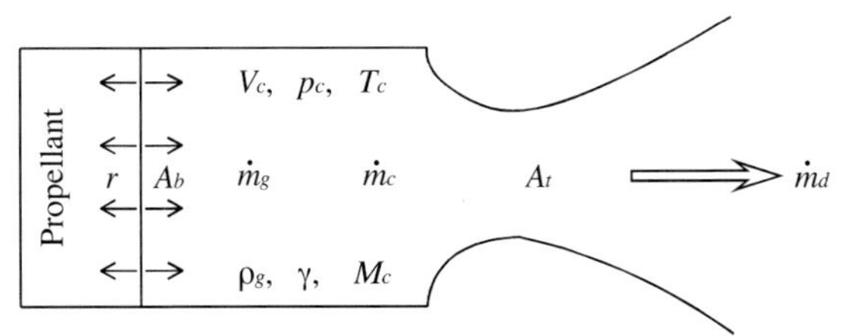

Fig. 14.8 Combustion parameters in a rocket motor.

where A_t is the nozzle throat area, c_D is the nozzle discharge coefficient given by Eq. (1.61), and p_c is the chamber pressure. The rate of mass accumulation in the chamber, $\dot{m}_c$, is given by

$$\dot{m}_c = d(\rho_g V_c)/dt \tag{14.5}$$

where V_c is the chamber free volume, ρ_g is the density of the gas, and t is time. The mass balance in the rocket motor is expressed by

$$\dot{m}_g = \dot{m}_c + \dot{m}_d \tag{14.6}$$

One assumes that V_c and T_c are constant and that the burning rate is expressed by Eq. (3.68) during the pressure transient in the rocket motor; Eq. (14.6) is then represented by[4]

$$(L^*/R_g T_c)\, dp_c/dt - K_n \rho_p a p_c{}^n + c_D p_c = 0 \tag{14.7}$$

where $L^* = V_c/A_t$, $K_n = A_b/A_t$, and R_g is the gas constant of the combustion gas given by Eq. (1.5). The transient pressure in the rocket motor is obtained by integration of Eq. (14.7) as

$$p_c(t) = \left[\left(p_i^{1-n} - \frac{a\rho_p K_n}{c_D}\right)\exp\left(\frac{(n-1)R_g T_c c_D}{L^*}\right)t + \frac{a\rho_p K_n}{c_D}\right]^{\frac{1}{1-n}} \tag{14.8}$$

where p_i is the initial pressure in the rocket motor. The final pressure when $t \to \infty$ is obtained as

(1) $n > 1$: $p_i < p_{eq}$ $\quad p_c \to 0$

$\qquad p_i > p_{eq}$ $\quad p_c \to \infty$

(2) $n = 1$ $\quad p_c \to p_i$

(3) $n < 1$ $\quad p_c \to p_{eq}$

where p_{eq} is the equilibrium pressure obtained when the time-dependent term shown in Eq. (14.6) is negligible and the mass balance is expressed by

$$\rho_p\, A_b\, r = c_D\, A_t\, p_c \tag{14.9}$$

Substituting Eq. (3.68) into Eq. (14.9), one gets the chamber pressure at a steady-state condition as

$$p_{eq} = p_c = \left(\frac{a\rho_p}{c_D}\frac{A_b}{A_t}\right)^{\frac{1}{1-n}} \tag{14.10}$$

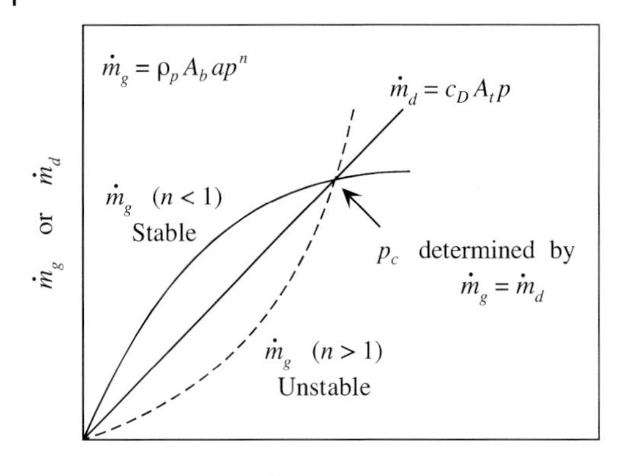

Fig. 14.9 Mass balance principle and stable burning point in a rocket motor.

As shown in Fig. 14.9, the chamber pressure is given by the crossover point of $\dot{m}_g$ and $\dot{m}_d$ in a plot of these parameters versus pressure. It is evident that stable combustion occurs only under the condition of

$$d(c_D\, A_t\, p)/dp > d(\rho_p A_b\, ap^n)/dp \tag{14.11}$$

Substituting Eq. (14.9) into Eq. (14.11) at $p = p_c$, one obtains the stability criterion as

$$n < 1 \tag{14.12}$$

Based on Eq. (14.8), the characteristic time in the combustion chamber, τ_c, is defined as

$$\tau_c = L^*/(n-1)R_g T_c c_D \tag{14.13}$$

If the pressure in a rocket motor changes during the time interval τ, the pressure transient is considered to represent a steady-state combustion when $\tau > \tau_c$ and a non-steady-state combustion when $\tau < \tau_c$.

14.1.4
Temperature Sensitivity of Pressure in a Rocket Motor

A change of burning rate resulting from a change of initial propellant temperature changes the equilibrium pressure, p_c, in a rocket motor. Thus, one needs to introduce the temperature sensitivity of pressure in the combustion chamber, π_k, which is defined similarly to σ_p:

$$\pi_k = \frac{1}{p}\frac{p_1 - p_0}{T_1 - T_0} \quad \text{at constant } K_n \tag{14.14}$$

where p_1 and p_0 are the pressures in the combustion chamber at T_1 and T_0, respectively, and p is an averaged pressure between p_1 and p_0.[1] Thus, the unit of tempera-

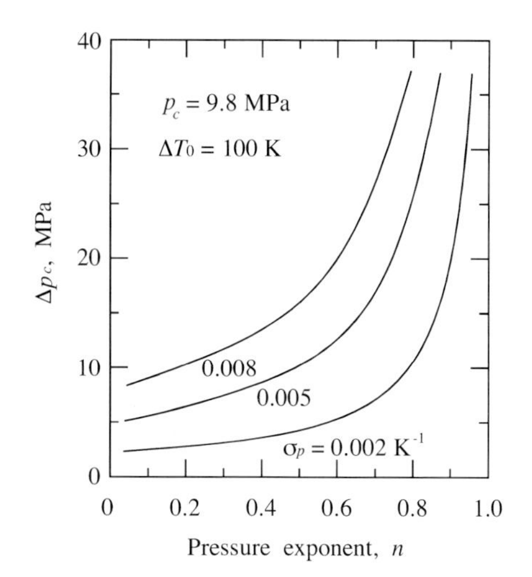

Fig. 14.10 Pressure increase in a rocket motor versus pressure exponent of burning rate as a function of temperature sensitivity.

ture sensitivity of pressure is K^{-1}. Since K_n is defined as the ratio of the burning surface area of the propellant and the nozzle throat area, $K_n = A_b/A_t$, constant K_n indicates a fixed physical dimension of a rocket motor. The differential form of Eq. (14.14) is expressed by

$$\pi_k = (\partial p/\partial T_0)_{Kn}/p = (\partial \ln p/\partial T_0)_{Kn} \tag{14.15}$$

Substituting Eq. (14.8) into Eq. (14.15), one gets

$$\pi_k = \frac{1}{1-n}\frac{1}{a}\left(\frac{\partial a}{\partial T_0}\right)_p = \frac{\sigma_p}{1-n} \tag{14.16}$$

Fig. 14.10 shows an example of calculated results of pressure increase (Δp_c) versus n when T_0 is increased by 100 K. It can be seen that Δp_c becomes large when a propellant having a large value of σ_p and/or n is used.[1,3,4]

Reduction of the temperature sensitivity of burning rate as defined in Eq. (3.73) increases the combustion stability in rocket motors, which results in improved ballistics. Fig. 14.11 shows the pressure versus burning time in a rocket motor with a propellant having $n = 0.5$ and $\sigma_p = 0.0030$ K^{-1}. When the initial propellant temperature T_0 is 233 K, the chamber pressure is low at about 9.5 MPa. However, the chamber pressure increases to 16 MPa when T_0 is increased to 333 K. The burning time is then reduced from 20 s to 13 s. The increase of the pressure in the rocket motor is significant as n becomes large and/or σ_p becomes large. Because of this change of pressure and burning time, the thrust generated by the rocket motor is also changed according to the relationship represented by Eq. (1.67).

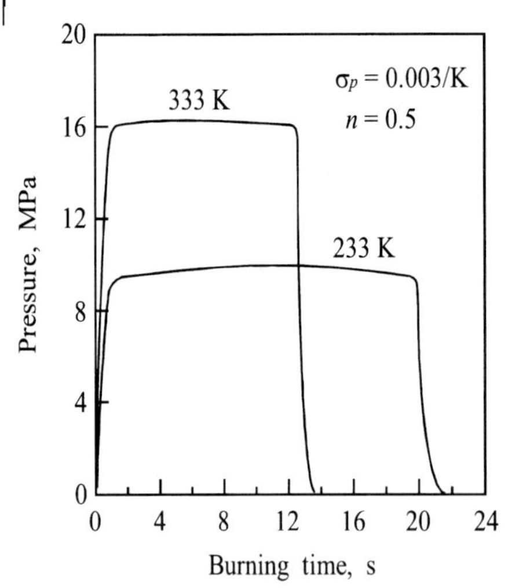

Fig. 14.11 Temperature sensitivity of a rocket motor.

14.2
Dual-Thrust Motor

14.2.1
Principles of a Dual-Thrust Motor

The purpose of a dual-thrust motor used for a rocket projectile is to generate two-stage thrust: booster-stage thrust and sustainer-stage thrust. The booster stage is used to accelerate the projectile from zero velocity to a certain flight-stabilized velocity, and then the sustainer stage is used to maintain a constant flight velocity. In general, two-stage thrust of a projectile is provided by means of two independent rocket motors, i. e., a booster motor and a sustainer motor, and the booster motor is mechanically separated from the projectile.

If separation of the booster motor is not favorable from the point of view of the design of the system, a dual-thrust motor composed of a single chamber is required. In order to increase the mass burning rate at the booster stage and to decrease the mass burning rate at the sustainer stage, the following grain design is required: the burning surface area of the propellant grain needs to be large for the booster stage and small for the sustainer stage. A dual-thrust motor may be realized by using a double-layered propellant grain. The first propellant grain burns at a high burning rate to generate booster thrust, and then the second grain burns at a low burning rate to generate sustainer thrust. Fig. 14.12 shows typical examples of cross-sections of propellant grains used for dual-thrust motors.[5]

14.2.2
Single-Grain Dual-Thrust Motor

The most conventional dual thrust is obtained by means of a single propellant grain of which the burning area varies with burning time. If one designs a dual-thrust motor so as to obtain a large thrust ratio of the booster stage and the sustainer stage, a

Fig. 14.12 Cross-sections of propellant grains used for dual-thrust motors.

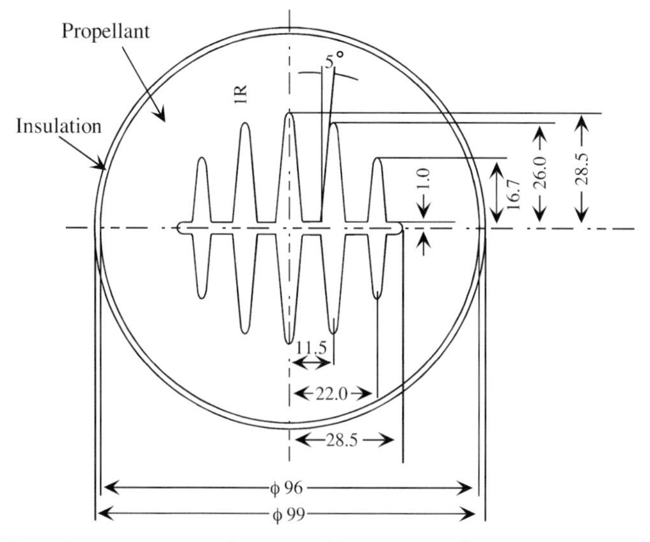

Fig. 14.13 Cross-section of an internal-burning propellant grain used in a dual-thrust motor.

large burning-area ratio between the booster stage and the sustainer stage is required. Based on Eqs. (1.69) and (14.10), the thrust of a rocket motor is represented by

$$F = c_p A_t p_c = c_F A_t \left(\frac{a\rho_p}{c_D} K_n \right)^{\frac{1}{1-n}} \tag{14.17}$$

where K_n is defined as the ratio of the burning surface area and the nozzle throat area, $K_n = A_b/A_t$. Since A_t remains constant during burning, a dual thrust is obtained by a change in A_b of the propellant grain.

Fig. 14.13 shows a typical example of a propellant grain for which the burning surface area ratio of the booster stage and sustainer stage is 2.18. The relationship

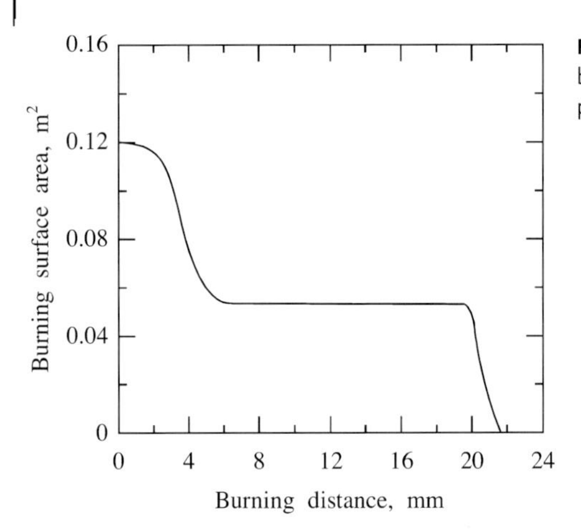

Fig. 14.14 Burning surface area versus burning distance for the dual-thrust propellant grain shown in Fig. 14–13.

between the burning area and the regressing grain from the initial surface to the final surface is shown in Fig. 14.14. A burning distance from the initial surface to 3 mm is used as the booster stage and that from 6 mm to 20 mm is used as the sustainer stage. The pressure and thrust transient stage corresponds to the burning distance range from 3 mm to 6 mm, and the remaining sliver of the propellant grain provides the end stage from 20 mm to 22 mm.[5]

The propellant grain is composed of $\xi_{AN}(0.84)$ as an oxidizer and $\xi_{PU}(0.16)$ as a fuel. The burning rate and $A_b/A_n = K_n$ characteristics are shown as a function of pressure in Fig. 14.15. The pressure exponent of burning rate, n, is 0.73. Combustion test results show that the average booster pressure from 0 s to 1.0 s is 11.0 MPa and the sustainer pressure from 2.5 s to 17.2 s is 1.7 MPa. The pressure ratio of the booster and sustainer stages is thus 6.47. Due to limitations imposed by the mechanical strength and elongation characteristics of propellant grains, the realistic internal port geometry of propellant grains that can be achieved is limited. The maximum pressure ratio obtainable from a single grain is considered to be no more than about 7.

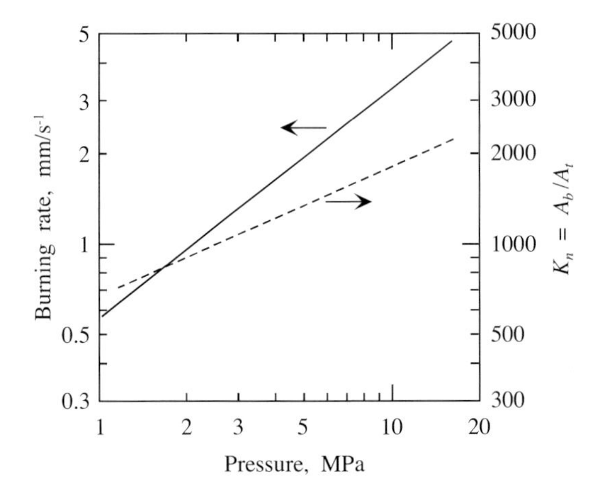

Fig. 14.15 Burning rate and K_n characteristics versus pressure.

14.2.3
Dual-Grain Dual-Thrust Motor

14.2.3.1 Mass Generation Rate and Mass Discharge Rate

When two different propellant grains, i. e., a high-burning-rate grain (propellant 1) and a low-burning-rate grain (propellant 2), are set in a single chamber along the chamber axis and burn simultaneously, a high booster thrust and a low sustainer thrust are generated. After the first stage of burning, the high-burning-rate grain burns out and the chamber pressure rapidly decreases. However, the low-burning-rate grain continues to burn as the second stage of burning, as shown in Fig. 14.16. Both the high- and low-burning-rate grains burn simultaneously under high chamber pressure during the first stage, and then the low-burning-rate grain continues to burn under low chamber pressure. This type of dual-thrust motor is used to obtain a high thrust ratio of the booster and sustainer stages.[5]

The mass generation rate at the booster stage is represented by $\dot{m}_{gB}$ and that at the sustainer stage is represented by $\dot{m}_{gS}$; these are the parameters that need to be addressed in designing a required dual-thrust motor. The chamber pressure p_B at the booster stage is determined by $\dot{m}_{gB}$, which is dependent on the burning rate characteristics of propellants 1 and 2. On the other hand, the chamber pressure p_S at the sustainer stage is determined by $\dot{m}_{gS}$, which is dependent only on the burning rate characteristics of propellant 2. The thrust F_B generated at the booster stage and the thrust F_S generated at the sustainer stage are represented by

$$F_B = I_{spB}\, \dot{m}_{gB}\, g \tag{14.18}$$

$$F_S = I_{spS}\, \dot{m}_{gS}\, g \tag{14.19}$$

where I_{spB} is the specific impulse at the booster stage, I_{spS} is the specific impulse at the sustainer stage, and g is the gravitational acceleration. The unit of specific impulse is seconds and it is given as a function of the chemical properties of the propellant and the nozzle expansion ratio. Since F_B is generated by the simultaneous

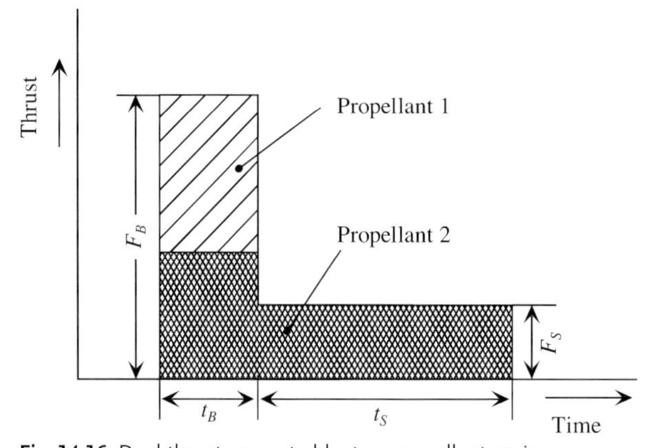

Fig. 14.16 Dual-thrust generated by two propellant grains.

combustion of propellants 1 and 2, I_{spB} is given by a mass-averaged value of $\dot{m}_{gB}$ composed of the mass generation rates of propellants 1 and 2 at pressure p_B. Table 14.1 shows the design parameters of dual-thrust motors using two different types of propellants, 1 and 2.

Table 14.1 Design parameters of the propellant grains required to obtain F_B and F_S.

	Booster stage	**Sustainer stage**
Thrust	F_B	F_S
Chamber pressure, MPa	p_B	p_S
Specific impulse	I_{spB}	I_{spS}
Propellant	1:2	2
Burning rate, m s^{-1}	r_1:r_2	r_2
Pressure exponent	n_1:n_2	n_2
Burning rate constant	k_1:k_2	k_2
Density, kg m^{-3}	ρ_1:ρ_2	ρ_2
Burning surface area, m^2	A_1:A_2	A_2
Burning time, s	t_B	t_S
Burning distance, m	L_1:L_2	L_2

14.2.3.2 Determination of Design Parameters

The linear burning rates of propellants 1 and 2 are expressed by

$$r_1 = k_1 p^{n_1} \quad \text{for propellant 1} \tag{14.20}$$

$$r_2 = k_2 p^{n_2} \quad \text{for propellant 2} \tag{14.21}$$

The mass generation rates are expressed by

$$\dot{m}_{g1} = \rho_1 A_1 r_1 = \rho_1 A_1 k_1 p^{n_1} \quad \text{for propellant 1} \tag{14.22}$$

$$\dot{m}_{g2} = \rho_2 A_2 r_2 = \rho_2 A_2 k_2 p^{n_2} \quad \text{for propellant 2} \tag{14.23}$$

Since both propellants 1 and 2 burn simultaneously during the booster stage at pressure p_B, and only propellant 2 burns during the sustainer stage at pressure p_S, the mass generation rates, $\dot{m}_{gB}$ for the booster stage and $\dot{m}_{gS}$ for the sustainer stage, are given by:[5]

$$\dot{m}_{gB} = \dot{m}_{g1} + \dot{m}_{g2} = \rho_1 A_1 k_1 p_B{}^{n_1} + \rho_2 A_2 k_2 p_B{}^{n_2} \quad \text{for the booster stage} \tag{14.24}$$

$$\dot{m}_{gS} = \dot{m}_{g2} = \rho_2 A_2 k_2 p_S{}^{n_2} \quad \text{for the sustainer stage} \tag{14.25}$$

The respective mass discharge rates of $\dot{m}_{dB}$ for the booster stage and $\dot{m}_{dS}$ for the sustainer stage are given by

$$\dot{m}_{dB} = c_{DB} A_t \, p_B \quad \text{for the booster stage} \tag{14.26}$$

$$\dot{m}_{dS} = c_{DS} A_t \, p_S \quad \text{for the sustainer stage} \tag{14.27}$$

where A_t is the nozzle throat area and c_{DB} and c_{DS} are the nozzle discharge coefficients for the booster stage and the sustainer stage, respectively. Since the mass generation rate and the mass discharge rate must be equal for a steady-state combustion, as shown in Fig. 14.9, the following mass balance equations are obtained for the booster stage and the sustainer stage:

$$\rho_1 A_1 k_1 p_B{}^{n_1} + \rho_2 A_2 k_2 p_B{}^{n_2} = c_{DB} A_t \, p_B \quad \text{for the booster stage} \tag{14.28}$$

$$\rho_2 A_2 k_2 p_S{}^{n_2} = c_{DS} A_t \, p_S \quad \text{for the sustainer stage} \tag{14.29}$$

The nozzle throat area, A_t, is determined by Eq. (14.28) as

$$A_t = (\rho_1 A_1 k_1 p_B{}^{n_1 - 1} + \rho_2 A_2 k_2 p_B{}^{n_2 - 1})/c_{DB} \tag{14.30}$$

The nozzle discharge coefficients, c_{DB} and c_{DS}, are determined by using Eq. (1.61). Using Eq. (14.29), the burning surface area of propellant 2 is obtained as

$$A_S = c_{DS} A_t / k_2 \rho_2 p_S{}^{n_2} \tag{14.31}$$

The equilibrium chamber pressure during the sustainer stage (p_S) is simply determined by Eq. (14.10) as

$$p_S = (\rho_2 k_2 K_{n2} / c_{DS})^{1/(1-n_2)} \tag{14.32}$$

where K_n is represented by $K_{n2} = A_S/A_t$. The pressure during the booster stage, p_B, is determined by means of Eq. (14.28). Based on the required pressure versus burning time characteristics, the geometrical grain shapes are obtained for propellants 1 and 2. Table 14.2 shows the physicochemical parameters of the propellants that

Table 14.2 Salient physical parameters of propellants 1 and 2.

Propellant	**1**	**2**
Combustion temperature, K	T_{g1}	T_{g2}
Gas constant, J kg^{-1} K^{-1}	R_{g1}	R_{g2}
Specific heat ratio	γ_1	γ_2
Nozzle discharge coefficient, s^{-1}	c_{DB}	c_{DS}

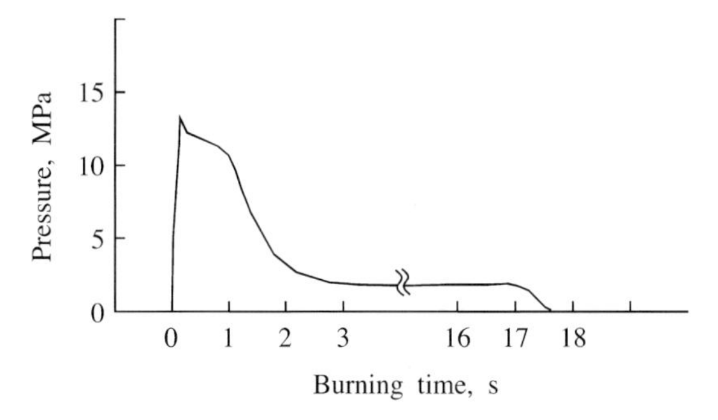

Fig. 14.17 Pressure versus burning time curve for a dual-grain dual-thrust motor.

need to be considered in order to obtain the requisite nozzle discharge coefficients-for the booster stage and the sustainer stage. The required thrust at the booster stage, F_B, and that at the sustainer stage, F_S, are determined by Eqs. (14.18) and (14.19) as a function of $\dot{m}_{gB}$ and $\dot{m}_{gS}$, respectively. Fig. 14.17 shows a typical example of a pressure versus burning time curve obtained for a dual-grain dual-thrust motor. The pressure ratio of the booster and the sustainer is approximately 6, which is difficult to attain with a single-grain dual-thrust motor.

In many cases, the rapid pressure decay at the combustion transition from the booster stage to the sustainer stage acts to extinguish the burning of propellant 2. No sustainer stage is then provided due to the burning interruption of propellant 2. Fig. 14.18 shows a set of photographs of such a burning interruption of propellant 2 at the combustion transition by a rate of pressure decay of $dp/dt = 10^4$ MPa s^{-1}. The burning interruption occurred after about 1 s of burning of the booster stage. Here, propellant 1 used at the booster stage was an NC-NG double-base propellant, and propellant 2 was an AN composite propellant. When an AP composite propellant was used as propellant 2, no combustion interruption occurred at dp/dt =

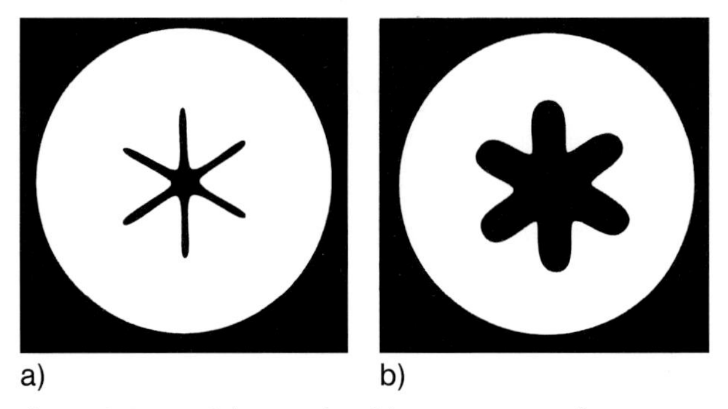

Fig. 14.18 A pair of photographs of the cross-section of propellant 2: (a) before combustion, and (b) after combustion interruption at the transition from booster stage to sustainer stage.

10^4 MPa s^{-1}. When a negative catalyst such as LiF or $SrCO_3$ was added to the AP composite propellant, combustion interruption occurred. As shown in Fig. 7.27, the pressure deflagration limit is lowered by the addition of the negative catalysts. Detailed design work of a dual-grain dual-thrust motor and the associated combustion test results are shown in Ref. [5].

14.3 Thrust Modulator

The thrust generated by a rocket motor is represented by $F = c_F A_t p_c$, as given by Eq. (1.69), where F is thrust, p_c is the chamber pressure, and c_F is the thrust coefficient defined by Eq. (1.70). When the chamber pressure is ideally expanded to the atmospheric pressure, c_F is given by the specific heat ratio of the combustion products, as given by Eq. (1.72). Accordingly, the thrust is dependent on A_t and p_c. The pressure can easily be regulated by adjusting the throat area mechanically. For example, if a pintle is inserted into the nozzle throat using a servo-mechanism, the throat area is decreased from A_{t1} to A_{t2}, and then the pressure is increased from p_{c1} to p_{c2}. As a result, the mass discharge rate is increased from $\dot{m}_{d1}$ to $\dot{m}_{d2}$, and the thrust is increased from $F_1 = c_{d1} A_{t1} p_{c1}$ to $F_2 = c_{d2} A_{t2} p_{c2}$.

Micro-rockets attached to the underneath of a pilot ejection seat of an aircraft are thrust modulators. Each thrust modulator generates variable thrust independently of the others in order to keep the seat horizontal. Each thrust modulator is controlled by signals transmitted from gyro-sensors. Thrust modulators are also used as side-thrusters attached to projectiles. Side-thrusters generate a pitching moment around the center of gravity of the projectiles, similarly to the aerodynamic fins of projectiles. Though aerodynamic fins are effective only when the projectiles are in the atmosphere, side-thrusters are useful in space or at high altitude. For example, a side-thruster is attached to the side-wall of a projectile and the combustion gas generated by the side-thruster is expelled from exhaust nozzles perpendicular to the projectile axis. The pitching moment is controlled by the thrust-modulated side-thruster. The thrust is modulated by adjusting the nozzle throat area by means of a pintle mechanism.

14.4 Erosive Burning in a Rocket Motor

14.4.1 Head-End Pressure

The combustion of a propellant grain in an internal burning occurs under the flow of the combustion gas along the burning surface, as shown in Fig. 14.1. Since the burning surface regresses perpendicularly to this flow, the flow velocity in the port of the propellant grain increases the heat flux transferred to the burning surface; the burning rate therefore increases and so-called erosive burning occurs.

Fig. 14.19 shows a typical set of pressure versus time curves obtained from tests on a rocket motor.[6] When the L/D ratio defined in Fig. 14.19 is increased, the head-end chamber pressure is increased drastically immediately after the ignition stage. These grains are seven-pointed-star-shaped neutral-burning grains (diameter D = 114 mm), and are made of an AP-Al-CMDB propellant with the composition $\xi_{NC}(0.25)$, $\xi_{NG}(0.31)$, $\xi_{TA}(0.08)$, $\xi_{AP}(0.27)$, and $\xi_{Al}(0.09)$. The ratio of the initial burning surface area (A_{b0}) to the nozzle throat area (A_t), $K_{n0} = A_{b0}/A_t$, and the ratio of the initial port area (A_{p0}) to the nozzle throat area, $J_0 = A_{p0}/A_t$, are shown in Table 14.3.

Table 14.3 Parameter values of an erosive-burning test motor.

L/D	10	12.5	15	16	17.5	18.5
K_{n0}	154	154	154	154	189	195
J_0	1.00	1.24	1.49	1.59	1.42	1.45

It is evident that erosive burning occurs only in the initial stage of combustion and diminishes about 0.5 s after ignition in each of the cases shown in Fig. 14.19; the burning pressure then returns to the designed pressure, p_{eq}. For example, the head-end pressure reaches more than 3.5 times the designed pressure of $p_{eq} = 5$ MPa just after ignition at $L/D = 16$, but the pressure decreases rapidly thereafter and the propellant continues to burn at constant pressure, p_{eq}.

Since the initial port area of the propellant grain, A_{p0}, is small, the flow velocity becomes large because the velocity at the nozzle throat is always at sonic level. Furthermore, when K_n is kept constant, A_b and A_t increase simultaneously with in-

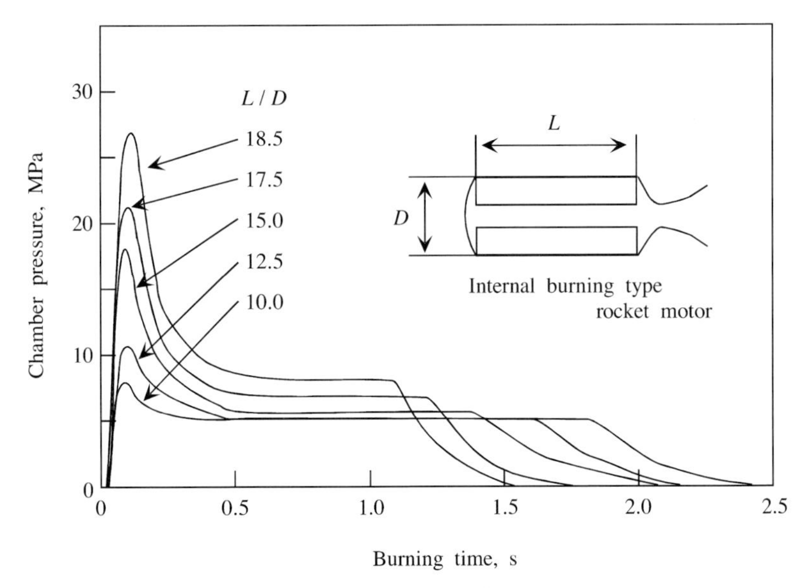

Fig. 14.19 Effect of L/D of a rocket motor on erosive burning.

creasing L/D. Since the port area of the propellant grain is kept constant, the flow velocity increases with increasing A_t in accordance with the mass conservation law given by Eq. (1.49). Therefore, the heat flux transferred from the gas flow increases and the erosive burning ratio increases. However, as the port area increases, the flow velocity decreases and so the erosive burning diminishes.

14.4.2
Determination of Erosive Burning Effect

The pressure or thrust versus burning time relationship in an internal burning of a rocket motor is computed by the semi-empirical equation, Eq. (13.26), proposed by Lenoir and Robillard. The maximum pressure, p_m, in the rocket motor is equivalent to the head-end pressure in the motor correlated with L/D (longitudinal length/outer diameter of propellant grain). The pressure in the motor is created by the burning of the propellant grain in accordance with Vieille's law, $r = ap^n$, and heat is transferred through the flow of burned gas perpendicular to the burning direction. The local mass generation rate at distance x from the head-end of the propellant grain is given by $\Delta A_{b,x} r_x \rho_p$, where $\Delta A_{b,x}$ is the burning surface area in an interval of length Δx at distance x and r_x is the burning rate at x. The burning rate, pressure, and flow velocity along the port of the propellant grain may be determined by the following iterative computations on the flow field:
Internal geometry of the propellant grain

↓

Nozzle throat area

↓

Assume the head-end pressure

↓

Compute velocity and pressure distributions along the flow direction in the port

↓

Compute burning rate as a function of pressure and determine p_{0i} at $t = t_0$

↓

Compute burning rate as a function of velocity

↓

Compute total mass flow rate at nozzle throat by $\dot{m}_g = \sum_{x-0}^{L} \Delta A_x r_x \rho_p$

↓

Compute head-end pressure by $p_0 = c^* A_t / \dot{m}_g$

↓

If $p_0 \neq p_{0i}$, assume another p_{0i} and continue to compute p_0 until $p_0 = p_{0i}$

↓

When $p_0 = p_{0i}$ is obtained, continue to compute internal propellant grain geometry at $t = t_1$

↓

Assume head-end pressure at $t = t_1$

↓

Repeat computation until the pressure reaches $p_0 = p_{eq}$.

↓

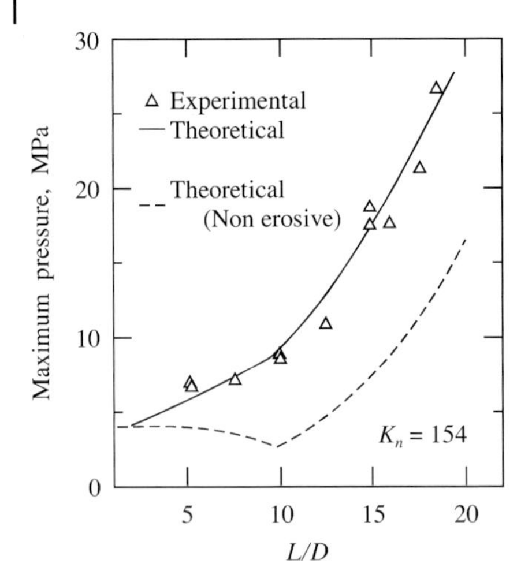

Fig. 14.20 Peak pressures during erosive burning.

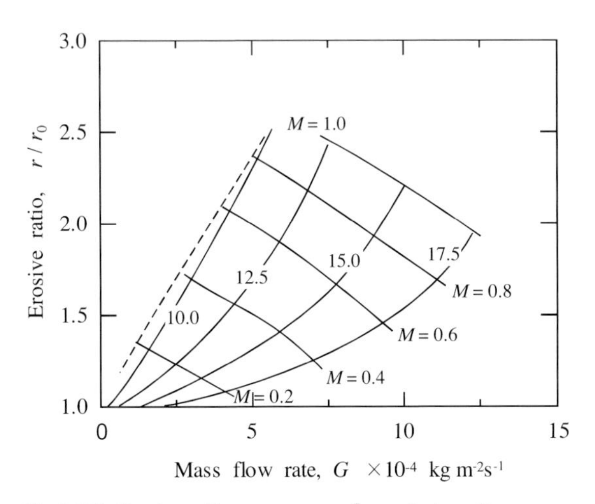

Fig. 14.21 Erosive ratio versus mass flow rate in unit cross-sectional area of the port.

The pressure peaks observed in the combustion tests shown in Fig. 14.19 are computed as a function of L/D as shown in Fig. 14.20. The peak pressures computed by means of the Lenoir–Robillard empirical equation are confirmed by the measured pressure at the head-end of the motor. It is evident that p_m values predicted without erosive burning are significantly lower than the measured maximum pressures. Fig. 14.21 shows the erosive ratio, $\varepsilon = r/r_0$, as a function of the mass flow rate per unit cross-sectional area in the port, G. The erosive ratio increases with increasing Mach number in the port at constant L/D.

In order to understand the erosive burning effect in a motor, an overall erosive burning ratio, ε^*, is defined as

$$\varepsilon^* = \dot{w}_e/\dot{w}_t = 1 - \dot{w}_p/\dot{w}_t \tag{14.33}$$

where $\dot{w}_t$ is the total mass flow rate exiting from the nozzle, $\dot{w}_e$ is defined as the mass burning rate caused by erosive burning, and $\dot{w}_p$ is defined as the mass burning rate caused by pressure. It is shown that $\dot{w}_p$ is given by

$$\dot{w}_p = \rho_p \sum_i r_i \Delta A_{b,i} \tag{14.34}$$

where r_i is the local burning rate and $\Delta A_{b,i}$ is the burning surface area at i along the port. The burning rate is given by a function of local pressure along the port, p_i at i, according to:

$$r_i = a p_i^n \tag{14.35}$$

The total mass flow rate is determined by Eqs. (1.60) and (1.69) as

$$\dot{w}_t = (c_D/c_F)F \tag{14.36}$$

where c_D is the nozzle discharge coefficient determined by the physical properties of the combustion gas and c_F is determined by the nozzle expansion ratio and the aft-end chamber pressure. The ratio ε^* is then determined from the computed values of $\dot{w}_p$ and $\dot{w}_t$ based on the measured data. Fig. 14.22 shows ε^* as a function of L/D. The erosive burning shown in Fig. 14.19 indicates that $\dot{w}_e$ occurring due to the gas-flow effect is approximately 20–35 % of the total mass flow rate, $\dot{w}_t$, and the remaining mass flow effect is caused by the burning pressure.[6]

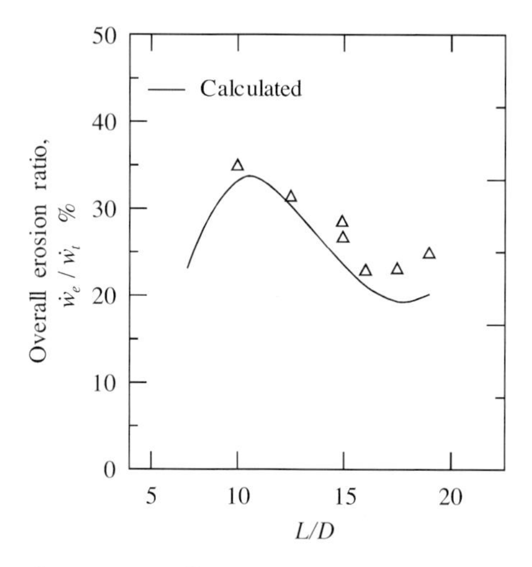

Fig. 14.22 Overall erosive ratio versus L/D.

14.5 Nozzleless Rocket Motor

14.5.1 Principles of the Nozzleless Rocket Motor

A nozzle used for a rocket is composed of a convergent section and a divergent section. The connected part of these two nozzle sections is the minimum cross-sectional area termed the throat. The convergent part is used to increase the flow velocity from subsonic to sonic velocity by reducing the pressure and temperature along the flow direction. The flow velocity reaches the sonic level at the throat and continues to increase to supersonic levels in the divergent part. Both the pressure and temperature of the combustion gas flow decrease along the flow direction. This nozzle flow occurs as an isentropic process.

The combustion gas of an internal burning of a propellant flows along the port of the propellant. If the nozzle attached to a rocket motor is removed, the pressure in the port becomes equal to atmospheric pressure and no sonic velocity is attained at the rear-end of the port. Then, no thrust is generated by the combustion of the propellant. However, if the mass burning rate of the propellant is high enough to choke the flow at the rear-end of the port, the pressure in the port is increased and the flow reaches sonic velocity. The increased pressure in the port is converted into thrust.

The thrust F is represented by

$$F = c_F A_t p_c \tag{1.69}$$

where p_c is the pressure in the combustion chamber, A_t is the cross-sectional throat area of the nozzle, and c_F is the thrust coefficient defined in Eq. (1.70). The pressure is given by

$$p_c = b K_n{}^m \tag{14.37}$$

where K_n is defined as $K_n = A_b/A_t$, and b and m are determined by the physicochemical properties of the propellant used. Substituting Eq. (14.37) into Eq. (1.69), one gets

$$F = c_F A_t b K_n{}^m \tag{14.38}$$

An adaptation of Eq. (14.38) for a nozzleless rocket indicates that the port area increases as the burning surface of the propellant regresses, K_n decreases and A_t increases, and so the choked condition is varied. Thus, the thrust generated by the nozzleless rocket is determined by the relationship of the mass generation rate in the port and the mass discharge rate at the rear-end of the port.[7–9]

14.5.2
Flow Characteristics in a Nozzleless Rocket

Let us consider a gas flow in a port of an internal-burning propellant. The cross-sectional area of the port is assumed to be constant throughout the port from the head-end to the rear-end, i. e., the port is one-dimensional along the flow direction.[7–9] The mass discharge rate from the rear-end of the port, $\dot{m}_d$, is given by

$$\dot{m}_d = \rho_g u_g A_p \tag{14.39}$$

where ρ_g is the density of the combustion gas, u_g is the flow velocity, and A_p is the cross-sectional area of the port. The mass generation rate in the port, $\dot{m}_g$, is represented by[7–9]

$$\dot{m}_g = \rho_p \int_0^e r(x) S(x) dx \tag{14.40}$$

where ρ_p is the density of the propellant, r is the linear burning rate, S is the internal burning surface of the port, and x is distance along the flow direction. The integration is from the head-end at $x = 0$ to the rear-end at $x = e$ of the propellant grain. Since the mass generation rate is equal to the mass discharge rate, the mass balance for a steady-state burning is obtained by

$$\dot{m}_g = \dot{m}_d \tag{14.41}$$

The momentum change in the port is represented by

$$p_0 A_p = p A_p + \dot{m}_d u_g \tag{14.42}$$

where p_0 is the pressure at the head-end of the port, at which $u_g = 0$. The energy equation is represented by

$$h_0 = h + u_g^2/2 \tag{14.43}$$

where h is the enthalpy and h_0 is the enthalpy at the head-end of the port. Since $u_g = 0$ at the head-end of the port, p_0 and h_0 are the stagnation pressure and the stagnation enthalpy, respectively. The enthalpy is defined by

$$h = c_p T \text{ and } h_0 = c_p T_0 \tag{14.44}$$

where T is temperature, T_0 is the stagnation temperature, which is equivalent to the adiabatic flame temperature of the combustion gas, and c_p is the specific heat at constant pressure.

Mach number, M, in the flow field is defined by

$$M = u_g/a = u_g/(\gamma R_g T)^{1/2} \tag{14.45}$$

where a is the sonic speed, γ is the specific heat ratio, and R_g is the gas constant of the combustion gas. Substituting Eqs. (14.39) and (14.45) into Eq. (14.42), one gets

$$p = p_0/(1 + \gamma M^2) \tag{14.46}$$

The temperature in the port is also given as a function of Mach number by substitution of Eqs. (14.44) and (14.45) into Eq. (14.43) as

$$T = T_0/\{1 + (\gamma - 1)M^2/2\} \tag{14.47}$$

The gas density in the port is given as a function of Mach number by the use of the equation of state, Eq. (1.5), and Eqs. (14.46) and (14.47) as

$$\rho_g = \rho_{g,0}\{1 + (\gamma - 1)M^2/2\}/(1 + \gamma M^2) \tag{14.48}$$

The pressure, temperature, and density of the gas at the rear-end of the port, i. e., at the exit of a nozzleless rocket motor, are obtained by Eqs. (14.46)–(14.48) as

$$p_e = p_0/(1 + \gamma) \tag{14.49}$$

$$T_e = 2T_0/(1 + \gamma) \tag{14.50}$$

$$\rho_{g,e} = \rho_{g,0}/2 \tag{14.51}$$

where the subscripts e and 0 denote the rear-end and the head-end, respectively. The choked condition at the rear-end is obtained when $p_0/p_e > 1 + \gamma$.

The mass flow rate at the rear-end of the port, $\dot{m}_{d,e}$, is given by

$$\dot{m}_{d,e} = u_{g,e}\rho_{g,e}A_e \tag{14.52}$$

where the subscript e denotes the rear-end of the port. The flow velocity at the rear-end of the port is represented by

$$u_{g,e} = (\gamma R_g T_e)^{1/2} \tag{14.53}$$

Substituting Eqs. (14.46)–(14.48) and the equation of state, Eq. (1.5), into Eq. (14.52), one gets

$$\dot{m}_{d,e} = A_e p_0 \{\gamma/2R_g T_0(\gamma + 1)\}^{1/2} \tag{14.54}$$

$$= c_{D,p}A_e p_0 \tag{14.55}$$

where $c_{D,p}$ is the discharge coefficient at the port given by

$$c_{D,p} = \{\gamma/2R_g T_0(\gamma + 1)\}^{1/2} \tag{14.56}$$

where T_0 is assumed to be equal to the adiabatic flame temperature of the propellant in this simplified analysis.[9] The nozzle discharge coefficient of a rocket nozzle, $c_{D,i}$, i. e., for an isentropic expansion nozzle, is represented by Eq. (1.61) as

$$c_{D,i} = \{(\gamma/R_g T_0)(2/(\gamma + 1))^{\zeta}\}^{1/2} \tag{14.57}$$

where $\zeta = (\gamma + 1)/(\gamma - 1)$. Thus, the ratio of the discharge coefficients without and with a nozzle, $c_{D,p}/c_{D,i}$, is given by

$$c_{D,p}/c_{D,i} = (1/2)\{(\gamma + 1)/2\}^{1/(\gamma - 1)} \tag{14.58}$$

Since the specific heat ratio of the combustion gas, γ, lies between 1.2 and 1.4, the ratio of the discharge coefficients, $c_{D,p}/c_{D,i}$, is 0.80 for $\gamma = 1.2$ and 0.78 for $\gamma = 1.4$. This result indicates that the mass flow rate from the port exit is approximately 20% lower than that from the isentropic nozzle exit used for a conventional rocket motor when the head-end pressure is equal to the chamber pressure in the rocket motor.[9]

It should be noted that the gas flow process in the port is not isentropic because mass and heat additions occur in the port. This implies that there is stagnation pressure loss and so the specific impulse is reduced for nozzleless rockets. When a convergent nozzle is attached to the rear end of port, the static pressure at the port exit, p_e, continues to decrease to the atmospheric pressure and the specific impulse of the nozzleless rocket motor is increased. The expansion process in a divergent nozzle is an isentropic process, as described in Section 1.2.

14.5.3 Combustion Performance Analysis

Nozzleless rockets are very simplified and low-cost rockets because no nozzles are used. Their specific impulse is lower than that of conventional rockets even when the same mass of propellant is used. Normally, a convergent-divergent nozzle is used to expand the chamber pressure to the atmospheric pressure through an isentropic change, which is the most effective process for converting pressure into propulsive thrust. The flow process without a nozzle increases entropy and there is stagnation pressure loss.

When a propellant used for a nozzleless rocket is perforated in the shape of a circular port, the flow velocity increases from $u_g = 0$ at the head-end of the port to the sonic velocity at the rear-end. The pressure decreases along the gas flow in the port from the head-end to the rear-end. There are two important factors with regard to the performance of the propellant burning: flow velocity and pressure. Erosive burning occurs due to the high flow velocity along the burning surface of the circular port. No erosive burning occurs in the upstream region of the port due to the low flow velocity, but there is high erosive burning in the downstream region due to the high flow velocity. Accordingly, the rate of port area increase in the downstream section becomes higher than that in the upstream section of the port. On the other hand, the burning rate in the upstream section is high due to the effect of high pressure, whereas that in the downstream section is low due to the lower pressure.

Though the effects of flow velocity and pressure in the port diminish soon after ignition, the pressure peak at the head-end must be limited due to the requirement of the motor case design and the thrust versus time should be adjustable in a controlled manner. Though the cross-sectional area of the port is small and so the burning area is also small at the initial stage of combustion, the head-end pressure is high due to the effect of erosive burning. Though the cross-sectional area of the port becomes large and hence the burning surface area becomes small at the final stage, the head-end pressure becomes low due to the disappearance of erosive burning. This successive combustion process provides a relatively constant thrust versus combustion time relationship for a nozzleless rocket motor. It is important to note that the pressure distribution along the gas flow in the port is dependent on the ratio of the port area to the port length, and is also dependent on the pressure exponent of burning rate, n, of the propellant. Combustion test data for nozzleless rockets are presented in Refs. [7–10].

14.6 Gas-Hybrid Rockets

14.6.1 Principles of the Gas-Hybrid Rocket

Hybrid rockets are intermediate between solid rockets and liquid rockets in terms of the nature of the combination of solid fuel and liquid oxidizer. Since the fuel and oxidizer components of a liquid rocket are physically separated, two mechanical systems are needed to feed these components into the combustion chamber. On the other hand, a hybrid rocket uses a polymeric inert material as a fuel and a liquid oxidizer, and so only one mechanical system is needed to feed this liquid oxidizer into the combustion chamber.

The liquid oxidizer is injected into the port of the perforated polymeric fuel grainin the combustion chamber. The burning occurs in the boundary layer formed along the perforated fuel surface. The polymeric fuel decomposes and gasifies due to the convective heating provided by the combustion gas of the mainstream flow. The mixing process of the fuel gas and the oxidizer gas determines the combustion performance of the hybrid rocket. The fuel surface regresses as the burning proceeds and the perforated port of the fuel grain becomes larger. The gas-flow velocity in the port therefore decreases and the heat flux transferred from the gas flow to the fuel surface is altered. In addition, the ratio of fuel to oxidizer for combustion is also altered during burning. Accordingly, the combustion performance of hybrid rockets is much lower than might be expected theoretically.

In order to improve the combustion performance of hybrid rockets, a new concept of a hybrid rocket is proposed: the gas-hybrid rocket.[4,11] Fig. 14.23 shows a schematic drawing of a gas-hybrid rocket composed of a combustion chamber, a gas generator, a liquid oxidizer tank, and oxidizer feeding pipes and a valve. Unlike conventional hybrid rockets, the gas-hybrid rocket is a two-stage combustion sys-

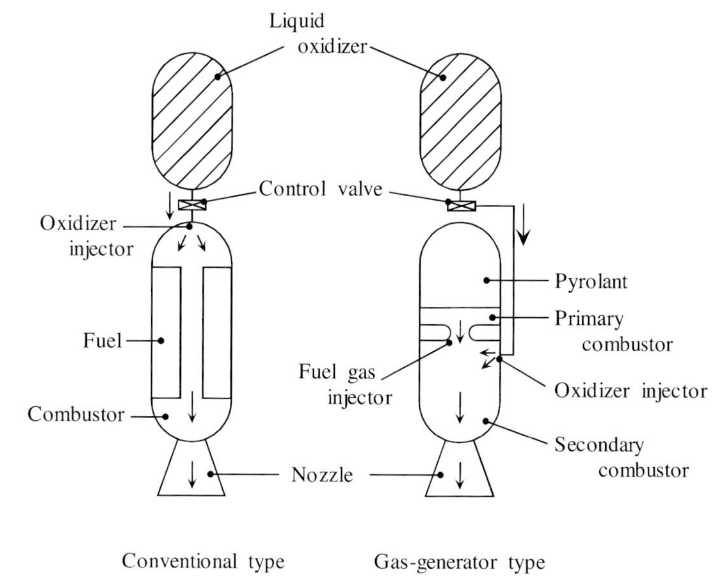

Fig. 14.23 A schematic diagram of a gas-hybrid rocket.

tem. A fuel-rich pyrolant is used to generate fuel-rich gas in the gas generator. When the pyrolant is ignited, fuel-rich gas is generated in the primary combustor, which is injected into the secondary combustor through a fuel gas injector. The liquid oxidizer is fed into the secondary combustor through a control valve and oxidizer injector. The mixture of fuel-rich gas and atomized oxidizer supplied by the oxidizer injector reacts to produce high-temperature combustion products, which are expelled from the nozzle attached to the rear-end of the secondary combustor. Through this simplified process, a gas-hybrid rocket is made. Since the temperature of the fuel-rich gas is high, the reaction rate in the secondary combustor is very high and high combustion performance is obtained. The mass generation rate in the secondary combustor is regulated by the mass flow rate of the liquid oxidizer, which is controlled by means of a valve attached to the oxidizer tank.

As shown in Fig. 14.24, a self-regulating oxidizer feeding mechanism is used to eliminate the liquid oxidizer pumping system. A flow of the pressurized fuel-rich gas generated in the primary combustor forces the oxidizer tank to supply the liquid oxidizer to the secondary combustor. Simultaneously, the fuel-rich gas is injected into the secondary combustor and reacts with the atomized oxidizer. The fuel-rich gas is injected from the primary combustor into the secondary combustor through the fuel gas injector under conditions of a choked gas flow. The pressure in the primary combustor is approximately double that in the secondary combustor. This system is termed a gas-pressurized system.

When a turbo pump is used to obtain high oxidizer fuel flow, this is also operated by the fuel-rich gas generated in the primary combustor. Since the fuel-rich gas is at a higher pressure than the pressure in the secondary combustor, it is used to operate the oxidizer pump and is then used as a fuel component in the secondary

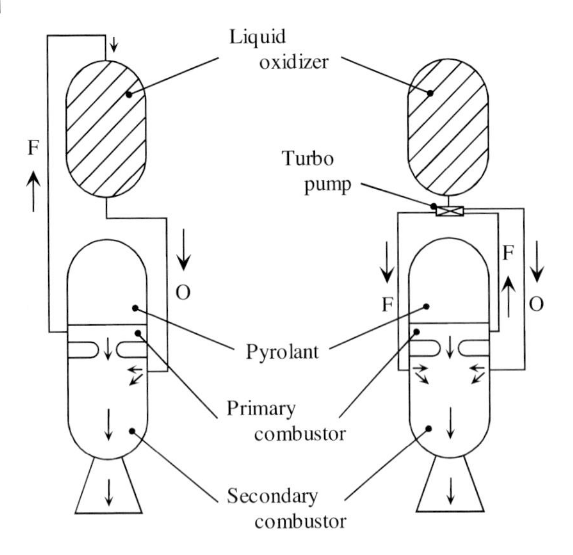

Fig. 14.24 Two types of gas-hybrid rocket: gas-pressurized system and turbo pump operation system.

combustor. Again, the fuel gas injector requires a choked gas flow and the pressure drop at the turbo pump must be taken into account to obtain a specified oxidizer flow rate.

14.6.2
Thrust and Combustion Pressure

As for a rocket motor, the combustion pressure is determined by the mass balance between the mass generation rate and the mass discharge rate according to

$$p_c = (\dot{m}_g + \dot{m}_o)/c_D A_t \tag{14.59}$$

where p_c is the pressure in the combustion chamber, $\dot{m}_g$ is the mass generation rate, $\dot{m}_o$ is the oxidizer flow rate, c_D is the nozzle discharge coefficient, and A_t is the nozzle throat area. The combustion temperature is determined by the mass flow rate ratio of oxidizer and fuel, O/F. The burning rate of the pyrolant used as the gas generator is given by Eq. (3.68), as in the case of rocket propellants. Equation (14.59) may then be rewritten as

$$p_c = a p_g{}^n \rho_p A_b / c_D A_t + \dot{m}_o / c_D A_t \tag{14.60}$$

where p_g is the pressure in the gas generator, A_b is the burning surface area of the pyrolant, and ρ_p is the density of the pyrolant.

When the gas injection nozzle is in an unchoked condition, the pressure in the gas generator is equal to the combustion chamber pressure, i. e., $p_g = p_c$. The gas generation rate is determined by the pressure in the combustion chamber. When the gas injection nozzle is in a choked condition, the burning rate is determined by the gas injection nozzle area, $A_{t,g}$, and then the pressure in the gas generator, p_g, is determined by

$$p_g = (a\rho_p K_{n,g}/c_{D,g})^{1/(1-n)} \tag{14.61}$$

where $c_{D,g}$ is the discharge coefficient of the gas injection nozzle of the gas generator and $K_{n,g} = A_b/A_{t,g}$. Thrust, F, is determined by

$$F = c_F A_t p_c \tag{14.62}$$

where c_F is the thrust coefficient, which is determined by the nozzle expansion ratio and the specific heat ratio of the combustion gas.

14.6.3
Pyrolants used as Gas Generators

Though the pyrolants used in gas-hybrid rockets burn in a similar manner as rocket propellants, their chemical compositions are fuel-rich. The pyrolants burn incompletely and the combustion temperature is below about 1000 K. When an atomized oxidizer is mixed with the fuel-rich gas in the secondary combustor, the mixture reacts to generate high-temperature combustion products. The combustion performance designated by specific impulse, I_{sp}, is dependent on the combination of pyrolant and oxidizer.

Typical pyrolants are GAP with or without metal particles, and mixtures of GAP and HTPB, BAMO, or AMMO. These are all self-deflagrating materials that generate fuel-rich products. Liquid oxygen, N_2O, N_2O_4, H_2O_2, and HNO_3 are typical oxidizers used in liquid rockets and hybrid rockets. These oxidizers are relatively stable liquid oxidizers at room temperature, except for liquid oxygen, which is a cryogenic oxidizer. Fig. 14.25 shows the specific impulse as a function of oxidizer/fuel (O/F) ratio when N_2O_4 is used as an oxidizer.[4] A combustion pressure, p_c, of 5 MPa and optimum expansion from the nozzle with $\varepsilon = 100$ are assumed for the computations. The maximum I_{sp} is obtained at O/F = 1.25 for HTPB40/AP60, and I_{sp} = 338 s at O/F = 1.60 for GAP. The specific impulse of a conventional N_2O_4/HTPB hybrid rocket is also shown in Fig. 14.25 for reference: I_{sp} = 340 s at N_2O_4/HTPB = 4.05. Tables 14.4, 14.5, 14.6, 14.7, and 14.8 show the maximum specific impulses obtained with various combinations of pyrolants and oxidizers. The oxidizer-to-fuel ratio (O/F by mass) and the density of the propellant system ($\rho_{O/F}$) are also important factors in the design of gas-hybrid rockets.

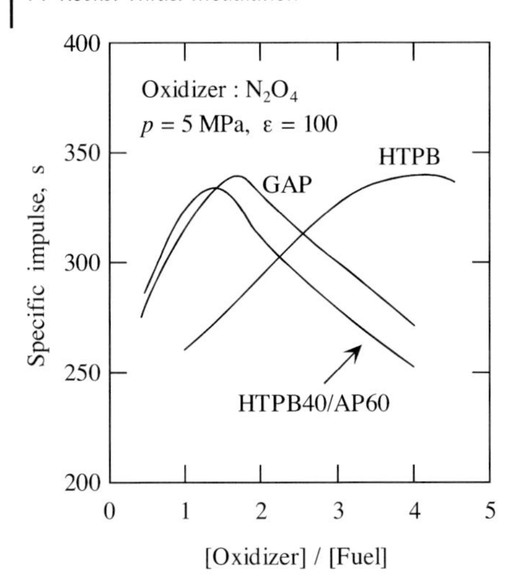

Fig. 14.25 Specific impulse of an N2O4/pyrolant gas-hybrid rocket as a function of oxidizer/fuel ratio.

Table 14.4 Specific impulse (oxidizer: liquid oxygen) at p_c = 5 MPa and ε = 100.

Pyrolant	I_{sp} (s)	O/F	$\varrho_{O/F} \times 10^3$ kg m^{-3}
$\xi_{GAP}(1.00)$	360.0	1.1	1.19
$\xi_{GAP}(0.80)/\xi_{Al}(0.20)$	361.9	0.8	1.28
$\xi_{GAP}(0.80)/\xi_{B}(0.20)$	361.4	1.1	1.25
$\xi_{GAP}(0.80)/\xi_{Zr}(0.20)$	350.1	0.9	1.31
$\xi_{GAP}(0.80)/\xi_{Mg}(0.20)$	357.1	0.8	1.25
$\xi_{GAP}(0.80)/\xi_{C}(0.20)$	353.1	1.3	1.24
$\xi_{GAP}(0.80)/\xi_{HTPB}(0.20)$	362.9	1.4	1.16
$\xi_{GAP}(0.60)/\xi_{HTPB}(0.40)$	365.3	1.7	1.13
$\xi_{BAMO}(1.00)$	365.1	1.1	1.21
$\xi_{BAMO}(0.80)/\xi_{Al}(0.20)$	366.9	0.7	1.30
$\xi_{AMMO}(1.00)$	366.0	1.5	1.11

Table 14.5 Specific impulse (oxidizer: N_2O) at p_c = 5 MPa and ε = 100.

Pyrolant	I_{sp} (s)	O/F	$\varrho_{O/F} \times 10^3$ kg m^{-3}
$\xi_{GAP}(1.00)$	321.8	3.3	1.24
$\xi_{GAP}(0.80)/\xi_{Al}(0.20)$	330.6	2.8	1.27
$\xi_{GAP}(0.80)/\xi_{B}(0.20)$	327.8	3.5	1.26
$\xi_{GAP}(0.80)/\xi_{Zr}(0.20)$	321.1	3.0	1.29
$\xi_{GAP}(0.80)/\xi_{Mg}(0.20)$	327.7	3.0	1.25
$\xi_{GAP}(0.80)/\xi_{C}(0.20)$	318.4	4.3	1.26
$\xi_{GAP}(0.80)/\xi_{HTPB}(0.20)$	321.6	4.3	1.22
$\xi_{GAP}(0.60)/\xi_{HTPB}(0.40)$	321.7	5.3	1.21
$\xi_{BAMO}(1.00)$	325.6	3.4	1.25
$\xi_{BAMO}(0.80)/\xi_{Al}(0.20)$	333.9	2.7	1.28
$\xi_{AMMO}(1.00)$	323.1	4.6	1.20

Table 14.6 Specific impulse (oxidizer: N_2O_4) at p_c = 5 MPa and ε = 100.

Pyrolant	I_{sp} (s)	O/F	$\varrho_{O/F} \times 10^3$ kg m^{-3}
$\xi_{GAP}(1.00)$	338.4	1.8	1.36
$\xi_{GAP}(0.80)/\xi_{Al}(0.20)$	344.8	1.3	1.43
$\xi_{GAP}(0.80)/\xi_{B}(0.20)$	341.6	1.8	1.42
$\xi_{GAP}(0.80)/\xi_{Zr}(0.20)$	333.2	1.4	1.43
$\xi_{GAP}(0.80)/\xi_{Mg}(0.20)$	340.9	1.3	1.39
$\xi_{GAP}(0.80)/\xi_{C}(0.20)$	332.6	2.1	1.42
$\xi_{GAP}(0.80)/\xi_{HTPB}(0.20)$	339.4	2.2	1.35
$\xi_{GAP}(0.60)/\xi_{HTPB}(0.40)$	340.1	2.7	1.33
$\xi_{BAMO}(1.00)$	343.1	1.6	1.38
$\xi_{BAMO}(0.80)/\xi_{Al}(0.20)$	349.4	1.2	1.44
$\xi_{AMMO}(1.00)$	341.5	2.3	1.29

Table 14.7 Specific impulse (oxidizer: H_2O_2) at p_c = 5 MPa and ε = 100.

Pyrolant	I_{sp} (s)	O/F	$\varrho_{O/F} \times 10^3$ kg m^{-3}
$\xi_{GAP}(1.00)$	339.3	2.7	1.39
$\xi_{GAP}(0.80)/\xi_{Al}(0.20)$	347.4	2.3	1.44
$\xi_{GAP}(0.80)/\xi_{B}(0.20)$	345.1	3.0	1.43
$\xi_{GAP}(0.80)/\xi_{Zr}(0.20)$	336.8	2.3	1.46
$\xi_{GAP}(0.80)/\xi_{Mg}(0.20)$	344.7	2.3	1.41
$\xi_{GAP}(0.80)/\xi_{C}(0.20)$	335.4	3.2	1.43
$\xi_{GAP}(0.80)/\xi_{HTPB}(0.20)$	340.1	3.3	1.37
$\xi_{GAP}(0.60)/\xi_{HTPB}(0.40)$	341.1	4.3	1.36
$\xi_{BAMO}(1.00)$	342.8	2.6	1.40
$\xi_{BAMO}(0.80)/\xi_{Al}(0.20)$	348.4	3.0	1.37
$\xi_{AMMO}(1.00)$	341.8	3.5	1.33

Table 14.8 Specific impulse (oxidizer: HNO_3) at p_c = 5 MPa and ε = 100.

Pyrolant	I_{sp} (s)	O/F	$\varrho_{O/F} \times 10^3$ kg m^{-3}
$\xi_{GAP}(1.00)$	324.3	1.9	1.41
$\xi_{GAP}(0.80)/\xi_{Al}(0.20)$	333.5	1.5	1.47
$\xi_{GAP}(0.80)/\xi_{B}(0.20)$	329.4	2.0	1.47
$\xi_{GAP}(0.80)/\xi_{Zr}(0.20)$	321.5	1.6	1.50
$\xi_{GAP}(0.80)/\xi_{Mg}(0.20)$	330.1	1.6	1.44
$\xi_{GAP}(0.80)/\xi_{C}(0.20)$	318.5	2.5	1.47
$\xi_{GAP}(0.80)/\xi_{HTPB}(0.20)$	324.4	2.6	1.40
$\xi_{GAP}(0.60)/\xi_{HTPB}(0.40)$	324.1	3.2	1.38
$\xi_{BAMO}(1.00)$	329.0	1.9	1.42
$\xi_{BAMO}(0.80)/\xi_{Al}(0.20)$	337.9	1.4	1.48
$\xi_{AMMO}(1.00)$	326.0	2.6	1.34

References

1 Kubota, N., Survey of Rocket Propellants and their Combustion Characteristics, Fundamentals of Solid-Propellant Combustion (Eds.: Kuo, K. K., and Summerfield, M.), *Progress in Astronautics and Aeronautics*, Vol. 90, Chapter 1, AIAA, New York (1984).

2 Glassman, I., and Sawyer, F., The Performance of Chemical Propellants, Circa Publications, New York (1970).

3 Sutton, G. P., Rocket Propulsion Elements, 6th edition, John Wiley & Sons, Inc., New York (1992), Chapter 11.

4 Kubota, N., Rocket Combustion, Nikkan Kogyo Press, Tokyo (1995).
5 Kubota, N., Principles of Solid Rocket Motor Design, Pyrotechnics Chemistry, J. of Pyrotechnics, Inc., Whitewater, CO (2004), Chapter 12.
6 Takishita, Y., Sumi, K., and Kubota, N., Experimental Studies on Erosive Burning of Rocket Motors, Proceedings of the 18th Symposium on Space Science and Technology, Tokyo (1974), pp. 197–200.
7 Procinsky, I. M., and McHale, C. A., Nozzleless Boosters for Integral-Rocket Ramjet Missile Systems, *J. Spacecraft and Rockets*, Vol. 18, 193 (1981).
8 Procinsky, I. M., and Yezzi, C. A., Nozzleless Performance Program, AIAA paper 82–1198 (1982).
9 Okuhara, H., Combustion Characteristics of Nozzleless Rocket Motors, Kogyo Kayaku, Vol. 48, No. 2, pp. 85–94 (1987).
10 Timnat, Y. M., Advanced Chemical Rocket Propulsion, Academic Press, New York (1987), Chapter 6.
11 Kuwahara, T., Mitsuno, M., Odajima, H., Kubozuka, S., and Kubota, N., Combustion Characteristics of Gas-Hybrid Rockets, AIAA-94–2880, New York (1994).

15
Ducted Rocket Propulsion

15.1
Fundamentals of Ducted Rocket Propulsion

15.1.1
Solid Rockets, Liquid Ramjets, and Ducted Rockets

Ducted rockets are intermediate between solid rockets and liquid ramjets in their propulsion characteristics. The propulsive force of solid rockets is generated by the combustion of propellants composed of oxidizer and fuel components. Thus, no additional fuels or oxidizers need to be introduced from the atmosphere into the rocket motor. The momentum change of the exhaust gas from the nozzle attached to the aft-end of the combustion chamber is converted into the thrust for propulsion. On the other hand, the propulsive force of liquid ramjets is generated by the combustion of a liquid hydrocarbon fuel with air introduced from the atmosphere.[1] The incoming air is compressed by a shock wave formed at the air-intake attached to the front end of the combustor. The air taken in from the atmosphere serves only as the oxidizer for the ramjets. The thrust is created by the momentum difference between the exhaust gas from the combustor and the air taken in from the atmosphere.

Similar to liquid ramjets, ducted rockets take in air from the atmosphere through an air-intake attached to the front end of the combustor. However, in contrast to liquid ramjets, the fuel components used for ducted rockets are fuel-rich pyrolants composed of fuel and oxidizer components. The products of incomplete combustion generated by a pyrolant in a gas generator burn with the air introduced from the atmosphere in the combustor.[1–6] As in the case of liquid ramjets, the thrust of ducted rockets is generated by the momentum difference between the exhaust gas from the combustor and the air taken in from the atmosphere.

Though air introduced from the atmosphere through an air-intake is used as an oxidizer in both ducted rockets and ramjets, the fuel used in ducted rockets is fundamentally different from that used in ramjets. The fuel supplied to the ramjet combustor is an atomized liquid that evaporates and mixes with the air for combustion. On the other hand, the fuel supplied to ducted rockets consists of high-temperature gaseous products. Thus, the ignition time and the reaction time of the ducted rocket fuel are much shorter than those of the ramjet fuel. Thus, high com-

Propellants and Explosives. Naminosuke Kubota

ISBN: 978-3-527-31424-9

bustion potential of ignitability and combustion stability are obtained for ducted rockets.

The compressed air introduced from the atmosphere through the air-intake is termed "ram air", and the associated pressure is termed "ram pressure". Ram pressure is built-up when the airflow velocity is decelerated in flow fields. The air-intake is designed as an aerodynamic tool to obtain maximum ram pressure.[1] Air-intakes are designed to decelerate supersonic flow to subsonic flow by the formation of shock waves in front of them. The combustor in which the fuel gas is burned with the ram air is termed a "ramburner".

15.1.2
Structure and Operational Process

Fig. 15.1 shows a typical model of a ducted rocket. The ducted rocket consists of a gas generator containing a gas-generating pyrolant, a gas-flow control system, air-intake, ramburner, and nozzle. Since the gas-generating pyrolant is composed of fuel-rich material, it burns incompletely in the gas generator and forms fuel-rich gaseous products.[1–7] These fuel-rich gaseous products are injected from the gas generator into the ramburner through the gas flow control system. The air introduced from the atmosphere into the ramburner is compressed by the shock wave formed at the front end of the air-intake. The fuel-rich gaseous products and the compressed air mix together in the ramburner and react to generate the products of complete combustion. These gaseous products are expelled through the exhaust nozzle attached to the aft end of the ramburner.

Unlike a solid rocket, a ducted rocket requires continuous airflow from the atmosphere to the ramburner through the air-intake. When a ducted rocket projectile is accelerated to a certain flight speed, compressed air is induced from the atmosphere. Once sufficient compressed air has been introduced into the ramburner, the ducted rocket starts to operate and generates thrust. The booster rocket attached

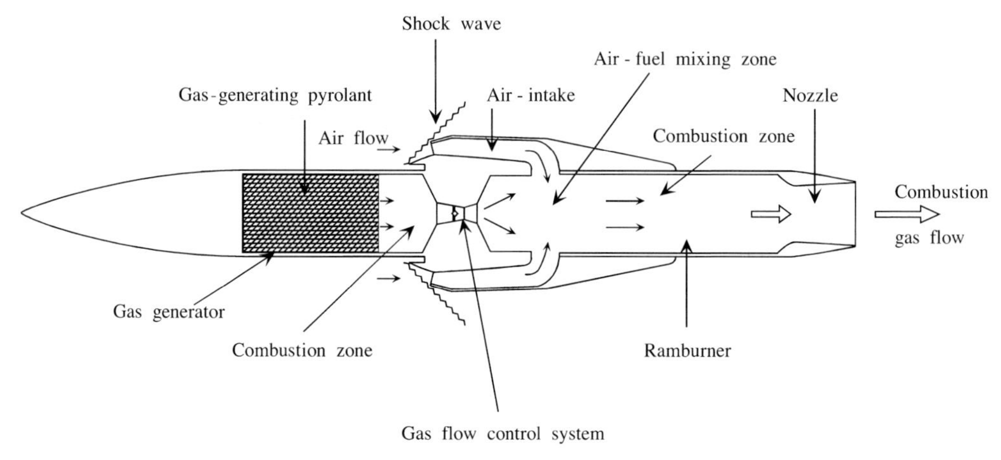

Fig. 15.1 Structure of a ducted rocket.

to the aft end of the ducted rocket is separated mechanically soon after burn-out of the booster propellant.

Since the ramburner is an empty pressure chamber, it is possible to integrate a booster propellant into the ramburner. The integrated booster rocket is composed of a booster propellant grain, a port cover attached to the air-intake, and a booster nozzle. The combustion gas of the booster propellant is expelled through the booster exhaust nozzle. The port cover is closed to maintain the chamber pressure during the booster propellant burning. When the booster rocket propellant is ignited, a booster thrust is generated, which accelerates the ducted rocket projectile. The projectile acquires a specified supersonic flight speed after burn-out of the booster propellant. The booster nozzle is then ejected to the outside and the port cover is opened. The compressed air resulting from the aforementioned shock wave is then introduced through the air-intake. The booster chamber becomes a ramburner and the gas-generating pyrolant is ignited to produce fuel-rich combustion products.

15.2 Design Parameters of Ducted Rockets

15.2.1 Thrust and Drag

The air-intake used to induce air from the flight-altitude atmosphere plays an important role in determining the overall efficiency of ducted rockets. The air pressure built up by the shock wave determines the pressure in the ramburner. The temperature of the compressed air is also increased by the heating effect of the shock wave. The fuel-rich gaseous products formed in the gas generator burn with the pressurized and shock-wave heated air in the ramburner. The nozzle attached to the rear-end of the ramburner increases the flow velocity of the combustion products through an adiabatic expansion process. This adiabatic expansion process is equivalent to the expansion process of a rocket nozzle described in Section 1.2.

Referring to Fig. 1.3, the momentum entering the air-intake is given by $\dot{m}_a v_a$ and that exiting from the nozzle is given by $(\dot{m}_a + \dot{m}_f)v_e$. The thrust created by the momentum change is fundamentally represented by Eq. (1.62). When the air-intake and the nozzle attached to the ducted rocket are designed to obtain maximum thrust efficiency, the pressures at the front end of the air-intake and at the aft end of the nozzle become $p_a = p_i = p_e$, and then Eq. (1.62) is represented by

$$F = (\dot{m}_a + \dot{m}_f)v_e - \dot{m}_a v_a \tag{15.1}$$

where F is thrust, $\dot{m}_a$ is the airflow rate at the air-intake, $\dot{m}_f$ is the fuel flow rate, and v_e is the exhaust velocity at the nozzle exit (at position e). The airflow rate at the air-intake is given by

$$\dot{m}_a = \rho_{ai} v_i A_i \tag{15.2}$$

where ρ_a is the density of air, v is velocity, and A is the cross-sectional area of the air-intake (at position i).

When a projectile assisted by a ducted rocket flies at velocity V along a trajectory with an angle θ with respect to the ground, the thrust F is represented by

$$F = D + M_p g \sin\theta \tag{15.3}$$

where D is the aerodynamic drag acting on the projectile, M_p is the mass of the projectile, and g is gravitational acceleration, and the flight velocity V is equal to the air-flow velocity v induced into the air-intake. The aerodynamic drag is obtained as a function of flight velocity according to:

$$D = (1/2) C_d \rho_a V^2 A \tag{15.4}$$

where C_d is the drag coefficient, and A is the cross-sectional area of the projectile. The drag coefficient depends on the shape of the projectile, the flight velocity, and the angle of attack of the projectile, α. Though C_d increases rapidly with increasing flight Mach number, M, in the region of transonic flight, $M = 0.7$–1.2, it decreases smoothly with increasing M in the high M region, $M = 1.8$–4, the region of practical operation for ducted rockets; for example, $C_d = 0.25$ at $M = 2$ and 0.18 at $M = 4$ for a flight of $\alpha = 0$. Accordingly, the aerodynamic drag is approximately proportional to the square of the flight velocity. If one assumes that a projectile has a level flight, i. e., $\theta = 0°$ at constant α, then the flight velocity is increased when $F > D$. The flight velocity becomes constant when $F = D$, and is decreased when $F < D$.

15.2.2
Determination of Design Parameters

In contrast to a rocket projectile, the flight speed of a ducted rocket projectile is dependent on the flight altitude since the density of the air induced into the air-intake varies with altitude, and hence the air-to-fuel ratio in the ramburner varies. This implies that the thrust generated by the ducted rocket varies, i. e., there is acceleration or deceleration of the level flight. The drag on the projectile also varies with flight velocity, which changes the airflow rate induced into the air-intake. It is highly difficult to obtain an optimized relationship between flight speed and flight altitude. The design parameters and procedure used for ducted rockets are shown in Table 15.1.

The parameters shown in Table 15.1 are used to determine the thrust and aerodynamic drag in a computational process as outlined in Table 15.2.

Table 15.1 Design parameters that need to be considered for ducted-rocket propulsion.

Flight altitude	h	m
Flight Mach number	M	
Fuel flow rate	$\dot{m}_f$	kg s^{-1}
Area ratios		
Area ratio of air-intakes	$\varepsilon_i = A_i/A_m$	
Nozzle expansion ratio	$\varepsilon_e = A_e/A_t$	
Area ratio of projectile	$\varepsilon_t = A_t/A_m$	
Area of air-intakes	A_i	m^2
Nozzle throat area	A_t	m^2
Area of nozzle exit	A_e	m^2
Cross-sectional area of projectile	A_m	m^2

Table 15.2 Computational process for determining the thrust of a ducted rocket.

1. Specify initial conditions		
Flight altitude	h	m
Atmospheric pressure	p_a	MPa
Air density	ϱ_a	kg m^{-3}
Air temperature	T_a	K
Flight velocity	v	m s^{-1}
Area of air-intake	A_i	m^2
Cross-sectional area of projectile	A_m	m^2
2. Compute $\dot{m}_a$		
$\dot{m}_a = \varrho_a v A_i$		
3. Determine $\dot{m}_f$ and A_t		
4. Compute ε_a		
$\varepsilon_a = \dot{m}_a/\dot{m}_f$		
5. Compute T_c, γ_c, and M_c		
Use thermochemical equilibrium calculation cord		
6. Compute the nozzle discharge coefficient c_D		
7. Determine p_c		
$p_c = (\dot{m}_f + \dot{m}_a)/c_D A_t$		
8. Determine the pressure recovery ratio		
9. Compute c_D		
10. Determine c_F		
11. Compute $c_F p_c A_t - D$	kN	
12. Determine F	kN	

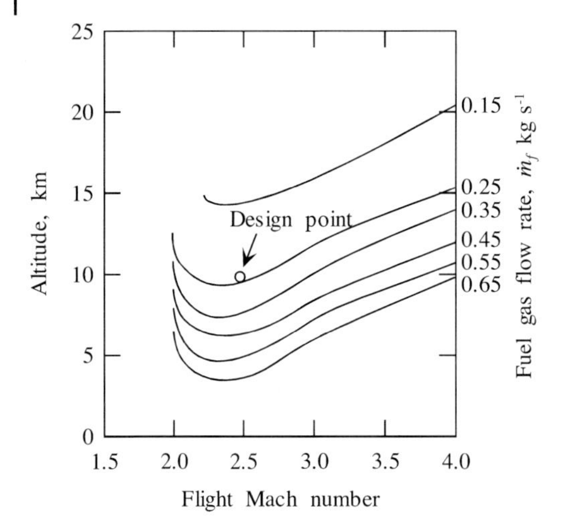

Fig. 15.2 Fuel flow rate and Mach number as a function of flight altitude.

15.2.3
Optimum Flight Envelope

Fig. 15.2 shows a typical result of a flight-model computation to establish the fuel-flow rate needed to sustain a required flight Mach number of a ducted rocket projectile at a flight angle of $\theta = 0°$. The design point is chosen at Mach 2.5 at altitude 10 km with $\dot{m}_f = 0.24$ kg s^{-1}, where $\dot{m}_f$ is the optimum fuel flow rate for the prerequisite flight Mach number and the specified altitude. The line at constant $\dot{m}_f$ indicates the possible flight zone. For example, when the flight altitude is changed from 15 km to 3 km at flight Mach number 2.5, $\dot{m}_f$ increases from 0.15 kg s^{-1} to 0.65 kg s^{-1}. The flight Mach number must be increased as the flight altitude is increased at constant $\dot{m}_f$. It is also shown that flight becomes impossible below Mach number 2.0 due to the inefficient airflow rate introduced from the air-intake.

It is evident that the fuel flow rate is decreased with increasing flight altitude because the airflow rate is decreased at constant flight Mach number. It must be noted that the air drag on the projectile decreases with increasing altitude because of the lower air density, and so the thrust can be decreased to maintain constant flight Mach number. The thrust is increased by increased fuel flow rate in order to increase the flight Mach number at constant altitude. The airflow rate is also increased by the increased flight Mach number.

15.2.4
Specific Impulse of Flight Mach Number

Though the specific impulse of a solid rocket is determined by the energetics of its propellant, the specific impulse of a ducted rocket is determined by the mixture ratio of the fuel flow rate from the gas generator and the airflow rate induced from the atmosphere. The ram pressure in the ramburner is raised by the shock wave formed at the air-intake. The combustion in the ramburner further increases the pressure therein. The specific impulse of a ducted rocket is determined not only by the energetics of the gas-generating pyrolant, but also by the aerodynamic efficiency of the air-intake. Since the density of air in the atmosphere decreases with increasing altitude and the airflow rate induced into the air-intake is dependent on the flight velocity, the specific impulse of a ducted rocket varies with flight velocity and altitude.

Though the thrust generated through the momentum change is as given by Eq. (15.1), the drag acting on the projectile must be matched by the thrust generated at the flight velocity. Flight analysis results indicate that high efficiency of ducted rocket operation is attained at supersonic flight in the range Mach 2–4.[1] The supersonic airflow is converted into pressure and temperature through the shock wave, accompanied by an entropy increase, and becomes a subsonic airflow as described in Section 1.2.1. The specific impulse of a ducted rocket, $I_{sp,d}$, is defined by:

$$I_{sp,d} = I_{sp} - \varepsilon u_{air}/g \tag{15.5}$$

where u_{air} is the flight velocity, g is gravitational acceleration, and I_{sp} is the specific impulse of the gas-generating pyrolant defined according to:

$$I_{sp} = (F/\dot{m}_g)/g \tag{15.6}$$

where F is the thrust of the ducted rocket and $\dot{m}_g$ is the fuel mass-flow rate. The momentum coming in through the air-intake is not included in the computation of I_{sp}. I_{sp} is determined by a similar method as the specific impulse of rocket propellants.

15.3 Performance Analysis of Ducted Rockets

15.3.1 Fuel-Flow System

The specific impulse of a ducted rocket is highly dependent on the airflow rate induced from the atmosphere. The fuel-rich gas generated by the combustion of a pyrolant in a gas generator burns with the air taken in from the atmosphere through the air-intake. Thus, the operational conditions are strongly dependent on the mass generation rate of the fuel-rich gas and the airflow rate induced from the atmosphere. In other words, the thrust of a ducted rocket is largely dependent on its flight speed and altitude.

When the flight trajectory of a ducted rocket projectile is given as a constant speed at a constant altitude, the thrust that needs to be given to the projectile is also constant throughout the flight trajectory. Since the airflow induced into the ramburner is kept constant, the fuel-flow rate generated in the gas generator is also kept constant. A fuel-flow control system for the gas generator is then no longer required and the ducted rocket system becomes a very simplified one. However, when a required flight trajectory from high altitude to low altitude or from high speed to low speed is specified, the airflow rate varies from point to point along the flight path. The fuel-flow rate then needs to be altered to obtain the required thrust at each position. Ducted rocket systems are divided into three types:

1. Non-choked fuel-flow system
2. Fixed fuel-flow system
3. Variable fuel-flow system.

15.3.1.1 Non-Choked Fuel-Flow System

The gas-generating pyrolant of a non-choked flow system is set at the forward part of the ramburner. The combustion pressure in the gas generator is then equivalent to the ramburner pressure. The pressure in the ramburner is approximately equal to the shock wave pressure formed at the air-intake attached to the ramburner. The pressure is varied by changing the flight speed and altitude of the projectile. Since the mass generation rate of the pyrolant burning is dependent on pressure, it is also varied by a change in the flight speed and altitude. This system is also termed a "solid ramjet".

When the flight speed is increased, the mass burning rate of the pyrolant in the gas generator is increased due to the increased ram pressure. The airflow rate induced from the air-intake is also increased by the increased flight speed. Thus, the thrust is increased in order to match the increased aerodynamic drag represented by Eq. (15.3). The non-choked fuel-flow system is a self-adjustable mass-flow system and its mechanical structure is a simple one without any moving parts. However, the thrust–drag relationship represented by Eq. (15.3) is adjustable in a flight envelope.

15.3.1.2 Fixed Fuel-Flow System

A fixed fuel-flow system is a simple set-up that is operated to maintain a constant fuel-flow rate. The fuel-rich gas flows out from the gas generator through a choked orifice that is attached at its aft-end. The mass generation rate of the fuel-rich gas is therefore independent of the pressure in the ramburner. When a projectile operated by a fixed-flow ducted rocket flies at a constant supersonic speed and at constant altitude, the airflow rate through the air-intake remains constant. Since the gas generation rate in the gas generator is kept constant, the air-to-fuel ratio also remains constant. Optimized combustion performance is thereby obtained. This class of ducted rocket is termed a "fixed fuel-flow ducted rocket".

However, a change in the flight speed and/or the flight altitude alters the airflow rate. Then, the air-to-fuel ratio in the combustion chamber is also altered, and the thrust produced by the ducted rocket is altered. Consequently, the flight envelope of the projectile becomes highly limited. These operational characteristics of the fixed fuel-flow ducted rocket restrict its application as a propulsion system.

15.3.1.3 Variable Fuel-Flow System

In order to overcome the difficulties associated with the non-choked fuel-flow system and the fixed fuel-flow system, a variable fuel-flow system is introduced: the fuel gas produced in a gas generator is injected into a ramburner. The fuel-flow rate is controlled by a control valve attached to the choked nozzle according to the airflow rate induced into the ramburner. An optimized mixture ratio of fuel and air, which is dependent on the flight altitude and flight velocity, is obtained by modulating the combustion rate of the gas-generating pyrolant. When a variable fuel-flow-rate system is attached to the choked nozzle of the gas generator, the fuel-flow rate is altered in order to obtain an optimized combustible gas in the ramburner. This class of ducted rockets is termed "variable fuel-flow ducted rockets" or "VFDR".

Since the mass generation rate in the gas generator is dependent on the pressure therein and the mass discharge rate is dependent on the throat area of the nozzle attached to the end of the gas generator, the mass generation rate is altered by changing the throat area. Thus, a throttable valve is attached to the end of the gas generator.

15.4 Principle of the Variable Fuel-Flow Ducted Rocket

15.4.1 Optimization of Energy Conversion

The supersonic air induced into the air-intake is converted into a pressurized subsonic airflow through the shock wave in the air-intake. The fuel-rich gas produced in the gas generator pressurizes the combustion chamber and flows into the ramburner through a gas flow control system. The pressurized air and the fuel-rich gas produce a premixed and/or a diffusional flame in the ramburner. The combustion gas flows out through the convergent-divergent nozzle and is accelerated to supersonic flow.

Since the airflow rate induced into the air-intake is dependent on the flight speed and altitude of the projectile, the mixture ratio of air and fuel gas must be adjusted accordingly. In some cases, the mixture may be too air-rich or too fuel-rich to burn in the ramburner, falling outside of the flammability limit (see Section 3.4.3), and no ignition occurs (see Section 3.4.1). In order to optimize the combustion in the ramburner under various flight conditions, a variable flow-rate system is attached to the gas flow control system.

15.4.2 Control of Fuel-Flow Rate

The mass generation rate in the gas generator is controlled by the variable flow system and the mixture ratio of fuel-rich gas to air in the ramburner is optimized. The burning rate is represented by the relationship $r = ap^n$, where r is the linear burning rate, p is the pressure, n is the pressure exponent of burning rate, and a is a con-

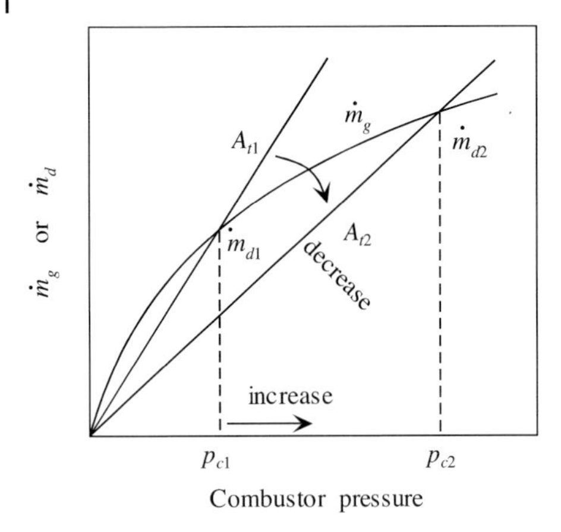

Fig. 15.3 Fundamental concept of a variable-flow system as a function of mass generation rate and mass discharge rate.

stant. Both n and a are dependent on the nature of the pyrolant. It is shown that the burning rate increases with increasing pressure since n is generally a positive value. As the pressure exponent increases, the burning rate becomes more pressure-sensitive.

The fundamental concept of the variable-flow system is based on the burning of the propellant in a rocket motor. The mass generation rate and the mass discharge rate must be equal to keep the fuel-flow rate constant, similarly to the mass balance in a rocket motor as described in Section 14.1. In Fig. 15.3, the mass balance between the mass generation rate, $\dot{m}_g$, and mass discharge rate, $\dot{m}_d$, in the gas generator is shown as a function of pressure. The mass generation rate is given by

$$\dot{m}_g = \rho_f A_b r \tag{14.3}$$

where A_b is the burning surface area and ρ_f is the density of the pyrolant. The mass discharge rate from the nozzle of the gas generator is given by Eq. (1.60) and is represented by

$$\dot{m}_d = c_D A_t p_c \tag{14.4}$$

where A_t is the nozzle throat area, c_D is the nozzle discharge coefficient given by Eq. (1.61), and p_c is the pressure in the gas generator. If the pressure variation in the gas generator with respect to time is negligibly small, the rate of mass accumulation in the gas generator is also negligibly small. Stable burning is obtained when $\dot{m}_g = \dot{m}_d$, as

$$\rho_f A_b r = c_D A_t p_c \tag{14.9}$$

Substituting Eq. (3.68) into Eq. (14.9), one obtains the gas generator pressure in a steady-state condition as

$$p_c = \{(a\rho_f/c_D)(A_b/A_t)\}^{1/(1-n)} \tag{14.10}$$

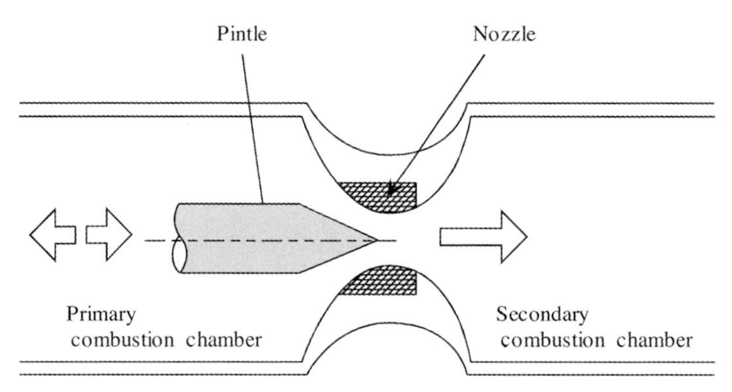

Fig. 15.4 A nozzle throat area controller based on insertion of a pintle into the throat.

When the nozzle throat area is A_{t1}, the mass discharge rate becomes $\dot{m}_{d1}$ and then the equilibrium pressure in the gas generator becomes p_{c1}, as shown in Fig. 15.3. The stable pressure is given as the crossover point of the $\dot{m}_g$ curve and the $\dot{m}_d$ line for A_{t1} in a plot of $\dot{m}_g$ or $\dot{m}_d$ versus pressure. When the nozzle throat area is decreased from A_{t1} to A_{t2}, the crossover point shifts to a mass discharge rate of $\dot{m}_{d2}$. The equilibrium pressure in the gas generator then shifts from p_{c1} to p_{c2}. The mass burning rate is increased and so the mass discharge rate is also increased by the increased gas generator pressure. The mass discharge rate from the gas generator is variable as long as the nozzle throat area is variable, and thereby the thrust generated by the VFDR may be varied.

Fig. 15.4 shows a schematic representation of a nozzle throat area controller used in a VFDR. The mass flow rate from the nozzle attached to the primary combustion chamber (gas generator) to the secondary combustion chamber (ramburner) is changed by inserting a pintle. The high-temperature gas produced in the gas generator flows into the ramburner through the pintled nozzle. The pintle inserted into the nozzle moves forward and backward in order to alter the nozzle throat area. As the nozzle throat area is made small, the mass flow rate increases according to the concept described above. The fuel-flow rate becomes throttable by the pintled nozzle.

As shown in Fig. 15.5, when a pyrolant of high pressure exponent is used, the variable flow range is increased. However, it is evident from Fig. 14.6 that the pressure exponent is required to be $n < 1$ for stable burning. It must also be noted that the temperature sensitivity of the pressure in the gas generator becomes high when a pyrolant of high pressure exponent is used.

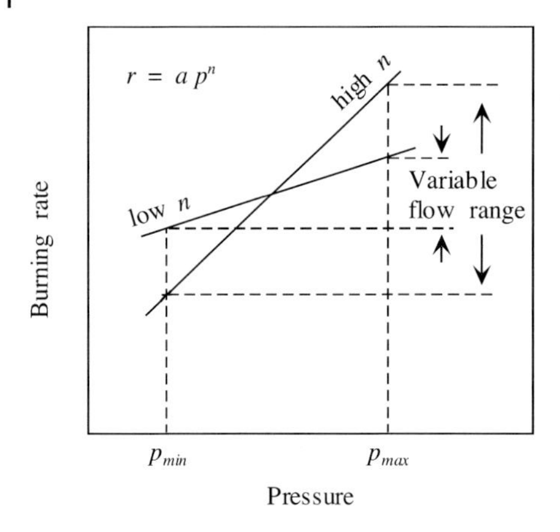

Fig. 15.5 Variable flow range of high and low pressure exponent pyrolants.

15.5 Energetics of Gas-Generating Pyrolants

15.5.1 Required Physicochemical Properties

The chemical compositions of gas-generating pyrolants used in ducted rockets are different from those of propellants used in rocket motors. The gas-generating pyrolants are fundamentally energetic materials that burn incompletely by themselves to generate fuel-rich combustion products. In general, the burning rates and the pressure exponents of burning rate of pyrolants decrease with decreasing energy contained within unit mass of the pyrolant.

Though the chemical components of the pyrolants are highly fuel-rich, the pyrolants are required to maintain self-sustaining combustion once ignited. A typical oxidizer used in pyrolants is ammonium perchlorate, which generates gaseous oxidizer fragments when thermally decomposed. Hydrocarbon polymers such as HTPB and HTPU are used as fuel components that produce gaseous and carbonaceous fragments upon thermal decomposition. Mixtures of ammonium perchlorate and hydrocarbon polymers form fuel-rich pyrolants. Their burning rate characteristics are dependent on the mixture ratio and the particle size of the crystalline ammonium perchlorate. This class of pyrolants is termed AP pyrolants.

Nitropolymers composed of -O–NO_2 functions and hydrocarbon structures are pyrolants that produce fuel-rich products accompanied by exothermic reaction. Typical nitropolymers are mixtures of nitrocellulose, nitroglycerin, trimethylolethane trinitrate, or triethylene glycol dinitrate, similar to the double-base propellants used in rockets and guns. Mixtures of these nitropolymers are formulated as fuel-rich pyrolants used in ducted rockets. This class of pyrolants is termed NP pyrolants.

The azide chemical bond, represented by -N_3, contains thermal energy, which is released when the bond is broken without oxidation. Typical chemicals containing azide bonds are glycidyl azide polymer, designated as GAP, BAMO, and AMMO. These polymers are copolymerized with hydrocarbon polymers to formulate fuel-rich pyrolants. Their exothermic decomposition products contain relatively high concentrations of fuel fragments such as H_2, CO, and $C_{(s)}$. This class of pyrolants is termed AZ pyrolants. When GAP is used as a major component of these AZ pyrolants, they are termed GAP pyrolants.

15.5.2 Burning Rate Characteristics of Gas-Generating Pyrolants

15.5.2.1 Burning Rate and Pressure Exponent

Typical gas-generating pyrolants include: (1) AP pyrolant composed of AP, $\xi_{AP}(0.50)$, and HTPB, $\xi_{HTPB}(0.50)$, which is cured with isophorone diisocyanate(IPDI); (2) NP pyrolant composed of NC, $\xi_{NC}(0.70)$ and NG, $\xi_{NG}(0.30)$, which is plasticized with diethyl phthalate (DEP); and (3) GAP pyrolant composed of glycidyl azide copolymer, $\xi_{GAP}(0.85)$, which is cured with hexamethylene diisocyanate(HMDI) and cross-linked with trimethylolpropane (TMP).

The AP pyrolant produces relatively high concentrations of solid carbon and hydrogen chloride (HCl). Though the mass fraction of the fuel components increases with increasing mass fraction of HTPB, self-sustaining burning of the AP pyrolant becomes impossible due to the lack of heat of decomposition, i. e., the thermal energy produced at the decomposing surface becomes too low to maintain thermal decomposition. On the other hand, the NP pyrolant burns to generate fuel-rich products even when the mass fraction of NC is increased. However, the burning rates of AP and NP pyrolants are low and their pressure exponents of burning rate are too low to permit their use as gas-generating pyrolants in VFDR.

Fig. 15.6 shows the burning rate characteristics of AP, NP, and GAP pyrolants in a gas generator. The burning rate of the GAP pyrolant is seen to be much higher than

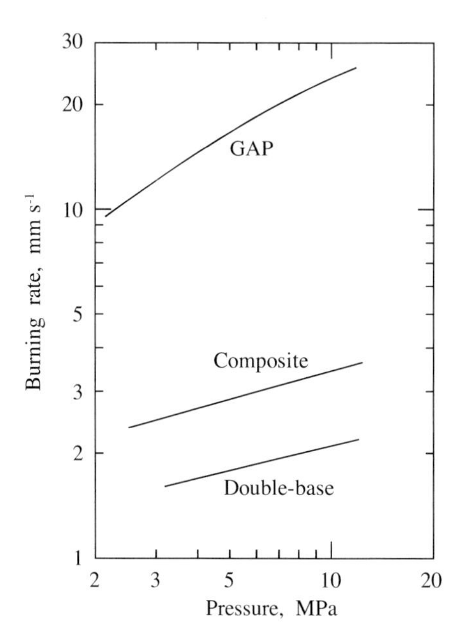

Fig. 15.6 Burning rate characteristics of gas-generating pyrolants.

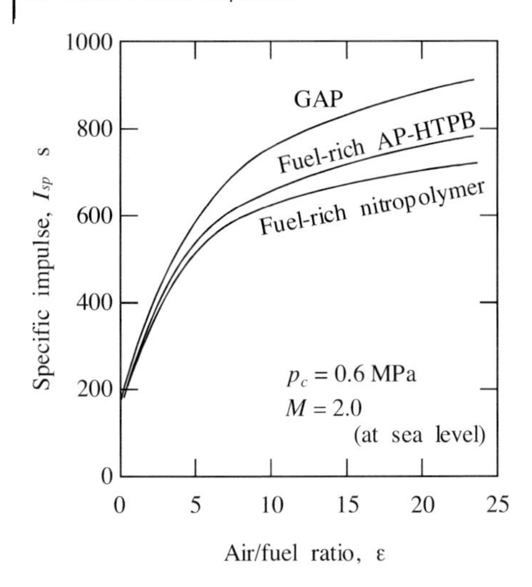

Fig. 15.7 Specific impulse of gas-generating pyrolants as a function of air-to-fuel ratio.

those of the AP and NP pyrolants. When the pressure is increased from 2.5 MPa to 10 MPa, the burning rate of the GAP pyrolant increases from 10 mm s^{-1} to 25 mm s^{-1}. The pressure exponents, n, of the AP and NP pyrolants are too low to permit their use as gas generators in VFDR. The pressure exponent of the GAP pyrolant is relatively high, about 0.7 in the pressure range between 1.0 MPa and 12 MPa.

The specific impulse of each pyrolant is computed as a function of air-to-fuel ratio, as shown in Fig. 15.7. In the computations, the pressure in the ramburner is assumed to be 0.6 MPa at Mach number 2.0 for a sea-level flight. When GAP pyrolant is used as a gas-generating pyrolant, the specific impulse is approximately 800 s at $\varepsilon = 10$. It is evident that AP pyrolant and NP pyrolant are not favorable for use as gas-generating pyrolants in VFDR. However, the specific impulse and burning rate characteristics of these pyrolants are further improved by the addition of energetic materials and burning rate modifiers.

15.5.2.2 Wired Gas-Generating Pyrolants

In general, the burning rates of fuel-rich gas-generating pyrolants are lower than those of rocket propellants due to the lower energy content of pyrolants compared to propellants. Thus, the design potential of the gas-generating pyrolants used in ducted rockets is very limited. Though the burning rates of propellants used for rockets are increased by the addition of catalysts, no catalysts are available for the gas-generating pyrolants due to the fuel-rich chemicals used.

The burning rates of rocket propellants are increased when fine metal wires are embedded within the propellants along the burning direction, as described in Section 13.6. The burning rate increases along the wire, and as a result the burning surface area increases. The increased burning surface area increases the mass burning rate. This method of burning rate augmentation is also effective for gas-generating pyrolants. As in the case of wired propellants, the burning rate along a fine wire embedded within a gas-generating pyrolant, r_w, is enhanced. The greater the thermal diffusivity of the wire, the higher the burning rate, r_w.

The burning-rate augmentation defined by r_w/r_0 is measured experimentally, where r_0 is the burning rate of the non-wired gas-generating pyrolant. Though r_w/r_0 increases with decreasing wire size, no burning rate augmentation effect is seen when the diameter of the wire is less than about 0.05 mm. The burning rates, r_0, of gas-generating pyrolants such as fuel-rich AP-HTPB and fuel-rich nitropolymer pyrolants are lower than those of rocket propellants such as AP-HTPB and nitropolymer propellants. The gas-phase temperature is low and hence the heat flux feedback through the wires is low for the gas-generating pyrolants as compared with propellants. However, r_w/r_0 appears to be approximately the same for both pyrolants and propellants. The obtained burning-rate augmentations are of the order of 2–5.

15.5.3
Pyrolants for Variable Fuel-Flow Ducted Rockets

A gas-generating pyrolant with a high pressure exponent of burning rate is a prerequisite for designing a VFDR offering a wide operational range of flight operation. The requirements for the pyrolants used in VFDR are summarized as follows:

1. Self-sustaining decomposition or combustion characteristics with higher fuel concentration and lower oxidizer concentration.
2. A high pressure exponent of burning rate; approximately 0.7–0.9, but must be less than 1.0.
3. A high burning rate when formed into an end-burning fuel grain: approximately 5×10^{-3} m s^{-1} in the low-pressure region around 1 MPa and 25×10^{-3} m s^{-1} in the high-pressure region around 10 MPa.
4. A combustion temperature that is low enough to protect the flow-rate control valve from heat and high enough to ignite the combustible gas when air is introduced into the ramburner; the combustion temperature of the gas-generating pyrolant is approximately 1400 K.

15.5.4
GAP Pyrolants

As described in Sections 4.2.4.1 and 5.2.2, GAP is a unique energetic material that burns very rapidly without any oxidation reaction. When the azide bond is cleaved to produce nitrogen gas, a significant amount of heat is released by the thermal decomposition. Glycidyl azide prepolymer is polymerized with HMDI to form GAP copolymer, which is crosslinked with TMP. The physicochemical properties of the GAP pyrolants used in VFDR are shown in Table 15.3.[4] The major fuel components are H_2, CO, and $C_{(s)}$, which are combustible fragments when mixed with air in the ramburner. The remaining products consist mainly of N_2 with minor amounts of CO_2 and H_2O.

Fig. 15.8 shows a typical set of burning rate versus pressure plots for GAP pyrolants composed of GAP copolymer with and without burning-rate modifiers. The burning rate decreases as the mass fraction of the burning-rate modifier, denoted by ξ(C), is increased. Graphite particles of diameter 0.03 μm are used as the burn-

Table 15.3 Physicochemical properties of GAP pyrolants.

Molecular mass	**1.98 kg mol^{-1}**
Density	**1.30×10^3 kg m^{-3}**
Heat of formation	**957 kJ kg^{-1} at 273 K**
Adiabatic flame temperature	**1465 K at 5 MPa**

Combustion products (mole fraction) at 5 MPa

H_2	CO	$C_{(s)}$	CH_4	N_2	H_2O	CO_2
0.3219	0.1395	0.2847	0.0215	0.2234	0.0071	0.0013

ing-rate modifier.[12,13] The pressure exponent of burning rate increases with increasing ξ(C) because of the suppression of the burning rate in the low-pressure region upon addition of the catalyst. Wide-ranging burning rate and pressure exponent modifications are possible by the addition of different mass fractions of ξ(C). For example, the pressure exponent is increased from 0.3 to 1.5 at a given pressure by the addition of ξ(C) = 0.10. In addition, the graphite particles act as fuel components in the ramburner.

The specific impulse of the GAP pyrolant is computed as a function of air-to-fuel ratio, ε, by means of Eq. (15.5) under conditions of a ramburner pressure of 0.6 MPa and a flight speed of Mach 2.0 at sea level. Fig. 15.9 shows the computed results; the specific impulse, I_{sp}, is seen to increase with increasing ε. The I_{sp} reaches approximately 800 s at ε – 15. The combustion temperature, T_f, in the ramburner is also computed as a function of ε. T_f increases with increasing ε in the range ε < 5 and reaches its maximum value, T_f = 2500 K, at ε = 5. T_f then decreases with increasing ε in the range ε > 5. It is important to note that I_{sp} increases continuously with increasing ε, even though T_f decreases in the range ε > 5.

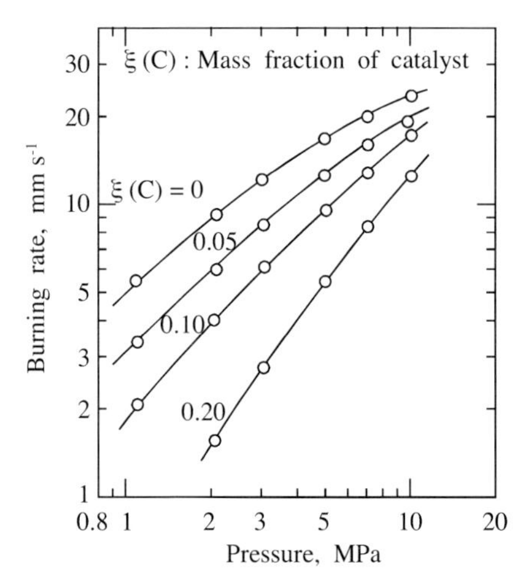

Fig. 15.8 Burning rate characteristics of a GAP copolymer with and without burning-rate modifier (graphite), showing that the pressure exponent increases with increasing amount of burning-rate modifier.

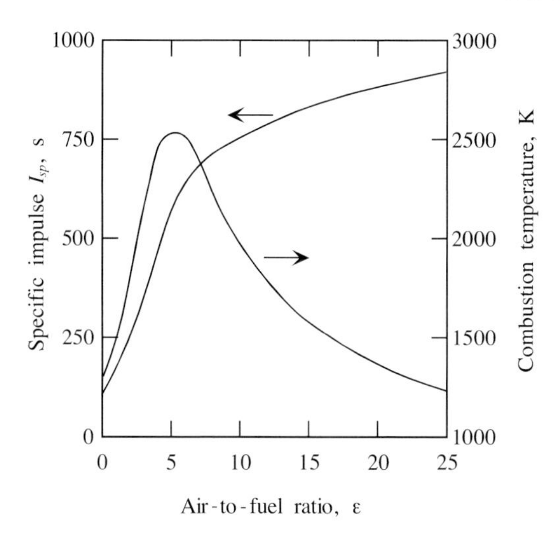

Fig. 15.9 Specific impulse and combustion temperature of GAP pyrolants as a function of air-to-fuel ratio; ramburner pressure 0.6 MPa and Mach number 2.0 at sea-level flight.

15.5.5 Metal Particles as Fuel Components

Metal particles represent energetic fuel components of gas-generating pyrolants. When metal particles are incorporated into a gas-generating pyrolant, they are heated by the combustion gas of the pyrolant without oxidation. This is because the pyrolants used in ducted rockets are fuel-rich and no excess oxidizing components are available to oxidize the metal particles. When these hot metal particles are ejected into the secondary combustion chamber, they are easily oxidized by the air introduced from the atmosphere to generate heat, and hence become hot metal oxide particles. Typical metal particles used are those of aluminum (Al), magnesium (Mg), titanium (Ti), or zirconium (Zr). Though boron (B) is a non-metal, the heat generated by its oxidation is much higher than that generated by the oxidation of metals.[1,5]

The molecular mass of the combustion products in the ramburner is increased by the formation of the oxidized metal particles. However, the temperature in the ramburner is also increased by the oxidation. The results of thermochemical calculations indicate that the specific impulse generated by the combustion in the ramburner is more dependent on the average combustion temperature than the average molecular mass of the products when metal particles are added. Table 15.4 shows the heats of combustion and the major oxidized products of the solid particles used in ducted rockets.

Al and Mg particles are favored metals in the formulation of pyrolants because of their high potential for ignitability and combustion. However, the combustion products of Al and Mg particles tend to agglomerate to form relatively large metal oxide particles. Since the densities and heats of combustion of Ti and Zr are higher than those of Al and Mg, Ti and Zr are more favorable for use as fuel metals in ducted rockets.

Table 15.4 Physicochemical properties of metal and non-metal particles used as fuel components in ducted rockets.

Solid particles	Density (kg m^{-3})	T_{mp} (K)	H_c (MJ kg^{-1})	Oxidized products
Metals				
Aluminum, Al	2700	934	16.44	Al_2O_3
Magnesium, Mg	1740	922	14.92	MgO
Titanium, Ti	4540	1998	8.50 (TiO)	TiO, Ti_2O_3, TiO_2
Zirconium, Zr	6490	2125	8.91	ZrO_2
Non-metals				
Boron, B	2340	2573	18.27	B_2O_3
Carbon, C	2260	3820	8.94 (CO_2)	CO, CO_2

T_{mp}: melting point temperature; H_c: heat of combustion

15.5.6 GAP-B Pyrolants

Boron particles are incorporated into GAP pyrolants in order to increase their specific impulse.[8–12] The adiabatic flame temperature and specific impulse of GAP pyrolants are shown as a function of air-to-fuel ratio in Fig. 15.10 and Fig. 15.11, respectively. In the performance calculation, a mixture of the combustion products of the pyrolant with air is assumed as the reactant. The enthalpy of the air varies according to the velocity of the vehicle (or the relative velocity of the air) and the flight altitude. The flight conditions are assumed to be a velocity of Mach 2.0 at sea level. An air enthalpy of 218.2 kJ kg^{-1} is then assumed.

The combustion temperature in a gas generator is maximized at the stoichiometric mixture of $\varepsilon = 5$. The specific impulse increases with increasing ε (fuel-lean mixture). However, the thrust decreases as ε is increased. The maximum thrust is obtained at the stoichiometric mixture ratio. It is evident that I_{sp} increases with increasing ξ_B. The I_{sp} of a pyrolant containing $\xi_B(0.2)$ is 1100 s and its T_f is about 1600 K at $\varepsilon = 20$.

Data for the combustion temperature in a gas generator are shown in Table 15.5. The GAP pyrolant without boron particles, $\xi_B(0.0)$, burns incompletely. The

Table 15.5 Theoretical and experimental combustion temperatures of GAP and GAP-B pyrolants.

GAP-B pyrolant	Combustion temperature, K Theoretical	Measured
$\xi_B(0.00)$	1347	1054
$\xi_B(0.10)$	2173	1139

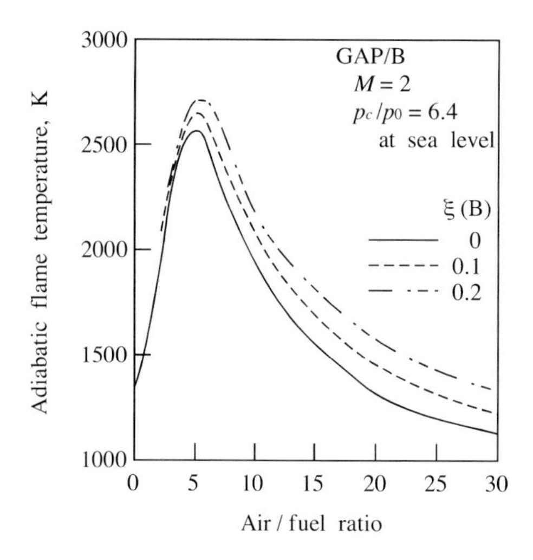

Fig. 15.10 Adiabatic flame temperature of a GAP-B gas-generating pyrolant as a function of air-to-fuel ratio.

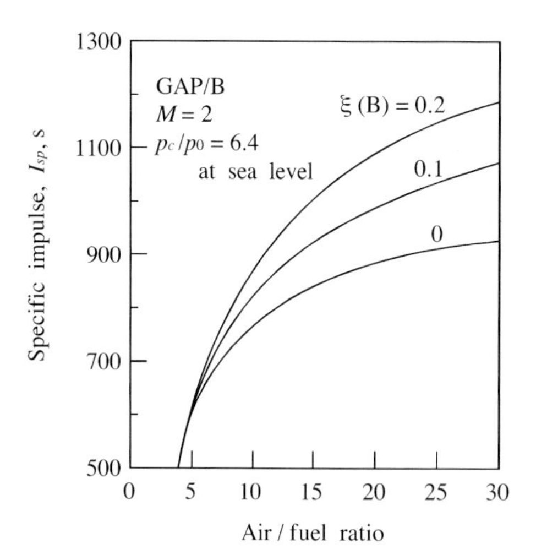

Fig. 15.11 Specific impulse of a GAP-B gas-generating pyrolant as a function of air-to-fuel ratio.

measured temperature is approximately 300 K lower than the theoretical temperature. A small temperature increase is observed when boron particles are incorporated into the GAP pyrolant. The difference between the theoretical combustion temperature and the measured temperature is approximately 1000 K when the mass fraction of boron is $\xi_B(0.1)$.

The temperature of the boron particles is raised by the heat generated by the decomposition of GAP. However, no combustion reaction occurs between the boron particles and the gaseous decomposition products of the GAP pyrolant. Thus, the temperature in the gas generator remains low enough to protect the attached nozzle from adverse heat.

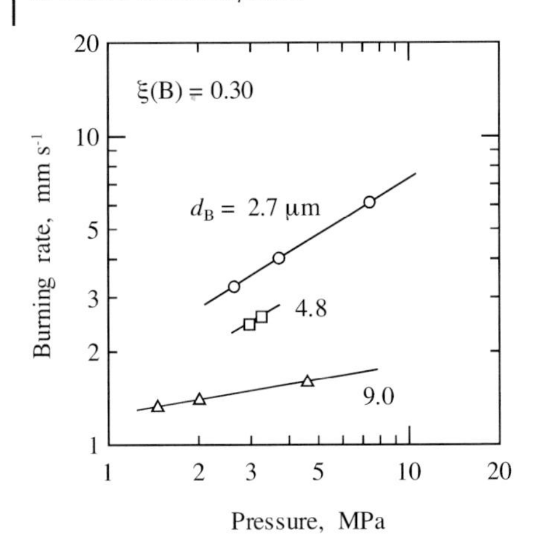

Fig. 15.12 Effect of boron particle size on the burning rates of an AP composite pyrolant.

15.5.7
AP Composite Pyrolants

Fig. 15.12 shows the burning rate characteristics of an AP composite pyrolant containing boron particles. The pyrolant is composed of $\xi_{AP}(0.40)$, $\xi_{CTPB}(0.30)$, and $\xi_{B}(0.30)$. The effect of boron particle size on burning rate is evident: both burning rate and pressure exponent decrease with increasing d_B. When d_B = 2.7 μm, the burning rate is 6 mm s^{-1} at 7 MPa and the pressure exponent is 0.65. However, when d_B = 9.0 μm, the burning rate is decreased drastically to 1.8 mm s^{-1} at the same pressure and the pressure exponent is decreased to 0.16. It is not possible to use such large boron particles in GAP-B pyrolants for VFDR.

15.5.8
Effect of Metal Particles on Combustion Stability

The reaction time required for complete combustion in a ramburner is determined by the flow velocity and the length of the ramburner. In order to obtain high combustion efficiency, the length of the ramburner should be as short as possible under high flow velocity. A combustible gas mixture is formed in the ramburner by mixing of the fuel-rich gas and the ram air induced through the air-intake. If the gas flow velocity in the ramburner is of the order of 500 m s^{-1} and the length of the ramburner is 1 m, for example, the residence time of the combustible gas in the ramburner is of the order of 2 ms. Since the flame spreading speed of the combustible gas is of the order of 10 m s^{-1}, the flame is expelled from the nozzle of the ramburner and it is difficult to maintain a flame in the ramburner. The combustible gas hence burns incompletely in the ramburner.

Metal particles incorporated into a gas-generating pyrolant act as flame holders to keep the flame in the ramburner. Each metal particle flows with the combustible gas and becomes a hot metal or metal oxide particle. Since the flow velocity of such a hot particle is lower than that of the combustible gas, the flow velocity of the combustible gas just downstream of the hot particle is decreased due to the aerodynamic effect between the gas flow and the surface of the hot particle. A micro-flame becomes attached to each hot particle and ignites the surrounding combustible gas, and the flame spreads into the combustible gas. In other words, the hot particle acts as an igniter. As a result, the hot particles dispersed in the ramburner ignite the combustible gas therein.

15.6 Combustion Tests for Ducted Rockets

15.6.1 Combustion Test Facility

Three types of combustion test facility are used to evaluate the combustion efficiencies of ducted rockets: direct-connect flow (DCF) test, semi-freejet (SFJ) test, and freejet (FJ) test, as shown in Fig. 15.13. Pressurized heated air or cooled air is supplied to the DCF, SFJ, and FJ test facilities. The pressure and temperature of the airflow are adjusted by means of an air control system to simulate the air conditions during flight of the ducted rocket projectile. In the case of the DCF test, the airflow is supplied to the ramburner from a pressurized air tank through a directly connected pipe. No air-intakes are used in the DCF test. Thus, the pressure and temperature of the air in the ramburner are as directly supplied from the pressurized air tank. No supersonic flow or shock waves are formed during the supply of air to the ramburner. In the DCF test, the combustion efficiency in the ramburner is measured as a function of the air-to-fuel flow ratio, ε. The combustion charac-

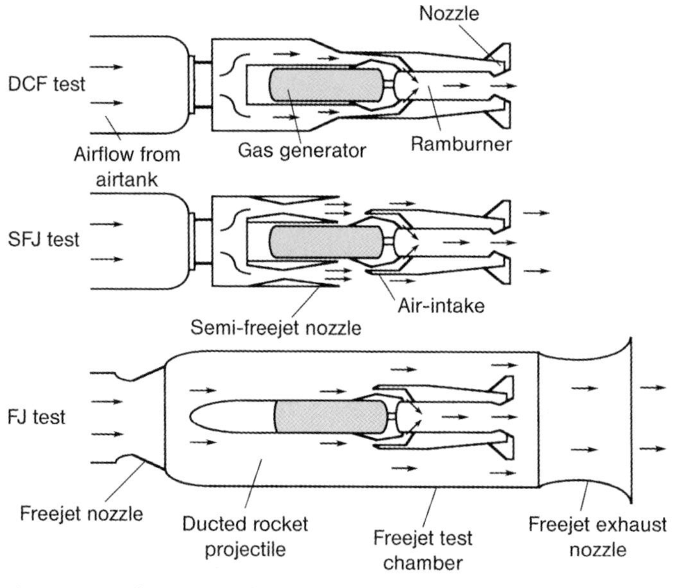

Fig. 15.13 Combustion test facilities for DCF, SFJ, and FJ tests.

teristics of ducted rockets are evaluated from the experimental data of ramburner pressure, airflow rate, and fuel-flow rate.

On the other hand, the airflow in the case of the SFJ test is supplied to the ramburner through the shock waves formed at the air-intake of a ducted rocket. The ducted rocket used for the combustion test is set on a thrust stand that is kept in a test chamber. A supersonic airflow, which simulates the flight conditions (flight altitude and flight Mach number), is supplied to the air-intake through a supersonic nozzle, which is directly connected to the pressurized air tank by a pipe. Mimicking the flight conditions of a ducted rocket projectile, air that is pressurized and heated by the shock waves is supplied to the ramburner. The air temperature of the supersonic airflow at the supersonic nozzle exit is adjusted to the pressure and temperature of the flight altitude. This airflow is produced by the air control system attached to the pressurized air tank. If two air-intakes are attached to the ducted rocket, two supersonic nozzles are needed for the SFJ test. The combustion gas in the ramburner is expelled through an exhaust pipe connected to the rear-end of the test chamber. Not only the combustion efficiency, but also the net thrust and the characteristics of the air-intake are measured by the SFJ test.

The FJ test is similar to an aerodynamic wind-tunnel test used for supersonic aircraft, except for the airflow condition. A ducted rocket projectile is mounted on a thrust stand and the projectile and thrust stand are placed in a test chamber. A supersonic airflow simulating the flight conditions is supplied to the projectile through a supersonic nozzle attached to the front-end of the test chamber. The pressure and temperature in the test chamber are kept equivalent to the flight altitude conditions. The aerodynamic drag on the projectile and the thrust generated by the ducted rocket are measured directly by the FJ test. The airflow surrounding the projectile and the combustion gas expelled from the ramburner flow out from the exhaust pipe attached to the rear-end of the test chamber.

15.6.2
Combustion of Variable-Flow Gas Generator

Fig. 15.14 shows a schematic drawing of a combustion test facility and DCF combustor. Pressurized, heated air is supplied to the DCF combustor through a regulator and a flow meter. The pressure and temperature of the air are adjusted in order to simulate designated flight altitude and speed conditions. The air tube directly connected to the plenum chamber is used to supply compressed air to the ramburner. The gas generator is attached to the forward end of the ramburner. The gas-generating pyrolant is an end-burning type grain and is ignited from the rear end. The outer surface and the forward end of the grain are insulated with polybutadiene rubber. The fuel gas flow rate is determined by the size of a choked orifice attached at the rear end of the gas generator. A pintle is inserted from the rear end of the gas generator into the choked orifice through the center axis of the gas-generating pyrolant. The pintle is moved back and forth by a DC servomotor through the ball screw as shown in Fig. 15.14. The gas flow rate is adjusted by changing the cross-sectional area of the choked orifice through displacement of the pintle.

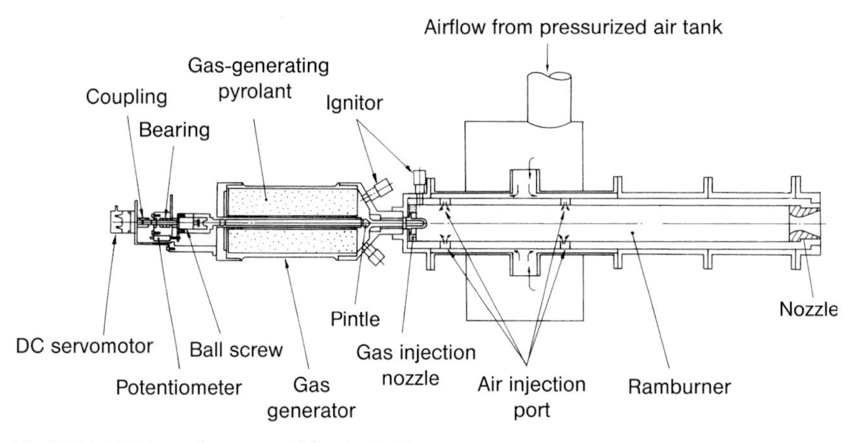

Fig. 15.14 VFDR combustor used for the DCF test.

The compressed, heated air is supplied to the ramburner through the air injection ports. Two types of air-injection ports, forming a so-called multi-port, are shown in Fig. 15.14: the forward port (two ports) and the rear port (two ports). The multi-port is used to distribute the airflow to the ramburner; 34 % is introduced via the forward port and the remaining 66 % via the rear port. The combustible gas formed by the combustion of the gas-generating pyrolant is injected through the gas injection nozzle and mixed with the air in the ramburner, and the burned gas is expelled form the ramburner exhaust nozzle. The pressures in the gas generator and the ramburner are measured by means of pressure transducers. The temperatures in the gas generator and the ramburner are measured with Pt-Pt/13 %Rh thermocouples.

Fig. 15.15 shows the specific impulse of a GAP pyrolant as a function of ε under variable flow conditions as obtained in a DCF test. The ramburner pressure ranges

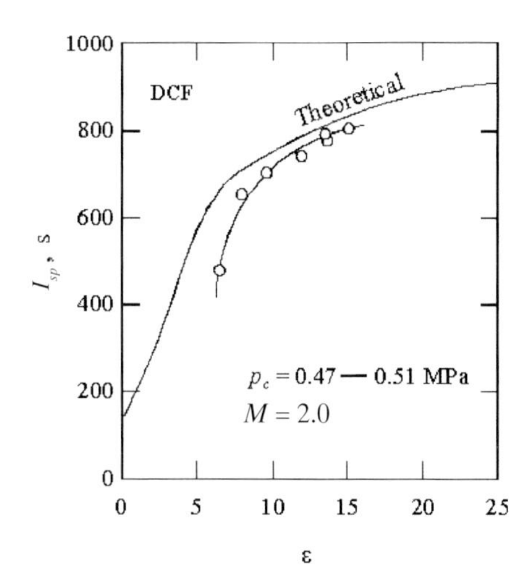

Fig. 15.15 Theoretical and experimental specific impulses of a GAP pyrolant of a VFDR as a function of air-to-fuel ratio obtained by a DCF test.

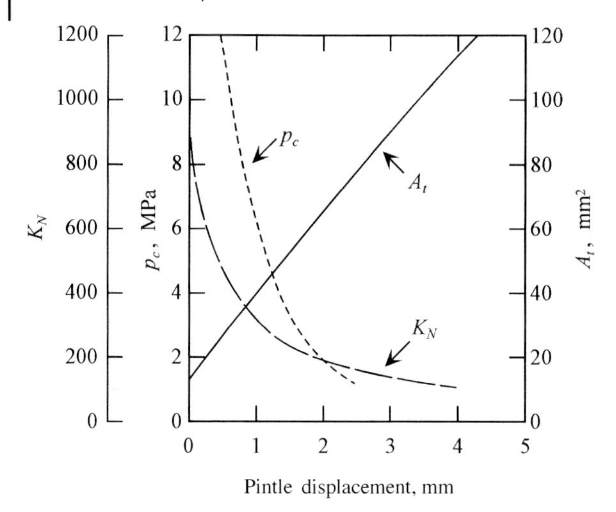

Fig. 15.16 Change of ramburner pressure with pintle displacement.

from 0.47 to 0.51 MPa and the characteristic length of the ramburner is 1.5 m. It is shown that the specific impulse increases with increasing ε and reaches 800 s at ε = 15. The experimental specific impulse is about 97% of the theoretical specific impulse at ε = 15. The flow velocity in the ramburner is about 250 m s^{-1}. The GAP pyrolant used for the combustion tests is composed of $\xi_{GAP}(0.85)$ without burning rate modifiers. The burning rate is 5.0×10^{-3} m s^{-1} at 1 MPa and 23×10^{-3} m s^{-1} at 10 MPa, and the pressure exponent is 0.76 at 1 MPa and 0.35 at 10 MPa.

Fig. 15.16 shows the characteristics of the gas-generating pyrolant and the pressure in the gas generator. The throat area of the gas generator, A_t, is changed by displacement of the pintle.

Fig. 15.17 (a) shows combustion test results obtained when the command signal-voltage applied to the servomotor of the flow-rate controller is in the form of a rectangular wave. The pintle displacement and thus the combustion pressure, p_c, are seen to emulate the form of the command signal. The combustion of the gas generator is completed after 16 s. Fig. 15.17 (b) shows combustion test results obtained when the command signal is applied as a triangular wave. It can again be seen that the pressure, p_c, follows the form of the command signal and the pintle displacement. Fig. 15.17 (c) shows combustion test results obtained when the command signal is applied in the form of a sine wave. The sine-wave frequency is incrementally increased from 1 Hz to 5 Hz. The results indicate that a pressure wave of frequency 2 Hz is generated when a command signal of 2 Hz is applied.

The response to pressure of the burning rate of the GAP pyrolant is sufficiently rapid to generate the required mass flux of combustible fuel-rich gas. The combustion temperature is low enough to protect the throttable flow-rate control valve from heat and is high enough to ignite the fuel gas when air is introduced into the ramburner. The combustion in the ramburner is very stable under high-velocity airflow and thus a high specific impulse is obtained. Fig. 15.18 shows a set of combustion

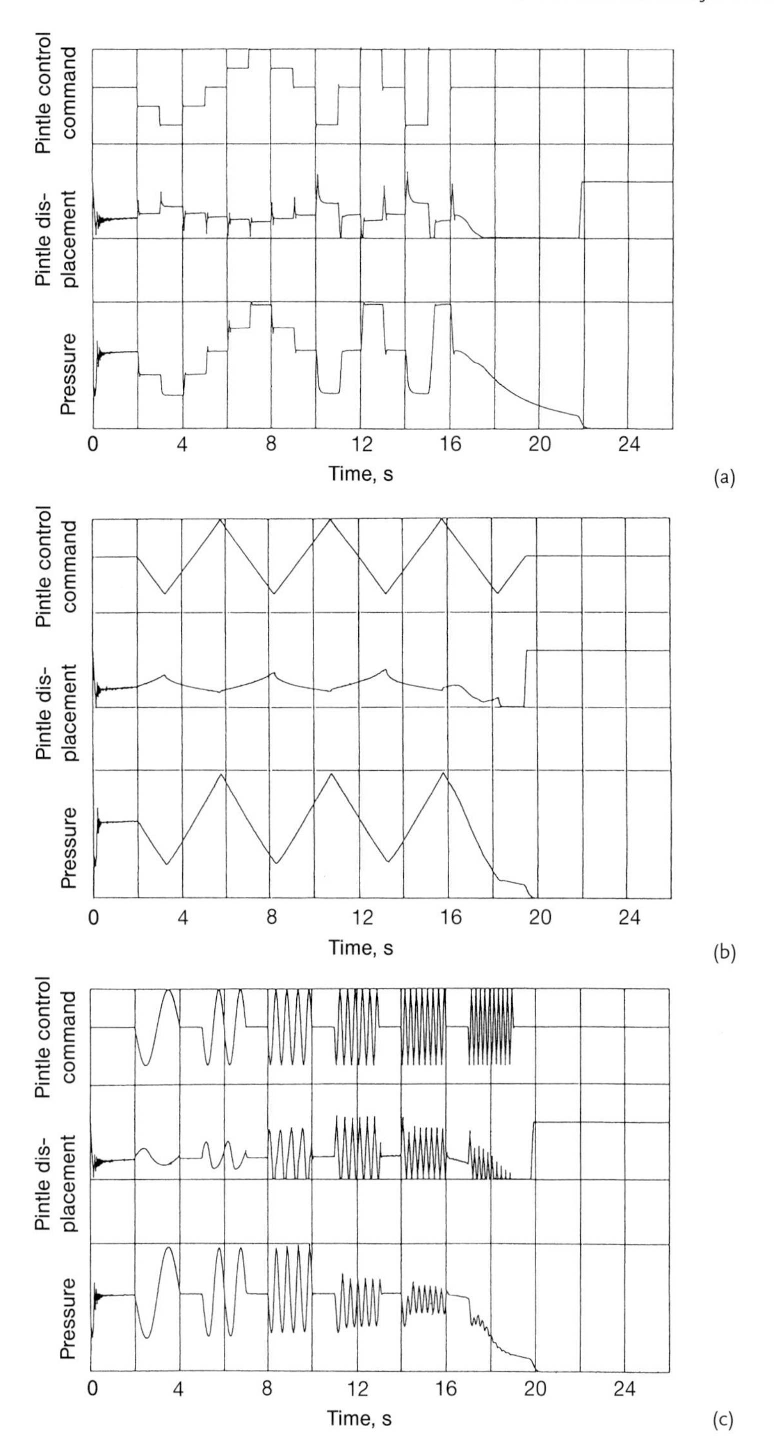

Fig. 15.17 DCF test results for a VFDR by applying command signals of different forms: (a) rectangular signal, (b) triangular wave, and (c) sine wave.

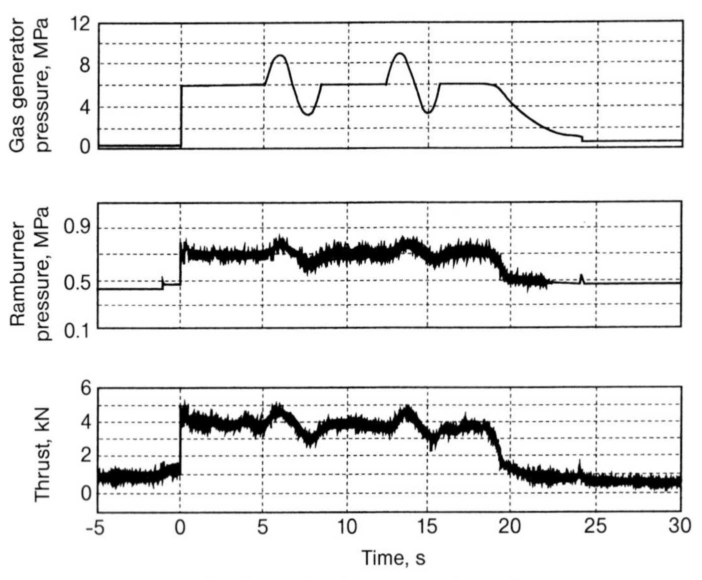

Fig. 15.18 SFJ test results obtained with a command signal composed of two sine waves, showing that the generated thrust follows the applied command signals through the gas generator pressure and the ramburner pressure.

results obtained from a DCF test. The thrust and the pressure in the ramburner are seen to follow the form of the command signal.

Fig. 15.18 shows SFJ test results for a VFDR obtained by applying a command signal composed of two sine waves. The gas generator pressure is 6.0 MPa and varies between 8.7 MPa and 3.3 MPa as a result of the sine-wave command signal. The generated ramburner pressure and hence the thrust are seen to respond to the variation of the pressure in the gas generator. Fig. 15.19 shows a pair of photographs of the combustion plume of a VFDR subjected to an SFJ test: (a) with an optimum air-to-fuel ratio, and (b) with a fuel-lean air-to-fuel ratio. Two side-air-intakes, as shown in Fig. D-5 of Appendix D, are attached to the VFDR.

15.6.3 Combustion Efficiency of Multi-Port Air-Intake

Boron is one of the essential materials for obtaining high specific impulse of a ducted rocket. However, the combustion efficiency of boron-containing gas-generating pyrolants is low due to incomplete combustion of the boron particles in the ramburner.[13–16] Fig. 15.20 shows the combustion temperature of a boron-containing pyrolant with and without boron combustion as a function of air-to-fuel ratio, ε. A typical boron-containing pyrolant is composed of mass fractions of boron particles $\xi_B(0.30)$, ammonium perchlorate $\xi_{AP}(0.40)$, and carboxy-terminated polybutadiene $\xi_{CTPB}(0.30)$. If the boron particles burn completely in the ramburner, the maximum combustion temperature reaches 2310 K at $\varepsilon = 6.5$ and v = Mach 2 (p_c =

(a)

(b)

Fig. 15.19 Combustion plumes of a VFDR subjected to an SFJ test: (a) with an optimum air-to-fuel ratio, and (b) with a fuel-lean air-to-fuel ratio.

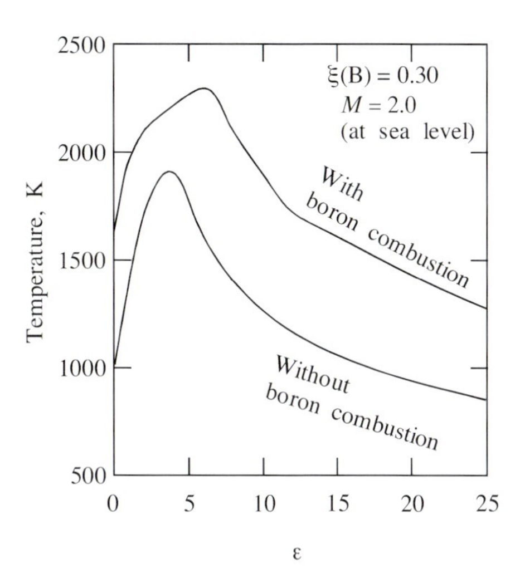

Fig. 15.20 Theoretical flame temperature versus air-to-fuel ratio of a boron-containing pyrolant with and without boron combustion.

0.6 MPa) under conditions of sea-level flight. However, if no boron combustion occurs, the temperature decreases to 1550 K under the same flight conditions. The combustion efficiency of boron particles is an important parameter for obtaining high specific impulse of ducted rockets.

The combustion efficiency of the ramburner, η_{c^*}, is defined by

$$\eta_{c^*} = c^*_{ex}/c^*_{th} \tag{15.7}$$

where c^*_{th} is the theoretical characteristic exhaust velocity and c^*_{ex} is the experimental characteristic exhaust velocity. It is shown that both c^*_{th} and c^*_{ex} are given by Eq. (1.73) or by Eq. (1.74) as

$$c^* = A_t p_c/(\dot{m}_a + \dot{m}_f) \tag{15.8}$$

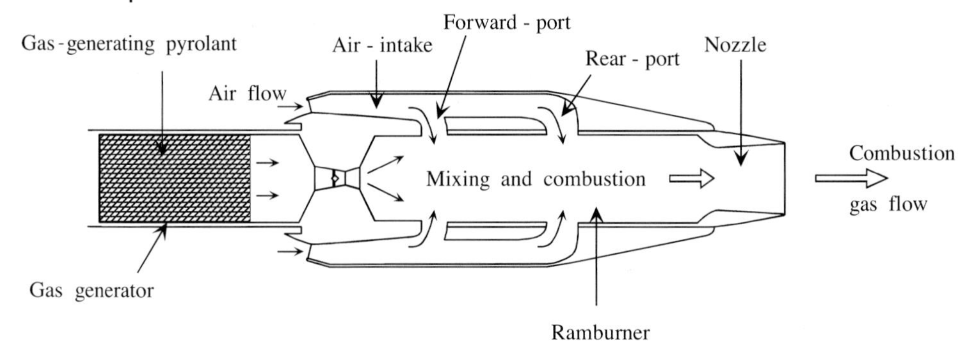

Fig. 15.21 Ducted rocket with multi-port air-intake.

where A_t is the nozzle throat area of the ramburner, p_c is the pressure in the ramburner, $\dot{m}_a$ is the airflow rate, and $\dot{m}_f$ is the fuel-flow rate. Experimental data on boron combustion indicate that the combustion efficiency is dependent on the mixing process of the air and fuel in the ramburner. Two types of air-intake, a single-port air-intake as shown in Fig. 15.1, and a multi-port air-intake as shown in Fig. 15.21, are used for DCF tests.[15,16] The single-port air-intake is composed of two air-intakes attached to the forward portion of the ramburner: one is on the right-hand side and the other is on the left. The multi-port intake is composed of four air-intakes: two are attached to the forward portion of the ramburner and the remaining two are attached to the rear portion.

When the airflow induced from the atmosphere is introduced through the single-port intake, the mixture formed in the forward part of the ramburner is fuel-lean because all the air induced from the single-port air-intake is introduced into the forward part. Thus, an excess-air mixture (fuel-lean mixture) is formed, the temperature of which becomes too low to initiate self-ignition. However, when a multi-port intake is used, the airflow is divided into two separate flows, entering at the forward part and the rear part of the ramburner. At the upstream flow, the air-to-fuel ratio can be made stoichiometric, which allows the mixture to ignite. At the downstream flow, the excess air is mixed with the combustion products and the temperature is lowered to increase the specific impulse.

When a gas-generating pyrolant containing boron particles $\xi(0.30)$ with d_B = 2.7 μm is burned in a ramburner, the combustion efficiency, η_{c^*} is approximately 79 % in the region of $\varepsilon = 5–22$ when a single-port air-intake is used, and is approximately 92 % at $\varepsilon = 12$ when a multi-port intake is used. The test results are shown in Fig. 15.22. The effect of the size of boron particles is evident, as shown in Fig. 15.23; η_{c^*} decreases rapidly with decreasing d_B when the single-port air-intake is used. No burning of the boron particles is observed in the ramburner when particles with d_B = 9.0 μm are incorporated into the gas-generating pyrolant. The same boron particles are partially burned when the multi-port air-intake is used, even though η_{c^*} is decreased to 81 %.

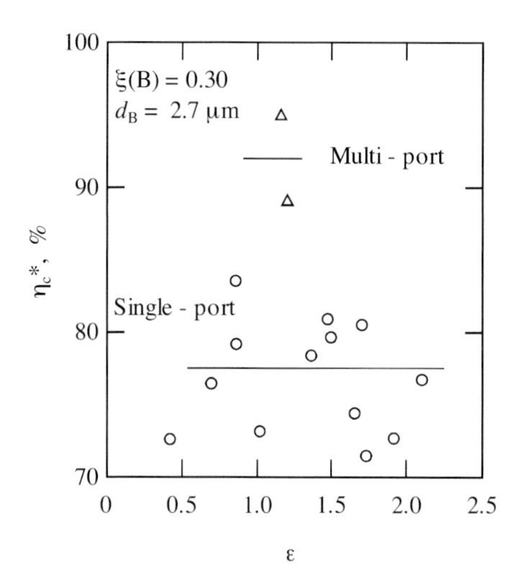

Fig. 15.22 Combustion efficiency of boron particles when two single-port intakes or two multi-port intakes are used.

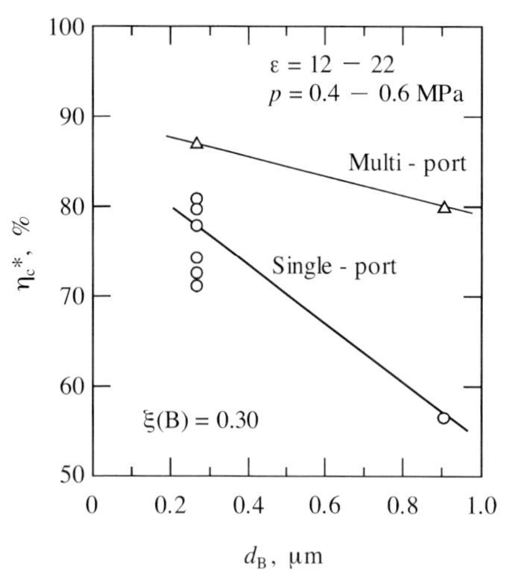

Fig. 15.23 Effect of air mixing process in the ramburner on combustion efficiency for two single-port intakes or two multi-port intakes as a function of boron particle size.

References

1 Kubota, N., and Kuwahara, T., Ramjet Propulsion, Nikkan Kogyo Press, Tokyo, 1997.

2 Besser, H.-L., "Solid Propellant Ducted Rockets", Messerschmitt-Bölkow-Blohm GmbH, Unternehmensbereich Apparate, München.

3 Technology of Ramjet and Ramrocket Propulsion at Bayern-Chemie, AY75, Jan. 1989.

4 Kubota, N., Yano, Y., Miyata, K., Kuwahara, T., Mitsuno, M., and Nakagawa, I., Energetic Solid Fuels for Ducted Rockets (II), *Propellants, Explosives, Pyrotechnics*, Vol. 16, 1991, pp. 287–292.

5 McClendon, S. E., Miller, W. H., and Herty III, C. H., Fuel Selection Criteria for Ducted Rocket Application, AIAA Paper No. 80–1120, June 1980.

6 Mitsuno, M., Kuwahara, T., Kosaka, K., and Kubota, N., "Combustion of Metallized Propellants for Ducted Rockets", AIAA Paper No. 87–1724, 1987.

7 Zhongqin, Z., Zhenpeng, Z., Jinfu, T., and Wenlan, F., Experimental Investigation of Combustion Efficiency of Air-Augmented Rockets, *J. of Propulsion and Power*, Vol. 4, 1986, pp. 305–310.
8 Kubota, N., and Sonobe, T. Combustion Mechanism of Azide Polymer, *Propellants, Explosives, Pyrotechnics*, Vol. 13, pp. 172–177, 1988.
9 Kubota, N., and Kuwahara, T., Energetic Solid Fuels for Ducted Rockets (1), *Propellants, Explosives, Pyrotechnics*, Vol. 16, pp. 51–54, 1991.
10 Kubota, N., and Sonobe, T., "Burning Rate Catalysis of Azide/Nitramine Propellants", 23 rd Symposium (International) on Combustion, The Combustion Institute, Pittsburgh, PA, 1990, pp. 1331–1337.
11 Ringuette, S., Dubois, C., and Stowe, R., On the Optimization of GAP-Based Ducted Rocket Fuels from Gas Generator Exhaust Characterization, *Propellants, Explosives, Pyrotechnics*, Vol. 26, 2001, pp. 118–124.
12 Limage, C., and Sargent, W., Propulsion System Considerations for Advanced Boron Powdered Ramjets, AIAA Paper No. 80–1283, June 1980.
13 Schadow, K., Boron Combustion Characteristics in Ducted Rockets, *Combustion Science and Technology*, Vol. 5, 1972, pp. 107–117.
14 Schadow, K., Experimental Investigation of Boron Combustion in Air-Augmented Rockets, *AIAA Journal*, Vol. 7, 1974, pp. 1870–1876.
15 Kubota, N., Miyata, K., Kuwahara, T., Mitsuno, M., and Nakagawa, I., Energetic Solid Fuels for Ducted Rockets (III), *Propellants, Explosives, Pyrotechnics*, Vol. 17, 1992, pp. 303–306.
16 Kubota, N., Air-Augmented Rocket Propellants, Solid Rocket Technical Committee Lecture Series, AIAA, Aerospace Sciences Meeting, Reno, Nevada, 1994.

Appendix A

List of Abbreviations of Energetic Materials

ADN	ammonium dinitramide
AMMO	3-azidomethyl-3-methyloxetane
AN	ammonium nitrate
AP	ammonium perchlorate
BAMO	bis-azide methyloxetane
CL-20	hexanitrohexaazatetracyclododecane (HNIW)
CTPB	carboxy-terminated polybutadiene
CuSa	copper salicylate
CuSt	copper stearate
DATB	diaminotrinitrobenzene
DBP	dibutyl phthalate
DDNP	diazodinitrophenol
DEGDN	diethylene glycol dinitrate
DEP	diethyl phthalate
DNT	dinitrotoluene
DOA	dioctyl adipate
DOP	dioctyl phthalate
DPA	diphenylamine
EC	ethyl centralite
GAP	glycidyl azide polymer
HMDI	hexamethylene diisocyanate
HMX	cyclotetramethylene tetranitramine
HNB	hexanitrobenzene
HNF	hydrazinium nitroformate
HNIW	hexanitrohexaazaisowurtzitane (CL-20)
HNS	hexanitrostilbene
HTPA	hydroxy-terminated polyacetylene
HTPB	hydroxy-terminated polybutadiene
HTPE	hydroxy-terminated polyether
HTPS	hydroxy-terminated polyester
IDP	isodecyl pelargonate

Propellants and Explosives. Naminosuke Kubota

ISBN: 978-3-527-31424-9

IPDI	isophorone diisocyanate
MAPO	tris(1-(2-methyl)aziridinyl) phosphine oxide
MT-4	adduct of 2.0 mol MAPO, 0.7 mol adipic acid, and 0.3 mol tartaric acid
NBF	*n*-butyl ferrocene
NC	nitrocellulose
NG	nitroglycerin
NP	nitronium perchlorate
NQ	nitroguanidine
2NDPA	2-nitrodiphenylamine
OXM	oxamide
PB	polybutadiene
PBAN	polybutadiene acrylonitrile
PbSa	lead salicylate
PbSt	lead stearate
Pb2EH	lead 2-ethylhexanoate
PE	polyether
PETN	pentaerythrol tetranitrate
Picric acid	2,4,6-trinitrophenol
PQD	paraquinone dioxime
PS	polyester
PS	polysulfide
PU	polyurethane
PVC	polyvinyl chloride
RDX	cyclotrimethylene trinitramine
SN	sodium nitrate
SOA	sucrose octaacetate
TA	triacetin
TAGN	triaminoguanidine nitrate
TATB	triaminotrinitrobenzene
TDI	toluene-2,4-diisocyanate
TEA	triethanolamine
TEGDN	triethylene glycol dinitrate
Tetryl	2,4,6-trinitrophenylmethylnitramine
TF	polytetrafluoroethylene, "Teflon"
TMETN	trimethylolethane trinitrate
TMP	trimethylolpropane
TNT	trinitrotoluene
Vt	vinylidene fluoride hexafluoropropane polymer, "Viton"

Appendix B

Mass and Heat Transfer in a Combustion Wave

Fig. B-1 presents a steady-state flow in a combustion wave, showing mass, momentum, and energy transfers, including chemical species, in the one-dimensional space of Δx between x_1 and x_2. The viscous forces and kinetic energy of the flow are assumed to be neglected in the combustion wave. The rate of heat production in the space is represented by ωQ, where ω is the reaction rate and Q is the heat release by chemical reaction per unit mass.

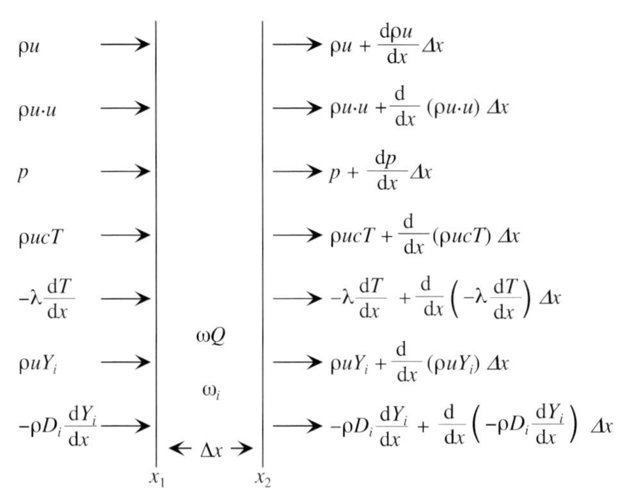

Figure B-1. Conservation of mass, momentum, energy, and chemical species in a combustion wave.

B.1
Conservation Equations at a Steady State in a One-Dimensional Flow Field

B.1.1
Mass Conservation Equation

The mass coming in through the cross-sectional area ΔA per unit time at x_1 is represented by

$$\rho u \Delta A$$

and the mass flowing out through ΔA per unit time at x_2 is represented by

$$\{\rho u + (\mathrm{d}/\mathrm{d}x(\rho u))\Delta x\}\Delta A$$

At a steady-state condition, the mass is conserved in the space according to

$$\rho u \Delta A = \{\rho u + \mathrm{d}/\mathrm{d}x(\rho u)\Delta x\}\Delta A$$

The mass conservation equation is then represented by

$$\frac{d}{dx}(\rho u) = 0 \tag{B.1}$$

B.1.2
Momentum Conservation Equation

The momentum coming in through ΔA per unit time at x_1 is represented by

$$\rho u u \Delta A$$

and that flowing out through ΔA per unit time at x_2 is represented by

$$\rho u \Delta A u + \mathrm{d}/\mathrm{d}x(\rho u u)\Delta A \Delta x$$

The pressures acting on ΔA at x_1 and x_2 are given by

$$p\Delta A$$

and

$$p\Delta A + \mathrm{d}p/\mathrm{d}x(\Delta A \Delta x)$$

respectively. The momentum in the space between x_1 and x_2 is given by

$$\rho u \Delta A + p\Delta A = \rho u \Delta A + \mathrm{d}/\mathrm{d}x(\rho u u)\Delta A \Delta x + p\Delta A + \mathrm{d}p/\mathrm{d}x(\Delta A \Delta x)$$

The momentum conservation equation is then represented by

$$\frac{d}{dx}(\rho u^2) + \frac{dp}{dx} = 0 \tag{B.2}$$

B.1.3
Energy Conservation Equation

The heat coming in through ΔA per unit time at x_1 is represented by

$$\rho u c T \Delta A$$

and the heat flowing out through ΔA per unit time at x_2 is represented by

$$\rho u c T \Delta A + d/dx(\rho u c T)\Delta A \Delta x$$

The heat transferred through ΔA per unit time at x_1 is represented by

$$(-\lambda dT/dx)\Delta A$$

and the heat transferred through ΔA per unit time at x_2 is represented by

$$(-\lambda dT/dx)\Delta A + d/dx(-\lambda dT/dx)\Delta A \Delta x$$

The heat produced by the chemical reaction in the volume $\Delta A \Delta x$ per unit time is given by

$$Q \Delta A \Delta x$$

The energy in $\Delta A \Delta x$ is conserved according to

$$\rho u c T \Delta A + (-\lambda dT/dx)\Delta A + \omega Q \Delta A \Delta x = \{\rho u c T \Delta A + d/dx(\rho u c T)\Delta A \Delta x\}$$
$$+ \{(-\lambda dT/dx)\Delta A + d/dx(-\lambda dT/dx)\Delta A + d/dx(-\lambda dT/dx)\Delta A \Delta x\}$$

The energy conservation equation is then represented by

$$\frac{d}{dx}\left(\lambda \frac{dT}{dx}\right) - \rho u \frac{d}{dx}(cT) + \omega Q = 0 \tag{B.3}$$

conductive heat transfer	convective heat transfer	heat produced by chemical reaction

B.1.4
Conservation Equations of Chemical Species

The chemical species i flowing in through ΔA per unit time at x_1 is given by

$$\rho u Y_i \Delta A$$

and the chemical species i flowing out through ΔA per unit time at x_2 is given by

$$\rho u Y_i \Delta A + \mathrm{d}/\mathrm{d}x(\rho u Y_i \Delta A)\Delta x$$

The chemical species i flowing in through ΔA per unit time at x_1 by diffusion is given by

$$(-\rho D_i \mathrm{d}Y_i/\mathrm{d}x)\Delta A$$

and the chemical species i flowing out through ΔA per unit time at x_2 by diffusion is given by

$$(-\rho D_i \mathrm{d}Y_i/\mathrm{d}x)\Delta A + \mathrm{d}/\mathrm{d}x(-\rho D_i \mathrm{d}Y_i/\mathrm{d}x)\Delta A \Delta x$$

The rate of mass loss of the chemical species i due to the chemical reaction in the space $\Delta A \Delta x$ is given by

$$\omega_i \Delta A \Delta x$$

The chemical species i conserved in $\Delta A \Delta x$ is given by

$$\rho u Y_i \Delta A + (-\rho D_i \mathrm{d}Y_i/\mathrm{d}x)\Delta A - \omega_i \Delta A \Delta x$$

$$= \rho u Y_i A + \mathrm{d}/\mathrm{d}x(\rho u Y_i)\Delta A \Delta x + (-\rho D_i \mathrm{d}Y_i/\mathrm{d}x)\Delta A + \mathrm{d}/\mathrm{d}x(-\rho D_i \mathrm{d}Y_i/\mathrm{d}x)\Delta A \Delta x$$

The conservation equation of species i is then represented by

$$\frac{d}{dx}\left(\rho D_i \frac{dY_i}{dx}\right) - \frac{d}{dx}(\rho u Y_i) + \omega_i = 0 \tag{B.4}$$

mass transfer by diffusion	mass transfer by convection	rate of mass production by chemical reaction

B.2
Generalized Conservation Equations at a Steady-State in a Flow Field

The conservation equations described in Section B.1 show the mass, momentum, energy, and chemical species equations at a steady state in a one-dimensional flow field. Similarly, the conservation equations at a steady-state in two- or three-dimensional flow fields can be obtained. The results can be summarized in a vector form as

mass conservation: $\nabla \cdot (\rho v) = 0$ (B.5)

momentum conservation: $\rho v \cdot \nabla v = -\nabla p$ (B.6)

energy conservation: $\nabla \cdot (\rho v c T - \lambda \nabla T) = -\Sigma \omega_i Q_i$ (B.7)

chemical species conservation: $\nabla \cdot (\rho v Y_i - \rho D \nabla Y_i) = \omega_i$ (B.8)

where v is the mass-averaged velocity vector.

Appendix C

Shock Wave Propagation in a Two-Dimensional Flow Field

The basic characteristics of a one-dimensional shock wave are described in Chapter 1 of this text. However, the shock waves in supersonic flow propagate not only one-dimensionally but also two- or three-dimensionally in space. For example, the shock waves formed at the air-intake of a ducted rocket are two- or three-dimensional in shape. Expansion waves are also formed in supersonic flow. The pressure downstream of an expansion wave is reduced and the flow velocity is increased. With reference to Chapter 1, brief descriptions of the characteristics of a two-dimensional shock wave and of an expansion wave are given here.[1–5]

C.1
Oblique Shock Wave

The formation of a shock wave is dependent on the objects that affect the flow field. The conservation of mass, momentum, and energy must be satisfied at any location. This is manifested in the formation of a shock wave at a certain location in the flow field to meet the conservation equations. In the case of a blunt body in a supersonic flow, the pressure increases in front of the body. The increased pressure generates a detached shock wave to satisfy the conservation equations in the flow field to match the conserved properties between the inflow and outflow in front of the body. The velocity then becomes a subsonic flow behind the detached shock wave. However, the shock wave distant from the blunt body is less affected and the detached shock wave becomes an oblique shock wave. Thus, the shock wave appears to be curved in shape, and is termed a bow shock wave, as illustrated in Fig. C-1.

Fig. C-2(a) shows an attached shock wave on the tip of a wedge. This is a weak shock wave formed when the associated pressure difference is small. On the other hand, as shown in Fig. C-2(b), a detached shock wave is formed when the pressure difference becomes large. An attached shock wave becomes a detached shock wave when the wedge angle becomes large.

When a two-dimensional wedge is placed in a supersonic flow, a shock wave that

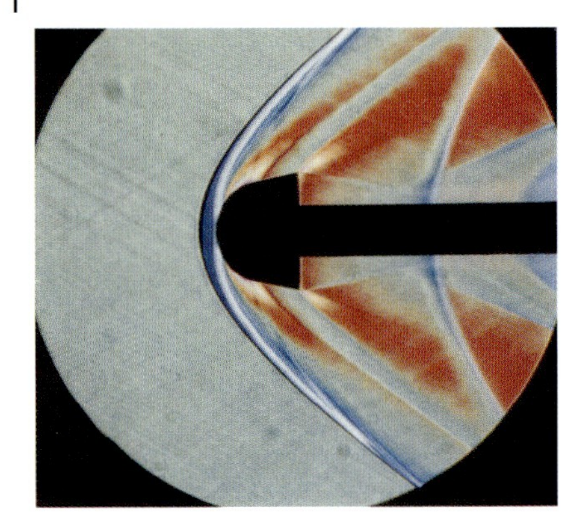

Figure C-1. A bow shock wave formed in front of a blunt body.

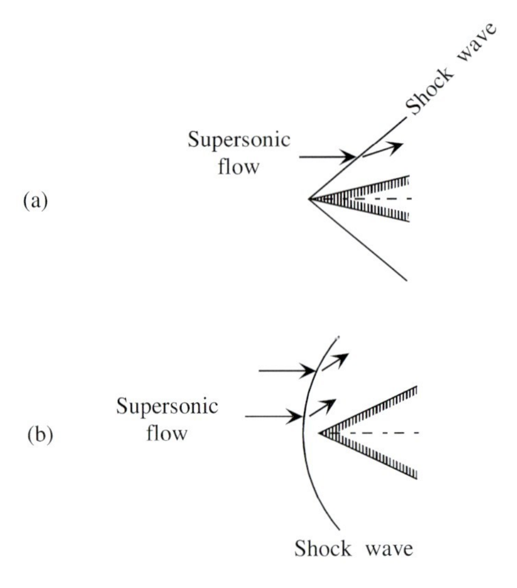

Figure C-2. Attached shock wave (a) and detached shock wave (b).

propagates from the tip of the wedge is formed. Unlike a normal shock wave, the streamline is not perpendicular to the shock wave, and this is termed an oblique shock wave. As shown in Fig. C-3, the shock wave is deflected by an angle, β, and the streamline is also inclined at an angle, θ. The velocity along the streamline changes from w_1 to w_2 through the oblique shock wave. The velocity component perpendicular to the shock wave changes from u_1 to u_2 and the velocity component parallel to the shock wave changes from v_1 to v_2. The velocity triangle shown in Fig. C-3 is expressed by

$$w_2^2 = u_2^2 + v_2^2 \tag{C.1}$$

Though the velocity component parallel to the shock wave remains unchanged, $v_1 = v_2$, the velocity component normal to the shock wave, $u_1 \rightarrow u_2$, changes through the shock wave. The change in the normal velocity component through the oblique

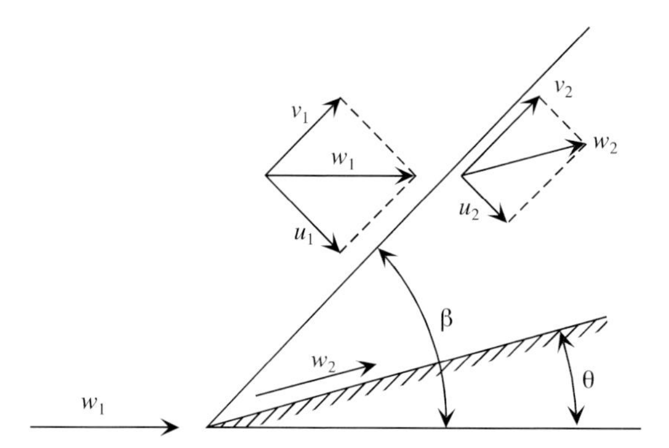

Figure C-3. Oblique shock wave formed by a two-dimensional wedge.

shock wave is equivalent to the velocity change through the normal shock wave. Then, u_1 shown in Fig. C-3 is equivalent to u_1 shown in Fig. 3.1 and the Rankine–Hugoniot relationship between pressure and density for an oblique shock wave becomes equivalent to that for a normal shock wave given by Eqs. (C.2) and (C.3) as follows:

$$p_2/p_1 = \{(\rho_2/\rho_1)\zeta - 1\}/(\zeta - \rho_2/\rho_1) \tag{C.2}$$

$$\rho_2/\rho_1 = \{(p_2/p_1)\zeta + 1\}/(p_2/p_1 + \zeta) \tag{C.3}$$

where ζ is given by $\zeta = (\gamma + 1)/(\gamma - 1)$.

The angle between the inflow streamline and the oblique shock wave, β, is expressed by

$$\beta = \tan^{-1}(u_1/v_1) \tag{C.4}$$

Since the Mach number of the inflow to the shock wave is given by $M_1 = w_1/a_1$ and that of the outflow from the shock wave is given by $M_2 = w_2/a_2$, the Mach number of the normal velocity perpendicular to the oblique shock wave, M_1^*, is represented by

$$u_1/a_1 = M_1 \sin\beta = M_1^* \tag{C.5}$$

Thus, the oblique shock-wave equations are obtained by replacing M_1 with M_1^* in the normal shock-wave equations, Eqs. (3.19)–(3.23), as follows:

$$\frac{p_2}{p_1} = \frac{2\gamma}{\gamma+1}M_1^{*2} - \frac{1}{\zeta} \tag{C.6}$$

$$\frac{p_{02}}{p_{01}} = \left[\frac{(\gamma+1)M_1^{*2}}{(\gamma-1)M_1^{*2}+2}\right]^{\frac{\gamma}{\gamma-1}} \left[\frac{2\gamma}{\gamma+1}(M_1^{*2}-1)+1\right]^{-\frac{1}{\gamma-1}} \tag{C.7}$$

$$M_1^* = \left[\frac{\gamma+1}{2\gamma}\frac{p_2-p_1}{p_1}+1\right]^{\frac{1}{2}} \tag{C.8}$$

$$\frac{T_2}{T_1} = \frac{2(\gamma-1)(M_1^{*2}-1)(\gamma M_1^{*2}+1)}{(\gamma+1)^2 M_1^{*2}} + 1 \tag{C.9}$$

$$\frac{\rho_2}{\rho_1} = \frac{(\gamma+1)M_1^{*2}}{(\gamma-1)M_1^{*2}+2} \tag{C.10}$$

The entropy change through the oblique shock wave is given by

$$s_2 - s_1 = c_p \ln(T_2/T_1) - R_g \ln(p_2/p_1) \tag{1.45}$$

Since $p_2/p_1 \geq 1$ in Eq. (C.8), the Mach number normal to the oblique shock wave is

$$M_1^{*2} \geq 1 \tag{C.11}$$

and then one obtains the relationship

$$\sin^{-1}(1/M_1) \leq \beta \leq \pi/2 \tag{C.12}$$

The incline angle along the streamline of the upstream defined by

$$\alpha = \sin^{-1}(1/M_1) \tag{C.13}$$

is termed the Mach angle, as defined in Chapter 2. It is shown the angle of the oblique shock wave, β, is larger than the Mach angle, α.

The angle η, between the shock-wave angle, β, and the angle of the streamline behind the shock wave, θ, given by

$$\eta = \beta - \theta \tag{C.14}$$

is represented by

$$\tan \eta = u_2/v_2 \tag{C.15}$$

Combining Eqs. (C.1), (C.2), and (C.8) with the relationships $v_1 = v_2$ and $u_2/u_1 = \rho_1/\rho_2$, one obtains

$$\tan \eta/\tan \beta = 2/\{(\gamma + 1)\, M_1^{*2}\} + 1/\zeta \tag{C.16}$$

It is shown that two β and two θ correspond to one M_1. When β is small, the static pressure ratio p_2/p_1 is small, and the shock wave is weak. On the other hand, when β is large, a strong shock wave is formed, for which p_2/p_1 is large. The Mach number behind the oblique shock wave becomes supersonic for weak shock waves and subsonic for strong shock waves.

Based on Eq. (C.15), the Mach number M_1^* is obtained as

$$M_1^* = \{(\gamma + 1)(M_1^2/2) \sin \beta \sin \theta / \cos \zeta + 1\}^{1/2} \tag{C.17}$$

When θ is small, Eq. (C.17) becomes

$$M_1^* = \{(\gamma + 1)(M_1^2/2)\theta \tan\theta + 1\}^{1/2} \tag{C.18}$$

C.2
Expansion Wave

Let us consider a supersonic flow along a wall surface with a corner of negative angle ($-\theta$). The flow is governed by the same conservation equations as for an oblique shock wave formed along a wall surface with a corner of positive angle ($+\theta$). The key difference is that an oblique shock wave is formed when the corner has a positive angle whereas an expansion wave is formed when the corner has a negative angle. The expansion wave is formed in a fan-shape with the corner at its center, as shown in Fig. C-4. The expansion wave consists of a multitude of Mach waves. The first Mach wave, with an angle of α_1, is formed at the front-end of the expansion wave, and the last Mach wave, with an angle of α_2, is formed at the rear-end of the expansion wave, these being represented by $\alpha_1 = \sin^{-1}(1/M_1)$ and $\alpha_2 = \sin^{-1}(1/M_2)$, respectively.

In the expansion wave, the flow velocity is increased and the pressure, density, and temperature are decreased along the stream line through the expansion fan. Since $\alpha_1 > \alpha_2$, it follows that $M_1 < M_2$. The flow through an expansion wave is continuous and is accompanied by an isentropic change known as a Prandtl–Meyer expansion wave. The relationship between the deflection angle and the Mach number is represented by the Prandtl–Meyer expansion equation.[1–5]

C.3
Diamond Shock Wave

When a supersonic flow emerges from a rocket nozzle, several oblique shock waves and expansion waves are formed along the nozzle flow. These waves are formed repeatedly and form a brilliant diamond-like array, as shown in Fig. C-5. When an under-expanded flow, i. e., having pressure p_e higher than the ambient pressure p_a, is formed at the nozzle exit, an expansion wave is formed to decrease the pressure. This expansion wave is reflected at the interface between the flow stream and the ambient air and a shock wave is formed. This process is repeated several times to form a diamond array, as shown in Fig. C-6 (a).

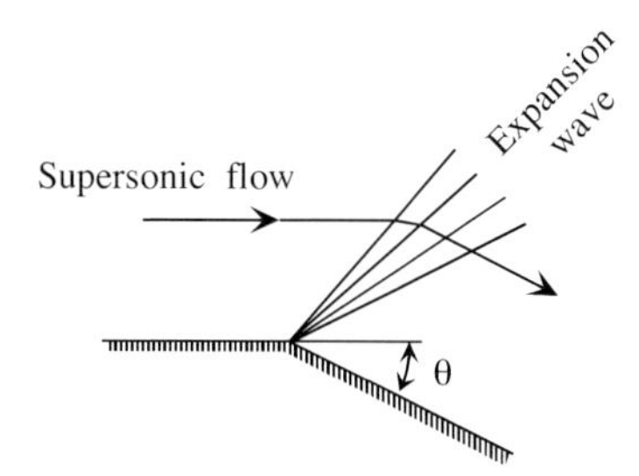

Figure C-4. Expansion wave formed in supersonic flow along a wall surface with a corner of negative angle.

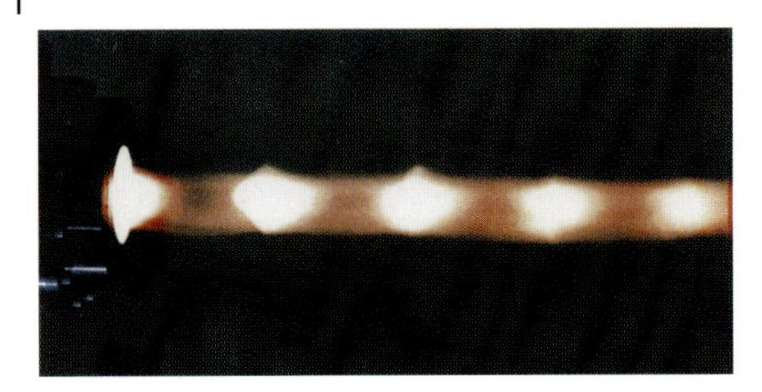

Figure C-5. Diamond shock wave array formed downstream of a rocket nozzle.

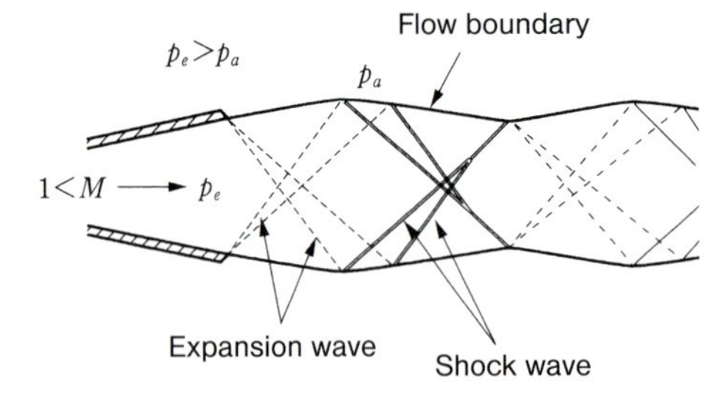

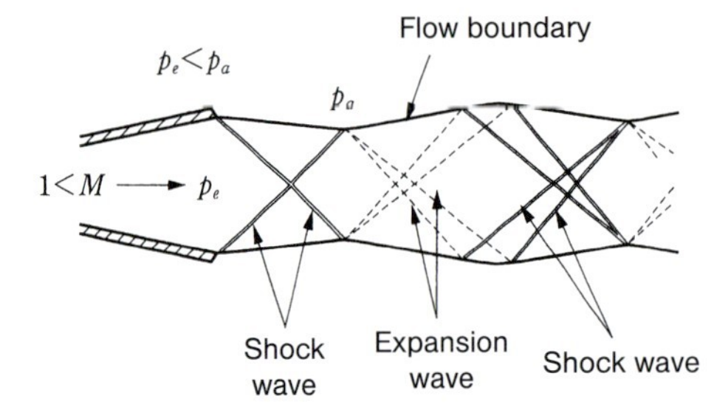

Figure C-6. Structures of (a) an under-expanded nozzle flow and (b) an over-expanded nozzle flow.

On the other hand, an over-expanded flow is formed at a nozzle exit when the pressure p_e is lower than that of the ambient atmosphere p_a, and a shock wave is formed to increase the pressure. This shock wave is reflected at the interface between the flow stream and the ambient air and an expansion wave is formed. As in the case of the under-expanded flow, this process is repeated several times to form a diamond array, as shown in Fig. C-6 (b).

References

1 Shapiro, A. H., The Dynamics and Thermodynamics of Compressible Fluid Flow, The Ronald Press Company, New York, 1953.
2 Liepman, H. W., and Roshko, A., Elements of Gas Dynamics, John Wiley & Sons, New York, 1957.
3 Kuethe, A. M., and Schetzer, J. D., Foundations of Aerodynamics, John Wiley & Sons, New York, 1967.
4 Zucrow, M. J., and Hoffman, J. D., Gas Dynamics, John Wiley & Sons, New York, 1976.
5 Kubota, N., Foundations of Supersonic Flow, Sankaido, Tokyo, 2003.

Appendix D
Supersonic Air-Intake

D.1
Compression Characteristics of Diffusers

D.1.1
Principles of a Diffuser

Air-intakes are important components of aero-propulsion engines such as turbojet, ramjet, and ducted rocket engines. The airflow velocity induced into the engines through air-intakes is converted as effectively as possible into static pressure in order to obtain a high momentum change. Diffusers are the main components of these air-intakes. The physical shape of the diffuser is fundamentally the same as that of a nozzle composed of convergent and/or divergent sections. The fundamental design principle of air-intakes is based on the aerodynamics of a shock wave.[1–6]

When a subsonic airflow is induced into a divergent nozzle, the flow velocity is reduced and the static pressure is increased. As shown in Fig. D-1, the air-intake used for a subsonic flow is composed of a subsonic diffuser, i. e., a divergent nozzle. On the other hand, when the airflow is supersonic, the airflow velocity is reduced and the static pressure is increased in the convergent part of a supersonic diffuser. After passing through the throat of the nozzle, the airflow velocity increases and the static pressure increases once more in the divergent part. A normal shock wave is formed downstream of the divergent part; the flow velocity is suddenly decreased and the pressure is increased. The supersonic air-intake is composed of a convergent-divergent nozzle. It must be noted that a normal shock wave is not formed in the convergent nozzle because such a shock wave would be unstable therein.

Fig. D-2 shows the shock-wave formation at a supersonic diffuser composed of a divergent nozzle. Three types of shock wave are formed at three different back-pressures downstream of the diffuser. When the back-pressure is higher than the design pressure, a normal shock wave is set up in front of the divergent nozzle and the flow velocity becomes a subsonic flow, as shown in Fig. D-2 (a). Since the streamline bends outwards downstream of the shock wave, some air is spilled over from the air-intake. The cross-sectional area upstream of the duct becomes smaller than the cross-sectional area of the air-intake, and so the efficiency of the diffuser is reduced. The subsonic flow velocity is further reduced and the pressure is increased in the divergent part of the diffuser.

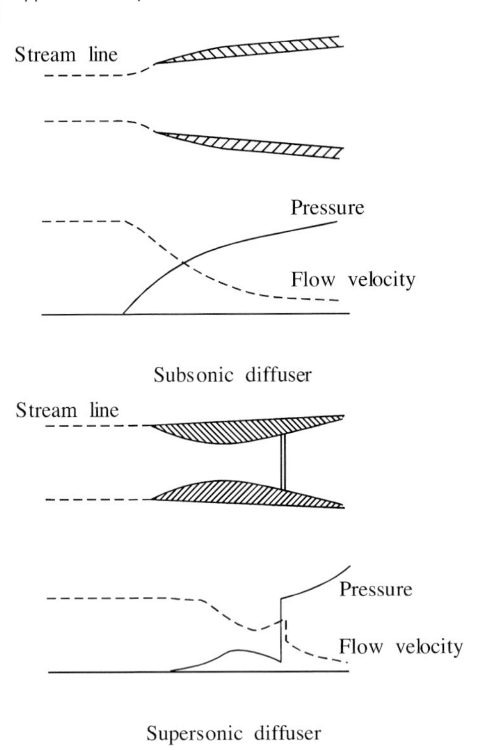

Figure D-1. Pressure and flow velocity in a subsonic diffuser and a supersonic diffuser.

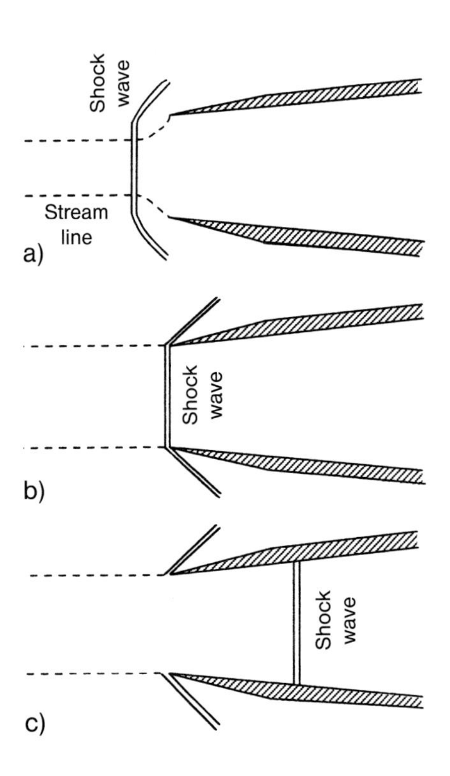

Figure D-2. Normal shock-wave formation of a supersonic diffuser at different back-pressures: (a) high back-pressure, (b) optimum back-pressure, and (c) low back-pressure.

When the back-pressure in the diffuser is optimized, a normal shock wave is set up at the lip of the diffuser and the pressure behind the shock wave is increased. No air spill-over occurs at the lip of the diffuser and the airflow velocity is as shown in Fig. D-2 (b). The pressure in the diffuser increases and the airflow velocity decreases along the flow direction. When the back-pressure is lower than the design pressure, a normal shock wave is "swallowed" inside of the diffuser, as shown in Fig. D-2 (c). Since the flow velocity in front of the normal shock wave in the diffuser is increased along the flow direction, the strength of the normal shock wave inside of the diffuser becomes higher than that in the case of the diffuser at the optimized back-pressure shown in Fig. D-2 (b). Thus, the pressure behind the shock wave is lowered due to the increased entropy.

D.1.2
Pressure Recovery

When a convergent-divergent nozzle is used in a ducted rocket, the convergent part reduces the flow velocity from supersonic flow to sonic flow at the nozzle throat and then the divergent part increases it once more from sonic flow to supersonic flow. When the supersonic flow in the divergent part generates a normal shock wave, the airflow becomes subsonic and the pressure increases. When a supersonic flow is reduced to a subsonic flow through isentropic compression, the pressure and the enthalpy of the airflow are increased from p_a to p_{0a} and from h_a to h_{0a}, respectively, as shown in Fig. D-3. However, when the supersonic flow is reduced to a subsonic flow by shock-wave compression, the pressure and the enthalpy are increased from p_a to p_{02} and from h_a to h_{02}, respectively. Though the enthalpy increase is the same for both compressions, i. e., $h_{0a} = h_{02}$, the pressure increase by the shock wave is less than that of the isentropic compression, i. e., $p_{02} < p_{0a}$.

The total pressure of p_{02}/p_{0a} is represented by

$$p_{02}/p_{0a} = (p_{02}/p_a)(p_a/p_{0a}) \tag{D.1}$$

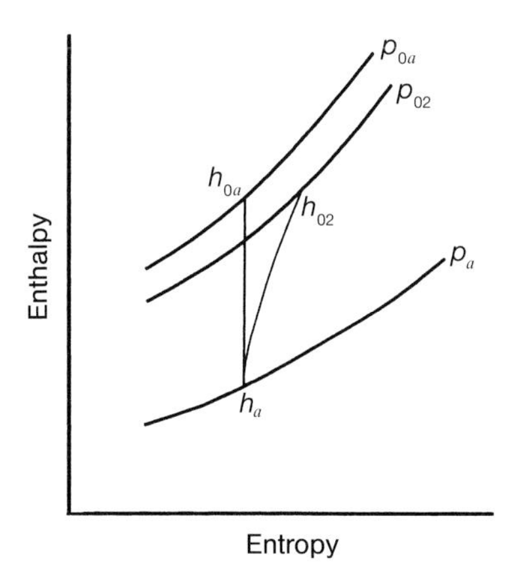

Figure D-3. Pressure recovery with isentropic and non-isentropic changes.

and p_{0a}/p_a is represented by

$$p_{0a}/p_a = \{1 + (\gamma - 1)M^2/2\}^{\gamma/(\gamma-1)} \tag{D.2}$$

The recovery factor, η_p, is defined by

$$\eta_p = p_{02}/p_{0a} \tag{D.3}$$

Substituting Eqs. (D.1) and (D.2) into Eq. (D.3), one obtains the pressure recovery factor as

$$\eta_p = \{1 + (\gamma - 1)M^2/2\}^{-\gamma/(\gamma-1)} (p_{02}/p_a) \tag{D.4}$$

If the flow process is an isentropic change, the total pressure p_{0a} remains unchanged throughout the nozzle flow. However, the process of the generation of a shock wave in the divergent part increases the entropy and the total pressure becomes p_{02}. It is evident that the inlet performance increases as p_{02} approaches p_{0a}.

When the compression process in the diffuser involves heat loss, the total enthalpy decreases from h_{0a} to h_{02}. The enthalpy recovery factor, η_d, is defined as

$$\eta_d = (h_{02} - h_a)/(h_{0a} - h_a) \tag{D.5}$$

The specific heat ratio of the air, γ, is considered to remain unchanged during the compression process in the diffuser,

$$\eta_d = (T_{02} - T_a)/(T_{0a} - T_a) \tag{D.6}$$

and the pressure and temperature changes during the compression process are given by

$$T_{02}/T_a = (p_{02}/p_a)^{(\gamma-1)/\gamma} \tag{D.7}$$

The temperature ratio of T_{0a}/T_a is given by

$$T_{0a}/T_a = 1 + (\gamma - 1)M^2/2 \tag{D.8}$$

Substituting Eqs. (D.7) and (D.8) into Eq. (D.6), one obtains the enthalpy recovery factor as

$$\eta_d = \{(p_{02}/p_a)^{(\gamma-1)/\gamma} - 1\}/(\gamma - 1)M^2/2 \tag{D.9}$$

The results of the analysis indicate that the pressure recovery factor is increased by the combination of several oblique shock waves and one weak normal shock wave in order to minimize the entropy increase at the air-intake.

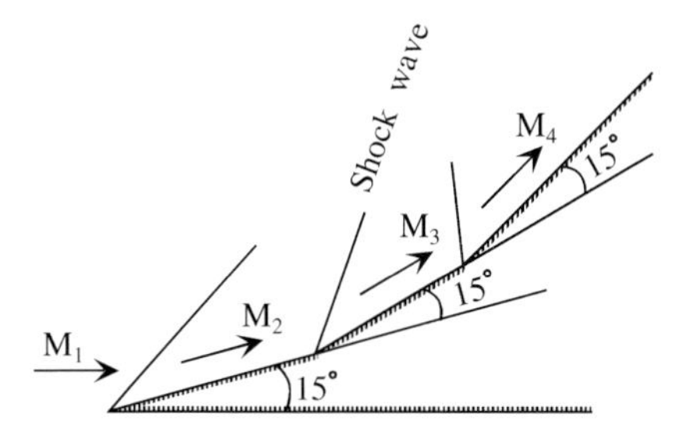

Figure D-4. Compression by the formation of three oblique shock waves with three ramps.

D.2 Air-Intake System

D.2.1 External Compression System

An external compression system used as a supersonic air-intake forms several oblique shock waves at its exterior and one normal shock wave at its lip. As a result of the formation of these oblique shock waves, the increase in entropy is reduced. For example, an oblique shock wave formed by the first ramp with an angle of 15°, as shown in Fig. D-4, decreases the Mach number from $M_1 = 3.5$ to $M_2 = 2.59$, and the stagnation pressure ratio of p_{02}/p_{01} becomes 0.848. The second ramp with an angle of 15° decreases the Mach number from $M_2 = 2.59$ to $M_3 = 1.93$, and the stagnation pressure ratio of p_{03}/p_{02} is 0.932. The third ramp with an angle of 15° decreases the Mach number from $M_3 = 1.93$ to $M_4 = 0.59$, and $p_{04}/p_{03} = 0.754$. The total stagnation pressure ratio between regions 1 and 4 is then obtained as:

$$p_{04}/p_{01} = (p_{02}/p_{01})(p_{03}/p_{02})(p_{04}/p_{03}) = 0.596$$

If the compression stems from one normal shock wave, $M_4 = 0.45$, $p_{04}/p_{01} = 0.213$, $p_4/p_1 = 14.13$, and $T_4/T_1 = 3.32$. It is evident that the pressure recovery factor obtained by the combination of oblique shock waves is significantly higher than that obtained by one normal shock wave.

D.2.2 Internal Compression System

An internal compression system forms several oblique shock waves and one normal shock wave inside the duct of the air-intake. The first oblique shock wave is formed at the lip of the air-intake and the following oblique shock waves are formed further downstream; the normal shock wave renders the flow velocity subsonic, as shown in the case of the supersonic diffuser in Fig. D-1. The pressure recovery factor and the changes in Mach number, pressure ratio, and temperature ratio are the same as in the case of the external compression system. Either external or internal air-intake systems are chosen for use in ramjets and ducted rockets, or combined systems are used to maximize the pressure recovery factor.

D.2.3
Air-Intake Design

Fig. D-5 shows an external compression air-intake designed for optimized use at Mach number 2.0. Fig. D-6 shows a set of computed airflows of an external compression air-intake designed for use at Mach number 2.0: (a) critical flow, (b) subcritical flow, and (c) supercritical flow. The pressures at the bottom wall and the upper wall along the duct flow are also shown. Two oblique shock waves formed at two ramps are seen at the tip of the upper surface of the duct at the critical flow shown in Fig. D-6 (a). The reflected oblique shock wave forms a normal shock wave at the bottom wall of the throat of the internal duct. The pressure becomes 0.65 MPa, which is the designed pressure. In the case of the subcritical flow shown in Fig. D-6 (b), the shock-wave angle is increased and the pressure downstream of the duct becomes 0.54 MPa. However, some of the airflow behind the oblique shock wave is spilled over towards the external airflow. Thus, the total airflow rate becomes 68 % of the designed airflow rate. In the case of the supercritical flow shown in Fig. D-6 (c), the shock-wave angle is decreased and the pressure downstream of the duct becomes 0.15 MPa, at which the flow velocity is still supersonic.

Fig. D-7 shows experimental and computed results for the pressure distribution from the tip along the bottom wall of the air-intake shown in Fig. D-5. The pressure is increased and decreased repeatedly due to the formation of three shock waves. The pressure downstream of the duct is effectively recovered from 0.1 MPa for the supersonic flow to 0.77 MPa for the subsonic flow.

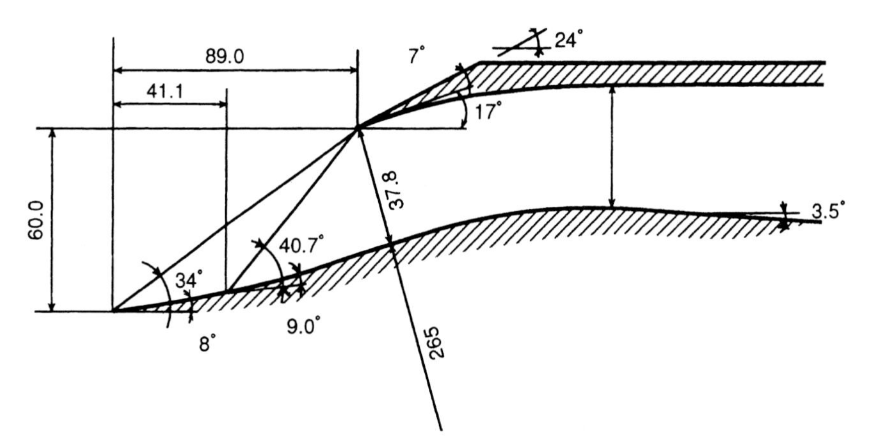

Figure D-5. External compression air-intake designed for use at Mach 2.0.

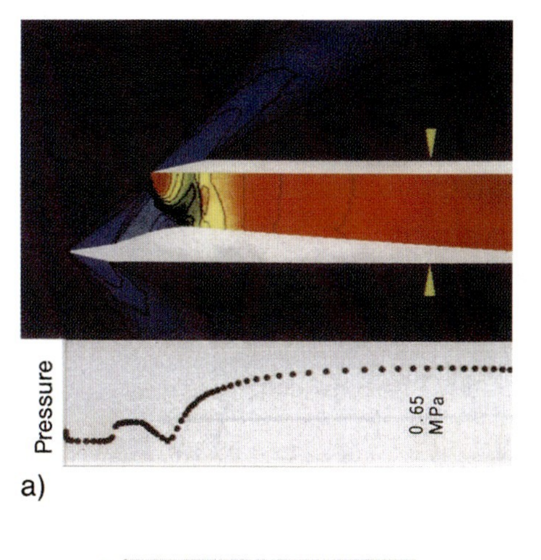

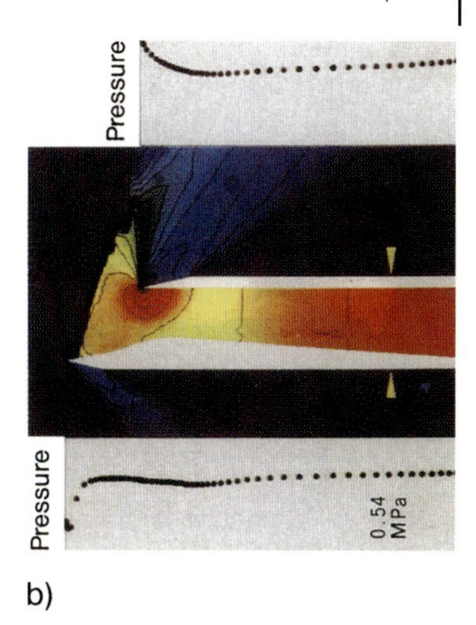

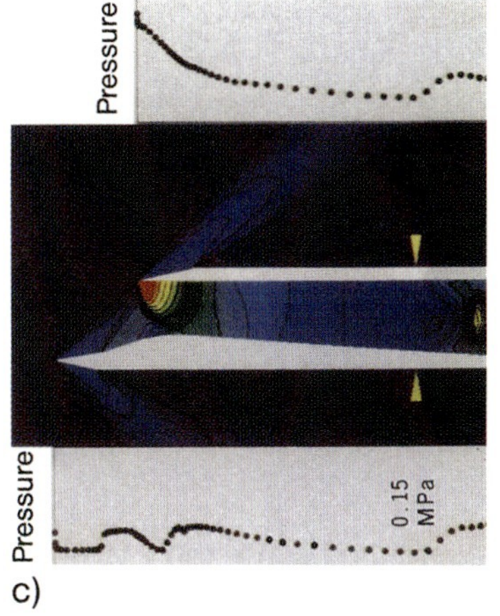

Figure D-6. Comparison of experimental and theoretical airflows under three types of operational conditions for the air-intake shown in Fig. D-5: (a) critical flow, (b) subcritical flow, and (c) supercritical flow.

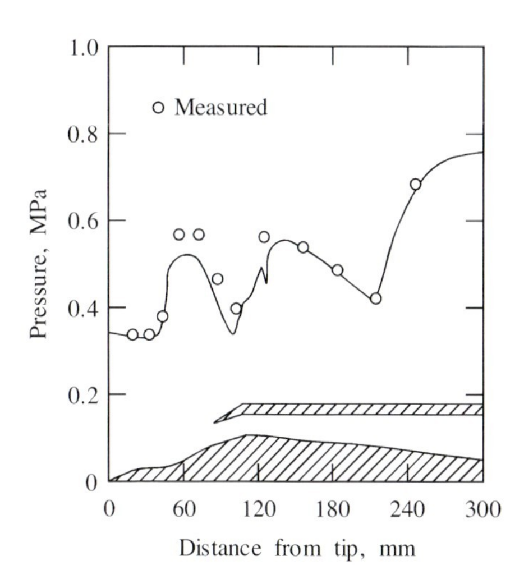

Figure D-7. Experimental and computed airflow data for the air-intake shown in Fig. D-5.

References

1 Shapiro, A. H., The Dynamics and Thermodynamics of Compressible Fluid Flow, The Ronald Press Company, New York, 1953.

2 Liepman, H. W., and Roshko, A., Elements of Gas Dynamics, John Wiley & Sons, New York, 1957.

3 Kuethe, A. M., and Schetzer, J. D., Foundations of Aerodynamics, John Wiley & Sons, New York, 1967.

4 Zucrow, M. J., and Hoffman, J. D., Gas Dynamics, John Wiley & Sons, New York, 1976.

5 Kubota, N., and Kuwahara, T., Ramjet Propulsion, Nikkan Kogyo Press, Tokyo, 1997.

6 Kubota, N., Foundations of Supersonic Flow, Sankaido, Tokyo, 2003.

Appendix E
Measurements of Burning Rate and Combustion Wave Structure

The burning rate of propellants is one of the important parameters for the design of rocket motors. The burning rate is obtained as a function of pressure and of initial temperature, from which pressure exponent of burning rate and temperature sensitivity of burning rate are deduced.

The combustion chamber used for measurements of burning rate is called a "strand burner". Fig. E-1 shows a drawing of a strand burner pressurized to the desired pressure with nitrogen gas. Purging nitrogen gas is introduced from the side at the bottom of the strand burner and is exhausted from the center-top of the burner through a choked orifice. The nitrogen gas flows around the propellant strand and then the burned gas flows together with the nitrogen gas towards the choked orifice. The nitrogen gas purge rate is adjusted so as to eliminate shear flow between the burned gas and the nitrogen gas above the burning strand, which en-

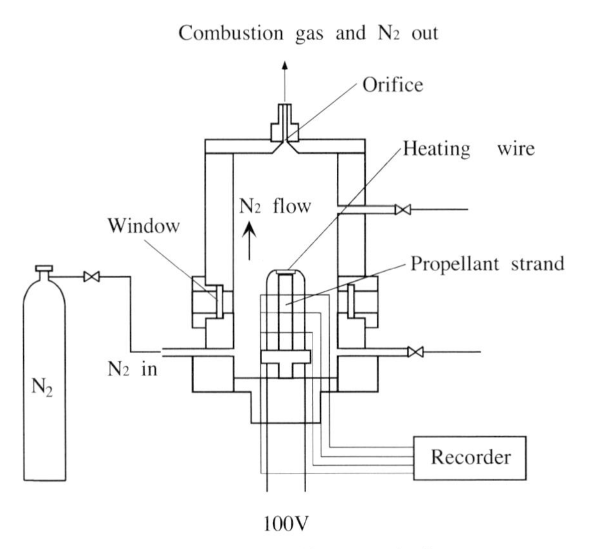

Figure E-1. Chimney-type strand burner with observation windows.

sures a stable combustion flame and a flat burning surface during burning. The described strand burner with a gas flow system is called a "chimney-type strand burner".

When a propellant strand is ignited under conditions of purging with nitrogen gas, the pressure in the strand burner is increased due to the additional gas emanating from the burning propellant. However, the pressure valve attached to the nitrogen gas supply is regulated automatically to reduce the nitrogen gas flow rate in order to maintain a constant pressure. In this way, the pressure in the burner is maintained at the desired level.

For burning rate measurements, propellants are formed into a strand shape, 7 mm × 7 mm in cross-section and 10 mm in length as standard dimensions. The strand is positioned vertically in the center of the burner. A fine metal wire (0.1 mm in diameter) is threaded through the top of the strand. The burning rate is measured by determining the instant of melting of each of five low melting point fuse wires made of lead metal, 0.25 mm in diameter, which are threaded through the strand at accurately known separation distances (15 mm). These five fuse wires, each in series with a resistor, form five parallel arms of an electrical circuit, and the output voltage changes discontinuously as soon as a fuse wire melts. The temperature of the strand is measured by means of a calibrated copper-constantan thermocouple threaded through the strand and the bead of the thermocouple is placed in the center of the strand.

Burning rates at different temperatures are measured by keeping the strand burner in a temperature-controlled environment. The purging nitrogen gas introduced into the strand burner is also cooled or heated to the desired temperature through a heat exchanger that is also kept within the same temperature-controlled environment. Thus, the temperatures of the strand burner, propellant strand, and nitrogen gas are all kept at the same desired level. The high-temperature gas generated by the burning of the strand does not affect the temperature because the burned gas flows in an upward direction. More than 5 minutes are needed to equilibrate the temperature of the strand under nitrogen-flow conditions.

A chimney-type strand burner for gas-phase and burning-surface observations consists of a chamber with four quartz windows built-in to its side-walls. A small cylinder 20 mm in diameter is mounted vertically inside the chamber and is connected to its base. Four transparent glass plates are mounted on the side of the cylinder. The cylinder is used to maintain a flow of nitrogen gas around the burning strand in order to keep the glass plates free of smoke deposits. The nitrogen gas is supplied through the base of the chamber and the flow rate is adjusted by changing the size of the orifice mounted on the top of the burner.

Photographs of the combustion wave structure in the gas phase are obtained using a high-speed video camera. The propellant strand is illuminated from outside of the strand burner by means of a tungsten filament lamp or a xenon lamp in order to observe the burning surface. Micro-photographs of the burning surface are obtained using a high-speed video camera fitted with micro-telescope equipment.

Index

Propellants and Explosives. Naminosuke Kubota

ISBN: 978-3-527-31424-9

i

o

p

s

Further Reading

Teipel, U. (Ed.)

Energetic Materials

Particle Processing and Characterization

643 pages with 417 figures and 70 tables
2005
ISBN 3–527-30240–9

Hattwig, M., Steen, H. (Eds.)

Handbook of Explosion Prevention and Protection

718 pages with 395 figures and 72 tables
2004
ISBN 3–527-30718–4

Meyer, R., Köhler, J., Homburg, A.

Explosives

434 pages with 20 figures and 27 tables
2002
ISBN 3–527-30267–0